sixth edition

Anesthesiology

Examination & Board Review

Mark Dershwitz, MD, PhD
Professor and Vice Chair of Anesthesiology
Professor of Biochemistry & Molecular Pharmacology
University of Massachusetts Medical School
Worcester, Massachusetts

J. Matthias Walz, MD
Assistant Professor of Anesthesiology
University of Massachusetts Medical School
Worcester, Massachusetts

McGraw-Hill
Medical Publishing Division

New York Chicago San Francisco Lisbon London Madrid Mexico City Milan
New Delhi San Juan Seoul Singapore Sydney Toronto

McGraw-Hill Specialty Board Review:
Anesthesiology Examination & Board Review, Sixth Edition

3 4 5 6 7 8 9 0 QPD/QPD 0 9 8

ISBN 0-07-144536-6

Notice

Medicine is an ever-changing science. As new research and clinical experience broaden our knowledge, changes in treatment and drug therapy are required. The authors and the publisher of this work have checked with sources believed to be reliable in their efforts to provide information that is complete and generally in accord with the standards accepted at the time of publication. However, in view of the possibility of human error or changes in medical sciences, neither the authors nor the publisher nor any other party who has been involved in the preparation or publication of this work warrants that the information contained herein is in every respect accurate or complete, and they disclaim all responsibility for any errors or omissions or for the results obtained from use of the information contained in this work. Readers are encouraged to confirm the information contained herein with other sources. For example and in particular, readers are advised to check the product information sheet included in the package of each drug they plan to administer to be certain that the information contained in this work is accurate and that changes have not been made in the recommended dose or in the contraindications for administration. This recommendation is of particular importance in connection with new or infrequently used drugs.

This book was set in Palatino by International Typesetting and Composition.
The editor was Marc Strauss.
The production supervisor was Sherri Souffrance.
Project management was provided by International Typesetting and Composition.
Quebecor World Dubuque was printer and binder.

This book is printed on acid-free paper.

Cataloging-in-Publication Data for this book is on file with the Library of Congress.

For Renée, Phil, Eli & Sally
—M.D.

For my parents for all their support
—J.M.W.

Contents

Contributors

Paul H. Alfille, MD
Instructor in Anaesthesia
Harvard Medical School
Associate Anesthetist
Massachusetts General Hospital
Boston, Massachusetts

Rae M. Allain, MD
Instructor in Anaesthesia
Harvard Medical School
Assistant Anesthetist
Massachusetts General Hospital
Boston, Massachusetts

Theodore Alston, MD, PhD
Assistant Professor of Anaesthesia
Harvard Medical School
Assistant Anesthetist
Massachusetts General Hospital
Boston, Massachusetts

Mark Dershwitz, MD, PhD
Professor and Vice Chair of Anesthesiology
Professor of Biochemistry & Molecular Pharmacology
University of Massachusetts Medical School
Worcester, Massachusetts

Peter J. Foley, MD, PhD
Associate Professor of Anesthesiology and Pediatrics
University of Massachusetts Medical School
Worcester, Massachusetts

Renée M. Goetzler, MD, MPH
Associate Program Director
Caritas Carney Hospital
Associate Clinical Professor of Medicine
Tufts University School of Medicine
Boston, Massachusetts

P. Grace Harrell, MD, MPH
Instructor in Anaesthesia
Harvard Medical School
Assistant Anesthetist
Massachusetts General Hospital
Boston, Massachusetts

John J. A. Marota, MD, PhD
Assistant Professor of Anaesthesia
Harvard Medical School
Assistant Anesthetist
Massachusetts General Hospital
Boston, Massachusetts

Ronald B. Rubin, MD, FACOG
Staff Anesthesiologist
The North Shore Medical Center
Salem, Massachusetts

J. Matthias Walz, MD
Assistant Professor of Anesthesiology
University of Massachusetts Medical School
Worcester, Massachusetts

Preface

The sixth edition of *McGraw-Hill Specialty Board Review: Anesthesiology Examination & Board Review* is actually the third edition in which the senior editor has been involved. Recognizing that the scope of anesthesiology has become too broad for one editor, the senior editor welcomes the participation of the junior editor to this edition and the expertise he brings in cardiopulmonary and critical care medicine. Three prior editions were written by Dr. Thomas P. Beach, formerly chief of the anesthesiology department at the Children's Hospital in Columbus, Ohio, and later of the department of anesthesia at the Massachusetts Eye and Ear Infirmary. This edition preserves the format of the last one which was so well received.

Once again, the subspecialty areas within anesthesiology were written by colleagues with special expertise in these fields. We are pleased by the addition of three new contributors. Dr. Ronald B. Rubin is one of the few physicians in the United States fully trained in both anesthesiology and obstetrics and gynecology. His expertise in both of these fields has facilitated a significant revision to Chapter 12. Dr. Peter J. Foley, our colleague and chief of pediatric anesthesiology at the University of Massachusetts, has assumed responsibility for Chapter 13. Dr. Renée M. Goetzler has brought her expertise in internal medicine and preventive medicine to Chapter 6. Finally, this edition once again includes an index in order to make it easier to locate questions on a specific topic. Every question is indexed, and most questions are listed in the index under two or three different keywords.

The reference list was expanded to include the popular text *Anesthesia and Coexisting Disease* by Stoelting and Dierdorf. The other references remain the same and the revision of this review book was timed to follow the revision of the text edited by Miller. There are three popular anesthesia reference books published in the United States, those edited by Barash et al., Longnecker et al., and Miller. Each of these books has its attributes and drawbacks. Longnecker and Miller have each edited two-volume reference works; however, they are organized quite differently. Longnecker's book emphasizes the preoperative evaluation and perioperative management of the patient organized by organ system. He and his coeditors acknowledge that the field of critical care medicine is now too broad to be covered within a few chapters in their text and chose not to attempt to do so. Miller's text begins with the basics: pharmacology, physiology, and instrumentation, and then considers anesthetic techniques for surgical procedures also organized by organ system. Miller's book includes several chapters devoted to critical care medicine. Barash decided to make a single-volume text. Although this makes many of the chapters more readable, it also means that it contains less of the minutiae sometimes needed to answer test questions. It is our belief that every practitioner of anesthesia should own one of the fat, two-volume texts, and the majority of questions in this review book are referenced to both of these books. Neither of these two-volume texts will fit in a briefcase and both are impractical to carry around when studying. Thus, most students of anesthesiology will also want a single-volume text, and for this reason many of the questions herein are referenced to Barash's book.

Electronic versions of all three anesthesiology reference texts are available. The approach taken by Miller is unique: it is available online and includes regular updates to the material. The drawback to this approach is that a continuous Internet connection is required to use the online version. The texts by Longnecker and Barash are available on CD-ROM. The latter CD-ROM includes in addition textbooks on pharmacology, regional anesthesia, and internal medicine.

Because of the supreme importance of pharmacology to the practice of anesthesiology, we believe every anesthesiologist should own a comprehensive reference. In the United States, the clear choice is Goodman and Gilman, now in its 10th edition and edited by Hardman et al. Thus the majority of the pharmacological questions in this book are referenced to it. Pediatric anesthesia remains an important part of every resident's training and a large portion of many persons' practices. Because the pediatric sections in the three references listed above are abbreviated, many of the pediatric questions are referenced to the textbook edited by Coté et al.

For this edition, the Introduction was also extensively rewritten to include new information from the American Board of Anesthesiology. The voluntary recertification program is being phased out. A new program, Maintenance of Certification in Anesthesiology (MOCA), is mandatory for diplomates with time-limited certificates issued after 2000 who wish to maintain their certification. MOCA is currently an alternative for diplomates certified prior to 2000 who wish to voluntarily recertify, and will become the only voluntary option after 2009.

This book contains 1800 questions and answers and more than double that number of references. There will be errors despite the best efforts of the authors and the editors to prevent them. We are very grateful to those persons who previously contacted the senior editor to report errors they had found in previous editions. Each one has been corrected in this edition. Once again, we would very much appreciate having any and all mistakes brought to our attention, and we have listed our e-mail addresses below to facilitate communication.

We would again like to thank our colleagues for their contributions to the subspecialty chapters in this book. We would also like to thank our families for tolerating our absences necessitated by the preparation of this latest edition.

Mark Dershwitz, MD, PhD
mark.dershwitz@umassmed.edu

J. Matthias Walz, MD
walzm@ummhc.org

Introduction

If you are planning to prepare for the American Board of Anesthesiology (ABA) written or oral examinations, including the recertification and MOCA (see below) examinations, then the *McGraw-Hill Specialty Board Review: Anesthesiology Examination & Board Review*, sixth edition, is designed for you. Here, in one package, is a comprehensive review resource with 1800 Board-type multiple-choice questions with referenced, paragraph-length discussions of each answer. In addition, the last 350 questions have been set aside as a practice test for self-assessment purposes.

Organization of this Book

Anesthesiology is divided into 17 chapters, all of which are designed to contribute to your review. The first seven chapters (Part I) are devoted to the basic science component of the ABA examination. The next nine chapters (Part II) review the clinical science component. The final chapter (Part III) is a two-part Practice Test which simulates an examination situation. Each of the components is discussed below in terms of what they contain and how you can use them in the most effective manner.

Questions

Each chapter in this book contains two different types of multiple-choice questions (or *items* in testing parlance). In general, about 50% of these are "one best answer–single item" questions and 50% are "multiple true–false (or K-type) items." In some cases, a group of two or three questions may be related to a situational theme. In addition, some questions have illustrative material (graphs or figures) that require understanding and interpretation on your part. Moreover, questions may be of three levels of difficulty: (1) rote memory questions, (2) memory questions that require more understanding of the problem, and (3) questions that require understanding and judgment. It is the judgment questions that we have tried to emphasize throughout this text. Finally, some of the items are stated in the negative. In such instances, we have printed the negative word in capital letters (e.g., "All of the following are correct EXCEPT"; "Which of the following choices is NOT correct"; and "Which of the following is LEAST correct").

One Best Answer-Single Item Question. This type of question presents a problem or asks a question and is followed by five choices, only one of which is entirely correct. The directions preceding this type of question will generally appear as below:

DIRECTIONS (Question 1): Each of the numbered items or incomplete statements in this section is followed by answers or by completions of the statement. Select the ONE lettered answer or completion that is BEST in each case.

An example for this item type follows:

1. A previously healthy 22-year-old male is undergoing surgical removal of a loose body in his knee under general anesthesia. Approximately 1 hour after beginning anesthesia, the patient becomes severely hypertensive. The most likely cause of the hypertension is

(A) pheochromocytoma
(B) malignant hyperthermia
(C) light anesthesia
(D) thyroid storm
(E) drug overdosage

TABLE 1. STRATEGIES FOR ANSWERING ONE BEST ANSWER–SINGLE ITEM QUESTIONS*

1. Remember that only one choice can be the correct answer.
2. Read the question carefully to be sure that you understand what is being asked.
3. Quickly read each choice for familiarity. (This important step is often not done by test takers.)
4. Go back and consider each choice individually.
5. If a choice is partially correct, tentatively consider it to be incorrect. (This step will help you lessen your choices and increase your odds of choosing the correct choice/answer.)
6. Consider the remaining choices and select the one you think is the answer. At this point, you may want to quickly scan the stem to be sure you understand the question and your answer.
7. Fill in the appropriate circle on the answer sheet. (Even if you do not know the answer, you should at least guess; you are scored on the number of correct answers, so do not leave any blanks.)

*Note that steps 2 through 7 should take an average of 70 seconds total. The actual examination is timed for an average of 70 seconds per question.

In this type of question, choices other than the correct answer may be partially correct, but there can only be one best answer. In the question above, the key word is *most*. Although all of the options listed may cause intraoperative hypertension, and all must be considered in the differential diagnosis, the most likely etiology of intraoperative hypertension is light anesthesia, or answer (C).

Multiple True–False Items. These questions are considered more difficult (or tricky), and you should be certain that you understand and follow the code that always accompanies these questions:

DIRECTIONS (Question 2): For each of the items in this section, ONE or MORE of the numbered options is correct. Choose answer

(A) if only 1, 2, and 3 are correct
(B) if only 1 and 3 are correct
(C) if only 2 and 4 are correct
(D) if only 4 is correct
(E) if all are correct

This code is always the same (i.e., "D" would never say "if 3 is correct"), and it is repeated throughout this book at the top of any page on which multiple true–false item questions appear.

SUMMARY OF DIRECTIONS

A	B	C	D	E
1, 2, 3 only	1, 3 only	2, 4 only	4 only	All are correct

TABLE 2. STRATEGIES FOR ANSWERING MULTIPLE TRUE–FALSE ITEM QUESTIONS

1. Carefully read and become familiar with the accompanying directions to this tricky question type.
2. Carefully read the stem to be certain that you know what is being asked.
3. Carefully read each of the numbered choices. If you can determine whether any of the choices are true or false, you may find it helpful to place a "+" (true) or a "–" (false) next to the number.
4. Focus on the numbered choices and your true/false notations, and use the following sequence to logically determine the correct answer:
   a. Note that in the answer code, choices 1 and 3 are always both either true or false together. If you are sure that either one is incorrect, your answer must be C or D.
   b. If you are sure that choice 2 and either choice 1 or 3 are incorrect, your answer must be D.
   c. If you are sure that choices 2 and 4 are incorrect, your answer must be B.*

*Remember, you only have an average of 70 seconds per question. Note that the following two combinations cannot occur: choices 1 and 4 both incorrect; choices 3 and 4 both incorrect.

A sample question follows:

2. Which of the following drugs has β-adrenergic antagonistic effects?

(1) propranolol
(2) ephedrine
(3) labetalol
(4) phenylephrine

You first need to determine which choices are right and wrong, and then which code corresponds to the correct numbers. In the example above, (1) and (3) are both β-adrenergic antagonists, and therefore (B) is the correct answer to this question.

Answers, Explanations, and References

At the end of the chapter, there is a section containing the answers, explanations, and references to the questions. This section (1) tells you the answer to each question, (2) gives you an explanation/review of why the answer is correct, background information on the subject matter, and why the other answers are incorrect, and (3) tells you where you can find more in-depth information on the subject matter in other reference books. We encourage you to use this section as a basis for further study and understanding.

If you choose the correct answer to a question, you can then read the explanation (1) for reinforcement and (2) to add to your knowledge about the subject matter (remember that the explanations usually tell not only why the answer is correct, but also why the other choices are incorrect). If you choose the wrong answer to a question, you can read the explanation for a learning/reviewing discussion of the material in the question. Furthermore, you can note the reference cited (e.g., 4:345), look up the full source in the References on page 387 (e.g., Longnecker DE, Tinker JH, Morgan GE, eds. *Principles and Practice of Anesthesiology*, 2nd ed. St. Louis, MO: Mosby Year Book; 1997), and refer to the page cited (p. 345) for a more in-depth discussion.

Practice Test

In the 350-question Practice Test at the end of the book, the questions are grouped according to question type (one best answer-single item, then multiple true–false items), with the subject areas integrated. This format mimics the actual examination and enables you to test your skill at answering questions in all of the areas under simulated examination conditions.

How to Use this Book

There are two logical ways to get the most value from this book. We shall call them Plan A and Plan B.

In Plan A, you go straight to the Practice Test and complete it according to the instructions. Analyze your areas of strength and weakness. This will be a good indicator of your initial knowledge of the subjects and will help to identify specific areas for preparation and review. You can now use the first 16 chapters of the book to help you improve your relative weak points.

In Plan B, you go through chapters 1 through 16 checking off your answers, and then comparing your choices with the answers and discussions in the book. Once you have completed this process, you can take the Practice Test to see how well prepared you are. If you still have a major weakness, it should be apparent in time for you to take remedial action.

In Plan A, by taking the Practice Test first, you get quick feedback regarding your initial areas of strength and weakness. You may find that you have a good command of the material indicating that perhaps only a cursory review of the first 16 chapters is necessary. This, of course, would be good to know early in your examination preparation. On the other hand, you may find that you have many areas of weakness. In this case, you could then focus on these areas in your review, not just with this book, but also with the cited references and with your current textbooks and journals.

It is, however, unlikely that you will not do some studying prior to taking the ABA examination (especially since you have this book). Therefore, it may be more realistic to take the Practice Test after you have reviewed the first 16 chapters (as in Plan B). This will probably give you a more realistic type of testing situation since very few of us just sit down to a test without studying. In this case, you will have done some reviewing (from superficial to in-depth), and your Practice Test will reflect this studying time. If, after reviewing the first 16 chapters and taking the Practice Test, you still have some weaknesses, you can then go back to the first 16 chapters and supplement your review with your texts.

The American Board of Anesthesiology

History

Anesthesiology is a relatively new specialty, and therefore the ABA is also relatively young.

The American Society of Anesthesiologists had its beginnings in 1911 in the New York area when two groups merged to form the New York Society of Anesthetists. The group grew and by 1935 was national in scope. In 1936 the name was changed to the American Society of Anesthetists, and the society was incorporated.

The American Society of Anesthesiology, Inc. grew out of a committee representing the American Society of Anesthetists, Inc., the American Society of Regional Anesthesia, Inc., and the Section on Surgery of the American Medical Association. In 1937, the American Board of Anesthesiology, Inc., was formed as an affiliate of the American Board of Surgery, Inc. In 1941, the board was approved as a separate entity.

The purposes of the Board are as follows*:

A. Maintain the highest standards of practice by fostering educational facilities and training in anesthesiology, which the ABA defines as the practice of medicine dealing with but not limited to:
   (1) Assessment of, consultation for, and preparation of, patients for anesthesia.
   (2) Relief and prevention of pain during and following surgical, obstetric, therapeutic and diagnostic procedures.
   (3) Monitoring and maintenance of normal physiology during the perioperative period.
   (4) Management of critically ill patients.
   (5) Diagnosis and treatment of acute, chronic and cancer related pain.
   (6) Clinical management and teaching of cardiac and pulmonary resuscitation.
   (7) Evaluation of respiratory function and application of respiratory therapy.
   (8) Conduct of clinical, translational and basic science research.
   (9) Supervision, teaching and evaluation of performance of both medical and paramedical personnel involved in perioperative care.
   (10) Administrative involvement in health care facilities and organizations, and medical schools necessary to implement these responsibilities.

B. Establish and maintain criteria for the designation of a Board certified anesthesiologist.

C. Inform the Accreditation Council for Graduate Medical Education ("ACGME") concerning the training required of individuals seeking certification as such requirements relate to residency training programs in anesthesiology.

D. Establish and conduct those processes by which the Board may judge whether a physician who voluntarily applies should be issued a certificate indicating that the required standards for certification or recertification as a diplomate of the ABA have been met.

   A Board certified anesthesiologist is a physician who provides medical management and consultation during the perioperative period, in pain medicine and in critical care medicine. A diplomate of the Board must possess knowledge, judgment, adaptability, clinical skills, technical facility and personal characteristics sufficient to carry out the entire scope of anesthesiology practice. An ABA diplomate must logically organize and effectively present rational diagnoses and appropriate treatment protocols to peers, patients, their families and others involved in the medical community. A diplomate of the Board can serve as an expert in matters related to anesthesiology, deliberate with others, and provide advice and defend opinions in all aspects of the specialty of anesthesiology. A Board certified anesthesiologist is able to function as the leader of the anesthesiology care team.

   Because of the nature of anesthesiology, the ABA diplomate must be able to manage emergent life-threatening situations in an independent and timely fashion. The ability to independently acquire and process information in a timely manner is central to assure individual responsibility for all aspects of anesthesiology care. Adequate physical and sensory faculties, such as eyesight, hearing, speech and coordinated function of the extremities, are essential to the independent performance of the Board certified anesthesiologist. Freedom from the influence of or dependency on chemical substances that impair cognitive, physical, sensory or motor function also is an essential characteristic of the Board certified anesthesiologist.

E. Establish and conduct those processes by which the Board may judge whether a physician who voluntarily applies should be issued a certificate indicating that the required standards for subspecialty certification or recertification in an ABA designated subdiscipline of anesthesiology have been met.

F. Serve the public, medical profession, health care facilities and organizations, and medical schools by providing the names of physicians certified by the Board.

Certification*

At the time of certification by the ABA, the candidate shall be capable of performing independently the entire scope of anesthesiology practice and must:

A. Hold an unexpired license to practice medicine or osteopathy in at least one state or jurisdiction of the United States or province of Canada that is permanent, unconditional and unrestricted. Further, every United States and Canadian medical license the applicant holds must be free of restrictions. Candidates for initial certification and ABA diplomates have the affirmative obligation to advise the ABA of any and all restrictions placed on any of their medical licenses and to provide the ABA with complete information concerning such restrictions within 60 days after their imposition. Such information shall include, but not be limited to, the identity of the State Medical Board imposing the restriction as well as the restriction's duration, basis, and specific terms and conditions. Candidates and diplomates discovered not to have made disclosure may be subject to sanctions on their candidate or diplomate status.

B. Have fulfilled all the requirements of the Continuum of Education in Anesthesiology.

C. Have on file with the ABA a Certificate of Clinical Competence with an overall satisfactory rating covering the final six-month period of Clinical Anesthesia training in each anesthesiology residency program.

D. Have satisfied all examination requirements of the Board.

E. Have a professional standing satisfactory to the ABA.

Recertification*

The ABA approved a policy of time-limited certification in 1995. All certificates issued by the ABA on or after January 1, 2000 will expire ten (10) years after the year the candidate passes the certification examination. The ABA took this step to reassure the public that the diplomate continues to demonstrate the attributes of a Board certified anesthesiologist.

The ABA Recertification Programs include a commitment to continuing education, an assessment of the quality of practice in the local environment, and an evaluation of knowledge. *Diplomates who hold a certificate that is not time-limited may voluntarily elect to apply to the ABA for recertification. The ABA will not alter the status of their certification if they do not recertify.*

*The American Board of Anesthesiology, Inc., Booklet of Information, February, 2005.

Only Diplomates certified in anesthesiology by the ABA before January 1, 2000, are eligible to apply for the recertification program. The soonest they may apply is seven years after their initial certification or recertification by the ABA. The ABA recertification program will not remain open indefinitely. Diplomates certified before 2000 who might have a future need to recertify should consider participating in the program before it closes in 2009. Participation will not jeopardize a participant's diplomate status.

Requirements:

The ABA recertification programs include two major components: an evaluation of the quality of current practice conducted at the local level and a secure written examination. To be admissible to an ABA recertification examination, the applicant shall be capable of performing independently the entire scope of specialty or subspecialty practice and must:

(1) Be a physician to whom the ABA previously awarded certification in the specialty or subspecialty.

(2) Have fulfilled the licensure requirement for certification. The applicant must inform the ABA of any conditions or restrictions in force on any active medical license he or she holds. When there is a restriction or condition in force on any of the applicant's medical licenses, the Credentials Committee of the ABA will determine whether, and on what terms, the applicant shall be admitted to the ABA examination system.

(3) Have on file in the ABA office documentation solicited by the ABA from the hospital/facility chief of staff, or equivalent, attesting to the applicant's current privileges where a substantial portion of the applicant's practice takes place. The documentation includes evaluations of various aspects of the applicant's current practice and verification that the applicant meets the Board's clinical activity requirement by practicing the medical discipline for which recertification is being sought, on average, at least one day per week during one of the previous three years. If the applicant's practice is entirely office-based, three letters of reference solicited by the ABA from referring physicians should be on file.

The ABA shall issue a recertification certificate to the applicant who is accepted for and satisfies the recertification examination requirement established by the ABA.

Maintenance of Certification in Anesthesiology (MOCA) Program*

The ABA issued diplomates certified on or after January 1, 2000 a certificate that is valid for 10 years. The voluntary recertification program is not open to holders of a time-limited anesthesiology certificate. They must satisfactorily complete the requirements of MOCA before their time-limited certificate expires to maintain diplomate status in the specialty.

MOCA is a 10-year program of ongoing self-assessment and lifelong learning, continual professional standing assessment, periodic practice performance assessments, and an examination of cognitive expertise. Each 10-year MOCA cycle begins the year after certification or the year the diplomate applies for MOCA, whichever occurs later. **Therefore, if a diplomate does not apply for MOCA before the end of the first calendar year following his or her certification, the diplomate's certification will expire before he or she can complete MOCA.**

Physicians should maintain competency in the following general areas: patient care, medical knowledge, practice-based learning and improvement, interpersonal and communication skills, professionalism, and systems-based practice. The MOCA requirements for Professional Standing, Lifelong Learning and Self-Assessment, Cognitive Expertise and Practice Performance are designed to provide assessments of these six general competencies.

A. PROFESSIONAL STANDING ASSESSMENT
ABA diplomates must hold an active, unrestricted license to practice medicine in at least one jurisdiction of the United States or Canada. Further, all US and Canadian medical licenses that a diplomate holds must be unrestricted at all times.

The ABA assesses a diplomate's Professional Standing continually. ABA diplomates have the affirmative obligation to advise the ABA of any and all restrictions placed on any of their medical licenses and to provide the ABA with complete information concerning such restrictions within 60 days after their imposition. Such information shall include, but not be limited to, the identity of the medical board imposing the restriction as well as the restriction's duration, basis, and specific terms and conditions. Diplomates discovered not to have made disclosure may be subject to sanctions on their diplomate status in the primary specialty. Professional Standing acceptable to the ABA is a prerequisite qualification for cognitive examination and for maintenance of certification.

B. PRACTICE PERFORMANCE ASSESSMENT
The ABA practice performance assessment process consists of attestations of clinical activity, acceptable clinical practice and participation in practice improvement activities. The ABA minimum clinical activity requirement is the practice of anesthesiology or a recognized anesthesiology subspecialty, on average, at least one day per week during one of the previous three years.

During the 5th and 9th years of a candidate's 10-year MOCA cycle, the ABA solicits attestations about a candidate's clinical practice participation in practice improvement activities from individuals identified by the diplomate as being familiar with his or her current practice of the specialty. Attestations of practice performance will be solicited more frequently if an assessment is not acceptable to the ABA.

Practice Performance assessments acceptable to the ABA are a prerequisite qualification for cognitive examination and for maintenance of certification.

C. LIFELONG LEARNING AND SELF-ASSESSMENT
ABA diplomates should continually seek to improve the quality of their clinical practice and patient care through self-directed professional development. This should be done through self-assessment and learning opportunities designed to meet the diplomate's needs and the MOCA requirement for Lifelong Learning and Self-Assessment ("LL-SA").

*The American Board of Anesthesiology, Inc., Booklet of Information, February, 2005.

The LL-SA requirement for maintenance of certification is 350 credits for continuing medical education ("CME") activities. Of the 350 credit total:

(1) At least 250 credits must be Category 1 credits for ACCME-approved programs or activities.

(2) At most 100 credits may be for programs and activities for which Category 1 credit is not awarded.

The prerequisite qualification for cognitive examination is at least 200 credits.

Diplomates should complete some LL-SA activity in at least 5 years of each 10-year MOCA cycle. They are encouraged to complete some LL-SA activity in each of the six general competencies for physicians.

MOCA candidates submit their CME activities and credits to the ABA electronically. CME activities are subject to audit and verification by the ABA within three years of their submission. **Therefore, diplomates must keep documentation of a CME activity for at least three years after they submit it to the ABA for LL-SA credit.**

D. COGNITIVE EXPERTISE ASSESSMENT

MOCA candidates must demonstrate their cognitive expertise by passing an ABA examination administered via computer under secure, standardized testing conditions. They may take the examination no earlier than the 7th year of their 10-year MOCA cycle. Examination prerequisites are:

(1) Professional standing (PS) acceptable to the ABA.

(2) Practice performance (PP) assessments acceptable to the ABA.

(3) At least 200 LL-SA credits submitted to the ABA prior to the examination year.

MOCA candidates may take the examination at most twice during a calendar year. There is no limit to the number of years they may take the examination.

E. MOCA CYCLE DURING AND AFTER TRANSITION PERIOD

The transition from a voluntary recertification examination program to MOCA began in January 2004. ABA diplomates whose date of initial certification or most recent recertification is before 2000 may voluntarily participate in either the recertification program until it ends in 2009 or the MOCA program. The first time they apply for MOCA they may complete the MOCA program in as soon as one year. They may expedite the completion of the MOCA program only once; thereafter, the 10-year MOCA program is their only option.

The MOCA Program is the only option for ABA diplomates certified in or after 2000 to maintain their certification. The ABA recommends that they apply for MOCA no later than the year after their certification. If a diplomate does not apply for MOCA before the end of the first calendar year following his or her certification, the diplomate's certification will expire before he or she can complete MOCA.

Diplomates issued a time-limited certificate in 2000, 2001, 2002, or 2003 have only 6, 7, 8 and 9 years, respectively, to complete their first MOCA cycle before their certification expires. For them, the LL-SA requirements for the secure examination prerequisite and for the awarding of maintenance of certification are prorated. They may visit the ABA website or contact the Board office for additional information regarding their MOCA program requirements.

Diplomates issued a time-limited certificate in or after 2004 have 10 years to complete all MOCA requirements before their certification expires. They have to complete 350 LL-SA hours with a minimum of 250 Category 1 hours. A prerequisite for them to take the secure exam is 200 LL-SA hours. PS assessment is continual, PP assessments occur in the 5th and 9th years of their MOCA cycle, and they can take the secure examination no sooner than the 7th year of their MOCA cycle.

Diplomates certified before 2000 have a certificate that is not time-limited. The first time they voluntarily apply for MOCA they can complete all of the MOCA requirements within one year of application. The Professional Standing assessment is continual. The Practice Performance assessment will be done within 6 months of application. They can satisfy LL-SA requirements on the basis of CME activities completed after certification and within the past 10 years. They can take the secure examination when they have satisfied all of the prerequisite requirements.

The ABA may audit and verify CME activities completed within three years of their submission. Therefore, diplomates should keep documentation of a CME activity for at least three years.

More PP assessments will be required if all MOCA requirements are not satisfied by the end of the fifth year following application. All requirements must be satisfied within the 10 years that precede the awarding of a certificate acknowledging completion of MOCA.

Diplomates with a certificate that is not time-limited may complete the expedited MOCA program only once. Thereafter, their next MOCA cycle will be 10 years and will begin the year they re-apply.

Information concerning the Board and its requirements may be obtained from:

The American Board of Anesthesiology, Inc.
4101 Lake Boone Trail, Suite 510
Raleigh, NC 27607-7506
Telephone: (919) 881-2570
FAX: (919) 881-2575
Web: *http://www.theaba.org*

Examination Preparation

Ideally, preparation for the examination should begin during training. During this period, time spent in a systematic approach to the body of knowledge will be well rewarded.

A Content Outline of the In-Training Examination, by the ABA and the American Society of Anesthesiologists, covers the knowledge that should be gained during the training period. It serves as a grid for the questions to be covered in the examination and also serves as an outline for study. Remember that the in-training examination, taken during the residency, and the written Board examination, taken after completing the residency, are identical. The scoring of the examination for the two groups is, however, different.

Textbooks provide a good background of information. Popular basic texts include those by Stoelting & Miller, Longnecker & Murphy, and Liu. There are comprehensive texts edited by Barash et al., Longnecker et al., and Miller. Many monographs are available on subspecialties, e.g., pain control, critical care, cardiothoracic, obstetric, and pediatric anesthesia.

The best sources of current information are journals and meetings. The newest data and latest techniques are reported in these two sources, and one must be familiar with them. Tape recordings of presentations at many major meetings are available.

An invaluable source of information is the annual refresher course given each year in conjunction with the American Society of Anesthesiologists meeting in October. A series of 1-hour lectures is given for 2 days. A syllabus is available covering all the lectures. If you cannot attend the lectures, the book may be purchased from the American Society of Anesthesiologists.

Format of Examination

Written Examination. The written examination of the ABA is of the objective, multiple-choice form. It is a comprehensive test that covers the wide field related to anesthesiology.

The questions are designed to test the information gained while in training and knowledge expected of physicians in general. *Trick* or ambiguous questions are avoided.

The test consists of a total of 350 questions to be answered in 7 hours. The test is split into two sessions of $3^1/_2$ hours, each consisting of 175 questions. You should plan your time accordingly. There are two types of questions used: one best answer-single item type and multiple true–false type. These are discussed earlier in this introduction.

In this book only these two types of questions are used.

Oral Examination. While the written examination is designed to measure knowledge, the oral examination is meant to measure the ability to use that knowledge in the clinical setting. This requires the ability to assimilate data and put it to use. One must be able to assess the information presented, sort out the priorities, and arrive at an anesthetic plan. This may also involve the ability to discuss the problems intelligently and convincingly. The ability to adapt to changing developments is also assessed.

The oral examination consists of two 35-minute examination periods with a 10-minute break between periods. Two examiners interview each candidate during each period. Only one examines at a time, and the time is divided between the examiners.

Each session starts with the presentation of a case report for discussion. All candidates receive the same report for discussion at that period. The approach of the candidate will determine, in part, further questions that may be asked.

A typical case report:

> A 26-year-old woman enters the obstetric suite with a diagnosis of abruptio placenta. She last ate one hour before her arrival at the hospital. She is actively bleeding. She states that she is a Jehovah's Witness and does not want any blood or blood products under any circumstances.

The discussion can take many routes: preparation, anesthetic of choice, complications, and so forth. Subsequent questions may result from the candidate's response.

It is important to think before you answer. Do not try to read the examiner's mind: do not try to figure out why he or she is asking that question. Answer the question as you understand it, and if you do not understand it, ask for clarification.

If you do not know much about regional anesthesia, do not take that path. You are bound to get further questions. If you do not know the answer, simply say "I do not know." You are better off moving to an area in which you have some knowledge than digging yourself deeper into a hole

PART I

Basic Sciences

CHAPTER 1

Physics and Chemistry

Questions

DIRECTIONS (Questions 1 through 28): Each of the numbered items or incomplete statements in this section is followed by answers or by completions of the statement. Select the ONE lettered answer or completion that is BEST in each case.

1. The statement that equal volumes of gases at the same temperature and pressure contain equal numbers of molecules is

   (A) Charles' law
   (B) Boyle's law
   (C) Lavoisier's law
   (D) Avogadro's hypothesis
   (E) Archimedes' hypothesis

2. The vapor pressure of a liquid is most dependent on the

   (A) atmospheric pressure
   (B) specific heat of the liquid
   (C) temperature
   (D) thermal conductivity of the container
   (E) molecular weight of the liquid

3. As a person ascends to an altitude of 10,000 ft, there is

   (A) no change in atmospheric pressure
   (B) a decrease in FIO_2
   (C) a decrease in PIO_2
   (D) a decrease in cerebrospinal fluid (CSF) pH
   (E) a decrease in minute ventilation

4. The premise that a wave train reflected from a moving surface will undergo a change in frequency is the basis of

   (A) the electrocardiogram (ECG)
   (B) the Doppler effect
   (C) thermodilution cardiac output monitors
   (D) force transducers
   (E) Raman spectroscopy

5. If three containers each contain 22.4 L of oxygen, nitrous oxide, and halothane vapor (each at standard pressure and temperature), the weights of the containers will

   (A) be equal
   (B) vary according to the molecular weights of the gases
   (C) vary according to how many moles of the gases are contained
   (D) be dependent on the individual vapor pressures of the gases
   (E) vary inversely with the density of the gases

6. The number of calories required to raise the temperature of 1 g of a substance by 1°C is

   (A) the heat of vaporization
   (B) the specific heat
   (C) the critical temperature
   (D) thermal conductivity
   (E) equal for all substances

7. The latent heat of vaporization

   (A) is equal for all liquids
   (B) is independent of the ambient temperature
   (C) varies with the temperature of the liquid
   (D) is very low for solids
   (E) for water is 1 cal/mL

8. Poiseuille's law describes the resistance to the flow of

(A) electrons in a conductor
(B) a fluid in relation to height
(C) a gas in a closed system
(D) a fluid in relation to the diameter of a tube
(E) a fluid through an orifice

9. In Figure 1-1, each side of the container holds gases with a total pressure of 760 mmHg. If the dividing wall were perforated to allow diffusion

(A) O_2 would diffuse from left to right
(B) no diffusion would occur
(C) the final partial pressure of O_2 would be 760 mmHg
(D) N_2O would diffuse from left to right
(E) the final partial pressure of N_2O would be 660 mmHg

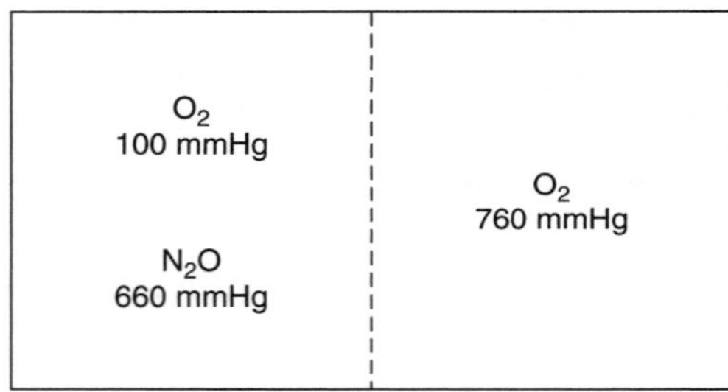

FIG. 1-1

10. The Bernoulli theorem states that

(A) pressure and volume in a pipe are related
(B) pressure and temperature in a pipe are related
(C) velocity of flow and lateral pressure in a pipe are related
(D) velocity of a fluid and its flow in a pipe are related
(E) flow of a fluid and its viscosity in a pipe are related

11. One milliliter of isoflurane liquid occupies what volume at 1 atm pressure and 20°C if all of the liquid is vaporized? The ideal gas constant is 0.082 (L · atm)/(K · mol), the specific gravity of isoflurane is 1.5, and its molecular weight is 184.5.

(A) 182 mL
(B) 195 mL
(C) 201 mL
(D) 207 mL
(E) 226 mL

12. A cylinder of oxygen has an internal volume of 6 L and a pressure of 1700 psi. How many liters of oxygen will this tank supply at sea level?

(A) 660 L
(B) 680 L
(C) 694 L
(D) 706 L
(E) 716 L

13. The electrical current, often called the *let-go* current, above which contraction of the finger flexors is unable to be overcome by voluntarily contracting the finger extensors, is approximately

(A) 0.15 mA
(B) 1.5 mA
(C) 15 mA
(D) 150 mA
(E) 1.5 A

14. Cellular telephones commonly cause interference with a medical device if

(A) used anywhere within a hospital
(B) used by a patient having an implanted cardiac defibrillator
(C) used while standing next to an anesthesia machine
(D) carried in a shirt pocket overlying a cardiac pacemaker
(E) used by a patient having an intracranial aneurysm clip

15. When desflurane, but not sevoflurane, in a carrier gas of 100% oxygen is passed through very dry carbon dioxide absorbent, the patient may be exposed to a toxic concentration of

(A) ozone
(B) phosgene
(C) carbon dioxide
(D) carbon monoxide
(E) fluoride

16. Current flowing in a loop of wire moving within a magnetic field is an example of

(A) inductance
(B) capacitance
(C) reactance
(D) resistance
(E) oscillation

17. A patient is undergoing operative repair of an arm fracture. The anesthesiologist is standing 4 ft from the patient's arm. The surgeon is using fluoroscopy to visualize the fracture. Assume the amount of radiation to which the anesthesiologist is exposed is x. If the anesthesiologist moves to a new position 8 ft from the arm, then the amount of radiation exposure will decrease to approximately

(A) $0.75x$
(B) $0.5x$
(C) $0.33x$
(D) $0.25x$
(E) $0.125x$

18. Laminar flow in a tube is proportional to the radius of the tube raised to the

(A) first power
(B) second power
(C) fourth power
(D) eighth power
(E) sixteenth power

19. A nerve cell has a resting potential of –90 mV. If the extracellular potassium concentration suddenly decreases to half its previous value, the resting potential will

(A) decrease to –180 mV
(B) decrease to –108 mV
(C) remain the same
(D) increase to –45 mV
(E) increase to +90 mV

20. A voltage of 100 V is applied to a resistor of 20 Ω. The amount of power dissipated by the resistor

(A) cannot be calculated without further information
(B) is 5 W
(C) is 80 W
(D) is 500 W
(E) is 2000 W

21. The osmolality of a solution of calcium chloride which is 0.1 molal is

(A) 50 mOsm/kg H_2O
(B) 100 mOsm/kg H_2O
(C) 200 mOsm/kg H_2O
(D) 300 mOsm/kg H_2O
(E) 400 mOsm/kg H_2O

22. The effect that occurs when a hypercalcemic patient is given a large dose of intravenous potassium phosphate is called

(A) synergy
(B) chelation
(C) precipitation
(D) excretion
(E) resorption

23. If the $PaCO_2$ is 40 mmHg and the bicarbonate concentration in the blood is 24 mEq/L, the blood pH is

(A) normal
(B) 7.3 and the mechanism is metabolic
(C) 7.3 and the mechanism is respiratory
(D) 7.5 and the mechanism is metabolic
(E) 7.5 and the mechanism is respiratory

24. At sea level, if a patient without any cardiopulmonary pathology is administered 70% O_2, after an appropriate time to reach equilibrium the PaO_2 will be approximately

(A) 415 mmHg
(B) 440 mmHg
(C) 476 mmHg
(D) 500 mmHg
(E) 532 mmHg

25. The absorption of one molecule of carbon dioxide by soda lime causes the net production of how many molecules of water?

(A) 0
(B) 1
(C) 2
(D) 3
(E) 4

26. A line isolation monitor is set to alarm at 2 mA. When an alarm condition exists, it means that

(A) 2 mA of current are flowing through the patient
(B) 2 mA of current could flow through the patient if the ungrounded power system were to become grounded
(C) the patient is likely to experience electrical injury if the alarm condition is not remedied
(D) there must be a faulty electrical device in the operating room
(E) surgery should stop immediately, and if it cannot stop, the patient should be relocated to a different operating room

27. For inhalational anesthetics, the anesthetic potency is proportional to

(A) lipid solubility
(B) boiling point
(C) vapor pressure
(D) critical temperature
(E) specific gravity

28. The current delivered to the patient by an electrocautery device differs from the current supplied by an electrical utility in its

(A) capacitance
(B) frequency
(C) amperage
(D) voltage
(E) power

DIRECTIONS (Questions 29 through 47): For each of the items in this section, ONE or MORE of the numbered options is correct. Choose answer

(A) if only 1, 2, and 3 are correct
(B) if only 1 and 3 are correct
(C) if only 2 and 4 are correct
(D) if only 4 is correct
(E) if all are correct

29. The boiling point of a liquid

(1) is the temperature where the vapor pressure equals atmospheric pressure
(2) will decrease if the atmospheric pressure is decreased
(3) is a temperature where the vapor and liquid are in equilibrium if enclosed within a container
(4) will increase if the ambient temperature is increased

30. Which of the following are colloidal solutions?

(1) 5% albumin
(2) 5% dextrose
(3) 6% hetastarch
(4) 7.5% sodium chloride

31. Specific gravity

(1) of water is 1
(2) applies only to liquids
(3) describes a ratio of the mass of one substance to the mass of a similar volume of water at the same temperature
(4) is measured in units of mass per unit volume

32. The density of a gas

(1) affects the rate of laminar flow through an orifice
(2) is independent of temperature
(3) is related to the explosive potential
(4) is important in the transition to turbulent flow

33. Microshock hazard is increased in patients

(1) with a temporary pacing wire in place
(2) in electrically operated beds
(3) receiving total parenteral nutrition
(4) who are in rooms with isolation transformers

34. The Venturi tube shown in Figure 1-2

(1) is the basis for injector systems
(2) to be effective must open out gradually distal to the constriction C
(3) has a delivery gas that is denoted by B
(4) has an entrained gas denoted by A

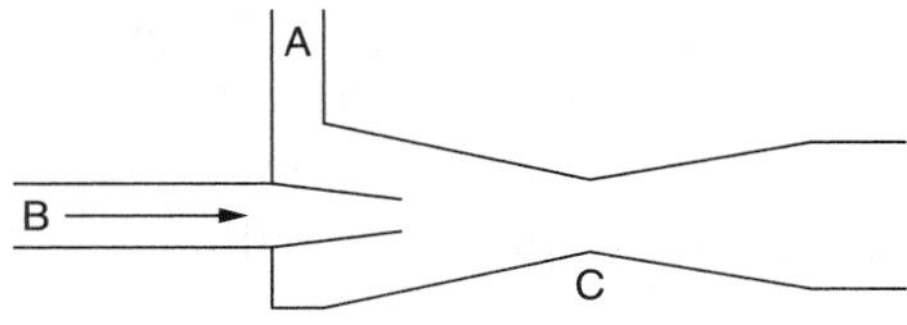

FIG. 1-2

35. Operating rooms use an isolated power supply because

(1) grounding cannot occur
(2) contact with both wires of the isolation transformer secondary would cause no shock
(3) leakage current is zero
(4) it affords added protection against high-amperage electrocution

36. Clinically significant amounts of heat loss from a patient in the operating room occurs via which of these mechanisms?

(1) conduction
(2) radiation
(3) convection
(4) evaporation

37. Leakage current

(1) is harmless
(2) occurs because of the magnitude of the inductance between electrical conductors supplying equipment
(3) is of the direct current (DC) type
(4) is unintentional flow of current from the internal wiring of a device

38. Safety rules of specific importance to anesthesiologists are contained in (a) publication(s) of the

(1) National Fire Protection Association (NFPA)
(2) American Society for Testing and Materials (ASTM)
(3) Joint Commission on Accreditation of Healthcare Organizations (JCAHO)
(4) Underwriters' Laboratories (UL)

39. Carbon dioxide and nitrous oxide have the same molecular weight. For this reason, these two gases

(1) have the same vapor pressure
(2) will occupy the same volume if 1 g of each is allowed to expand at the same temperature and pressure
(3) can be used in the same calibrated flowmeter
(4) cannot be distinguished by conventional mass spectroscopy

40. Turbulent flow

(1) in a tube is effective in purging the contents of the tube
(2) increases as the viscosity of the gas increases
(3) is noted at branches, obstructions, or roughness of the tube
(4) rates are calculated by the Hagen-Poiseuille equation

41. The viscosity of a fluid

(1) depends on its density
(2) affects its specific gravity
(3) does not affect resistance to flow
(4) determines its rate of laminar flow

42. A system for measuring intra-arterial blood pressure consists of a force transducer connected to an intra-arterial cannula by a length of fluid-filled tubing. The amount of damping in the system may be decreased by

(1) removing a bubble from the tubing connecting the transducer and cannula
(2) decreasing the length of the tubing connecting the transducer and cannula
(3) using less compliant material for the tubing connecting the transducer and cannula
(4) filling the tubing connecting the transducer and cannula with a less viscous fluid

43. Which of the following solutions has approximately (within 10%) the same osmolality as human plasma?

(1) 5% glucose
(2) 2.5% glucose + 0.2% sodium chloride
(3) 0.9% sodium chloride
(4) 5% glucose + 0.45% sodium chloride

44. The normal value for total serum calcium is approximately 9 mg/dL. If a patient has a normal total serum calcium and a measured serum ionized calcium of 5 mg/dL, the patient

(1) will probably have a positive Chvostek's sign
(2) will have a high circulating level of parathyroid hormone
(3) may have severe vitamin D deficiency
(4) probably does not have a serum albumin level of 2.5 g/dL

45. A ground fault circuit interrupter

(1) is almost never used in operating rooms
(2) is a foolproof method of preventing serious electrical shock
(3) may be used in a grounded electrical system
(4) has both audio and visual alarms to indicate the presence of a ground fault

46. Which of the following statements regarding carbon dioxide transport in the blood is (are) true?

(1) Most of the carbon dioxide entering the blood is converted to bicarbonate ion in the red cell.
(2) Carbon dioxide entering the red cell in peripheral tissues causes enhanced oxygen release from hemoglobin.
(3) Conversion of carbon dioxide to carbonic acid proceeds at a negligible rate nonenzymatically compared to that catalyzed by carbonic anhydrase.
(4) Metabolic alkalosis facilitates oxygen delivery to peripheral tissues.

47. The solution property (ies) which the Lambert-Beer law relate(s) is (are)

(1) light absorbance
(2) length of light path
(3) solute concentration
(4) temperature

Answers and Explanations

1. **(D)** The statement that 1 mole of anything contains an equal number of molecules is Avogadro's hypothesis. The number, 6.023×10^{23}, is Avogadro's number. *(4:945, 1015, 1030)*

2. **(C)** The vapor pressure of a liquid is most dependent on the ambient temperature, and is independent of atmospheric pressure. *(1:574; 4:1019; 5:284–5)*

3. **(C)** At an increased altitude, there is a decrease in both atmospheric pressure and PIO_2, although FIO_2 remains constant at 21%. The hypoxic drive to ventilation is stimulated, increasing minute ventilation, and causing CSF alkalosis. *(1:798–9; 5:2688)*

4. **(B)** The Doppler effect depends on the fact that frequency of a source is changed by the relative motions of the source and the observer or receiver. Doppler technology is being used to an increasing degree in medicine, although not in the ECG, thermodilution cardiac output monitors, force transducers, or Raman spectrometers for determination of gas composition. *(1:680; 4:831–2; 5:1217)*

5. **(B)** At standard temperature and pressure, 1 mole of a gas occupies 22.4 L. Therefore, the weight of these containers would vary according to the molecular weights of the gases. *(1:379; 4:1030)*

6. **(B)** The number of calories required to raise the temperature of 1 g of a substance by 1°C is its specific heat. The specific heat varies for different substances. *(1:574; 4:1021; 5:285)*

7. **(C)** The latent heat of vaporization is the number of calories needed to convert 1 g of liquid into vapor at a constant temperature. This value is dependent on the ambient temperature. The colder the liquid, the more calories needed to vaporize a given amount of liquid. *(1:574; 4:1020; 5:285)*

8. **(D)** Poiseuille's law relates resistance to laminar flow of a fluid through a tube to the length and radius of the tube and the viscosity of the fluid. *(4:439, 1034; 5:2842)*

9. **(D)** A gas will diffuse from a place where it is at a high partial pressure to a place where it is at a low partial pressure. This is the reason nitrous oxide diffuses into air-containing pockets. The final partial pressure of O_2 will be 430 mmHg and that of N_2O will be 330 mmHg. *(1:800; 4:1020; 5:112–3)*

10. **(C)** The Bernoulli theorem states that the velocity of flow and lateral pressure are related for an ideal fluid, i.e., a fluid with no viscosity. If a fluid is flowing through a pipe, the lateral wall pressure decreases as the flow velocity increases. *(4:848; 5:1202)*

11. **(B)** One milliliter of isoflurane liquid is 1.5 g or 0.00813 mole (1.5 g/184.5 g/mole). Thus, by the ideal gas law, $V = nRT/P = (0.00813) \times (0.082) \times (273 + 20) = 0.195\ \text{L} = 195\ \text{mL}$. *(1:379; 4:1021)*

12. **(C)** At constant temperature, the product of the pressure and volume of a gas is constant (Boyle's law). Thus, the volume of gas at sea level where the atmospheric pressure is 14.7 psi will be $(1700\ \text{psi}) \times (6\ \text{L})/(14.7\ \text{psi}) = 694\ \text{L}$. *(4:1013; 5:2667)*

13. **(C)** When electrical currents of 10–20 mA are applied to the upper extremity, sustained muscle contraction occurs of a magnitude that cannot be overcome. If the individual is holding onto a wire, he or she probably will not be able to let go. *(1:145–6; 5:3144–5)*

14. **(D)** Cellular telephones are often prohibited in many locations without adequate scientific justification. Interference with the majority of types of medical equipment occurs extremely rarely. There is a significant likelihood of interference with cardiac pacemakers but only when the telephone was placed right over the pacemaker. When used next to the ear of a person with a pacemaker or defibrillator, there is essentially no risk. *(1:160–1)*

15. **(D)** Desflurane, but not sevoflurane, may react with dry soda lime to yield a potentially toxic concentration of carbon monoxide. The reaction of trichloroethylene, an obsolete anesthetic agent, yielded phosgene. *(1:403–5; 5:297–8)*

16. **(A)** A moving magnet will generate an electric current in a stationary wire and vice versa. This is an example of inductance and is the principle behind electric generators. *(1:144; 4:702)*

17. **(D)** Radiation intensity decreases as a function of the square of the distance. Thus, by doubling the distance between the patient and the anesthesiologist, the amount of the radiation exposure will decrease to one-fourth its previous value. *(1:1329; 5:3153)*

18. **(C)** Poiseuille's law relates laminar flow to the fourth power of the radius:

$$\text{Flow} = \frac{\pi p r^4}{8vl}$$

where p is the pressure, r is the radius of the tube, v is the viscosity of the gas, and l is the length of the tube. *(4:1016; 5:2842)*

19. **(B)** According to the Nernst equation, the resting potential of the cell is proportional to the logarithm of the ratio of the potassium concentrations inside and outside of the cell. This ratio is normally about 30–40. If the ratio were to double to 60–80, the resting potential would decrease (i.e., become more negative) by approximately 20%. *(1:743; 5:578)*

20. **(D)** By Ohm's law, if a voltage of 100 V is applied to a resistance of 20 Ω, a current of 100 V/20 Ω = 5 A will flow. Power dissipation is the product of voltage and current in the circuit, or in this case 5 A × 100 V = 500 W. *(1:143; 4:701–2)*

21. **(D)** A molecule of calcium chloride dissociates in solution to yield three particles: one calcium ion and two chloride ions. For each mole of calcium chloride dissolved in a solution, an osmotic concentration of 3 Osm results. Therefore, a 0.1 molal solution of calcium chloride yields a 0.3 osmolal solution, or 300 mOsm/kg H_2O. *(1:176, 180; 4:947–8; 5:1104)*

22. **(C)** Calcium phosphate is an insoluble salt; therefore, the administration of phosphate ion to a patient with hypercalcemia will cause precipitation of calcium phosphate. This therapy is immediately effective in treating hypercalcemia, but at the cost of the deposition of insoluble calcium phosphate throughout the body. *(1:192; 4:970–1; 5:1774)*

23. **(A)** By the Henderson-Hasselbalch equation, $pH = pK + \log [HCO_3^-]/[H_2CO_3]$. $[H_2CO_3]$ may be replaced by $PaCO_2 \times 0.03$. With the values given, this yields a normal pH of 7.4. *(1:165; 4:967; 5:1600)*

24. **(B)** The alveolar gas equation relates PAO_2 to FIO_2 as follows:

$$PAO_2 = FIO_2(P_{atm} - P_{H_2O}) - \frac{PaCO_2}{RQ}$$

where P_{H_2O} is the value for saturated water vapor pressure at body temperature, and RQ is the respiratory quotient, which is normally approximately 0.8. Thus, in this patient, PAO_2 is calculated to be 449 mmHg. Since the normal alveolar-arterial oxygen difference is less than 10 mmHg, the PaO_2 should be approximately 440 mmHg. *(1:804; 5:1010; 1439)*

25. **(B)** A molecule of carbon dioxide reacts with one molecule of water to form carbonic acid. The reaction of one molecule of carbonic acid with two molecules of sodium hydroxide produces two molecules of water. Therefore, there is a net production of one molecule of water. *(1:582; 4:1046; 5:296)*

26. **(B)** When the line isolation monitor alarms, the threshold amount of current could flow through the patient if the ungrounded power system were to become grounded. A current of 2 mA will not injure a patient unless it is applied directly to the heart. If a large number (e.g., 20) of properly functioning electrical devices are plugged in simultaneously in an operating room, each having an acceptable leakage current (100 μA), it is still possible to exceed the 2 mA threshold for sounding the line isolation monitor. When the line isolation monitor alarms, noncritical electrical equipment should be disconnected. If the alarm condition is not corrected, surgery may proceed, but the remaining equipment should be checked before being used again. *(1:152–5; 5:3142–3)*

27. **(A)** According to the Meyer-Overton rule, anesthetic potency is proportional to lipid solubility. *(1:131; 4:1106–7; 5:115)*

28. **(B)** The standard frequency used by electrical utilities in the United States is 60 Hz, while the current delivered by an electrocautery device is in the range of 10^5 to 10^6 Hz. *(1:157–8; 5:3146)*

29. **(A)** The boiling point of a liquid is a function of the atmospheric pressure, and will decrease with decreasing atmospheric pressure. At the boiling point, the vapor pressure of the liquid is equal to atmospheric pressure. Liquid enclosed within a container will be at equilibrium with vapor above the liquid at any temperature below the critical temperature and above the freezing point. Ambient temperature has no effect on the boiling point of a liquid. *(1:379; 4:1019–20; 5:285)*

30. **(B)** The albumin and hetastarch solutions are colloidal, meaning that they contain molecules that do not diffuse freely across capillary membranes. Although the 7.5% sodium chloride solution is hyperosmolar and can expand plasma volume, capillary membranes are freely permeable to sodium chloride. *(1:175–7; 4:986–91; 5:1786)*

31. **(B)** Specific gravity is defined as the ratio of the mass of one substance to the mass of a similar volume of water at the same temperature. Specific gravity may be specified for solids, liquids, or gases; however, for gases, the mass of the similar volume of air is used in the denominator. Both substances must be compared at the same conditions of temperature and pressure. Mass per unit volume is the unit used for density. *(4:1370–1; 5:1667)*

32. **(D)** The density of a gas determines its rate of turbulent flow through an orifice; however, laminar flow is independent of gas density but is a function of viscosity. Density is a function of temperature and is not related to the explosive potential of a gas. *(1:795–6; 4:1016–7; 5:278)*

33. **(B)** Any direct entry into the heart, such as with a temporary pacing wire or a central venous catheter for total parenteral nutrition, is a potential pathway for microshock. Electrically operated beds may increase the patient's risk of macroshock if the bed were to malfunction. Isolation transformers serve to protect the grounded patient from contact with an electrical wire. *(1:155–6; 4:708–10; 5:3145–6)*

34. **(E)** The Venturi tube is the basis of injector systems. The gradual opening distal to the constriction maintains the increased flow through the constriction. Two gases are noted in the diagram: the delivery gas B and the entrained gas A. *(4:2203; 5:1203–4, 2813)*

35. **(D)** An isolated power supply affords protection against macroshock. Grounding and leakage current can still occur. An isolation transformer provides no protection against shock if a person contacts both wires from the transformer secondary. *(1:150–2; 4:705–6; 5:3141–2)*

36. **(E)** Clinically significant amounts of heat loss in the operating room occur via all four processes. *(1:684; 5:1576)*

37. **(D)** Leakage current is the unintentional flow of alternating current from the internal wiring of a device. It may be a hazard to the patient with direct lines to the heart at levels over 10 μA. The most common source of leakage current is the capacitance between electrical conductors supplying equipment. *(1:152–4)*

38. **(E)** Safety regulations for operating room design are in publications from the NFPA and the JCAHO. ASTM publishes specifications on medical devices, including anesthesia machines. UL publishes standards for electrical equipment used in medicine. They also test many electrical devices, and those that meet their specifications may carry the "UL" seal. *(1:94–5, 157, 161, 567; 4:710–5; 5:274, 2987, 3140)*

39. **(C)** If pressure and temperature are constant, the volume of a gas is proportional to the number of gas molecules. The same mass of two gases with the same molecular weight will have the same number of molecules. Mass spectroscopy determines compounds on the basis of their molecular weight. Some sophisticated mass spectrometers can differentiate two compounds of identical molecular weight by their fragmentation patterns. Two gases can be used in the same flowmeter if they had the same densities and viscosities. *(1:379, 570–1, 669–70; 4:871–2, 1013)*

40. **(B)** Turbulent flow is characterized by irregular lines of flow and is likely at branches, obstructions, and roughness of the tube. When turbulent flow is present in a tube, the gas contents of the tube are effectively purged. Turbulent flow is inversely related to the density of the gas and independent of its viscosity. The Hagen-Poiseuille equation defines the flow rate in a tube when the flow is laminar. *(1:795; 4:1016)*

41. **(D)** The viscosity of a fluid is an intrinsic property that affects its laminar flow characteristics. It is not dependent on density and does not affect its specific gravity. It does affect resistance to flow. *(1:795; 4:1016; 5:1217)*

42. **(E)** All of these measures will decrease the damping in the system. *(1:673; 4:807–8; 5:1276–7)*

43. **(B)** The osmolality of plasma is about 285 mOsm/kg H_2O. The osmolality values for 5% glucose and 0.9% sodium chloride are 278 and 308, respectively; therefore, these solutions are approximately isosmotic (or isotonic) with plasma. The osmolality of 2.5% glucose + 0.2% sodium chloride is 207; therefore, it is a hypotonic solution. The osmolality of 5% glucose + 0.45% sodium chloride is 437; therefore, it is a hypertonic solution. *(1:175–6; 4:948–50; 5:2463–4)*

44. **(D)** This patient has a normal value for ionized calcium, which is normally approximately 50% of the total serum calcium. Therefore, the patient is unlikely to be hypoalbuminemic, i.e., having a serum albumin level less than 3.5 g/dL, since albumin is the molecule to which most calcium is bound. If the patient has a low value for ionized calcium, which may result from severe vitamin D deficiency, a positive Chvostek's sign and an elevated parathyroid hormone level are likely. *(1:189–90; 4:434,969–72; 5:1771)*

45. **(B)** A ground fault circuit interrupter prevents current from flowing in a circuit when there is an imbalance in the current flowing in the two sides of the circuit. This device provides a high degree (although not foolproof) of protection but is rarely used in operating rooms because it might disconnect power from life-support equipment. It does not provide either an audio or visual alarm but disconnects the power when it detects a fault. It is usually used in grounded electrical systems. *(1:155; 5:3140)*

46. **(A)** Metabolic alkalosis shifts the oxyhemoglobin dissociation curve to the left, causing an increased affinity of hemoglobin for oxygen, and impairing oxygen transport to peripheral tissues. The other statements accurately describe the physiology of carbon dioxide transport. *(1:165; 5:703)*

47. **(A)** The Lambert-Beer law states that absorbance of light through a solution is equal to the products of the length of the light path, the solute concentration, and the molar extinction coefficient of the solute. *(5:1212–3)*

CHAPTER 2

Anesthesia Equipment

Questions

DIRECTIONS (Questions 48 through 72): Each of the numbered items or incomplete statements in this section is followed by answers or by completions of the statement. Select the ONE lettered answer or completion that is BEST in each case.

48. A patient is being monitored with a bispectral index (BIS) monitor. When the value for BIS is 60, it means that the patient

(A) has about a 60% probability of being awake
(B) has a very small (less than 1%) probability of having recall of intraoperative events
(C) does not require additional opioid
(D) if not pharmacologically paralyzed will not move in response to surgical incision
(E) is less likely to be awake than if the BIS value were 50

49. The lifetime of a canister of soda lime

(A) depends on the method of filling
(B) is independent of the volume of CO_2 exhaled
(C) is independent of the location of the relief valve
(D) is prolonged by low gas flows
(E) is shortened by channeling

50. The oxygen tanks on an anesthesia machine are

(A) G tanks
(B) M tanks
(C) E tanks
(D) D tanks
(E) B tanks

For Questions 51 and 52, please refer to Figure 2-1

51. The system depicted in the figure is the

(A) Georgia valve
(B) Jackson-Rees system
(C) T-piece
(D) Bain circuit
(E) Mapleson A system

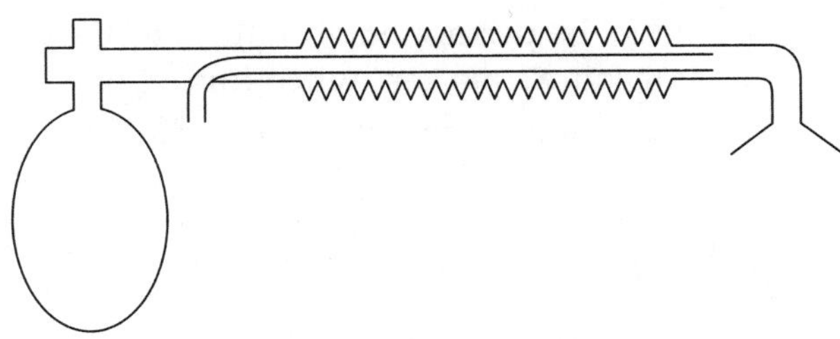

FIG. 2-1

52. A disadvantage of the circuit in Figure 2-1 is the

(A) inability to use spontaneous ventilation with the system
(B) requirement for low flow
(C) inability to scavenge waste gases
(D) presence of overflow valve farther from the patient
(E) kinking of inner delivery tube

53. In closed-circuit anesthesia, the one parameter that must be met is the

(A) tidal volume
(B) minute volume
(C) respiratory rate
(D) oxygen consumption
(E) needed anesthesia

54. If a variable bypass vaporizer is tipped over

(A) the wick will become saturated
(B) the vaporizer must be returned to the factory
(C) the concentration of the vapor will be higher than calculated
(D) a low concentration will result
(E) the vaporizer should be righted and then may be put into use

55. A negative-pressure leak test

(A) is accomplished by having the clinician apply mouth suction to the tubing connected to the common gas outlet
(B) reliably finds leaks in the carbon dioxide absorber
(C) is only appropriate for anesthesia machines containing a check valve downstream from the vaporizers
(D) must be performed with the anesthesia machine's master switch turned on
(E) may detect internal vaporizer leaks

56. The position best tolerated by the surgical patient is the

(A) lithotomy position
(B) prone position
(C) horizontal supine position
(D) Trendelenburg position
(E) Fowler position

57. The pumping effect in reference to a vaporizer

(A) leads to a decreased concentration of the agent
(B) refers to a method of filling the vaporizer
(C) refers to a method of cleaning the vaporizer
(D) leads to a decreased flow through the vaporizer
(E) is more pronounced at low flow rates

58. An open waste gas scavenging system

(A) must have a negative-pressure relief valve
(B) must have a positive-pressure relief valve
(C) must be connected to a source of vacuum
(D) does not need a reservoir
(E) cannot be simultaneously connected to the APL (adjustable pressure limiting) and ventilator relief valves

59. A mechanism used to reduce the pressure of a gas from a compressed gas cylinder to a usable, nearly constant pressure is

(A) a gauge
(B) a flowmeter
(C) an indicator
(D) a regulator
(E) a check valve

60. Two identical syringes are fitted with needles, one of which is twice the diameter of the other. The same force is applied to the plungers. The volume ejected in unit time is

(A) not related to the needle size
(B) four times greater through the large needle
(C) eight times greater through the large needle
(D) sixteen times greater through the large needle
(E) thirty-two times greater through the large needle

61. If a nitrous oxide tank is contaminated with water vapor, ice will form on the cylinder valve as a result of the

(A) latent heat of vaporization
(B) specific heat
(C) vapor pressure
(D) low pressure of the nitrous oxide
(E) ambient temperature

62. Electrocautery machines do not cause ventricular fibrillation because the current they deliver differs from the electric current supplied by wall electrical outlets primarily by being

(A) direct current instead of alternating current
(B) lower in voltage
(C) lower in frequency
(D) higher in frequency
(E) lower in amperage

63. The Tec-5 vaporizers are

(A) temperature and pressure compensated
(B) temperature compensated only
(C) pressure compensated only
(D) neither pressure nor temperature compensated
(E) most accurate at 30°C

64. The line isolation monitor

(A) measures leakage current flowing from the patient to ground
(B) measures leakage current flowing from the electrical equipment to the patient
(C) sounds an alarm if the leakage current exceeds 50 mA
(D) measures the impedance between the AC wiring and ground
(E) cuts off power to the circuit if a faulty piece of equipment is connected

65. The anesthesia system depicted in Figure 2-2 is a

(A) to-and-fro system
(B) Magill system
(C) circle system
(D) nonrebreathing system
(E) T-piece

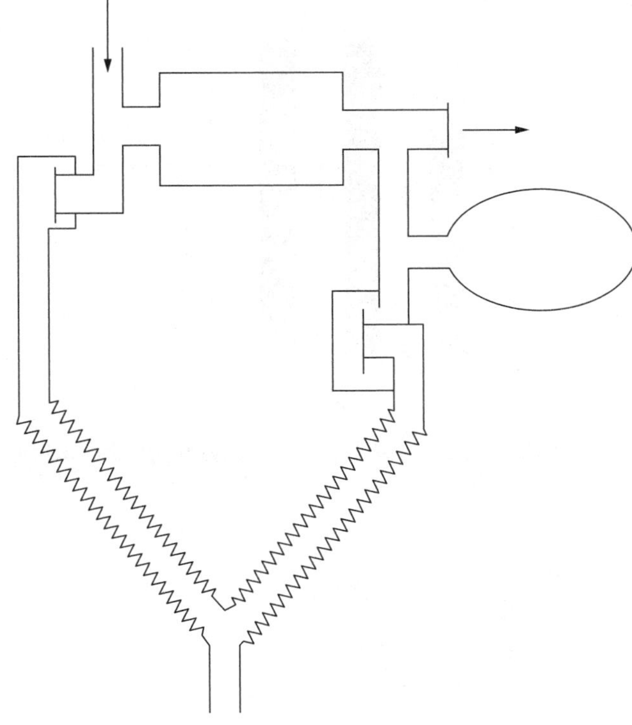

FIG. 2-2

66. Which of the following statements is TRUE regarding blood pressure cuffs?

(A) The bladder length should be 50% of the limb circumference.
(B) The bladder width should be 40% of the limb circumference.
(C) A cuff designed for a thigh cannot be used on a large arm.
(D) A cuff designed for an arm cannot be used on a small thigh.
(E) A cuff whose bladder is too narrow for the limb will give an erroneously low blood pressure.

67. The indicator in carbon dioxide absorbent is

(A) methylene blue
(B) ethyl violet
(C) bromthymol blue
(D) phenolphthalein
(E) cresol purple

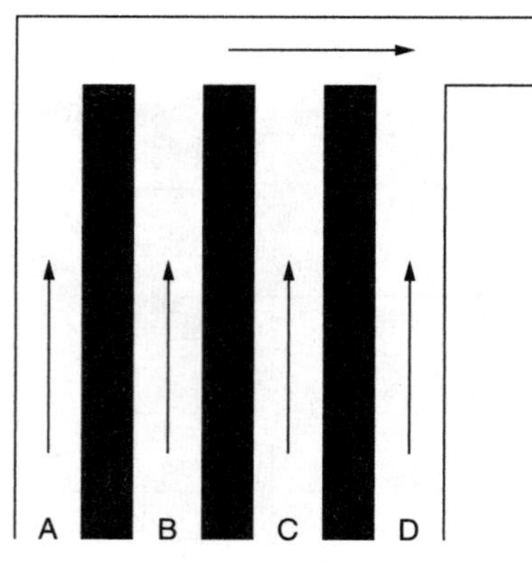

FIG. 2-3

68. In Figure 2-3, the oxygen flowmeter should be at position

(A) A
(B) B
(C) C
(D) D
(E) position is not critical

69. If the skin resistance between a person's arms is 6000 Ω, and the person's right arm is connected to an earth ground, and 120 V are applied to the left arm, the current flowing from arm to arm will be

(A) nearly 0 if the 120 V are supplied by a grounded electrical system
(B) 20 mA if the 120 V are supplied by a grounded electrical system
(C) 20 mA if the 120 V are supplied by an isolated electrical system
(D) 50 mA if the 120 V are supplied by an isolated electrical system
(E) 50 A if the 120 V are supplied by an isolated electrical system

70. Overfilling of a vaporizer

(A) will cause the vaporizer not to function
(B) will yield a concentration of agent lower than that dialed
(C) is not likely with modern vaporizers
(D) is usually manifested by the agent leaking out of the vaporizer
(E) will decrease oxygen delivery

71. All of the following are true of the lithotomy position EXCEPT

(A) the back is supine
(B) the knees are flexed
(C) the hips are flexed
(D) the legs are internally rotated
(E) there are few hemodynamic disadvantages to the position

72. The Vapor 19.1 vaporizer is temperature compensated as a result of a(n)

(A) bimetallic band
(B) copper element
(C) expansion element in the vaporizing chamber
(D) sintered disc
(E) heating element that keeps the vaporizer at a constant temperature

DIRECTIONS (Questions 73 through 100): For each of the items in this section, ONE or MORE of the numbered options is correct. Choose answer

(A) if only 1, 2, and 3 are correct
(B) if only 1 and 3 are correct
(C) if only 2 and 4 are correct
(D) if only 4 is correct
(E) if all are correct

73. Which of the following flowmeter indicators are read at the center of the indicator?

(1) plumb bob float
(2) skirted float
(3) floats which do not rotate in the rotameter tube
(4) ball float

74. An anesthesiologist is using an anesthesia machine equipped with an oxygen proportioning system, as well as an extra flowmeter which is supplied by a helium tank. A hypoxic gas mixture may be administered to the patient if

(1) the gas supplied by the wall oxygen outlet is not oxygen
(2) there is a leak in the oxygen flowmeter
(3) the helium flowmeter is adjusted to an excessively high flow rate
(4) the nitrous oxide flowmeter is adjusted to an excessively high flow rate

75. When using a Macintosh blade for laryngoscopy, the

(1) epiglottis is lifted directly
(2) laryngoscope is held in the right hand
(3) blade enters the mouth on the left side
(4) tip is advanced into the vallecula

76. Nasopharyngeal airways are contraindicated in the patient with

(1) nasal polyps
(2) infection of the nasopharynx
(3) a very low platelet count
(4) the jaw wired shut

77. The suction port of a fiber-optic bronchoscope may be used to

(1) administer oxygen
(2) spray local anesthetic
(3) aspirate tracheal secretions
(4) ventilate the patient during bronchoscopy

78. Anesthesia equipment that is safe to use in a room containing a functioning magnetic resonance imaging (MRI) machine includes

(1) laryngeal mask airway
(2) fiber-optic oximeter probe
(3) blood pressure cuff
(4) electrocardiogram (ECG) cable and electrodes

79. A capnometer may detect

(1) exhaustion of carbon dioxide absorbent
(2) malfunctioning of an open scavenger system
(3) added mechanical dead space
(4) disconnection of wall oxygen supply

80. The oxygen analyzer

(1) is a reliable detector of endotracheal tube disconnection
(2) works by infrared spectroscopy
(3) should be calibrated using a mixture of oxygen and nitrous oxide
(4) can detect if the patient is administered a hypoxic gas mixture

81. When an anesthesia machine equipped with an outlet check valve is tested for leaks, the correct procedure involves

(1) applying back pressure at the junction of the circle absorber system and the common gas outlet
(2) testing for vaporizer leaks with the vaporizer turned on
(3) disconnecting the machine from the gas outlets on the wall
(4) attaching a suction bulb to the common gas outlet

82. Pulse oximeters

(1) are based on Beer's law
(2) can differentiate carboxyhemoglobin from oxyhemoglobin
(3) cannot differentiate methemoglobin from oxyhemoglobin
(4) use a single wavelength of red light

83. Soda lime granules

(1) are manufactured to be smooth and round
(2) are of a size 4 to 8 mesh
(3) of a small size give a lower resistance
(4) of a small size give better absorption function

84. The isolated power supply system used in operating rooms requires that

(1) the metal portions of the operating table be connected to earth ground
(2) the patient be insulated from the metal portions of the operating table
(3) conductive flooring be used in the operating room
(4) a transformer be connected between electrical equipment in the operating room and the electric power supplied by the utility company

85. The standard measurement(s) for anesthesia breathing circuit and ventilator fittings is (are)

(1) 12 mm
(2) 15 mm
(3) 19 mm
(4) 22 mm

86. In a nonrebreathing system

(1) inspired gas tensions may be held constant
(2) humidification is optimal
(3) change of flow produces rapid changes in inspired concentrations
(4) one adds oxygen in amounts just to satisfy metabolic requirements

87. Gases and vapors that may be measured by infrared spectrometry include

(1) oxygen
(2) isoflurane
(3) nitrogen
(4) nitrous oxide

88. The fail-safe valves on anesthesia machines

(1) prevent delivery of a hypoxic mixture
(2) prevent flow unless the oxygen flowmeter is on
(3) require that oxygen tanks be full
(4) are on if oxygen pressure is present

89. Venous air embolism may be detected by

(1) alteration in heart sounds detected by precordial Doppler
(2) decrease in expired carbon dioxide on the capnometer
(3) increase in expired nitrogen measured by mass spectroscopy
(4) decrease in oxygen saturation indicated by the oximeter

90. The BIS monitor reliably indicates the depth of hypnosis when it is achieved via the administration of

(1) isoflurane
(2) midazolam
(3) propofol
(4) ketamine

91. A nitrous oxide tank contains gas at a pressure of 750 psi. When the last drop of liquid nitrous oxide evaporates

(1) the pressure will fall rapidly
(2) the pressure will be zero
(3) the pressure will remain at 750 psi until the tank is empty
(4) the rate of pressure fall is dependent on the rate of flow

92. The float in a flowmeter

(1) may stick at a high level if the tube is dirty
(2) is made to remain static during flow
(3) may give an inaccurate reading if back pressure builds up
(4) may be replaced with another if it is lost

93. The oxygen flush valve

(1) can apply oxygen at the pressure of the compressed oxygen cylinders to the breathing system
(2) can cause barotrauma
(3) may introduce high fresh gas flows through an open vaporizer
(4) should not be pressed during the inspiratory phase of positive-pressure ventilation

94. Which of the following situations regarding intraoperative monitoring is mandated by the American Society of Anesthesiologists (ASA) Standards for Basic Intraoperative Monitoring?

(1) A patient having a total knee replacement under spinal anesthesia with intravenous sedation and receiving supplemental oxygen must have the expired carbon dioxide measured.
(2) A patient having intraoperative radiation therapy under general anesthesia must have a resident or attending anesthesiologist or nurse anesthetist present in the operating room at all times during the procedure.
(3) A woman in labor with an indwelling epidural catheter and receiving a continuous epidural infusion of local anesthetic must have her blood pressure taken and recorded every 5 minutes.
(4) A patient having an inguinal herniorrhaphy under local anesthesia with monitored anesthesia care must have a resident or attending anesthesiologist or nurse anesthetist present in the operating room at all times during the surgical procedure.

95. The diameter index safety system (DISS)

(1) prevents attachment of gas-administering equipment to the wrong type of gas
(2) prevents incorrect yoke-tank connections
(3) is based on matching specific bores and diameters
(4) is used on scavenger hoses

96. In the supine position, pressure points which should be padded are the

(1) occiput
(2) elbows
(3) heels
(4) knees

97. The American Society for Testing and Materials standard number F1850-00 requires newly manufactured anesthesia machines to have a(n):

(1) oxygen analyzer
(2) pulse oximeter
(3) pressure monitor in the breathing system
(4) ventilator with a disconnect alarm

98. Factors which increase the difference between arterial and measured end-tidal carbon dioxide include

(1) mismatch of ventilation and perfusion
(2) wheezing
(3) high fresh gas flow rates
(4) high cardiac output

99. True statements regarding ECG lead placement include

(1) Lead I displays the ECG signal recorded from the left arm and right arm.
(2) Lead II displays the ECG signal recorded from the left arm and right leg.
(3) Lead V6 displays the ECG signal recorded from a unipolar electrode placed in the fifth intercostal space at the left midaxillary line.
(4) Lead V1 displays the ECG signal recorded from a unipolar electrode placed in the second intercostal space just to the right of the sternum.

100. The laryngeal mask airway

(1) protects against aspiration as well as an endotracheal tube
(2) may prevent airway obstruction during monitored anesthesia care
(3) requires a rigid or flexible fiber-optic laryngoscope for proper placement
(4) permits positive-pressure ventilation

Answers and Explanations

48. **(B)** The BIS monitor measures the depth of hypnosis and the BIS value can be used to predict the probability that a patient will be awake and responsive or will have recall at a given point in time. The BIS value is not a measure of the likelihood of movement or of the magnitude of noxious stimuli requiring opioid therapy. During surgery, the typical BIS target value is between 40 and 60. At a BIS value of 60, the probability of responsiveness to verbal command is about 20%, while the probability of recall is less than 1%. *(1:682; 5:1250–7)*

49. **(E)** Channeling of the airflow shortens the life of soda lime by allowing passage of the exhaled gas along channels of low resistance. This expends the soda lime along the channels, and carbon dioxide will flow through the system without being absorbed. The method of filling does not affect the length of soda lime usefulness. The relief valve position does have an effect. The valve should be placed to vent the air that has the highest concentration of carbon dioxide. The higher the gas flow, the longer the soda lime will last. *(1:582; 4:1046–7; 5:296)*

50. **(C)** Gas cylinders are designated by letters from A to M. The size increases as one advances in the alphabet. *(1:568; 4:1013; 5:276)*

51. **(D)** The circuit shown is a Bain circuit, which is a modification of the Mapleson D system. *(1:580–1; 4:1044; 5:293–4)*

52. **(E)** The inner tube may become disconnected or it may kink, leading to excessive rebreathing of exhaled gases or disruption of fresh gas flow. The Bain circuit permits spontaneous ventilation, requires relatively high fresh gas flows, permits easy waste gas scavenging, and positions the pop-off valve away from the patient for easier adjustment. *(1:580–1; 4:1044; 5:293–4)*

53. **(D)** In any anesthesia system, the one factor that must be provided for is oxygen. In a closed-circuit system, the inflow of fresh gas matches that which is taken up by the patient. *(5:143–6)*

54. **(C)** If the vaporizer tips over, liquid agent may find its way into the bypass chamber. This will cause an increased concentration of anesthetic agent in the system. If tipping occurs, it is not necessary to return the vaporizer to the factory, but it must be righted and flushed with fresh gas at high flow rates for at least 30 minutes with the vaporizer set for a low concentration of agent. It should not be put into immediate use. *(1:574–5; 4:1027; 5:285–8)*

55. **(E)** The negative-pressure leak test is a universal test for leaks in the low-pressure circuit of anesthesia machines, regardless of whether the low-pressure circuit contains a check valve. Performing the test requires applying negative pressure with a suction bulb to the common gas outlet (from which the carbon dioxide absorber has been disconnected) with the machine's master switch turned off. With the suction bulb collapsed, each vaporizer is individually turned on and reinflation of the suction bulb indicates an internal leak in that vaporizer. *(1:590; 5:307–8)*

56. **(C)** The horizontal supine position is the one best tolerated, but even that one has its problems. Merely because the patient does not have to be moved should not cause one to be complacent with positioning. Pressure points should be padded, the superficial nerves and the eyes protected. *(1:640–5; 4:692; 5:1158–63)*

57. **(E)** The pumping effect refers to the increase in anesthetic concentration from a vaporizer when there is reversed gas flow through the vaporizer as may occur during assisted or controlled ventilation. This effect is more pronounced at low flow rates. *(1:575–6; 4:1034; 5:287)*

58. **(C)** An open waste gas scavenging system requires a reservoir but no valves. It must be connected to a source of vacuum to actively withdraw waste gases. It is usually connected to both the APL and ventilator relief valves via a "Y" connector. *(1:586–8; 5:304–6)*

59. **(D)** The mechanism to reduce pressure of a gas to a useful pressure is a regulator. A gauge is a device to measure the pressure. The flowmeter is a device to measure the flow being delivered. A check valve is a device to allow flow in one direction only. *(1:568; 4:1014; 5:276)*

60. **(D)** According to Poiseuille's law, flow is proportional to the radius of the tube raised to the fourth power. Since the internal radii are in a ratio of 1:2, the volume ejected in that time will be 16 times greater through the larger needle. *(4:1016; 5:2842)*

61. **(A)** The nitrous oxide tank contains a liquid, and in order for it to become vaporized, heat must be supplied. As the cylinder is opened, heat is removed from the cylinder and from the air in the immediate vicinity. The temperature falls, causing condensation. *(4:1020; 5:285)*

62. **(D)** Wall electrical outlets supply alternating current at a frequency of 60 Hz (in the United States; 50 Hz in some other countries). This frequency is actually the one to which the heart muscle is most susceptible. Electrocautery machines use alternating current of a much higher frequency, approximately 10^5 to 10^6 Hz. *(1:157–8; 5:3146)*

63. **(B)** The Tec-5 series of vaporizers are temperature compensated. Variable bypass vaporizers are not pressure compensated. In terms of anesthetic potency, operating a variable bypass vaporizer under hypobaric conditions (e.g., high altitude) delivers a more potent than expected anesthetic concentration to the patient. Conversely, operating the vaporizer under hyperbaric conditions (e.g., in a hyperbaric chamber) delivers a less potent than expected anesthetic concentration to the patient. *(1:574–5; 4:1025–6; 5:285–9)*

64. **(D)** The line isolation monitor measures the impedance from the AC wiring of an isolated power system to ground, and sounds an alarm when the current that could flow to ground exceeds 5 mA. Unlike a fuse or a circuit breaker, it does not cut off power to the circuit. *(1:152–3; 4:705–6; 5:3140–3)*

65. **(C)** The system in the picture is a circle system. This system has an absorber that contains soda lime or a similar absorbent, a fresh gas inlet, valves to direct flow, a relief valve, pressure gauge, breathing tubes, reservoir bag, and Y-piece. An oxygen analyzer is also a part of the system. *(1:581–2; 4:1046; 5:295–6)*

66. **(B)** The bladder length should be at least 80%, and the width 40% of the limb circumference. If the bladder is too small, a larger pressure will be needed to occlude flow and an erroneously high blood pressure reading will result. The labeling of a cuff as "adult" or "thigh" is a general guideline; the cuff size should be chosen to match a particular limb size, regardless of whether the limb is the arm or leg. *(1:671–2; 4:804; 5:1169–70)*

67. **(B)** The indicator in carbon dioxide absorbent is ethyl violet. *(1:582; 4:1047; 5:297)*

68. **(D)** Placement of the oxygen in the position nearest to the common gas outlet will avoid delivery of a hypoxic mixture should a crack occur in the flowmeter tubing. If a leak occurs in

this position, all components of the gas will leak out of the crack in the tubing. *(1:572; 4:1016; 5:280)*

69. (B) Current will flow only if both the patient and the electrical system are grounded. The magnitude of the current is given by Ohm's law, which states that current (in amps) equals the voltage (in volts) divided by the resistance (in ohms). Thus, the current will equal 120 V/ 6000 Ω or 0.02 A, which is 20 mA. *(1:143–4; 4:701–2; 5:1209, 3139–43)*

70. (C) Overfilling of the vaporizer is mainly a problem of the past. Modern vaporizers are designed to avoid this hazard. When overfilling of a vaporizer occurs, the hazard of high delivered concentrations is present due to the presence of liquid agent in the bypass chamber. *(1:574–6; 4:1027; 5:285–9)*

71. (D) In the lithotomy position, internal rotation of the legs stretches the common peroneal nerve around the head of the fibula, causing a nerve palsy to be more likely. Hemodynamic stability is preserved because the legs are elevated above the heart. *(1:642–3; 4:692–3; 5:1155, 1158–9)*

72. (C) The Vapor 19.1 vaporizer accomplishes temperature compensation by virtue of an expansion element that extends into the vaporizing chamber. No heating elements are involved. *(1:575–6; 4:1025–6; 5:287)*

73. (D) The ball float is read at the midpoint of the ball. If one reads the flow at the bottom of a float designed to be read at the top, there is a considerable difference in flows. This may not be important in all machines or in all cases, but the effect is magnified at low flow rates. *(1:571; 5:279)*

74. (A) An oxygen proportioning system mechanically connects the oxygen and nitrous oxide flow controls so that as the nitrous oxide flow is increased, or the oxygen flow is decreased, the flow of the other gas is changed to maintain a minimum oxygen concentration of 25%. The system depends on oxygen being correctly supplied to the oxygen flow control. Therefore, increasing the nitrous oxide flow will not, by itself, result in a hypoxic mixture. Any additional flow controls (e.g., helium and carbon dioxide) are not linked to the oxygen flow control, and increasing their flow could lead to a hypoxic mixture. Oxygen leaking from the oxygen flowmeter is not detected by the proportioning system and could also result in a hypoxic mixture. *(1:573; 1017–8; 5:283)*

75. (D) The Macintosh blade is used by placing it in the vallecula (the space between the base of the tongue and the epiglottis). The laryngoscope is held in the left hand, and the blade enters the mouth on the right. *(1:607; 4:1084–5; 1634)*

76. (A) Any instrumentation of the nose should be done after taking a history to ascertain the absence of a condition likely to precipitate problems. Nasal polyps that may bleed or be sheared off, a low platelet count, and infection of the nasal area are contraindications to the use of a nasopharyngeal airway. A wired jaw, in itself, is not a contraindication to a nasopharyngeal airway. *(1:599; 4:1076; 5:1619, 1625)*

77. (A) The suction port of a fiber-optic bronchoscope is designed primarily to aspirate secretions; however it may also be used to administer oxygen to a spontaneously breathing patient or to spray local anesthetic in the airway or on the vocal cords. The patient cannot be ventilated via this port. *(1:621; 5:1645–6)*

78. (E) Administering anesthesia and/or monitoring a patient during an MRI scan is a challenge. Laryngeal mask airways are safe. An oximeter probe with a fiber-optic cable to the monitor is safe, while an oximeter probe connected with an electrical cable may cause a burn at the probe site. Specially shielded ECG cables are required and burns may still occur during very long MRI scans. Blood pressure cuffs are safe as long as the tubing connectors are not ferromagnetic. *(1:759–60; 5:2644–6)*

79. (B) Exhaustion of the carbon dioxide absorbent and added mechanical dead space will increase the amount of rebreathing of exhaled air and also increase the inspired carbon dioxide concentration. This result would be displayed as an increasing baseline on the capnometer waveform. An open scavenging system cannot apply positive or negative pressure to the breathing circuit, and therefore a malfunction cannot be detected within the breathing circuit. If the wall oxygen supply is disconnected, oxygen will be supplied by the gas cylinders on the anesthesia machine, if their valves are open. If the cylinders are closed or empty, all fresh gas flow from the common gas outlet will cease. *(1:821; 4:1073, 1046; 5:1458–9)*

80. **(D)** The oxygen analyzer is the only monitor designed to quickly and reliably detect when a hypoxic gas mixture is administered to the patient, regardless of the reason. It is usually placed in the inspired limb of the breathing circuit, where it cannot detect a disconnection of the endotracheal tube. Available oxygen analyzers use electrochemical or galvanic detectors; infrared spectroscopy is used to measure carbon dioxide, nitrous oxide, and volatile anesthetics. The oxygen analyzer is calibrated to 21% using room air or to 100% oxygen. *(1:668; 4:863–6; 5:305, 311, 1454)*

81. **(C)** In an anesthesia machine with an outlet check valve, the application of pressure to the circle absorber system will check the absorber and breathing system tubing for leaks, but it will also close the outlet check valve and not detect leaks upstream from this valve. Therefore, a negative-pressure leak test is performed by attaching a suction bulb to the common gas outlet. The presence of vaporizer leaks is determined by turning on each vaporizer and determining if the suction bulb remains collapsed. *(1:589–90; 4:1039–40; 5:307–8)*

82. **(B)** Pulse oximeters use two wavelengths of light—one red and the other infrared—to estimate the percentage of hemoglobin in the oxygenated form by applying Beer's law which relates absorbance of light by a chemical to its concentration. Both carboxyhemoglobin and methemoglobin are misinterpreted as oxyhemoglobin by pulse oximeters. *(1:670–1; 4:877–9; 5:1213–4, 1448–9)*

83. **(C)** Soda lime granules are in the size of 4 to 8 mesh. Smaller granules give a larger absorptive surface but cause more resistance. The irregular shape is better to give more absorptive area. *(1:582, 4:1046–7; 5:296)*

84. **(D)** In an isolated power supply system, an isolation transformer is connected between the electric power supplied by the utility company and the electrical outlets in the operating room. The safety provided by this system depends on the patient and any wires or metallic objects in contact with the patient not being connected to earth ground. Conductive flooring was used in the past to avoid static electricity sparks which could cause fires or explosions in the presence of flammable anesthetics. *(1:150–2; 4:705–6; 5:3141–3)*

85. **(C)** Standard fitting sizes used in modern anesthesia delivery systems are 15 mm (endotracheal tube fitting) and 22 mm (breathing system fitting). The fitting for the scavenger hose is 19 mm. *(4:1059; 5:305)*

86. **(B)** In a nonrebreathing system, gas concentrations can be held constant. Changes can be made rapidly also, since there is no reservoir that must be equilibrated. Humidification is poor because no rebreathing occurs and exhaled gases are scavenged. Oxygen is always provided in amounts sufficient to provide a wide latitude in consumption. *(4:1041–6; 5:293–5, 2393)*

87. **(C)** Nitrous oxide and all of the volatile anesthetics absorb infrared light. Nitrogen and oxygen may be determined by mass spectroscopy or Raman spectroscopy. Oxygen may also be measured by electrochemical or galvanic detectors. *(1:669–70; 4:869–70; 5:1213, 1454)*

88. **(D)** The fail-safe valves on anesthesia machines are on if oxygen pressure is present. It is important to understand that this valve is strictly pressure related. As long as there is oxygen pressure in the machine, the valves will be on, permitting flow of other gases such as nitrous oxide, carbon dioxide, air, helium, nitrogen, and so forth. It is therefore possible to still deliver a hypoxic mixture to the patient even with a functioning fail-safe valve. In contrast, newer machines have proportioning systems which ensure a minimum concentration of oxygen delivered at the common gas outlet. *(1:569–70; 4:1014; 5:277)*

89. **(A)** A decrease in oxygen saturation may be a late manifestation of venous air embolism, and should not be the method on which detection is relied. The precordial Doppler, capnometer, and mass spectrometer will all indicate the presence of a venous air embolism quickly and with high sensitivity. *(1:766; 4:1635–7; 5:2290)*

90. **(A)** The BIS monitor is useful for measuring the depth of hypnosis produced by volatile anesthetics, propofol, thiopental, or midazolam. Ketamine typically produces activation in the electroencephalogram (EEG) even when the patient is unconscious thus the BIS monitor is not able to assess ketamine-induced depth of hypnosis. *(1:682; 5:1255–7)*

91. **(D)** When the last drop of nitrous oxide liquid evaporates, the tank is approximately 16% full. The pressure fall will depend on the size of the tank and the rate of flow. *(4:1015)*

92. **(B)** The float may stick if the tubing is dirty. If back pressure builds up in the system, the reading may be inaccurate. The float should rotate to show that gas is flowing. If one component of the flowmeter set is broken or lost, the entire set must be replaced. *(1:572; 4:1015–7; 5:280)*

93. **(C)** The oxygen flush valve can apply oxygen to the breathing system at the pressure of the pipeline supply (e.g., approximately 50 psi) causing barotrauma. If oxygen is being supplied by the gas cylinders mounted on the anesthesia machine, the cylinder pressure is decreased to approximately 45 psi in the anesthesia machine, and this pressure is applied to the breathing circuit when the oxygen flush valve is pressed. During the inspiratory phase of positive-pressure ventilation, both the ventilator relief and pop-off valves are closed, and pressing the oxygen flush valve will apply high pressure to the breathing circuit. The oxygen flush valve is positioned distally to the vaporizers in an anesthesia machine. *(1:567; 4:1014; 5:283–4)*

94. **(D)** Standard I of the ASA Standards for Basic Intraoperative Monitoring mandates that "qualified anesthesia personnel shall be present in the room throughout the conduct of all general anesthetics, regional anesthetics, and monitored anesthesia care. In the event there is a direct known hazard (e.g., radiation) to the anesthesia personnel that might require intermittent remote observation of the patient, some provision for monitoring the patient must be made." The standards "are not intended for application to the care of the obstetric patient." Patients having spinal anesthesia with intravenous sedation may have their ventilation "evaluated, at least, by continual observation of qualitative clinical signs," according to Standard II. *(1:31; 4:796)*

95. **(B)** The DISS prevents attachment of gas administration equipment to the wrong gas. The system is based on matching specific bores and diameters that are assigned to the specific gases. This is not the protective system for cylinder-yoke attachments, which uses the pin index safety system. Suction hoses are not included in the DISS. *(1:568; 4:1014; 5:276)*

96. **(A)** Pressure damage may occur on the occiput, elbows, and heels after prolonged procedures. Padding the popliteal fossae may decrease postoperative stiffness, but there is little risk of a pressure injury at this location. *(1:641; 4:682–4,692; 5:1158)*

97. **(A)** The American Society of Testing and Materials (ASTM) standard number F1850-00 mandates that all new anesthesia machines incorporate an oxygen analyzer, a pulse oximeter, and a pressure monitor in the breathing system (among others). A ventilator is optional. *(5:274)*

98. **(A)** The difference between arterial and measured end-tidal carbon dioxide will be increased by ventilation-perfusion mismatch (e.g., emboli,

decreased cardiac output), prolonged expiratory phase (e.g., wheezing), and high fresh gas flow rates which dilute expired carbon dioxide. *(1:668–9; 4:866–71; 5:1459–61)*

99. **(B)** Leads I and V6 are properly described. Lead II is recorded from the right arm and left leg. Lead V1 is placed just to the right of the sternum in the fourth intercostal space. *(1:1507; 5:1390–1)*

100. **(D)** The laryngeal mask airway permits positive-pressure ventilation if the fit and seal of the airway are adequate for a particular patient. It is positioned blindly in a patient under general anesthesia. If anesthesia depth is inadequate, laryngospasm may result. There is no protection against aspiration. *(1:599–602; 4:1076; 5:1627)*

CHAPTER 3

Circulation

Questions

DIRECTIONS (Questions 101 through 172): Each of the numbered items or incomplete statements in this section is followed by answers or by completions of the statement. Select the ONE lettered answer or completion that is BEST in each case.

101. Hypoxic vasoconstriction is prominent in the

(A) lungs
(B) brain
(C) heart
(D) liver
(E) skin

102. An anticoagulant originally derived from leech saliva is

(A) abciximab
(B) clopidogrel
(C) hirudin
(D) eptifibatide
(E) aprotinin

103. Torsades de pointes

(A) may be treated with procainamide
(B) is a type of ventricular tachycardia
(C) can be caused by hypermagnesemia
(D) is characterized by a short QT interval
(E) is more frequent in men than women

104. The predominant cardiovascular effect of propofol as an intravenous induction agent is

(A) venodilation and mild depression of myocardial contractility
(B) pulmonary hypertension
(C) decreased heart rate
(D) increased systemic vascular resistance (SVR)
(E) increased cardiac output

105. At equianalgesic doses, the opioid most likely to increase heart rate is

(A) morphine
(B) meperidine
(C) fentanyl
(D) sufentanil
(E) alfentanil

106. Myocardial oxygen demand may be decreased by

(A) tachycardia
(B) decreased preload
(C) increased afterload
(D) increased contractility
(E) increased wall tension

107. Coronary perfusion pressure (CPP) is increased as a result of

(A) increased diastolic blood pressure (DBP)
(B) increased left ventricular end-diastolic pressure (LVEDP)
(C) systolic hypertension
(D) tachycardia
(E) hypocapnia

108. Baroreceptor reflex activity is

(A) unimportant as a circulatory regulator
(B) responsive to oxygen content
(C) located primarily in the carotid sinus
(D) inversely proportional to age
(E) responsive to CO_2 concentration

109. At a constant stroke volume, cardiac output is

(A) proportional to resistance
(B) a linear function of heart rate
(C) dependent on blood volume
(D) related to potassium concentration
(E) inversely proportional to arterial blood pressure

110. The intrinsic rhythmicity of the sinoatrial node is affected by the metabolism of the pacemaker cells themselves and may be increased by

(A) digoxin
(B) hypothyroidism
(C) increased body temperature
(D) hyperkalemia
(E) calcium

111. Stroke volume is primarily a function of

(A) the extent of myocardial fiber shortening
(B) heart rate
(C) peripheral resistance
(D) arterial oxygen content
(E) serum potassium level

112. Afterload of the left ventricle is

(A) related only to the aortic valve
(B) dependent on the distensibility of the large arteries
(C) best measured by estimating the mean arterial pressure
(D) increased by vasodilators
(E) unrelated to ventricular size

113. The intrinsic rate of the sinoatrial node is generally

(A) 20 to 40 beats per minute
(B) 40 to 60 beats per minute
(C) 70 to 80 beats per minute
(D) 80 to 100 beats per minute
(E) 100 to 120 beats per minute

114. Slow diastolic depolarization (phase 4) is

(A) a property of all cells
(B) independent of potassium flux
(C) the process underlying intrinsic rhythmicity in pacemaker cells
(D) more pronounced in the bundle of His than in the sinoatrial node
(E) unrelated to rhythmicity

115. Cardiac output will be decreased by

(A) raising the heart rate from 50 to 72 beats per minute
(B) increased preload
(C) decreased afterload
(D) myocardial disease
(E) digoxin

116. The second heart sound coincides with all of the following EXCEPT

(A) closing of the aortic valve
(B) isometric relaxation of myocardial fibers
(C) closure of the mitral valve
(D) the T wave of the electrocardiogram (ECG)
(E) closure of the pulmonic valve

117. As a flow-directed pulmonary artery catheter is inserted, you would expect to observe all of the following EXCEPT a

(A) right atrial pressure of 8 mmHg
(B) right ventricular systolic pressure of 25 mmHg
(C) pulmonary artery systolic pressure of 25 mmHg

(D) pulmonary artery diastolic pressure lower than right ventricular diastolic pressure
(E) pulmonary capillary wedge pressure of 10 mmHg

118. The cardiac conduction system includes all of the following EXCEPT

(A) the sinoatrial node
(B) atrial conduction pathways
(C) right bundle branch
(D) left bundle branch
(E) coronary sinus

119. The standard ECG records the correct electrical potentials when leads are placed such that

(A) lead I reads left arm-left leg
(B) lead I reads right arm-right leg
(C) lead II reads right arm-left leg
(D) lead III reads right arm-left leg
(E) lead II reads right arm-left arm

120. Calcium ion

(A) decreases myocardial contractile force
(B) decreases duration of systole
(C) decreases vascular tone
(D) decreases ventricular automaticity
(E) enters the cardiac cell, causing excitation

121. Electrocardiographic monitoring is indicated during anesthesia to detect all of the following EXCEPT

(A) efficacy of pump function
(B) arrhythmias
(C) ischemia
(D) electrolyte disturbance
(E) pacemaker function

122. An endogenous inhibitor of intravascular coagulation is

(A) factor V Leiden
(B) activated protein C
(C) bradykinin
(D) Hageman factor
(E) argatroban

Questions 123 through 126 refer to Figure 3-1:

123. LVEDP is most closely approximated by

(A) arterial blood pressure
(B) central venous pressure (CVP)
(C) pulmonary artery systolic pressure
(D) pulmonary capillary wedge pressure
(E) right atrial pressure

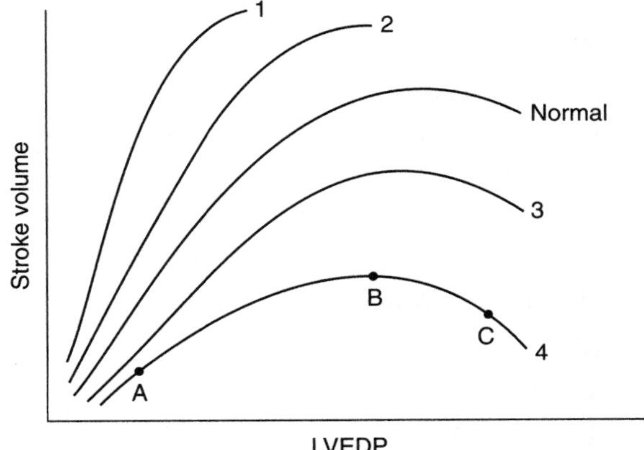

FIG. 3-1

124. The curves marked 3 and 4 in the graph shown could be seen in patients

(A) with increased contractility
(B) on digoxin or epinephrine
(C) who are in heart failure
(D) who have been treated with glucagon
(E) who respond to increased LVEDP with increased stroke volume

125. If a patient's status is represented by point A on curve 4 in the graph shown, then administration of intravenous fluid may

(A) shift the patient to the normal curve
(B) shift the patient to point B or C
(C) take the patient to point B only
(D) have no effect
(E) shift the patient to curve 2

126. Administration of an inotrope to a patient at point B on curve 4 is intended to

(A) shift the patient to curve 3
(B) shift the patient to point A
(C) shift the patient to point C
(D) maintain the patient at point B
(E) shift the patient to curve 1

127. Clinical manifestations of fat emboli include all of the following EXCEPT

(A) petechiae
(B) hypoxemia
(C) confusion
(D) bradycardia
(E) cyanosis

128. The minimum number of platelets needed for surgical hemostasis is approximately

(A) $10{,}000/mm^3$
(B) $30{,}000/mm^3$
(C) $80{,}000/mm^3$
(D) $100{,}000/mm^3$
(E) $1{,}000{,}000/mm^3$

129. The average unit of packed red cells, processed by centrifugation, has a hematocrit of

(A) 40%
(B) 50%
(C) 60%
(D) 70%
(E) 90%

130. Morbid obesity is associated with

(A) decreased cardiac output
(B) hypertension
(C) decreased pulmonary artery pressure
(D) decreased blood volume
(E) decreased cardiac workload

131. Technical difficulties in treating patients with morbid obesity include all of the following EXCEPT

(A) spuriously low blood pressure cuff readings
(B) difficult venous access
(C) difficult intubation
(D) difficult airway maintenance with mask
(E) difficulty with nerve blocks

132. Systolic flow reversal occurs normally in the

(A) abdominal aorta
(B) right coronary artery
(C) left coronary artery
(D) hepatic veins
(E) pulmonary veins

133. Left ventricular afterload reduction and augmentation of CPP can be achieved by means of

(A) intra-aortic balloon counterpulsation
(B) dopamine together with nitroprusside
(C) epinephrine
(D) β-adrenergic blockade
(E) dobutamine

134. Patent ductus arteriosus (PDA) may close in response to

(A) hypoxemia
(B) alprostadil (prostaglandin E_1)
(C) heparin
(D) indomethacin
(E) nitrous oxide

135. When an echocardiography probe is in the midesophageal position, the cardiac structure that is closest to the probe is the

(A) right atrium
(B) left atrium
(C) right ventricle
(D) left ventricle
(E) coronary sinus

136. The coronary circulation responds to hyperventilation with

(A) no change
(B) an increase in flow
(C) a decrease in flow

(D) a transient increase followed by an intense vasodilatation
(E) intense vasoconstriction

137. Myocardial oxygen consumption is generally tied most closely to

(A) heart rate
(B) blood viscosity
(C) cardiac output
(D) stroke volume
(E) FIO_2

138. With the arm at atrial level, venous pressure in the median basilic vein is

(A) equal to right atrial pressure
(B) above right atrial pressure
(C) below right atrial pressure
(D) not related to blood volume
(E) independent of arm position

139. If the recorded blood pressure is 160/100, the mean pressure is approximately

(A) 110 mmHg
(B) 120 mmHg
(C) 130 mmHg
(D) 140 mmHg
(E) 150 mmHg

140. The presence of a bubble in an arterial line

(A) is not significant
(B) leads to an artificially high reading
(C) affects only the diastolic pressure
(D) leads to a damping of the tracing
(E) has a greater effect on the mean blood pressure

141. Under normal circumstances, the plasma protein (colloid) osmotic pressure is

(A) 0 mmHg
(B) 5 mmHg
(C) 20 mmHg
(D) 50 mmHg
(E) 100 mmHg

142. Pulmonary edema may result from all of the following EXCEPT

(A) altered permeability
(B) decreased pulmonary capillary pressure
(C) decreased oncotic pressure
(D) increased negative airway pressure
(E) head injury

143. Pulmonary edema in the recovery room

(A) usually occurs as a late sign
(B) is always associated with a rise of CVP
(C) will be detected because of distended neck vessels
(D) may be due to airway obstruction
(E) usually occurs with normal breath sounds present

144. During transesophageal echocardiography, the velocities of blood flow are measured by means of the

(A) Bernoulli equation
(B) Venturi effect
(C) Stewart-Hamilton equation
(D) Doppler effect
(E) continuity equation

145. An inotrope which may increase myocardial contractility in pulmonary edema is

(A) oxygen
(B) digoxin
(C) morphine
(D) furosemide
(E) nitroglycerin

146. Mechanisms by which patients with failing hearts compensate for low cardiac output include all of the following EXCEPT

(A) increased sympathetic drive to the heart
(B) myocardial hypertrophy
(C) renal loss of salt and decreased blood volume
(D) secondary hyperaldosteronism
(E) increased ventricular filling pressure

147. The Frank-Starling mechanism refers to the relationship of

(A) the left and right ventricles
(B) the left and right atria
(C) preload and stroke volume
(D) afterload and ventricular volume
(E) stroke volume and afterload

148. If the size of the ventricle increases

(A) wall tension needed to pump the same amount of blood is less
(B) less oxygen is needed to pump blood
(C) the heart becomes more efficient
(D) wall tension will increase proportionally with the radius
(E) wall tension will be proportional to wall thickness

149. Coronary blood flow

(A) is independent of the systolic pressure
(B) is not affected by humoral agents
(C) is increased by a slow heart rate
(D) is increased by a fast heart rate
(E) occurs almost entirely in systole

150. Pulmonary artery pressure

(A) increases passively with increases of cardiac output
(B) remains constant with change of cardiac output
(C) is not important to pulmonary vascular resistance (PVR)
(D) is not dependent on the radius of the vessels
(E) depends entirely on cardiac output

151. The transplanted heart

(A) has an abnormal Frank-Starling mechanism
(B) responds as does an innervated heart to atropine
(C) bears functional α- and β-adrenoceptors
(D) does not respond to isoproterenol
(E) does not increase cardiac output with increased preload

152. Signs and symptoms associated with an increased incidence of postoperative congestive heart failure include all of the following EXCEPT

(A) New York Heart Association (NYHA) functional class II
(B) previous pulmonary edema
(C) presence of an S3 gallop
(D) jugular venous distention
(E) signs of left heart failure

153. Increased sympathetic tone in congestive heart failure may be indicated by

(A) memory loss
(B) weakness
(C) fatigue
(D) confusion
(E) anxiety

154. Right heart failure soon after heart transplant is managed with

(A) measures to reduce PVR
(B) measures to reduce SVR
(C) positive end-expiratory pressure (PEEP)
(D) reduced FIO_2
(E) permissive hypercarbia

155. Hypotension noted on initiation of cardiopulmonary bypass may be due to all of the following EXCEPT

(A) hemodilution
(B) decreased catecholamines
(C) aortic stenosis
(D) persistent systemic to pulmonary shunt
(E) aortic dissection

156. Myocardial preservation during cardiopulmonary bypass may include all of the following EXCEPT

(A) maintaining the ventricle in a distended state
(B) cardioplegia
(C) nearly continuous perfusion
(D) avoidance of ventricular fibrillation
(E) maintaining the heart at a low temperature

157. You are called to see your patient in the recovery room because of elevated blood pressure readings. Your approach should be to

(A) treat the blood pressure with small doses of an antihypertensive medication
(B) do nothing, but wait to see if the hypertension is a transient problem associated with emergence from anesthesia
(C) examine the patient for evidence of hypoxia or hypercarbia
(D) recheck the blood pressure yourself to make sure the cuff is the correct size
(E) arrange a cardiology consultation

158. Cardiovascular changes that occur with advancing age are

(A) decreasing blood pressure
(B) increase in cardiovascular reserve
(C) loss of elasticity of the vascular tree
(D) increased number of myofibrils
(E) increase in cardiac output

159. The right ventricle

(A) has little function in the adult patient
(B) is very easily cooled in the patient undergoing cardiopulmonary bypass
(C) can overcome PVR very effectively, and, therefore, is not of concern in terminating cardiopulmonary bypass
(D) is more likely to be injured by intracoronary air than is the left ventricle during cardiopulmonary bypass
(E) is unaffected by PEEP

160. A cause of systemic hypertension that is amenable to surgical correction is

(A) essential hypertension
(B) secondary aldosteronism
(C) renal parenchymal disease
(D) pheochromocytoma
(E) long-standing renal artery stenosis

161. The Allen's test

(A) is used to assess adequacy of radial artery perfusion
(B) has little predictive value
(C) is positive if good flow occurs
(D) is negative if good flow occurs
(E) is independent of hand position

162. Central venous cannulation is indicated in all the following procedures EXCEPT

(A) a surgical procedure in which there is an unusual position, e.g., head-down position
(B) patients in shock
(C) hyperalimentation
(D) intravenous administration of vasopressors
(E) procedures in which large volume shifts may occur

163. The precordial thump is indicated at the time of cardiac arrest

(A) in all patients with arrest
(B) only in pediatric patients
(C) since it can do nothing but good
(D) only in those with unmonitored arrhythmias
(E) and should be delivered to the midpart of the sternum

164. Defibrillation is

(A) indicated in all instances of cardiac arrest
(B) delivered at an energy level of 400 J in all patients
(C) always successful if used early
(D) the initial step in resuscitation of the unwitnessed arrest
(E) initially attempted with an energy level of 200 J for ventricular fibrillation in an adult

165. In advanced life support

(A) epinephrine is used primarily for its β-adrenergic effect
(B) bretylium is preferable to lidocaine because of its less severe side effects
(C) atropine is a first-line drug used to reduce vagal tone
(D) isoproterenol is used first for the treatment of bradycardia
(E) sodium bicarbonate should be started immediately

166. Administration of drugs during advanced life support

(A) may be by intramuscular injection
(B) should preferably be by peripheral line
(C) should be by femoral line if possible
(D) should be via central line if possible
(E) may be via the endotracheal tube, but only epinephrine and bicarbonate can be given by this route

167. All of the following may be seen in the patient with hypovolemia EXCEPT

(A) increased heart rate
(B) wide pulse pressure
(C) decreased urine volume
(D) flat neck veins
(E) pale mucous membranes

168. When monitoring the CVP

(A) a waveform should be evaluated for the most accurate reading
(B) the catheter should be placed in the right atrium
(C) an accurate reflection of fluid status is obtained in all patients
(D) analysis of waveforms is no better than a digital readout
(E) a flow-directed pulmonary artery catheter should always be used, since it will give the CVP and other information

169. When using arterial pressure monitoring

(A) it is best to use a catheter that approximates the size of the vessel to avoid pressure artifact
(B) one should never aspirate the catheter during decannulation
(C) a flush system should be employed
(D) the duration is not related to complications
(E) caution is needed if using the radial artery, since it has no collaterals

170. A heart lesion associated with high pulmonary flow is

(A) pulmonic stenosis
(B) tetralogy of Fallot
(C) coarctation of the aorta
(D) ventricular septal defect (VSD)
(E) tricuspid atresia

171. A true statement in reference to sites of vessel cannulation is that

(A) chylothorax is common with right internal jugular cannulation
(B) brachial plexus trauma is common with antecubital approach for central venous cannulation
(C) pneumothorax is not common with the subclavian approach
(D) the Allen's test rules out ischemic complications
(E) the dorsalis pedis is not the artery of choice in a patient with diabetes

172. The patient with aortic stenosis has

(A) a rapid downhill course once symptoms are present
(B) a large left ventricular cavity
(C) low electrocardiographic voltages
(D) protection against ischemia due to the large ventricle
(E) a very compliant ventricle

DIRECTIONS (Questions 173 through 262): For each of the items in this section, ONE or MORE of the numbered options is correct. Choose answer

(A) if only 1, 2, and 3 are correct
(B) if only 1 and 3 are correct
(C) if only 2 and 4 are correct
(D) if only 4 is correct
(E) if all are correct

173. Inhaled nitric oxide (NO)

(1) may cause methemoglobinemia
(2) often lowers mean pulmonary artery pressure
(3) exhibits virtually no systemic vasodilation
(4) often decreases systemic oxygenation

174. Nitric oxide (NO)

(1) is released from endothelial cells on cholinergic stimulation
(2) relaxes vascular smooth muscle on activation of guanylate cyclase
(3) is a likely metabolite of sodium nitroprusside and nitroglycerin
(4) increases cyclic-guanosine monophosphate (GMP) levels in vascular smooth muscle

175. Agents that are inhibitors of fibrinolysis include

(1) tissue plasminogen activator
(2) aprotinin
(3) streptokinase
(4) ε-aminocaproic acid

176. Vitamin K

(1) is required for synthesis of functional clotting factors VII, IX, X, and II (prothrombin)
(2) antagonizes heparin
(3) antagonizes warfarin
(4) antagonizes protamine sulfate

177. Erythropoietin

(1) is produced by the kidney
(2) increases platelet production
(3) increases red blood cell production
(4) increases white blood cell production

178. In the Wolff-Parkinson-White (WPW) syndrome the

(1) heart rate is usually 60 to 80 beats per minute during sinus rhythm
(2) PR interval is less than 0.12 seconds during sinus rhythm
(3) QRS duration is usually longer than 0.12 seconds
(4) the upstroke of the QRS complex is often slurred

179. Drugs exhibiting α-adrenergic agonist activity include

(1) clonidine
(2) labetalol
(3) phenylephrine
(4) vasopressin

180. Agents which may reduce right-to-left shunt in tetralogy of Fallot include

(1) sodium nitroprusside
(2) labetalol
(3) propranolol
(4) phenylephrine

181. The transplanted, denervated heart responds to

(1) circulating catecholamines
(2) circulating acetylcholine
(3) adrenergic blocking agents
(4) muscarinic blocking agents

182. Causes of pulseless electrical activity (PEA) include

(1) pulmonary embolism
(2) atrial fibrillation
(3) tension pneumothorax
(4) atrial flutter

183. Vasopressin

(1) causes vasoconstriction on reacting with V_2 receptors
(2) exhibits a half-life of less than 1 minute
(3) is a catecholamine
(4) may be used as an alternative pressor to epinephrine in the treatment of electroshock-refractory ventricular fibrillation

184. Inhibitors of platelet glycoprotein IIb/IIIa include

(1) eptifibatide
(2) tirofiban
(3) abciximab
(4) aspirin

185. Milrinone

(1) decreases SVR
(2) is a catecholamine
(3) is a positive inotrope
(4) is inactivated by monoamine oxidase

186. The platelet glycoprotein IIb/IIIa

(1) participates in aggregation of platelets
(2) is inhibited by abciximab
(3) is inhibited by eptifibatide
(4) is activated by diverse stimuli

187. Verapamil may exacerbate

(1) sick sinus syndrome
(2) atrioventricular block
(3) WPW syndrome
(4) congestive heart failure

188. Digoxin

(1) activates sodium-potassium adenosine triphosphatase (ATPase)
(2) inhibits atrioventricular conduction
(3) decreases intracellular calcium levels
(4) increases contractility

189. Pulmonary capillary wedge pressure will be decreased by

(1) right ventricular failure
(2) left ventricular failure
(3) severe pulmonary vasoconstriction
(4) severe mitral stenosis

190. Prominent v waves in the pulmonary capillary wedge trace may indicate

(1) mitral regurgitation
(2) ischemic papillary muscle dysfunction
(3) fluid overload
(4) left ventricular failure

191. Amiodarone

(1) suppresses SA node function
(2) slows the ventricular rate in atrial fibrillation
(3) can slow polymorphic ventricular tachycardia
(4) is a vasoconstrictor

192. Sinus bradycardia

(1) is a serious arrhythmia when seen during anesthesia
(2) is an irregular rhythm
(3) always requires treatment
(4) is characterized by a rate of 40 to 60 beats per minute

193. The tetralogy of Fallot includes

(1) atrial septal defect (ASD)
(2) VSD
(3) PDA
(4) pulmonary outflow obstruction

194. The preoperative ECG can be used to detect

(1) chamber enlargement
(2) ischemic heart disease
(3) electrolyte disturbances
(4) heart block

195. Cardiac output is

(1) identical in the left and right heart
(2) usually 5 to 6 L/min in a 70-kg man
(3) unaffected by blood volume
(4) the product of heart rate and stroke volume

196. Left ventricular compliance may be decreased in

(1) aortic stenosis
(2) ischemia
(3) inotrope usage
(4) the patient on vasodilators

197. Preload is affected by

(1) total blood volume
(2) increased heart rate
(3) atrial function
(4) venous tone

198. Myocardial contractility is decreased by

(1) sympathetic stimulation
(2) parasympathetic tone
(3) inotrope administration
(4) myocardial ischemia

199. The coronary circulation is composed of the right and left coronary arteries

(1) the right supplying the right ventricle
(2) which are remarkably constant in their configuration and areas they supply
(3) the left dividing into the left anterior descending and the circumflex
(4) both of which empty into the right ventricle

200. When inserting a CVP line

(1) air embolism may occur
(2) the tip should be placed in the right atrium
(3) use of the external jugular vein avoids many of the major complications
(4) the left side of the neck is preferred

201. Calcium administration

(1) decreases myocardial contractility
(2) decreases excitability
(3) can be antagonized by calcium chelating agents, but the effects are too long for clinical use
(4) may potentiate development of digoxin toxicity

202. The Allen's test

(1) should be performed, since only 40% of patients have a complete ulnar arch
(2) should be done with the wrist in extension
(3) is described as positive if good flow is demonstrated
(4) should be described in terms of return of color and time

203. To obtain an accurate CVP

(1) the catheter tip should be within the thoracic portion of the vena cava
(2) the tip should be at the vena cava-atrium junction
(3) the pressure should be demonstrated to fluctuate with respiration
(4) one must be able to aspirate blood

204. Factors of importance when choosing the cephalic versus the basilic vein for insertion of a central venous catheter are

(1) the cephalic vein is usually smaller
(2) the cephalic vein is more superficial in the upper arm
(3) the cephalic vein enters the axillary vein at a 90° angle
(4) the basilic vein lies lateral to the cephalic vein

205. Determination of cardiac output with the thermodilution technique

(1) requires a pulmonary artery catheter with a thermistor
(2) uses the same principle as the dye-dilution technique
(3) requires the use of an exact amount of fluid with a known temperature
(4) requires the measurement of the temperature in the pulmonary artery

206. Pressure values in a normal adult might include

(1) right atrium: mean 5 mmHg
(2) right ventricle: 25/5 mmHg
(3) pulmonary artery: 23/9 mmHg
(4) pulmonary capillary wedge pressure: 10 mmHg

207. The effect of hyperventilation on coronary blood flow

(1) is to cause increased flow
(2) leads to coronary spasm
(3) does not respond to nitroglycerin
(4) can be suppressed by diltiazem

208. Factors that affect myocardial ischemia are

(1) anemia, which aggravates it
(2) adrenergic blockade, which aggravates it
(3) digoxin treatment in the failing heart, which alleviates it
(4) nitroglycerin, which aggravates it

209. Cardioplegia solution is an important adjunct in heart surgery since

(1) oxygen consumption increases with ventricular fibrillation
(2) oxygen consumption is reduced to one-fourth with arrest
(3) a quiet field is provided
(4) myocardial oxygen consumption varies with rhythm

210. Air embolism seen with cardiopulmonary bypass

(1) is common after aortic valve procedures
(2) is alleviated with the use of ventricular venting
(3) is alleviated by the use of an aortic vent
(4) comes in part from air that accumulates in the pulmonary veins during bypass

211. Methods of detecting intraoperative air embolism include

(1) mass spectrometry
(2) end-tidal carbon dioxide determination
(3) Doppler monitoring
(4) transesophageal echocardiography

212. The position of the tip of the CVP catheter can be confirmed by

(1) electrocardiography
(2) x-ray
(3) pressure monitoring
(4) Doppler monitoring while injecting 1 mL of air

213. The phlebostatic axis

(1) is located at the tip of the left ventricle
(2) is defined at the level at which a change of position will cause a pressure change of less than 1 mmHg
(3) has no clinical importance
(4) is used for pressure references

214. Baroreceptor structures include

(1) the walls of the large arteries of the neck and thorax
(2) the carotid sinus
(3) the aortic arch
(4) the glossopharyngeal nerve

215. The renin-angiotensin mechanism includes

(1) renin, which is in the liver
(2) converting enzyme, which is in the lung
(3) the conversion of renin substrate to angiotensin I by the converting enzyme
(4) initiation of the entire process by a decrease in blood pressure

216. Cerebral function changes after cardiopulmonary bypass have been attributed to

(1) hypotension while on pump
(2) barbiturate administered while on pump
(3) microemboli from the pump
(4) pulmonary hypertension

217. A pacemaker with the letter code AVTPN

(1) paces the atrium
(2) senses the ventricle
(3) triggers a pacemaker spike on sensing an impulse
(4) is not programmable

218. Pacemaker threshold changes that can occur by various agents and chemical changes include

(1) an increase with stress
(2) a decrease with decrease in carbon dioxide
(3) an increase with increase in potassium
(4) an increase with hypoxia

219. Thebesian veins

(1) always drain into the left atrium
(2) are part of the pulmonary circulatory system
(3) carry oxygenated blood
(4) which empty into the left ventricle contribute to shunt

220. Cardiac index

(1) is synonymous with cardiac output
(2) is calculated by dividing the cardiac output by the patient's weight
(3) is lower in an adult than in a child
(4) is calculated to give a better comparison among patients of varying size

221. After insertion of a flow-directed pulmonary artery catheter, you note the presence of large v waves in the pulmonary capillary wedge trace. This may be due to

(1) tricuspid atresia
(2) aortic valve stenosis
(3) aortic regurgitation
(4) mitral regurgitation

222. Coarctation of the aorta

(1) may be the cause of hypertension
(2) may be associated with congestive heart failure
(3) may be associated with ventricular hypertrophy
(4) in the neonate is always associated with other heart lesions

223. Transesophageal echocardiography

(1) cannot be used in a continuous mode
(2) can only be used in the closed chest, therefore is not useful in the cardiac surgery patient
(3) gives poor delineation of the left ventricle
(4) gives good visualization of the right heart

224. Lymph

(1) is essentially interstitial fluid
(2) enters the venous circulation through the external jugular vein
(3) flow is increased by exercise
(4) from the intestine is low in fat

225. Factors that increase the risk of anesthesia include

(1) an S3 gallop
(2) rhythm other than sinus
(3) age greater than 70 years
(4) emergency operations

226. Methods of detecting ventricular dysfunction include

(1) determination of ejection fraction
(2) wall motion studies
(3) determination of cardiac output
(4) determination of end-diastolic pressure

227. During cardiopulmonary resuscitation (CPR)

(1) a central line is preferable for drug administration
(2) peripheral intravenous lines are not suitable for administration of epinephrine
(3) lidocaine, atropine, and epinephrine may be administered via the endotracheal tube
(4) intracardiac injections should be attempted before an attempt is made to secure a central line

228. The baroreceptors

(1) are stretch receptors
(2) respond to absolute pressure change as well as rate of change in pressure
(3) function through the cardioinhibitory and vasomotor centers
(4) are reset by chronic hypertension

229. The arterial supply to the spinal cord includes

(1) the anterior spinal arteries
(2) the posterior spinal arteries
(3) radicular arteries
(4) artery of Adamkiewicz

230. The pericardium

(1) serves as a buffer to the heart at times of acceleration
(2) contains nerve fibers that, when stimulated, will slow the heart rate
(3) minimizes cardiac dilatation
(4) is not essential to cardiac function

231. The right and left ventricles

(1) are completely independent of one another assuming the septum is intact
(2) are affected identically by a deep inspiration
(3) are not affected by septal position
(4) are each affected by the systolic pressure of the other

232. Hypertrophic obstructive cardiomyopathy (HOCM) is

(1) treated with inotropic agents
(2) a form of primary myocardial disease
(3) always responsive to medical treatment
(4) treated with β-adrenergic blocking agents

233. Afterload will decrease after

(1) a reduction of blood viscosity
(2) the opening of an arteriovenous fistula
(3) administration of a vasodilator
(4) development of mitral regurgitation

234. Stroke volume is dependent on myocardial fiber shortening, which is determined by

(1) preload
(2) contractility of the myocardium
(3) afterload
(4) heart rate

235. Histamine elicits

(1) intense pulmonary vasoconstriction
(2) an effect on the circulation that is the opposite of the effect of hypoxia
(3) systemic vasodilatation
(4) bronchodilation

236. An increase in mean diastolic ventricular pressure will

(1) not be reflected in the pulmonary circulation
(2) necessarily increase cardiac output

(3) follow the use of diuretics
(4) be transmitted to the atrium and proximal vascular structure

237. Morphine is useful in the treatment of acute pulmonary edema because it

(1) has sedative effects
(2) increases venous pressure
(3) decreases venous return
(4) increases respiratory activity

238. Acute pulmonary edema may occur with

(1) sudden onset of dyspnea
(2) moist rales throughout both lungs
(3) cyanosis
(4) respiratory acidosis

239. Ventricular hypertrophy

(1) helps by decreasing oxygen demand
(2) improves perfusion
(3) allows a decrease in left ventricular filling pressure
(4) makes the ventricle more susceptible to ischemia

Questions 240 and 241: A 55-year-old male is admitted with a history of hypertensive heart disease and evidence of acute myocardial infarction. Vital signs are blood pressure 170/100, heart rate 124 beats per minute, body temperature 101°F, and respiratory rate 24 beats per minute.

240. Efforts to improve myocardial oxygenation should include

(1) decreasing arterial blood pressure
(2) decreasing body temperature
(3) slowing heart rate
(4) administration of oxygen

241. Chest pain worsens as the heart rate and blood pressure are reduced with metoprolol and nitroglycerin. Additional measures might include

(1) an inhibitor of cyclic-GMP phosphodiesterase
(2) intra-aortic balloon counterpulsation
(3) antifibrinolytics
(4) anticoagulation

Questions 242 through 245

A 24-year-old female was admitted for evaluation of chest pain. The pain was sharp and aggravated by breathing. A chest x-ray showed an enlarged heart, ECG showed decreased voltage, and there was a pericardial friction rub. Pericardiocentesis demonstrated purulent fluid. Serial x-rays showed widening of the cardiac shadow, and the patient was scheduled for pericardial drainage.

242. You would expect to find

(1) a water-hammer pulse
(2) an increased cardiac output
(3) distended neck veins that flatten in the sitting position
(4) decreased arterial pressure

243. A pulsus paradoxus is a pulse that

(1) reduces in amplitude on inspiration
(2) shows alternate strong and weak beats
(3) is merely an exaggeration of the normal respiratory effect on the arterial pulse
(4) is stronger on inspiration

244. In monitoring this patient, the

(1) arterial blood pressure is of no help
(2) Swan-Ganz catheter is mandatory
(3) x-ray evaluation is of utmost importance
(4) CVP will provide useful information

245. This patient is brought to the operating room for the procedure. In administering anesthesia

(1) a propofol induction should always be used
(2) vasodilators should be avoided or used very cautiously
(3) efforts should be made to increase cardiac output
(4) cardiac depressants must be avoided or used cautiously

246. Coronary artery circulation is influenced by

(1) oxygen saturation of the blood
(2) cardiac output
(3) autoregulation
(4) adenosine

247. Myocardial oxygen consumption is

(1) increased by propranolol
(2) decreased by digoxin
(3) increased by nitroglycerin
(4) increased by isoproterenol

248. Calculation of SVR requires determination of

(1) mean arterial pressure
(2) mean CVP
(3) cardiac output
(4) pulmonary artery pressure

249. Venous return is

(1) the major determinant of cardiac preload
(2) impeded by muscular contractions
(3) increased by a decrease in right atrial pressure
(4) impeded by the valves of the superior vena cava

250. Oxygen consumption under anesthesia

(1) is related to body temperature
(2) will fall more in a patient who is excited before induction
(3) generally decreases
(4) varies with tidal volume

251. Goals for the anesthetic management of a patient with coronary artery disease are to

(1) maintain high afterload
(2) maintain contractility
(3) maintain high heart rate
(4) maintain sinus rhythm

252. The capillaries are

(1) the site of gas exchange between blood and tissue
(2) seldom open to full capacity
(3) regulated by local physical and chemical agents
(4) primarily controlled by the autonomic nervous system

253. The patient with mitral stenosis usually has

(1) left ventricular hypertrophy
(2) increased left atrial pressure
(3) increased cardiac output
(4) right ventricular hypertrophy

Questions 254 and 255 refer to Figure 3-2:

254. The relationship between the transmembrane action potential and the ECG shown in the graph is such that

(1) phase 0 corresponds to the QRS
(2) phase 1 is responsible for the T wave
(3) phase 2 is a plateau and represents a recovery phase
(4) phase 4 is continued repolarization

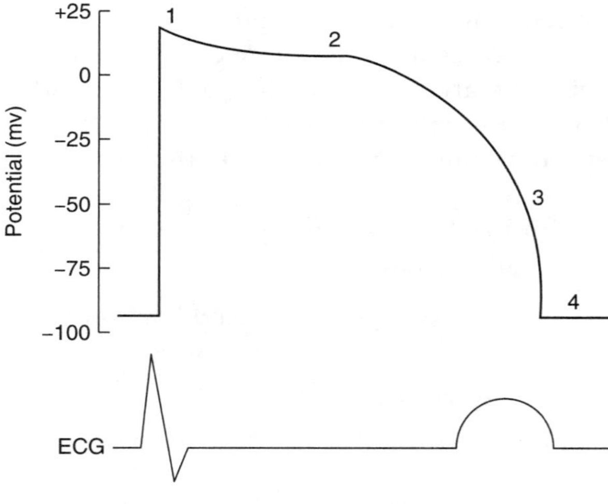

FIG. 3-2

255. Phase 4 of the action potential

(1) is the same for all cells
(2) is denoted for a slow diastolic depolarization in some cells
(3) is related to flux of calcium and magnesium
(4) is responsible for pacemaking activity

256. Stimulation of the vagus

(1) slows the heart
(2) decreases contractile force
(3) slows conduction
(4) increases excitability

257. Stimulation of the sympathetic fibers to the heart

(1) increases rate
(2) prolongs systole
(3) enhances conduction
(4) weakens contractions

258. The coronary sinus

(1) represents a minor drainage system of the heart
(2) receives the entire coronary venous blood flow
(3) drains into the pulmonary artery
(4) contains blood with an oxygen saturation of less than 30%

259. The bronchial arteries

(1) are part of the systemic circulation
(2) arise from the pulmonary artery
(3) drain into the azygos vein
(4) have no anastomoses with the pulmonary artery

260. Pulmonary circulation

(1) is very expansile, being able to accept large changes in cardiac output
(2) time is about 11 seconds
(3) totals about 400 to 600 mL of blood at any one time
(4) from the right ventricle equals the output from the left ventricle

261. During the Valsalva's maneuver, there is

(1) decreased blood return to the right ventricle
(2) increased venous pressure in the head
(3) decreased cardiac output
(4) a reflex increase in heart rate when the aortic pressure falls

262. During blood storage, there is an increased plasma concentration of

(1) potassium
(2) dextrose
(3) ammonia
(4) bicarbonate

Answers and Explanations

101. **(A)** The vasoconstrictor response of the pulmonary circulation differs from that of other tissues in order to match lung perfusion to ventilation. *(1:831; 3:394; 4:1765; 5:166)*

102. **(C)** Hirudin, the anticoagulant from the leech, is now produced through recombinant DNA technology (lepirudin). A synthetic peptide, bivalirudin, is modeled on hirudin. Abciximab is a monoclonal antibody, clopidogrel is synthetic, eptifibatide was discovered in snake venom, and the three are antiplatelet drugs. Aprotinin is a bovine protein which has some anticoagulant activity but generally reduces bleeding. *(1:229; 3:863; 3:1535; 5:2649)*

103. **(B)** The phrase was taken from a ballet-dancing maneuver. It is a type of polymorphic ventricular tachycardia characterized by a long QT interval. Procainamide can prolong the QT interval and elicit torsade. Magnesium can be therapeutic. The arrhythmia occurs more frequently in women than in men. *(1:194; 3:963; 5:360; 5:2934; 6:90)*

104. **(A)** Cardiac output falls mainly through decreased preload. *(1:334; 3:345; 5:325)*

105. **(B)** Meperidine tends to decrease cardiac output and increase heart rate. Other μ-receptor agonists tend to decrease heart rate. *(1:353; 2:147; 3:358; 4:1272; 5:380)*

106. **(B)** Myocardial oxygen consumption is increased by tachycardia. It causes increased demand for oxygen at the same time that it leads to decreased oxygen supply by decreasing coronary blood flow. Increased afterload leads to increased wall tension, both of which require more oxygen. Increases in contractility require more oxygen. *(1:873; 3:90; 4:1664; 5:2106; 6:19)*

107. **(A)** Coronary perfusion occurs for the most part during diastole. CPP = DBP – LVEDP. As DBP rises or LVEDP falls, the flow will increase. Tachycardia decreases time of perfusion, since diastole is shortened. Systolic hypertension decreases perfusion, since the ventricle contracts harder, allowing less perfusion during systole. *(1:892; 4:166; 5:680)*

108. **(C)** The baroreceptors are located principally in the carotid sinus. These are important to the regulation of circulation. The baroreceptors respond to pressure, not to carbon dioxide or oxygen. *(1:280–1; 4:2090; 5:738)*

109. **(B)** At a constant stroke volume, cardiac output is a linear function of heart rate. The other factors are important in the generation of cardiac output, but they are direct or indirect determinants of stroke volume: resistance (afterload), arterial pressure, and blood volume (preload). Cardiac output is not directly related to potassium concentration. *(1:867; 3:903; 4:821; 5:1333)*

110. **(C)** Intrinsic rhythmicity may be altered by anything that affects diastolic depolarization. Increased temperature increases heart rate. Digoxin decreases rhythmicity. Hypothyroidism does not increase rhythmicity. Calcium does not alter the resting membrane potential but does alter the threshold potential. Hyperkalemia decreases rhythmicity. *(1:856; 5:1406; 6:81)*

111. **(A)** Stroke volume is related to myocardial fiber shortening. Heart rate and peripheral resistance do not have a direct effect on stroke volume. Serum potassium and arterial oxygen saturation are not factors in regulation of stroke volume. *(1:870; 3:903; 5:2834)*

112. **(B)** Afterload is dependent on the distensibility of the arteries into which the ventricle ejects the blood. The best measure of afterload is the SVR. SVR is a function of mean arterial blood pressure, CVP, and cardiac output. Afterload is a function of the aortic valve resistance, but not exclusively. Vasodilators decrease afterload. Ventricular size is related to afterload but is not an important consideration. *(1:291; 3:903; 4:842; 5:725; 6:81)*

113. **(C)** The intrinsic rate of the sinoatrial node is generally 70 to 80 beats per minute. Since it has a faster rate, it is the dominant pacemaker. The more caudal a pacemaker cell is located in the conduction system, the slower its intrinsic rate. The atrioventricular node has an intrinsic rate of 40 to 60 beats per minute. *(1:856; 3:933; 4:93; 5:202)*

114. **(C)** The process of slow diastolic depolarization is a property of pacemaker cells and is the process causing intrinsic rhythmicity. The more cephalad the pacemaker, the more pronounced is the process. Therefore, it is seen less in the His bundle. *(1:862; 3:933; 4:1564; 5:730)*

115. **(D)** Myocardial disease will decrease cardiac output. Rate increases will increase output assuming that the stroke volume is constant. An increase in preload or a decrease in afterload will increase output. Digoxin, through an increase in contractility, will increase output. *(1:867; 3:903; 5:726; 5:733)*

116. **(C)** At the time of the second heart sound, the mitral valve is opening. The closing of the aortic and pulmonic valves, isometric relaxation, and the T waves are coincident. *(1:861; 4:275; 5:1078; 6:35)*

117. **(D)** The step-up in diastolic pressure is indicative of the entry into the pulmonary artery. *(1:675; 4:823; 5:679)*

118. **(E)** The conduction system includes the sinoatrial node, the principal pacemaker. The impulse is transmitted through the atrial conduction pathways to the atrioventricular node and then to the bundle of His before dividing into the right and left branch bundles. The coronary sinus is not involved in impulse conduction but receives the venous drainage of the heart. *(1:853; 2:746; 3:937; 4:93; 5:1406; 6:77)*

119. **(C)** The correct lead placement is right arm-left leg in lead II. The other leads are right arm-left arm in lead I and left arm-left leg in lead III. *(1:1507; 5:1390–5; 6:18)*

120. **(E)** Excitation of the cardiac cell membrane and depolarization are accompanied by calcium entering the cell. Calcium ions increase myocardial contractile force, prolong duration of systole, increase vascular tone, and increase ventricular automaticity. *(1:1123; 2:370; 3:1718; 4:1691; 5:1770; 6:113)*

121. **(A)** The ECG cannot detect efficacy of pump function. The ECG may show a normal sinus rhythm during PEA (electromechanical dissociation) when the pump is not working effectively at all. The ECG can also be observed for changes indicative of electrolyte disturbance, especially of calcium and potassium. Ischemia may be detected by changes in the tracing. Pacemaker function may be checked by observing the tracing while turning the pacemaker on and off (e.g., with a magnet). Arrhythmias are detected by the ECG. *(1:887; 5:1389; 6:18)*

122. **(B)** Endothelial thrombomodulin converts intravascular thrombin from a clotting enzyme into an activator of protein C. The activated protein C has anticoagulant activity through its destruction of factors V and VIII. Factor V Leiden is a single point mutation that makes the molecule resistant to destruction by activated protein C and therefore increases the risk of intravascular thrombosis. Argatroban is a synthetic inhibitor of thrombin. Activated Hageman factor initiates the intrinsic blood clotting cascade and activates prekallikrein to produce bradykinin. *(1:217; 3:1526; 5:1116, 2796; 6:495)*

123. **(D)** Pulmonary capillary wedge pressure is the best estimate of LVEDP. By measuring changes in cardiac output and pressures, a ventricular function curve can be drawn and therapeutic interventions evaluated. Right heart pressures do not necessarily provide accurate estimates of left heart pressures and function. Therefore, CVP gives the worst approximation of left side function. *(1:676; 3:903; 4:815; 5:1320; 6:35)*

124. **(C)** The curves are representative of patients who are not responding to increases in LVEDP with increases in stroke volume. These do not show any effect of increased contractility or any effect of inotropic intervention. In such a patient, inotrope administration may be the next logical step. *(1:291; 3:903; 4:841; 5:726)*

125. **(B)** A patient on such a low curve may respond to fluid by advancing to point B, may also react adversely and fall even farther down on the curve. This would be represented by point C. *(1:291; 3:903; 4:841; 5:726)*

126. **(A)** The inotrope ought to improve contractility. *(1:291; 3:903; 5:726)*

127. **(D)** Fat emboli may be seen with fractures. Fat embolism syndrome is associated with petechiae, hypoxemia, confusion, and cyanosis. Tachycardia is often present. *(1:1114; 4:295; 5:2425; 6:175)*

128. **(B)** The minimum number of platelets needed for normal coagulation is controversial, but is frequently cited as 30,000 to 50,000/mm^3. *(1:219; 5:1116)*

129. **(D)** Most units of red cells have a hematocrit of 70%. *(1:1201; 2:237; 4:2416; 5:1812)*

130. **(B)** Severe hypertension is seen in 5 to 10% of patients with morbid obesity. Moderate hypertension is seen in 50% of patients. The cardiac output is increased, as is the pulmonary artery pressure. Blood volume is increased, and the cardiac workload is increased. *(1:1035; 2:462; 4:507; 5:1033; 6:443)*

131. **(A)** Blood pressure readings usually are falsely high due to the difficulty in obtaining a suitable cuff. The cuff bladder length should be at least 80%, and the width 40% of the limb circumference. Venous and arterial access may be difficult. Intubation may be difficult. It may be impossible to get a suitable mask fit. Nerve blocks may be difficult because of the problem in finding landmarks. *(1:1038; 2:462; 4:507; 5:1033, 1270; 6:449)*

132. **(C)** Because of intramyocardial pressure, systolic flow reversal is most likely to occur in the left coronary artery. Hence, left ventricular perfusion occurs mainly during diastole. Aortic flow reversal might indicate, for instance, severe aortic valve insufficiency. Hepatic vein and pulmonary flow reversals occur during atrial contraction, but systolic flow reversal in those veins would indicate severe tricuspid or mitral regurgitation, respectively. These flows may be analyzed by means of Doppler echocardiography. *(1:865; 4:1684; 5:1378)*

133. **(A)** Pharmacologic interventions will tend to improve one parameter at the expense of the other. *(1:916; 4:191; 5:1991; 6:9)*

134. **(D)** The ductus normally closes in response to elevated arterial oxygen tension and patency may be maintained deliberately with alprostadil (PGE_1). *(1:921; 2:421; 3:679; 4:1709; 5:2009; 6:49)*

135. **(B)** The left atrium is closest to the probe and appears in the apex of the sector display in the midesophageal short axis view. *(1:886; 4:830; 5:1363–85; 6:28; 6:396)*

136. **(C)** Coronary circulation reacts to carbon dioxide in a manner similar to cerebral circulation. Although this may be salutary in the instance of head trauma, in the patient with chest pain and ischemia, hyperventilation may lead to further decrease of coronary flow. The magnitude of the decrease is not large. *(1:865; 4:1663; 5:1079)*

137. (A) Myocardial oxygen consumption is closely and directly related to cardiac rate over wide ranges. The rate-pressure product is also closely related to myocardial oxygen consumption. An increase in stroke volume does not cause as much of an increase in oxygen consumption as an increase in pressure. *(1:884; 3:903; 4:1664; 5:1942)*

138. (B) Since blood flow to the heart runs along pressure gradients in the upper extremities, the blood in the basilic vein runs from a site of higher pressure to a site of lower pressure, the right atrium. In the lower extremities, blood flow is assisted by valves and muscular activity. *(1:856; 4:141; 5:1196)*

139. (B) $\text{MAP} = \text{diastolic} + \frac{(\text{systolic} - \text{diastolic})}{3}$. *(1:860; 4:805; 5: 1281)*

140. (D) The pressure waveform may be affected by the presence of even small bubbles. The bubble, being very compliant, leads to a damping of the trace and a reading that hovers around the mean. Both systolic and diastolic pressures will be affected. *(1:673; 5:1278)*

141. (C) The plasma oncotic pressure is important in maintaining the fluid balance in the capillaries. The balance between the driving pressure and the oncotic pressure prevents tissue edema. The delicate balance can be disturbed by either increases or decreases in pressure or increases or decreases in oncotic pressure. *(1:176; 4:948; 5:753; 6:343)*

142. (B) Pulmonary edema results from an increase in pulmonary capillary pressure. Pulmonary edema may also result from neurologic injury. Negative airway pressure pulmonary edema has been reported in the patient with airway obstruction who is breathing against a closed glottis or an occluded endotracheal tube. *(1:676; 4:1827; 5:680; 5:704; 6:207)*

143. (D) The patient who develops pulmonary edema in the recovery room usually does so within the first hour. Distended neck veins or increased CVP may be absent, but the patient is usually wheezing. An occluded endotracheal tube may be the problem. *(1:991; 4:2356; 5:2713; 6:207)*

144. (D) The Doppler effect is the change in frequency that occurs when an echo is coming from a reflector that is moving toward or away from the probe. The frequency shift is directly proportional to the velocity of the reflector in the radial direction. The Doppler-measured velocity of a jet of blood can be converted to an estimate of pressure gradient by means of the Bernoulli equation. The Stewart-Hamilton equation pertains to thermodilution cardiac output measurement. *(1:680; 4:831; 5:1334)*

145. (B) The drug of choice to improve contractility is an inotropic agent. Digoxin is the only appropriate choice of the listed medications. Oxygen should be administered to improve oxygenation. Furosemide may be required to decrease venous return. Morphine may decrease venous return and provide sedation. Administration of nitroglycerin can improve coronary blood flow and decrease pre and afterload. *(1:307; 2:370; 3:916; 4:1563; 5:1125; 6:110)*

146. (C) The failing heart attempts to compensate by salt retention and increasing blood volume. Myocardial hypertrophy occurs, and there is an increased sympathetic outflow to improve output. Ventricular filling pressure increases as the heart decompensates. *(1:295; 3:779; 4:275; 5:633)*

147. (C) The Frank-Starling "law of the heart" is the underlying principle in the use of ventricular function curves. It relates preload and stroke volume. As more volume enters the heart, a larger cardiac output will occur. *(1:291; 1:872; 2:357; 4:476; 5:726; 5:733)*

148. (D) As the size of the ventricle increases, the radius and the wall tension increases. The law of Laplace states that more tension is needed to generate the same pressure as the radius increases. This makes the heart more inefficient and requires more oxygen to pump blood. *(1:861; 5:1376)*

149. **(C)** Coronary perfusion is improved by slow heart rates, since most perfusion occurs during diastole. The coronary flow is not independent of systolic pressure, since areas of the subendocardium are poorly perfused during systole. *(1:883; 4:1664; 5:204)*

150. **(A)** As the cardiac output increases, pulmonary artery pressure increases. Other factors also are involved in pulmonary artery pressure, e.g., the radius of the vessels. Vasoconstriction and vasodilatation of the pulmonary vessels occur with changes of cardiac output to regulate PVR. *(1:877; 4:1744; 5:679)*

151. **(C)** The transplanted heart has intact α- and β-adrenoceptors. The Frank-Starling effect is intact. Atropine will not have any effect since there is no autonomic innervation nor circulating cholinergic agonist. *(1:281; 2:544; 5:2253; 6:21)*

152. **(A)** An increased incidence of postoperative congestive heart failure is seen in patients with NYHA functional class IV. Patients with less severe heart disease do not have a higher incidence of postoperative congestive heart failure. The remaining options are all associated with signs of heart failure and should be sought and corrected in the preoperative period. *(1:475; 4:183; 5:908; 6:114)*

153. **(E)** The anxiety seen in patients with congestive heart failure is due to increased sympathetic activity. The increased sympathetic activity is a compensatory mechanism, and it is not directly related to the low flow state, as are the other options. Memory loss, weakness, fatigue, and confusion are directly caused by low output. *(1:295; 4:275; 5:633; 6:110)*

154. **(A)** Hypoxia, hypercarbia, and PEEP increase the afterload of the failing right heart. In transplantation, a healthy right ventricle may fail when suddenly challenged with increased afterload due to chronic or fixed pulmonary hypertension. Inhaled nitric oxide can be useful to reduce PVR. *(1:914; 2:544; 3:394; 4:1142; 5:165; 6:20)*

155. **(C)** Hypotension persisting after initiation of bypass may be a sign of aortic insufficiency. The initial hemodilution may cause decreased catecholamines and decreased viscosity. These should be transient. If persistent, one must look for other problems, e.g., a persistent shunt or aortic dissection. *(1:912; 4:1659; 5:1971–82)*

156. **(A)** Allowing the ventricle to overdistend can be detrimental, and most centers use a ventricular vent to decompress it. The ventricle should be kept in a nonbeating state and kept cold to lower oxygen consumption. Fibrillation should be avoided, since the oxygen cost is great. *(1:906; 4:1684; 5:1977)*

157. **(C)** The recovery room patient with hypertension should be examined to rule out hypoxia or hypercarbia. If present, these must be treated, and this takes precedence over other considerations. After you have ruled out those problems, you can proceed with a methodical assessment of the problem: retake the pressure, and if it is not life-threatening, you may want to watch it or treat it with a drug. In some cases, a consultation may be in order. *(1:1383; 4:141; 5:1217)*

158. **(C)** The vascular tree becomes less elastic with age. Blood pressure rises with age, which may be a reflection of the loss of elasticity. The number of myofibrils decreases and the cardiac output decreases. The cardiovascular reserve decreases as a reflection of the other changes. *(1:1206; 4:212; 5:2436; 6:740)*

159. **(D)** Right ventricular function is receiving much more attention, and the interdependence of the two ventricles is being appreciated to a greater degree. Blood flow occurs during systole and diastole. Since there is good collateral flow, it is harder to cool the right ventricle. PVR is hard to overcome, since the ventricle muscle is not as well developed. PEEP will be transmitted to the right ventricle, decreasing venous return. *(1:853; 4:1825; 5:2033)*

160. **(D)** Pheochromocytoma is amenable to surgical correction. Other causes of secondary hypertension—secondary aldosteronism and renal parenchymal disease—may not be amenable. The patient with renal artery stenosis may be treated surgically if the condition is diagnosed early. If later, surgery may be of no help. *(1:876; 4:163; 5:2072; 6:364)*

161. (B) Allen's test is used to assess the adequacy of the ulnar artery flow when cannulating the radial artery. Studies have shown poor correlation with subsequent problems. The reason for this is that many of the problems are embolic in nature, which cannot be predicted with an Allen's test. In addition to the lack of data showing correlation, there is confusion as to what is positive or negative. It is better to describe the result of the test. Since the position of the wrist is important, one must ascertain that the wrist is not hyperextended when performing the test. *(1:673; 1:817; 2:751; 4:810; 5:1272; 5:2069)*

162. (A) A central venous line is not indicated because the patient is in an abnormal position. It is indicated in patients with shock. Hyperalimentation, inotropic drugs, and irritating solutions should be infused through a central venous catheter. When performing a procedure where large fluid shifts are expected, measurement of CVP may be useful to determine the patient's volume status. *(1:675; 2:740; 4:811; 5:1286)*

163. (E) The precordial thump is administered to the midpart of the sternum with the fleshy part of the fist from a distance of about 8 to 12 in. It is recommended in monitored situations. It is not recommended in children. Indiscriminate use may lead to worse arrhythmias. *(1:1485; 2:265; 4:657; 5:2929–35)*

164. (E) The current recommendation is to attempt defibrillation with an energy level of 200 J in an adult. If that is unsuccessful, a second attempt is made with 200 to 300 J; if that is unsuccessful, a third attempt with 360 J is made. It is not always indicated and it is not always successful. *(1:1492; 2:265; 4:659; 5:2929–35)*

165. (C) Atropine is the first-line drug for reducing vagal tone. Epinephrine is used for its α-adrenergic effect. Lidocaine is preferred over bretylium because of lidocaine's fewer side effects. Isoproterenol is not used first in cases of bradycardia. *(1:1485; 2:265; 3:957; 4:647; 5:2937)*

166. (D) Drug administration should be by central line, if possible. If no central line is available, a peripheral line can be used. Femoral lines are not favored, since circulation below the diaphragm is not as good. If no access is available, the endotracheal tube may be used for administration of epinephrine, atropine, and lidocaine. Sodium bicarbonate cannot be used via the endotracheal route. *(1:1485; 2:265; 5:647)*

167. (B) In hypovolemic states, the pulse pressure is narrowed. Heart rate is increased to maintain cardiac output, and the neck veins are flat. Urine volume is decreased to preserve volume. The mucous membranes are pale, reflecting lower blood flow to peripheral areas. *(1:116; 2:706; 5:1326)*

168. (A) The most accurate interpretation of CVP is by analysis of waveform. The catheter should not be placed in the right atrium, since perforation may occur. CVP gives an indication of fluid status but is not accurate in all patients, especially in those with disparate right and left ventricular function. It is not necessary to use a flow-directed catheter in every patient. Some can be monitored by CVP alone, and the extra risk is not warranted. *(1:675; 2:740; 4:811; 5:1289)*

169. (C) A flush system is indicated in pressure monitoring. When decannulating the vessel, gentle aspiration is advised to aspirate any clots or debris that may be present. The arterial catheter should be the smallest that is compatible with the need, and it should be removed as soon as possible, since the complication rate rises with the duration of placement. The radial artery is a favorite site, since it has good collaterals. *(1:673; 2:749; 4:809; 5:1279)*

170. (D) VSD is associated with high pulmonary flow. The options of pulmonic stenosis, tetralogy of Fallot, and tricuspid atresia are associated with decreased blood flow to the pulmonary circuit. Coarctation of the aorta is not associated with a change in pulmonary flow. *(1:919; 2:428; 4:2105; 5:2009; 6:48)*

171. **(E)** Pneumothorax is a common problem with subclavian cannulation. Chylothorax is a complication of left internal jugular cannulation. There should be no brachial plexus trauma with the antecubital approach. Allen's test is not a foolproof method to determine the patency of the ulnar arch. The dorsalis pedis artery should be avoided in the patient with diabetes or peripheral vascular disease. *(1:107; 2:479; 5:1294)*

172. **(A)** The patient with aortic stenosis has a rapid course once angina, syncope, and congestive heart failure occur. The size of the ventricular muscle mass increases, but the cavity size does not change. The increased size also renders the muscle more prone to ischemia. Fast heart rates will cause a low cardiac output, since ventricular filling will be compromised. In addition, coronary perfusion will suffer. Low heart rates can also be devastating, since the stroke volume is fixed. The most common problem is that of tachycardia. *(1:894; 2:450; 4:18; 5:1954)*

173. **(A)** Inhaled nitric oxide dilates blood vessels in ventilated regions of the lungs and can thereby improve the matching of perfusion to ventilation. While intravenous vasodilators can worsen ventilation:perfusion matching and so worsen systemic arterial oxygenation, inhaled nitric oxide often improves systemic oxygenation. Inhaled nitric oxide exhibits little or no systemic vasodilation because the molecule is rapidly destroyed in the systemic circulation. Reaction with heme is one mechanism of NO destruction, and hemoglobin oxidation occurs concomitantly. In most clinical situations, NO benefits occur at doses that do not cause significant methemoglobin levels. *(1:1172; 2:357; 3:394; 5:1454)*

174. **(C)** Nitric oxide is normally produced by endothelial cells but can be delivered pharmacologically. It relaxes smooth muscle by stimulation of guanylate cyclase. *(1:279; 2:357; 3:394; 5:685)*

175. **(C)** ε-aminocaproic acid inhibits conversion of plasminogen to plasmin and aprotinin inhibits plasmin. Tissue plasminogen activator and streptokinase convert plasminogen to plasmin and thus promote fibrinolysis. *(1:232; 2:401; 3:1533; 5:1811; 6:497)*

176. **(B)** Protamine antagonizes heparin. *(1:228; 2:20; 3:1783; 5:749; 6:543)*

177. **(B)** The anemia of chronic renal failure stems from diminished production of erythropoietin, a protein available for therapeutic purposes. *(1:1012; 2:248; 3:1488; 4:2410; 5:1826)*

178. **(E)** An accessory conduction pathway is evidenced by the wide QRS complex and its slurred upstroke (delta wave). Patients are prone to develop atrial fibrillation or paroxysmal supraventricular tachycardia. *(1:1513; 4:1570; 5:1400)*

179. **(B)** Clonidine lowers blood pressure by stimulating presynaptic α_2-adrenergic receptors in the central nervous system and periphery. Phenylephrine raises blood pressure by stimulating vascular α_1-adrenergic receptors. Labetalol blocks α- and β-adrenoceptors. The action of vasopressin is mediated through V_1 and V_2 receptors. *(1:316; 2:180; 3:789; 3:879; 4:747; 5:650; 6:431)*

180. **(D)** Increased peripheral resistance may reduce cyanosis in tetralogy of Fallot. *(1:919; 2:433; 3:239; 4:1714; 5:2009; 6:55)*

181. **(B)** Because of esterase activities of the blood, there is no circulating acetylcholine. β-Adrenergic receptor antagonists slow the transplanted heart, but antimuscarinic agents do not speed the transplanted heart. *(1:1366; 2:544; 5:2253; 6:20)*

182. **(B)** Possible causes of PEA include pulmonary embolism, acidosis, tension pneumothorax, cardiac tamponade, hypovolemia, hypoxia, abnormal temperature, abnormal potassium level, myocardial infarction, and drug overdose. *(1:1493; 4:661; 5:2939)*

183. **(D)** Vasopressin is a peptide that vasoconstricts on binding to V_1 receptors and is an approved alternative to epinephrine for CPR purposes.

Its half-life is about 15 minutes. *(1:1495; 4:315; 5:2938)*

184. **(A)** The IIb/IIIa inhibitors are indicated in the management of unstable angina/non-ST-segment elevation myocardial infarction. They are often given with aspirin and heparin, which anticoagulate via other receptors. *(1:930; 4:924; 5:1985)*

185. **(B)** Milrinone is a bipyridine derivative. It inhibits the phosphodiesterase that breaks down cyclic-adenosine monophosphate (AMP). *(1:306; 2:370; 4:1688; 5:2032)*

186. **(E)** The glycoprotein is a receptor for fibrinogen or von Willebrand's factor, proteins which, in turn, provide cross-links to hold platelets together in an aggregate. Diverse platelet-activating stimuli operate via the glycoprotein. Abciximab is an inhibitory monoclonal antibody, and eptifibatide is an inhibitor peptide originally found in snake venom. *(1:213; 3:1535; 5:1985)*

187. **(E)** Verapamil may increase the heart rate in case of supraventricular tachycardia due to WPW syndrome. *(1:314; 2:379; 3:938; 5:2935)*

188. **(C)** Digoxin inhibits the sodium-potassium ATPase. *(1:306; 2:379; 3:916; 4:1654; 5:1125; 6:112)*

189. **(B)** Left ventricular failure and mitral stenosis tend to increase the wedge pressure. *(1:676; 4:1673; 5:1311)*

190. **(E)** Correlation with degree of regurgitation is poor. *(1:900; 4:849; 5:1312; 6:28)*

191. **(A)** Amiodarone slows the SA node, slows AV conduction, and decreases the ventricular response to atrial fibrillation. It is a coronary and peripheral vasodilator. *(1:1496; 4:1571; 5:1404)*

192. **(D)** Sinus bradycardia is characterized by rates of 40 to 60 beats per minute. There are many causes of bradycardia, and most do not require treatment. An exception to this general statement is the bradycardia occurring in the newborn. Any rate under 100 in the newborn is bradycardia and treatment is appropriate, since the newborn has a rate-dependent cardiac output. Sinus bradycardia is regular. The etiology of symptomatic bradycardia must be established; if the bradycardia occurs in a patient with a sick sinus, treatment may be appropriate. *(1:1384; 2:316; 4:1378; 5:1398; 6:80)*

193. **(C)** The tetralogy of Fallot includes a VSD, pulmonary outflow obstruction, overriding of the aorta, and right ventricular hypertrophy. An ASD or PDA is not part of the complex. *(1:919; 2:433; 4:1712; 5:2009; 6:53)*

194. **(E)** The preoperative ECG can be used for the detection of all the options. Chamber enlargement is readily seen in the tracing, as are some electrolyte disturbances, especially changes in calcium and potassium. Heart block is seen as are ischemic changes. *(1:485; 4:211; 5:951; 6:18)*

195. **(C)** Cardiac output is a function of many components of the circulatory system. The typical output is 5 to 6 L/min in an adult. Output is affected by blood volume. Cardiac output is the product of heart rate and stroke volume. Communications of the bronchial artery capillaries with those of the pulmonary veins is a reason for left ventricular output exceeding the right. Left-to-right shunting through an ASD could increase right output over left. *(1:678; 2:367; 5:683)*

196. **(A)** Left ventricular compliance is decreased in the patient with aortic stenosis or ischemia and in the patient receiving inotropes. It is increased in the patient on vasodilators. The typical aortic stenosis patient has a normal-sized cavity with a large myocardial mass. Small increases in fluid lead to a large increase in pressure. *(1:853; 4:1685; 5:728)*

197. **(E)** Preload is affected by all the options. Volume changes are important for preload, since the volume is the chief determinant of the amount of blood returning to the heart. Venous tone plays a part by assisting venous return. Heart rate is important, since large increases in rate decrease filling time. The atrial contribution is important to ventricular filling. *(1:291; 2:357; 4:841; 5:725)*

198. **(C)** Myocardial contractility is affected by both neural and humoral factors. Parasympathetic tone and myocardial ischemia both decrease contractility, whereas sympathetic stimulation and inotrope administration increase contractility. *(1:853; 4:104; 5:1327)*

199. **(B)** There are many variations of the blood supply to the heart, but the usual configuration is for a right coronary artery supplying the right side, and a left coronary that divides into the left anterior descending and the circumflex. Most of the venous drainage empties into the coronary sinus, but there are also direct connections of the thebesian veins into all chambers. *(1:853; 2:443; 5:2648; 6:18)*

200. **(B)** When inserting a CVP line, air embolus may occur if the catheter is left open to the atmosphere while the patient creates negative pressure in the thorax. The tip should be out of the atrium to avoid perforation of that chamber. The external jugular vein is less prone to complications than the internal. The right side of the neck is preferred because of the presence of the thoracic duct on the left. *(1:1192; 2:740; 4:1673; 5:1289)*

201. **(D)** Calcium administration increases myocardial contractility and increases excitability. The effect of calcium can be antagonized by calcium chelating agents, but the effect is too short for clinical use. Calcium may predispose to digoxin toxicity. *(1:307; 2:370; 3:1718; 4:1691; 5:731)*

202. **(D)** The Allen's test should be described in terms of return of flow and time to return to flow, since there is confusion in terminology regarding positive and negative. There is controversy concerning the usefulness of the test in general. Extension of the wrist can itself affect perfusion. Eighty percent of patients have a complete volar arch. *(1:673; 1:817; 2:751; 4:810; 5:1272)*

203. **(E)** A correctly positioned CVP line should be in the thoracic portion of the vena cava or great veins, preferably at the cava-atrium junction. It should not be in the heart chambers. To demonstrate patency and proper placement, there should be variation of pressure with respiration, and aspiration of blood should be possible. *(1:675; 2:740; 4:1673; 5:1289)*

204. **(A)** Reasons for choosing the basilic vein in preference to the cephalic are the larger size of the basilic and a straighter shot with the basilic. The basilic vein becomes larger as it proceeds proximally. The basilic vein lies medial to the cephalic vein. *(1:856; 2:739; 5:1286)*

205. **(E)** Cardiac output determination with the thermodilution technique uses a dilution calculation just as does the dye-dilution technique. A thermistor measures the change in temperature in the pulmonary artery. In the dye-dilution technique, changes in dye concentration are determined. Attention to detail is important, including the injection of an exact amount with a known temperature. *(1:678; 4:1608; 5:1222)*

206. **(E)** The values are all within normal limits. *(1:675; 4:820; 5:1302)*

207. **(C)** Hyperventilation decreases coronary flow and can cause spasm of the coronary arteries. This effect can be suppressed by diltiazem, and the decreased flow can be treated with nitroglycerin. *(1:865; 4:1729; 5:717)*

208. **(B)** Many factors affect the ischemic heart. Anemia aggravates ischemia by decreasing the amount of oxygen available. Adrenergic blockade alleviates ischemia by decreasing oxygen demand. Digitalis will increase oxygen demand, but it increases the pumping function of the heart, making it more efficient. This relieves some of the ischemia. Nitroglycerin increases blood flow, which alleviates the ischemia. *(1:884; 3:852; 4:2310; 5:1946)*

209. **(E)** Cardioplegia has improved the success rate of cardiac procedures, since it provides a relaxed heart. This factor leads to a decrease in oxygen consumption. Oxygen consumption approximately doubles in the presence of ventricular fibrillation. *(1:906; 4:1684; 5:1977)*

210. **(E)** Air embolism is a factor in many procedures but is most problematic in procedures on the aortic valve. Vents that are placed in the ventricle and aorta allow the air to escape. Some air accumulates in the pulmonary veins. This air can be evacuated by expansion of the lungs while the vents are still in place. *(1:910; 2:400; 5:2672)*

211. **(E)** All the options are methods of detecting air embolism. Doppler monitoring will detect small amounts of air. Mass spectrometry measures levels of nitrogen. End-tidal carbon dioxide, which decreases with air embolism, is sensitive but requires some decrease in pulmonary blood flow before a change is noted. *(1:803; 2:729; 4:1635; 5:1381; 6:244)*

212. **(E)** The position of the tip of the CVP catheter can be detected by the injection of 1 mL of air, but that is a dangerous method; carbon dioxide should be used. Electrocardiography can be used, looking for large P waves as the catheter is advanced. X-ray with contrast medium in the catheter is very useful. Pressure monitoring on insertion, similar to that used when inserting a pulmonary artery catheter, also is helpful. *(1:818; 2:740; 4:811; 5:1286)*

213. **(C)** The phlebostatic axis is a reference for measuring pressures. It is the level of the heart at which there is little pressure change with change of position. It is of obvious clinical importance since monitors must have a good reference level. The site is at the lower aspect of the atrium. *(1:673; 5:1286)*

214. **(E)** All of the structures are involved with baroreceptor activity. The carotid sinus is the most important structure involved, but baroreceptors are also present in the arch of the aorta and the wall of the great vessels of the neck. Impulses are carried to the vasomotor center by the glossopharyngeal nerve. *(1:280; 4:157; 5:738)*

215. **(C)** The renin-angiotensin system includes renin, which is secreted by the kidney. Renin acts on plasma angiotensinogen to form angiotensin I, which is converted to angiotensin II by a converting enzyme in the lung. The entire process is set into motion by decreased blood volume. *(1:876; 3:809; 4:1592; 5:1486; 6:425)*

216. **(B)** Postbypass neurologic changes have been attributed to hypotension and decreased perfusion while on the pump. Others have postulated that debris caused by the pump may be at fault. Attempts to improve the outcome have led to the use of barbiturates during bypass. The pulmonary artery pressure should have no direct effect. *(1:113; 2:402; 5:1541; 6:5)*

217. **(A)** There is a standard code for pacemakers—first letter: the chamber paced; second letter: the chamber sensed; third letter: mode of response; fourth letter: programmable features; and fifth letter: dysrhythmia treatment. When treating a patient who has a pacemaker, it is important to know what kind of pacer is involved. *(1:1514; 5:1431; 6:72)*

218. **(C)** The pacemaker threshold can respond to many bodily changes, including decreased threshold with stress. This may be related to hyperventilation and decrease in carbon dioxide, which also decreases the threshold. Increased potassium values cause a lowered threshold. Hypoxia increases the threshold. *(1:1514; 5:1419; 6:72)*

219. **(D)** Thebesian veins form part of the venous drainage system of the heart. The veins carry desaturated blood and they empty into any chamber. They are part of the coronary circulatory system. Since they carry desaturated blood, they increase shunt when they empty into the left side of the heart. *(1:856; 5:701)*

220. **(D)** Cardiac index is an attempt to match data from various sized patients. It is calculated by dividing the patient's cardiac output by his/her body surface area. Thus, when comparing two patients of varying sizes, comparison can be done on the basis of surface area. Cardiac index in an adult is about 3.5 L/min/m^2, and in a 2-year-old it is about 2.1 L/min/m^2. *(1:860; 2:357; 4:521; 5:2834)*

221. **(D)** The presence of large v waves is due to some influence affecting left atrial filling. Large v waves may show evidence of mitral regurgitation. Aortic valve and tricuspid valve abnormalities should not be reflected in the measurements. It would be difficult to insert a pulmonary artery catheter into a patient with tricuspid atresia. *(1:900; 2:432; 4:847; 5:1312; 6:28)*

222. **(A)** Coarctation of the aorta can be the cause of hypertension. In the infant, it may be associated with congestive heart failure due to left ventricular failure. Coarctation of the aorta may occur as an isolated lesion in the newborn. *(1:916; 2:424; 4:1712; 5:2010; 6:52)*

223. **(D)** Transesophageal echocardiography has the advantage of being able to delineate the left ventricle very well. The right ventricular structures are also well seen. It can be used in a continuous mode. With the esophageal location, monitoring can take place with the chest open or closed. *(1:886; 2:396; 4:1676; 5:1363–85)*

224. **(B)** Lymph is essentially an interstitial fluid. The flow is increased by exercise. The lymph flowing from the intestine is high in fat. Lymph is returned to the systemic circulation via the thoracic duct. *(1:176; 4:1739; 5:705)*

225. **(E)** The Goldman study showed a correlation between all the factors listed and increased risk of anesthesia. There is not complete agreement among anesthesiologists concerning the implications of the Goldman study, but it has stimulated many studies on outcome. *(1:1213; 4:171; 5:908; 5:1067)*

226. **(E)** Ventricular dysfunction may be detected by the several methods outlined. *(1:889; 4:820; 5:1222)*

227. **(B)** Injection of medications at the time of cardiac arrest usually is problematic. If a deep line is in place, it should be used. If no line is in place, an antecubital line should be attempted first. This will allow drug administration and not interrupt resuscitation procedures. If no line is available, lidocaine, atropine, and epinephrine can be administered through the endotracheal tube. *(1:1485; 2:265; 4:647; 5:2923–48)*

228. **(E)** Baroreceptors are essentially stretch receptors. They respond to actual pressure and rate of change of the pressure. Through connections to the cardioinhibitory and vasomotor centers, the receptors are able to regulate cardiac output. In the patient with chronic hypertension, the baroreceptors are reset to ignore the baseline pressure and respond only to pressures above that level. *(1:280; 4:157; 5:738)*

229. **(E)** The blood supply to the spinal cord is tenuous at best. There are anterior and posterior arteries that are branches of the vertebral arteries. In addition, there are radicular arteries at several levels, which supply the cord. The largest radicular artery is the artery of Adamkiewicz. These arteries are not constant, and low perfusion pressure or ligation of a large artery may jeopardize the perfusion of the cord. *(1:952; 2:508; 4:2119; 5:2088; 6:146)*

230. **(E)** The pericardium is not necessary for life, as is evidenced in the many patients who function after open heart surgery without one. The function of the pericardium is to buffer the heart from sudden movement, provide lubrication between the myocardium and adjacent structures, and minimize cardiac dilatation. *(1:879; 4:383; 5:1316; 6:136)*

231. **(D)** There is an interdependence of the two ventricles, and changes in the pressures on one side will have an effect on the other. The interventricular septum can shift with changes in pressure and compromise the other side. In normal inspiration, right ventricular end-diastolic volume increases, while that of the left ventricle decreases. *(1:819; 4:1670; 5:724)*

232. **(C)** HOCM is a form of primary myocardial disease in which there is hypertrophy of the muscle cells. This may occur in an asymmetric fashion. Contraction leads to obliteration of the ventricular cavity and diastolic filling is also impaired. Inotropic agents cause increased functional obstruction of the outflow tract and should be avoided. β-Adrenergic receptor antagonists can be employed. Medical treatment is not always successful. *(1:895; 4:274; 5:1956; 6:120)*

233. (E) Afterload will decrease in any situation where less pressure is required of the left ventricle. Decreases in blood viscosity will decrease afterload. An arteriovenous fistula will shunt blood from the arterial side. Vasodilators make the arterial bed more expansile. Mitral regurgitation decreases the force necessary to pump blood from the left ventricle. *(1:291; 2:546; 3:901; 4:842; 5:683)*

234. (A) Preload, contractility, and afterload are important for determination of stroke volume. Stroke volume and heart rate are the determinants of cardiac output. *(1:870; 2:357; 3:903; 5:724–9)*

235. (B) Histamine injection causes pulmonary vasoconstriction and systemic vasodilatation. There is also bronchoconstriction. The effect seen on the circulation is the same, in general, as that seen with hypoxia. *(1:207; 2:147; 3:645; 4:389; 5:688; 6:286)*

236. (D) An increase in mean diastolic ventricular pressure is usually reflected in the pulmonary circulation, assuming the mitral valve is competent. The increases may not have an effect on cardiac output depending on where the patient is on the Starling's curve; an increase may lead to decompensation. Treatment with a diuretic leads to a decrease in volume. The increase usually is transmitted to the atrium and proximal vascular structures. *(1:676; 4:1664; 5:1322)*

237. (B) Morphine has beneficial effects in the patient with acute pulmonary edema. These include its sedative effects and venous dilatation. There will be a decrease in venous pressure with its administration and a decrease in respiratory activity. *(1:351; 3:584; 5:680; 6:110)*

238. (E) Acute pulmonary edema can occur with all of the options. Treatment is aimed at the cause and oxygenation of the patient. Positive pressure ventilation may be necessary. *(1:676; 2:466; 5:1468; 6:207)*

239. (D) Ventricular hypertrophy presents additional problems to an already injured heart. Hypertrophy increases oxygen demand. Perfusion is decreased since diastolic pressure is decreased. The enlargement of the chamber requires an increased filling pressure. The ventricle becomes more susceptible to ischemia even in the absence of coronary artery disease. *(1:896; 4:35; 5:1946)*

240. (E) This patient has the typical list of problems. One must do what is possible to limit the inequities of myocardial oxygen supply and demand. Lowering the pressure will decrease demand, as will lowering the temperature and heart rate. Administration of oxygen will increase the supply. *(1:884; 4:139; 5:1946)*

241. (C) A cyclic-GMP phosphodiesterase inhibitor has a positive inotropic effect and would lower SVR. It would only be used in a patient with a low flow state due to low cardiac index or cardiogenic shock. A balloon pump is expected to reduce myocardial demand and improve supply. As a coronary thrombosis has probably occurred, fibrinolytics (not antifibrinolytics) and anticoagulants may help *(1:884; 1:932; 3:850; 5:1946; 6:5)*

242. (D) A water-hammer pulse is seen in the patient with aortic insufficiency. The patient with pericardial effusion will exhibit pulsus paradoxus. The cardiac output is decreased. The neck veins are distended and stay distended in the supine position, since little drainage can occur. Blood pressure will decrease. *(1:918; 4:384; 5:1967)*

243. (A) Pulsus paradoxus is an exaggeration of the normal variation in the pulse with inspiration. The pulse reduces in amplitude on inspiration. In the exaggerated state of pericardial effusion, the pulse may show alternate strong and weak beats. *(1:875; 4:614; 5:1285)*

244. (D) Monitoring this patient does not require a pulmonary artery catheter. The right side of the heart should be monitored, and this can be done with a central venous catheter. An arterial line would be helpful. X-ray, once the diagnosis is established, is of little value. *(1:978; 4:385; 5:1967)*

245. **(D)** The anesthetic management of the patient with pericardial effusion is very demanding. One must maintain the cardiac output that is present. There is very little to do to improve it. Any vasodilator may cause circulatory collapse. Cardiodepressants must be avoided. Some believe that ketamine or etomidate is the drug of choice. *(1:978; 4:385; 5:1966)*

246. **(E)** Coronary circulation is influenced by oxygen saturation and cardiac output. The coronary flow is autoregulated to maintain flow in the pressure range of 50 to 120 mmHg. Autoregulation is through oxygen levels, but functions through the action of adenosine as an intermediary. *(1:866; 2:379; 4:1664; 5:1946)*

247. **(D)** Myocardial oxygen consumption is increased by isoproterenol, due in part to increased rate and in part to increased inotropy. Consumption is increased by digoxin and decreased by nitroglycerin. *(1:866; 3:861; 4:1664; 5:1946; 6:19)*

248. **(A)** The calculation requires mean arterial pressure, mean CVP, and cardiac output. Pulmonary pressure is not needed. *(1:676; 1:860; 4:1660; 5:1333)*

249. **(B)** Venous return is the chief determinant of preload. As the right atrial pressure increases, return will fall. Muscular contractions, especially in the lower extremities, aid in venous return. The venous valves also assist in venous return. *(1:879; 4:804; 5:725)*

250. **(A)** Oxygen consumption under anesthesia will fall about 15%. It will fall more in the patient who is excited before going to sleep, since that patient has a higher consumption. The fall in consumption is not affected by premedication or tidal volume. *(1:395; 4:397; 5:702)*

251. **(C)** Goals for anesthetic management of the patient with coronary artery disease include maintaining contractility and maintaining sinus rhythm. One wants to avoid a high afterload and a high heart rate. These factors apply to the patient for cardiac or noncardiac surgery. *(1:884; 4:1665; 5:1946)*

252. **(A)** The capillary bed is the site of gas and nutrient exchange between blood and tissues. The bed is seldom open to full capacity but is regulated by the metarterioles and precapillary sphincters. Local physical and chemical factors play a role in the flow through the capillary bed, but capillaries themselves are not regulated by the autonomic nervous system. *(1:873; 5:723)*

253. **(C)** The mitral stenosis patient usually has increased left atrial pressure because of poor outflow from the atrium. The right ventricle is enlarged and the left ventricle is not enlarged. Cardiac output is decreased due to the decreased left ventricular preload. *(1:898; 4:848; 5:1959; 6:29)*

254. **(B)** Depolarization of the cardiac muscle cell (phase 0) corresponds to the QRS. Phase 1 is the beginning of the repolarization. Phase 2 is a plateau and represents a recovery phase. Phase 3 is repolarization and is responsible for the T wave. Phase 4 is diastolic depolarization in the automatic cell. In the nonpacemaker cell, phase 4 restores resting membrane potential pending another depolarization. *(1:862; 3:933; 4:1330; 5:730)*

255. **(C)** Phase 4 of the action potential is responsible for pacemaking activity in some cells. The changes are due to fluxes of sodium and potassium. Diastolic depolarization is seen in pacemaker cells only. *(1:862; 3:933; 4:1330; 5:730)*

256. **(B)** Stimulation of the vagus slows the heart but does not affect contractility, since the vagus does not affect the ventricle directly. Vagal stimulation slows conduction and decreases excitability. *(1:282; 2:513; 4:726; 5:723)*

257. **(B)** Stimulation of the sympathetic fibers to the heart increases the rate and enhances the conduction. The sympathetic fibers are more directed to the ventricle. The duration of systole is decreased, and the contractions are strengthened. *(1:296; 5:723)*

258. **(D)** The coronary sinus receives blood that is very desaturated, having passed through the myocardial tissues that extract most of the oxygen. The sinus is the major venous drainage system of the heart. It receives about 75% of the coronary blood flow. The blood from the coronary sinus drains into the right atrium. *(1:853; 5:1303)*

259. (B) The bronchial arteries are part of the systemic circulation, although they are rich in anastomoses with the pulmonary artery. The bronchial arteries arise from the aorta. They drain into the azygos vein. *(1:794; 4:1739; 5:723)*

260. (E) The pulmonary circulation from the right side of the heart is very expansile, being able to absorb large increases in blood due to increases in cardiac output. The circulation time in the pulmonary circuit is about 11 seconds. The pulmonary circuit holds about 400 to 600 mL at any one time. *(1:801; 4:1711; 5:723)*

261. (E) The maneuver was described by Valsalva as a means to force air into the middle ear via the Eustachian tubes. The maneuver causes a decrease of blood return to the right ventricle due to increased intrathoracic pressure. This also leads to an increased venous pressure in the head. Cardiac output falls during the intrathoracic pressure increase. There is a reflex increase in heart rate when the blood pressure falls. *(1:280; 2:81; 5:739)*

262. (B) During blood storage, there is an increasing plasma concentration of potassium and ammonia. Dextrose and bicarbonate levels decrease. *(1:204; 2:242; 4:2410; 5:1804)*

CHAPTER 4

Respiration

Questions

DIRECTIONS (Questions 263 through 351): Each of the numbered items or incomplete statements in this section is followed by answers or by completions of the statement. Select the ONE lettered answer or completion that is BEST in each case.

263. If one measured pleural pressure in a standing human, one would find that the pressure was

(A) highest at the apex of the lung
(B) highest at the base of the lung
(C) equal at all levels
(D) unrelated to body position
(E) completely unpredictable from one level to another

264. The Laplace law is important in pulmonary physiology because

(A) it describes the properties of gas mixtures
(B) it describes the angles of the bronchi
(C) it describes the bucket handle movement of the ribs during ventilation
(D) it describes the pressure relationships within the alveoli
(E) it describes resistance in large airways

265. Surfactant is a substance that

(A) is produced in the liver of the newborn
(B) is important in newborns but has little importance in the adult
(C) is produced by the basement membrane of the lung
(D) lowers surface tension in the alveoli
(E) is a long-chained carbohydrate molecule

266. The functional residual capacity (FRC) is defined as the combination of

(A) tidal volume and residual volume
(B) tidal volume and expiratory reserve volume
(C) tidal volume and inspiratory reserve volume
(D) residual volume and expiratory reserve volume
(E) vital capacity less the closing volume

267. The maximum volume of air inhaled from the end of a normal inspiration is referred to as

(A) tidal volume
(B) inspiratory capacity
(C) inspiratory reserve volume
(D) vital capacity
(E) functional residual capacity

268. The metabolic functions of the lung do NOT include inactivation of

(A) 5-hydroxytryptamine
(B) epinephrine
(C) bradykinin
(D) prostaglandin
(E) norepinephrine

269. Postoperative pulmonary function

(A) is characterized by increased airway resistance
(B) is worse after upper-abdominal as compared to thoracic or lower-abdominal incisions
(C) is characterized by air trapping and increased FRC
(D) is unaffected by a laparoscopic approach
(E) returns to baseline after 48 hours of smoking cessation

270. Factors contributing to increased airway pressure under anesthesia include all of the following EXCEPT

(A) muscle paralysis of the chest wall
(B) a decrease in FRC
(C) the supine position
(D) the presence of an endotracheal tube
(E) controlled ventilation

271. Humidification of inhaled air or gases is

(A) more efficient with an endotracheal tube in place
(B) increased with the administration of atropine
(C) more efficient in an open system than a closed system
(D) at its optimum with the patient breathing through the nose
(E) at its optimum with the patient breathing through a tracheostomy

272. The principal muscle(s) of respiration is (are) the

(A) diaphragm
(B) intercostals
(C) abdominals
(D) sternomastoids
(E) scalenes

273. Distribution of ventilation in the lung is such that

(A) the apical portions are better ventilated
(B) the dependent areas are better ventilated
(C) the central or hilar areas are better ventilated
(D) all areas are ventilated equally
(E) ventilation is not affected by position

274. Inspiratory force

(A) is a measurement of reserve of ventilatory effort
(B) is dependent on conscious effort
(C) correlates with force of cough
(D) is difficult to measure
(E) normally is between 5 and 10 cm H_2O above atmospheric pressure

275. A patient is admitted for bariatric surgery for treatment of morbid obesity. His height is 65 in., and his weight is 134 kg. He is diagnosed as having the obesity-hypoventilation (Pickwickian) syndrome. His preoperative evaluation would show

(A) alveolar hyperventilation, anemia, and hypoxemia
(B) alveolar hyperventilation, erythrocytosis, and hypoxemia
(C) decreased expiratory reserve volume and higher intra-abdominal pressure when supine
(D) lower work of breathing
(E) unimproved respiratory condition after weight reduction

276. In the normal adult lung

(A) small airways cause the majority of airway resistance
(B) gas exchange occurs only in terminal alveoli
(C) tracheal cartilage is incomplete anteriorly
(D) terminal bronchioles have no cartilage in their walls
(E) the pores of Kohn connect the right and left sides

277. All of the following components contribute to physiologic shunt EXCEPT

(A) bronchial veins
(B) pleural veins
(C) thebesian veins
(D) abnormal arterial-venous communications in the lung
(E) nonperfused alveoli

278. During one-lung anesthesia with a double lumen tube

(A) no shunting occurs in the unventilated lung
(B) no alveoli are in continuity with the proximal airways
(C) CO_2 removal is difficult
(D) continuous positive airway pressure (CPAP) with oxygen improves oxygenation
(E) it is not possible to administer oxygen to the collapsed lung

279. After bilateral carotid endarterectomy, a patient will

(A) have no respiratory changes
(B) show no change in arterial carbon dioxide
(C) respond to hypoxia with hyperventilation
(D) be more susceptible to hypoxemia
(E) always develop hypertension

280. Adequacy of alveolar ventilation is determined by measuring

(A) the oxygen gradient
(B) oxygen tension
(C) oxygen saturation
(D) cardiac output
(E) carbon dioxide tension

281. Measurement of arterial oxygen tension

(A) depends on hemoglobin concentration
(B) uses the Clark electrode
(C) uses the Severinghaus electrode
(D) is not affected by temperature
(E) is performed on clotted blood

282. Mechanisms that may cause hypoxemia under anesthesia include all of the following EXCEPT

(A) hypoventilation
(B) hyperventilation
(C) increase in FRC
(D) supine position
(E) increased airway pressure

283. In a patient with a fixed shunt, a decrease in cardiac output will

(A) decrease the arterial oxygenation
(B) decrease the arterial carbon dioxide tension
(C) have little effect on oxygenation
(D) decrease dead space
(E) increase urinary output

284. After a 2-hour anesthetic with hyperventilation, a patient breathing room air will

(A) return to normal parameters within 30 minutes
(B) remain hypocarbic for 2 hours
(C) possibly become hypoxemic if not treated with oxygen
(D) become hypoxemic and hypercarbic
(E) be well oxygenated if the air exchange is unimpaired by drugs

285. Carbon dioxide retention in chronic obstructive lung disease is

(A) due primarily to low cardiac output
(B) due primarily to inspiratory obstruction
(C) due primarily to low minute ventilation
(D) due primarily to increased V_D/V_T
(E) treated best by increasing inspired oxygen tension

286. Tracheal mucus flow is

(A) impeded by deviation from ambient oxygen tension
(B) decreased in patients with chronic obstructive lung disease
(C) not important as a cleansing mechanism
(D) unrelated to ciliary activity
(E) more active after a patient smokes a cigarette

287. All of the following are true of closing volume EXCEPT that it is

(A) measured by a single-breath nitrogen technique
(B) useful in determining disease of small airways
(C) decreased at the extremes of age
(D) unchanged in obesity
(E) measured at phase IV on the nitrogen washout curve

288. The ventilatory response to $PaCO_2$

(A) is independent of hypoxemia
(B) has a major peripheral component
(C) is depressed by metabolic acidemia
(D) is unaffected by opioid antagonists
(E) is augmented by norepinephrine

289. Pulmonary vascular resistance can be decreased by all of the following EXCEPT

(A) leukotriene
(B) bradykinin
(C) isopreterenol
(D) nitric oxide
(E) prostacyclin (prostaglandin I_2)

290. The peripheral chemoreceptors are

(A) located in the medulla oblongata
(B) poorly perfused and, therefore, respond slowly to changes in the oxygen content of the blood
(C) responsible for the hypoxic drive to respiration
(D) influenced by oxygen content rather than oxygen tension
(E) maximally stimulated by arterial oxygen tension over 250 mmHg

291. When considering oxygen transport in the lung, the LEAST important cause of hypoxemia is

(A) ventilation/perfusion mismatch
(B) diffusion barrier
(C) venous admixture
(D) bronchial artery blood flow
(E) altitude

292. In pulmonary function testing, diffusing capacity for carbon monoxide (DLCO)

(A) is greater than FRC
(B) is unchanged in anemia
(C) is decreased in pulmonary fibrosis
(D) estimates the gas transfer ability of the lung
(E) estimates the dead space ratio (V_D/V_T) of the lung

293. General anesthesia alters the mechanical properties of the chest wall leading to

(A) increased FRC
(B) increased lung volume
(C) decreased elastic recoil of the lung
(D) decreased compliance
(E) no change in lung physiology

294. The oxyhemoglobin dissociation curve describes the relationship of oxygen saturation to oxygen tension. All of the following are true EXCEPT that

(A) at an oxygen tension of 60 mmHg, the saturation is approximately 90%
(B) the curve is shifted to the left with a more acid pH
(C) the curve is shifted to the right with an increase in carbon dioxide tension
(D) the curve is shifted to the left with a decrease in temperature
(E) the curve is shifted to the right with increased levels of 2,3-DPG

295. A 35-year-old man enters the hospital with a diagnosis of restrictive pulmonary disease. The pulmonary function test result compatible with this diagnosis is

	FVC	FEV_1	$\frac{FEV_1}{FVC}$
(A)	normal	decreased	decreased
(B)	decreased	decreased	normal
(C)	decreased	decreased	increased
(D)	increased	increased	normal
(E)	normal	decreased	normal

where FVC: forced vital capacity.
FEV_1: forced expiratory volume in 1 second.

Questions 296 through 298

An 18-year-old male was involved in an automobile accident that resulted in a cervical cord injury at the C5 level. His signs and symptoms include paresthesias, motor weakness, a tender abdomen with an equivocal abdominal tap for blood, and a fracture of the femur. He is being treated with oxygen, 40% by mask, and skeletal traction. He is being evaluated for splenic injury. Initial arterial blood gas results are pH 7.40, PCO_2 42 mmHg, and PO_2 96 mmHg. Over the next 2 hours, his weakness becomes more profound, he becomes agitated, and repeat arterial blood gas results are pH 7.32, PCO_2 49 mmHg, and PO_2 79 mmHg.

296. At this time, the appropriate management is to

(A) observe for another hour and reevaluate
(B) intubate and ventilate
(C) administer anxiolytics cautiously, withhold opioids to preserve respiratory drive and not cloud the abdominal findings
(D) obtain an immediate chest film and evaluate it before making a decision
(E) increase oxygen delivery by mask

297. His respiratory changes may be caused by all of the following EXCEPT

(A) hypoventilation due to cord injury
(B) hypoventilation due to upper abdominal injury
(C) fat embolus
(D) respiratory depression secondary to oxygen administration
(E) pulmonary contusion

298. This patient would be a prime candidate for respiratory failure even in the absence of a leg fracture because

(A) he has a decreased ability to cough
(B) his lesion is high enough to predispose to aspiration
(C) his treatment may lead to pulmonary oxygen toxicity
(D) he may have blood loss from his other injuries
(E) he is being kept supine for evaluation

Questions 299 and 300

Pulmonary function test results for a 25-year-old, 70-kg man are (at sea level, breathing room air)

vital capacity: 3600 mL
inspiratory capacity: 2500 mL
tidal volume: 500 mL
dead space: 300 mL
O_2 saturation: 93%
O_2 tension: 82 mmHg
CO_2 tension: 48 mmHg

299. These values show that the patient has

(A) a normal dead space
(B) an abnormal inspiratory capacity
(C) a low PCO_2
(D) a decreased vital capacity
(E) an abnormal O_2 saturation

300. Treatment of this patient should be directed toward

(A) decreasing dead space
(B) improving tidal volume
(C) maintaining the status quo
(D) decreasing atelectasis
(E) slowing ventilation

Questions 301 and 302 refer to Figure 4-1:

301. In the graph shown, the inflation and deflation curves differ. This is

(A) due to the pull of the thoracic cage
(B) referred to as hysteresis
(C) an abnormal condition in humans
(D) seen only in patients with obstructive lung disease
(E) a measurement of dead space

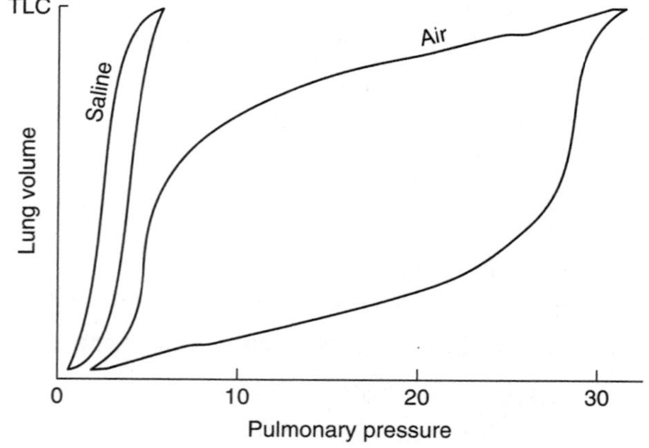

FIG. 4-1

302. In the graph, there is a marked difference between the curve for air and the curve for saline. This can be attributed to the

(A) effect of pollutants in the air
(B) effect of surface tension on lung elasticity
(C) method of measurement
(D) position of the esophageal balloon used in measurement
(E) presence of diseased lung tissue

303. The definitive test of adequacy of ventilation is

(A) listening to the esophageal stethoscope
(B) watching the rise and fall of the chest
(C) analyzing arterial blood gases
(D) measuring tidal volume with a spirometer
(E) using an apnea monitor

304. Pulmonary vascular resistance

(A) is entirely dependent on the cardiac output
(B) is entirely dependent on the pressure in the pulmonary artery
(C) is equal to pressure divided by radius of the artery
(D) depends on the state of vasomotor tone, flow, and pressure
(E) is not affected by cardiac output

305. Hypoxic pulmonary vasoconstriction

(A) is not important in the intact human being
(B) is active only at high altitude
(C) causes more blood flow to the base of the lung
(D) causes higher dead space/tidal volume ratio (V_D/V_T) than in the nonhypoxic lung
(E) diverts blood flow from hypoxic to non-hypoxic lung areas

306. The work of breathing

(A) is increased in the anesthetized patient breathing spontaneously
(B) is solely due to airway resistance
(C) is solely due to elastic forces
(D) is at its lowest at a respiratory rate of 25 breaths per minute
(E) is increased in the patient with restrictive disease if the respiratory rate is increased

307. Anatomic dead space

(A) is independent of lung size
(B) is about 1 mL/kg body weight
(C) is not affected by equipment
(D) combined with alveolar dead space constitutes physiologic dead space
(E) is of less importance in the newborn than the adult

308. The term P_{50} in reference to the oxyhemoglobin dissociation curve

(A) refers to the position on the curve at which the PO_2 is 50 mmHg
(B) normally has a value of 27 mmHg
(C) describes an enzyme system in hemoglobin
(D) is constant
(E) is affected only by type of hemoglobin

309. All of the following are frequently found in carbon monoxide poisoning EXCEPT

(A) seizures and coma
(B) lactic acidosis
(C) desaturation by pulse oximetry
(D) carboxyhemoglobin
(E) normal PaO_2

310. The carotid bodies primarily

(A) respond to elevated PCO_2
(B) respond to SvO_2
(C) respond to PaO_2
(D) signal the medulla via the vagus nerve
(E) respond to hydrogen ions

311. The Bohr effect refers to the

(A) effect of carbonic anhydrase on the uptake of CO_2
(B) ability of the blood to pick up more CO_2 when the oxygen tension is low
(C) amount of hydrogen ion in solution in the blood
(D) effect of PCO_2 on the saturation of oxygen
(E) presence of catalytic enzymes in the blood

312. Hypercapnia under anesthesia may be a result of

(A) hyperventilation
(B) decreased dead space ventilation
(C) decreased carbon dioxide production
(D) use of an Ayre's T-piece at less than peak inspiratory flow rate
(E) increased pulmonary artery flow

Questions 313 through 316 refer to Figure 4-2:

313. In Zone 1 of the lung

(A) no air is moving
(B) circulation is highest
(C) venous pressure is high
(D) dead space is high
(E) shunting is high

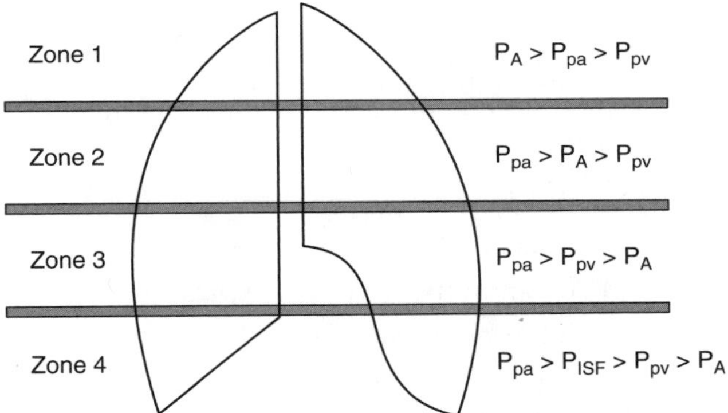

FIG. 4-2

314. In Zone 2 of the lung

(A) there is good blood flow regardless of ventilation
(B) venous pressure is high
(C) dead space is high
(D) the pulmonary vessels are collapsed
(E) the blood flow is determined primarily by pulmonary artery pressure and alveolar pressure

315. In Zone 3 of the lung

(A) blood flow is governed by the arteriovenous pressure difference
(B) dead space is high
(C) there is high alveolar pressure
(D) venous pressure is very low
(E) little blood flow occurs

316. All of the following are true EXCEPT

(A) Zone 2 is the "waterfall" region of the lung
(B) Zone 1 will increase in hypovolemic shock
(C) Zone 4 will increase with lymphatic blockage
(D) these zones would matter less if we could breathe in water
(E) these zones are independent of gravitational effects

317. Which statement about the diaphragm is NOT true

(A) it is innervated via the vagal nerve
(B) it has no fixed insertion
(C) it is mainly active in inspiration
(D) it has an equal mix of slow twitch and fast twitch fibers
(E) it is deficient in stretch receptors

318. Maximum voluntary ventilation (MMV)

(A) is measured over 3 minutes
(B) is effort independent
(C) is reduced in restrictive lung disease
(D) is a measure of oxygen transport
(E) is reduced in obstructive lung disease

319. The main buffering system in the circulation is

(A) hemoglobin
(B) bicarbonate
(C) phosphate
(D) sulfate
(E) plasma proteins

320. Hemoglobin as a buffer

(A) is the primary noncarbonic buffer
(B) is unimportant
(C) acts only on respiratory pH changes
(D) acts as the intact hemoglobin molecule, not as a salt
(E) is a stronger acid than carbonic acid

321. Recognition of hypoxemia in the recovery room

(A) depends on the detection of cyanosis
(B) depends on the detection of apnea
(C) depends on the detection of circulatory responses
(D) is best monitored with pulse oximetry
(E) is done better with a transcutaneous oxygen monitor than with a pulse oximeter

322. Hypoventilation in the recovery room

(A) should always be treated with opioid reversal
(B) is common after inhalation anesthesia
(C) is uncommon after upper abdominal procedures
(D) is best detected by pulse oximetry
(E) is always accompanied by increases in blood pressure

323. A patient who is breathing 100% oxygen

(A) completely eliminates the nitrogen in the lung in two breaths
(B) eliminates nitrogen rapidly, dependent on the volume of the breaths
(C) eliminates nitrogen regardless of presence of disease
(D) eliminates nitrogen faster if he or she is a smoker
(E) eliminates nitrogen faster if he or she is older

324. Signs of hypoxemia include all of the following EXCEPT

(A) decreased ventilatory effort due to chemoreceptor stimulation
(B) increased heart rate due to sympathetic stimulation
(C) cyanosis
(D) decreased heart rate due to direct effect of hypoxemia on the heart
(E) increased blood pressure due to sympathetic stimulation

325. Which of the following are true about Type II alveolar cells

(A) they produce surfactant
(B) they are the major components of gas exchange
(C) they line the capillary endothelium
(D) they can be replaced by Type I cells
(E) they are migratory and phagocytic

326. Diffusion nonequilibrium or diffusion impairment is

(A) important in CO_2 elimination
(B) important in oxygenation
(C) the gradient between alveolar and capillary gas tensions
(D) greater in low flow states
(E) greater in respiratory obstruction

327. The alveolar-arterial oxygen difference (A-aDO_2)

(A) in healthy adults is about 40 to 80 mmHg
(B) can be measured directly
(C) increases with age because of a decrease in arterial oxygen tension
(D) is a good screening tool for detecting dead space/vital capacity (V_D/V_T) changes
(E) is a good screening tool for detecting ventilation/perfusion (V/Q) changes

328. All of the following are involved in oxygen delivery EXCEPT

(A) glucose concentration
(B) carbon dioxide concentration
(C) cardiac output
(D) hemoglobin concentration
(E) dissolved oxygen

329. The Haldane effect refers to

(A) the effect of the cytochrome system on oxygen uptake
(B) the shift in the CO_2 dissociation curve caused by altered levels of oxygen
(C) the shift of the CO_2 dissociation curve due to temperature
(D) the shift of the oxyhemoglobin curve due to blood pressure
(E) the shift of the CO_2 response curve due to drugs

330. Carbon dioxide is transported in the circulation primarily

(A) as dissolved CO_2
(B) bound to proteins as carbamino compounds
(C) hydrolyzed to carbonic acid
(D) bound to hemoglobin as carboxyhemoglobin
(E) excluded from the red cell

331. Intrapulmonary shunting refers to

(A) anatomic dead space
(B) alveolar dead space
(C) ventilation without perfusion
(D) perfusion without ventilation
(E) wasted ventilation

332. The cuff of an endotracheal tube should be inflated to what pressure to prevent aspiration while not causing underlying ischemia?

(A) 5 mmHg
(B) 20 mmHg
(C) 40 mmHg
(D) 60 mmHg
(E) 80 mmHg

333. If a patient is allowed to breathe 100% oxygen under anesthesia

(A) atelectasis will disappear
(B) bowel distention will decrease
(C) the PO_2 will rise due to increased dead space
(D) lung units with low ventilation/perfusion (V/Q) ratios may become shunt units
(E) the oxygen tension will rise due to an increase in FRC

334. Calculation of physiologic dead space requires

(A) the Bohr equation
(B) the measurement of pulmonary artery oxygen tension
(C) the measurement of mixed venous CO_2
(D) the measurement of inspired CO_2 tension
(E) a body plethysmograph

335. All of the flowing statements about ciliated respiratory epithelium are true EXCEPT

(A) it extends to the alveoli
(B) is not involved in gas transport
(C) its function is inhibited by low humidity
(D) its function is inhibited by high FIO_2
(E) its function is inhibited by inhaled anesthetics

336. All of the following nerves are important in the motor supply of the larynx EXCEPT

(A) the superior laryngeal nerve
(B) the inferior laryngeal nerve
(C) the internal laryngeal branch of the superior laryngeal nerve
(D) the external laryngeal branch of the superior laryngeal nerve
(E) the recurrent laryngeal nerve

337. The musculature of the bronchioles

(A) runs along one side of the bronchus only
(B) becomes thinner as it proceeds distally
(C) becomes thinner relative to the thickness of the bronchiole as it proceeds distally
(D) has little effect on the diameter of the bronchiole
(E) has no relationship to the elastic fibers

338. Sequelae of unilateral recurrent laryngeal nerve injury include

(A) airway obstruction
(B) hoarse voice
(C) Horner's syndrome
(D) dysphagia
(E) masseter spasm

339. A patient is admitted for treatment of injuries suffered in an auto accident. After induction of anesthesia with thiopental and maintenance of anesthesia with halothane, nitrous oxide, and oxygen, the blood is noted to be very dark. The cause of the dark blood may be all of the following EXCEPT

(A) carbon monoxide
(B) positional compromise of venous drainage of the operated site
(C) sulfhemoglobin
(D) methemoglobin
(E) profound hypoxemia

340. All of the following statements about blood preservation are correct EXCEPT

(A) adding adenine to whole blood preservatives (CPD vs. CPDA-1) extends to storage life from 3 to 5 weeks
(B) 2,3-DPG is maintained above 70% of initial level
(C) more than 70% of red cells survive 24 hours after transfusion
(D) pH of the stored blood falls
(E) plasma potassium level in the stored blood rises

341. The pulmonary manifestations of scleroderma are characterized by all of the following EXCEPT

(A) increased compliance
(B) diffuse fibrosis
(C) decreased vital capacity

(D) hypoxemia
(E) increased V_D/V_T ratio

342. The Hering-Breuer reflex

(A) has little importance in the adult human
(B) causes a deep inspiration after a total expiration in the laboratory animal
(C) causes a deep inspiration after a cough
(D) has a major role in the control of ventilation
(E) is stimulated by barbiturates

343. The carbon dioxide response curve describes the pattern of ventilation following challenge by various concentrations of carbon dioxide. Which of the following statements is TRUE?

(A) The patient under anesthesia will show the same effects regardless of the agent.
(B) The slope of the response may change, but the position of the curve remains the same.
(C) The position and slope are the same in adults and infants.
(D) Increased work of breathing leads to a steeper slope.
(E) The slope of the line measures the patient's sensitivity to carbon dioxide.

344. Airway resistance

(A) is greatest in the terminal bronchioles
(B) is reduced by surfactant
(C) is increased in pulmonary fibrosis
(D) determines the distribution of ventilation
(E) is increased in anaphylaxis

345. All of the following will influence the carbon dioxide response curve EXCEPT

(A) individual variation
(B) hypoxemia
(C) chronic bronchitis
(D) opioids
(E) oxygen at 50% concentration

346. All of the following statements are true about alveoli EXCEPT

(A) they are 100 to 300 μm in diameter
(B) they are mostly lined with Type I alveolar cells
(C) they are partially lined with Type II alveolar cells
(D) they are partially lined with Type III alveolar cells
(E) they are surrounded by capillaries

347. Circulation to the lung

(A) consists entirely of the pulmonary artery and its branches
(B) consists of the pulmonary artery and the bronchial arterial system, which are of equal importance
(C) is under high pressure relative to the left side of the heart
(D) consists of two systems, but the bronchial system is expendable
(E) consists of the pulmonary artery, which nourishes the bronchioles, and the bronchial arteries, which progress only to the level of the second branching of the bronchioles

Questions 348 and 349 refer to Figure 4-3:

348. All of the following are TRUE of the flow/volume loop shown in the figure EXCEPT

(A) the x-axis is volume
(B) the y-axis is pressure
(C) vital capacity is the distance from points 1 to 3
(D) a breath proceeds through the points 1 to 4 in that order
(E) there is no air leak

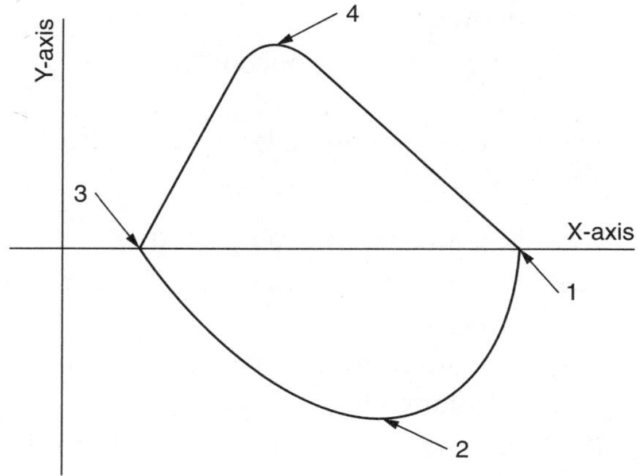

FIG. 4-3

349. In this flow/volume loop

(A) there is evidence of tracheal stenosis
(B) point 2 is at maximum expiratory volume
(C) point 3 is at maximum inspiratory volume
(D) one can determine inspiratory reserve volume
(E) one can determine FRC

Questions 350 and 351 refer to Figure 4-4:

350. As compared to the normal Curve 2, all of the following are TRUE about Curve 3 EXCEPT

(A) residual volume is elevated
(B) late expiratory flow is reduced
(C) vital capacity is relatively normal
(D) total lung capacity is elevated
(E) it is characteristic of restrictive lung disease

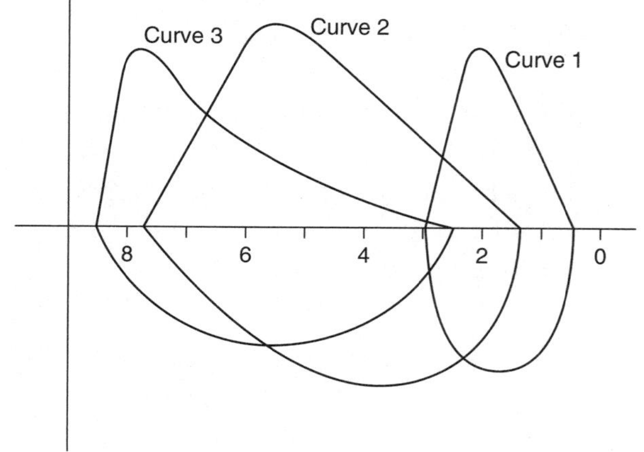

FIG. 4-4

351. As compared to the normal Curve 2, all of the following are TRUE about Curve 1 EXCEPT

(A) residual volume is reduced
(B) late expiratory flow is reduced
(C) vital capacity is reduced
(D) total lung capacity is reduced
(E) it is characteristic of restrictive lung disease

DIRECTIONS (Questions 352 through 424): For each of the items in this section, ONE or MORE of the numbered options is correct. Choose answer

(A) if only 1, 2, and 3 are correct
(B) if only 1 and 3 are correct
(C) if only 2 and 4 are correct
(D) if only 4 is correct
(E) if all are correct

352. FRC can be measured by

(1) use of an esophageal balloon
(2) helium wash-in
(3) spirometry
(4) body plethysmography

353. Under controlled ventilation, an increase in dead space would be characterized by

(1) decreased end-tidal CO_2
(2) decreased PaO_2
(3) increased $PaCO_2$
(4) increased peak airway pressure

354. In the lateral decubitus position, the lung relationships

(1) are the same as in the semirecumbent position, i.e., the apex is in Zone 1 and the bases are in Zone 3
(2) ventilation is highest at the apex
(3) perfusion is greater in the upright lung
(4) compliance is unequal in the two lungs

355. Hypoxia at high altitude is caused by

(1) low FIO_2
(2) low ambient pressure

(3) low $PaCO_2$
(4) high-altitude pulmonary edema in some people

356. Positive end-expiratory pressure (PEEP)

(1) always improves oxygenation
(2) increases lung volume at end expiration (FRC)
(3) decreases lung compliance
(4) increases ventilation/perfusion (V/Q) ratio

357. Physiologic dead space

(1) is increased in areas of high ventilation/perfusion (V/Q) ratio
(2) is increased in old age
(3) is increased in the lung apex when standing
(4) is increased after administration of atropine

358. Which of the following anesthetic agents have been shown to cause bronchodilation?

(1) halothane
(2) sevoflurane
(3) isoflurane
(4) ketamine

359. The work of breathing

(1) involves both resistive and elastic work
(2) is expended mostly in expiration
(3) to overcome elastic forces is increased when breathing is deep and slow
(4) to overcome resistive forces is decreased when breathing is fast and shallow

360. Carbon dioxide is transported

(1) in physical solution
(2) as carbonic acid
(3) as bicarbonate
(4) as carbaminohemoglobin

361. The respiratory pattern seen with various levels of anesthesia with halothane includes

(1) irregular respirations at light (less than minimum alveolar concentration) levels
(2) faster, more shallow respirations at moderate levels
(3) jerky, gasping respirations at deep levels
(4) a rocking-boat pattern at deep levels due to the loss of diaphragmatic movement

362. A decrease in FRC under anesthesia may be due to

(1) the supine position
(2) increased intra-abdominal pressure
(3) a cephalad shift in the diaphragm
(4) bronchial dilation

363. Hypoxic pulmonary vasoconstriction is decreased by

(1) mitral stenosis
(2) intravascular volume overload
(3) hypothermia
(4) inhalational anesthetics

364. In the patient with an intrapulmonary shunt

(1) decreased cardiac output will increase PO_2
(2) increased oxygen consumption will decrease oxygen content
(3) decreased cardiac output will increase oxygen content
(4) decreased cardiac output will decrease oxygen content

365. A patient with hypercapnia will have

(1) decreased levels of epinephrine
(2) increased levels of norepinephrine
(3) respiratory alkalosis
(4) increased plasma potassium

366. The patient who is hyperventilated to a PCO_2 of 20 mmHg under anesthesia will have

(1) increased cerebral blood flow
(2) increased ionized calcium
(3) increased oxygen delivery to the tissues
(4) increased ventilation/perfusion (V/Q) mismatch due to inhibition of hypoxic pulmonary vasoconstriction

367. Pulse oximeter readings

(1) are not affected by low perfusion states
(2) of 90% correspond to a PaO_2 of 60 mmHg
(3) are equally reliable with oxyhemoglobin, carboxyhemoglobin, and methemoglobin
(4) may be inaccurate due to light interference

368. Specific effects of anesthesia on control of breathing include a decreased response to

(1) carbon dioxide
(2) hypoxemia
(3) metabolic acidemia
(4) added airway resistance

369. Factors leading to pulmonary edema include

(1) increased capillary pressure
(2) decreased oncotic pressure
(3) lymphatic insufficiency
(4) increased capillary permeability

370. The respiratory quotient

(1) depends on the CO_2 output
(2) is independent of the metabolic substrate
(3) depends on the O_2 uptake
(4) is always 0.8

371. PEEP usually

(1) increases FRC
(2) decreases compliance
(3) decreases work of breathing
(4) decreases lung volume

372. The change in FRC with body position is

(1) marked reduction in the supine position
(2) head-down is little different from supine
(3) lateral is greater than supine
(4) prone is greater than supine

373. Therapy directed at the hypoxemic patient in the postoperative period may include

(1) supplemental oxygen
(2) upright position
(3) prevention of atelectasis
(4) administration of carbon dioxide to stimulate ventilation

374. The adult trachea is

(1) 10 to 15 cm in length
(2) half in the chest
(3) in the midline in the neck
(4) bifurcates at the level of the third thoracic vertebra

375. Airway resistance

(1) is a dynamic measurement
(2) is measured by dividing volume per time by pressure
(3) depends on the caliber of the airway
(4) is independent of the pattern of airflow

376. In comparing closing capacity (CC) and FRC,

(1) obesity increases both CC and FRC
(2) increasing FRC relative to CC results in areas of low ventilation/perfusion (V/Q)
(3) anything that decreases CC below FRC results in areas of atelectasis
(4) increasing CC above the tidal volume plus FRC results in areas of atelectasis

377. Dead space to tidal volume ratio (V_D/V_T)

(1) is a measure of ventilatory efficiency
(2) is normally under 0.4
(3) is increased in the patient with chronic lung disease
(4) increases require increased minute volumes to maintain the same PCO_2

378. During anesthesia, the diaphragm assumes a more cephalad position because of

(1) paralysis from muscle relaxants
(2) increased end-expiratory tone of the abdominal wall
(3) surgical retraction
(4) second gas effect

379. Forced exhaled vital capacity (FVC)

(1) is measured in the first second
(2) is a measure of inspiratory reserve volume
(3) is affected by restrictive disease in the first second
(4) may vary with patient cooperation

380. The medullary chemoreceptors are maximally stimulated by

(1) low oxygen tension
(2) reflex activity from the diaphragm
(3) carbon dioxide
(4) hydrogen ion

381. In a mixture of gases (such as air)

(1) each gas obeys Dalton's law
(2) each gas contributes a partial pressure that is proportional to its solubility
(3) each gas contributes a partial pressure that is proportional to its molecular fraction
(4) the total pressure is equal to the sum of the gases divided by the number of gases

382. The composition of alveolar gases differs from that of inhaled gases because

(1) oxygen is being absorbed from the alveoli
(2) carbon dioxide is being added to the alveoli
(3) water vapor is being added
(4) nitrogen is taken up by the alveolar capillaries

Questions 383 through 387

A 26-year-old man is admitted to the emergency room with first-, second-, and third-degree burns on his chest, face, and arms. He is alert but in severe pain. His respirations are shallow and rapid. Arterial blood gases are obtained on room air, and the results are pH 7.48, PCO_2 30 mmHg, PO_2 78 mmHg.

383. This patient's laboratory results demonstrate

(1) an uncompensated respiratory acidosis
(2) a respiratory alkalosis
(3) an oxygen tension that is normal
(4) an oxygen tension that is low but does not require immediate treatment

384. The patient's pulmonary status

(1) can be compromised by upper airway swelling
(2) should be observed carefully for at least 24 hours
(3) is dependent, to a degree, on the type of burn
(4) will always require a tracheostomy

385. His oxygenation may be adversely affected by

(1) pulmonary edema
(2) sickle cell disease
(3) carbon monoxide poisoning
(4) alkalosis

386. The patient's shallow, rapid respirations

(1) move air more efficiently
(2) are due to increased compliance
(3) allow him to oxygenate better with little effort
(4) magnify the effect of dead space on total ventilation

387. The patient's carboxyhemoglobin level is found to be 20%. Which of the following statements is (are) TRUE?

(1) This abnormality will be manifest by cyanosis.
(2) Pulse oximetry may be falsely elevated.
(3) FIO_2 should be adjusted to 100% O_2 saturation by pulse oximetry.
(4) This concentration is below the lethal range.

Questions 388 through 390

A 75-year-old woman undergoes an exploration and removal of her right kidney and adrenal gland for a large tumor. The procedure is performed in the left lateral decubitus position, with flexion of the table and use of the kidney rest. The patient is intubated, and ventilation is controlled. Anesthesia is maintained with isoflurane, nitrous oxide, oxygen, and vecuronium.

388. The blood pressure, initially 136/80 falls to 80/40 after positioning and incision. This change may be due to

(1) poor ventilation
(2) poor cardiac output
(3) isoflurane anesthesia
(4) blood loss

389. The initial response(s) to this situation is (are) to

(1) listen to the chest to ascertain good ventilation
(2) decrease isoflurane concentration
(3) lower kidney rest
(4) insert an arterial catheter

390. After blood pressure is restored and exploration proceeds, there is a rise in blood pressure to 180/100 and the appearance of multiple ectopic ventricular beats on the electrocardiogram. These signs may be due to

(1) poor ventilation
(2) pneumothorax
(3) hormone released from the adrenal tumor
(4) hypocapnia

391. DLCO will decrease with

(1) pulmonary embolism
(2) exercise
(3) pulmonary fibrosis
(4) increased hemoglobin

392. When inhaling from FRC to maximal inspiration, which of the following is (are) true?

(1) pulmonary vascular resistance increases
(2) pleural pressure decreases
(3) residual volume stays constant
(4) basilar alveoli become larger than apical alveoli

393. A laryngeal mask airway (LMA) is an appropriate anesthetic technique for

(1) fiberoptic bronchoscopy
(2) prevention of aspiration
(3) spontaneous ventilation
(4) prevention of laryngospasm

394. Nerve block for placement of an LMA should include the

(1) superior laryngeal nerve
(2) anterior ethmoid nerve
(3) glossopharyngeal nerve
(4) recurrent laryngeal nerve

395. An endotracheal tube position is evaluated by the following: examining the patient, the numerical markings on the tube, and the chest x-ray. Which of the following is (are) consistent with the proper position?

(1) the left side is ventilated better than the right side
(2) the tip of the tube is 30 cm from the upper front teeth
(3) the tip of the tube overlies the sixth thoracic vertebra
(4) both sides ventilate equally

396. In describing laminar airflow through a tube

(1) Henry's law is important
(2) radius is a factor

(3) tube length is more important than tube diameter
(4) viscosity is more important than density

397. A 34-year-old woman with myasthenia gravis is ventilated with 100% oxygen. This may cause

(1) injury to her lungs
(2) PaO_2 values in excess of 760 mmHg
(3) absorption atelectasis
(4) retrolental fibroplasia

398. The factors that influence arterial oxygen tension under anesthesia include

(1) inspired oxygen fraction (FIO_2)
(2) barometric pressure
(3) position
(4) age

399. A 12-year-old boy with muscular dystrophy is scheduled to have an operation to correct scoliosis. Evaluation of this patient should

(1) include x-rays of the spine to determine the amount of curvature
(2) include pulmonary function testing
(3) include a cardiac examination
(4) address the possibility of autologous blood donation

400. Factors which increase the P_{50} of hemoglobin include

(1) increased temperature
(2) decreased 2,3-DPG
(3) increased hydrogen ion concentration
(4) methemoglobin

401. Factors which increase the incidence of postoperative pulmonary complications include

(1) upper versus lower abdominal surgery
(2) obesity
(3) smoking
(4) longer surgical duration

402. Mixed venous oxygen tension is

(1) normally 60 mmHg
(2) a guide for determining tissue hypoxia
(3) independent of metabolic conditions and body temperature
(4) best sampled from the pulmonary artery

403. A pulse oximeter can give falsely low readings in the presence of

(1) blue nail polish
(2) hyperbilirubinemia
(3) methylene blue administration
(4) high altitude

404. The pulmonary vascular system

(1) has no α-adrenergic or β-adrenergic receptors
(2) is well endowed with smooth muscle
(3) has no connections to autonomic ganglia
(4) responds to norepinephrine by vasoconstriction

405. Neurogenic pulmonary edema

(1) may follow head injury
(2) is not associated with hypoxemia
(3) is treated with reduction in intracranial pressure
(4) develops only in a denervated lung

406. Closing capacity is

(1) the lung volume at which the onset of airway closure is detected
(2) increased in smoking
(3) greater than residual volume
(4) greater than closing volume

407. Transpulmonary pressure

(1) measures intralung pressure
(2) is zero at FRC
(3) is a gradient between the airway opening and the alveolar pressure
(4) increases with increasing lung volume

408. The effect of a deliberate hypotensive technique on pulmonary gas exchange is increased by

(1) chronic obstructive lung disease
(2) decreased cardiac output
(3) nicardipine as opposed to nitroprusside administration
(4) nitroprusside as opposed to isoflurane administration

409. In interpreting pulmonary function tests, it is important to remember that

(1) vital capacity measurement is not a timed measurement
(2) spirometry fails to detect early disease in small airways
(3) maximal breathing capacity is dependent on cooperation
(4) the FEV_1 will detect restrictive disease

Questions 410 and 411

A patient with chronic obstructive lung disease is admitted to the hospital with the following arterial blood gases: pH 7.10, PCO_2 92 mmHg, and PO_2 52 mmHg.

410. These blood gas values reveal

(1) respiratory acidosis
(2) O_2 administration
(3) acute CO_2 retention
(4) chronic hypoxemia

411. Treatment should consist of

(1) improvement of ventilation
(2) assessment of vital signs and state of consciousness
(3) supplemental oxygen and treatment of the underlying cause of the crisis
(4) sodium bicarbonate, 90 mEq intravenously

412. The immediate preoperative preparation of a patient with asthma may include

(1) β-adrenergic agonists
(2) corticosteroids
(3) anticholinergic agents
(4) cromolyn sodium

413. A patient who arrives in the recovery room after a general anesthetic should be

(1) allowed to awaken without disturbance
(2) encouraged to lie on the back for easier access to the airway
(3) given opioids at fixed intervals
(4) closely observed for respiratory depression

414. A 14-gauge catheter is inserted through the cricothyroid membrane and attached to a wall oxygen source in such a way that oxygen can be delivered intermittently. With this technique

(1) pneumothorax is inevitable
(2) adequate oxygenation is possible
(3) gastric dilatation is a hazard
(4) pulmonary aspiration is unlikely

Questions 415 and 416

A 35-year-old man is admitted to the emergency room following an automobile accident. It is noted that there is a contusion over the anterior thorax, he is tachypneic, and has a scaphoid abdomen. Auscultation reveals poor breath sounds on the left side. Chest x-ray shows a large air cavity in the left side of the thorax. Blood pressure is 80/60 and heart rate is 120 beats per minute.

415. Diagnoses that must be considered include

(1) ruptured spleen
(2) pneumothorax
(3) diaphragmatic hernia
(4) cardiac contusion

416. A diagnosis of diaphragmatic hernia is made, and the patient is transported to the operating room. In transport, the patient becomes apneic and is ventilated with a bag and mask unit. One might expect

(1) the patient's blood pressure to fall
(2) pulmonary compliance to decrease

(3) a shift of the mediastinum
(4) the abdomen to become distended

Questions 417 and 418

A 74-year-old former smoker is scheduled for resection of a giant bulla of the left upper lobe.

417. Which of the following findings would be expected in the preoperative evaluation?

(1) increased residual volume
(2) elevated arterial CO_2
(3) decreased diffusing capacity (DLCO)
(4) abnormal chest x-ray

418. After induction of anesthesia, prudent airway management includes

(1) controlled ventilation with large tidal volumes to reduce atelectasis
(2) addition of nitrous oxide to reduce potent inhaled agents and improve blood pressure
(3) prophylactic chest tube placement to prevent bronchopleurocutaneous fistulae
(4) lung isolation with a double lumen tube

419. Findings in fat emboli syndrome include

(1) fever
(2) tachypnea
(3) petechiae
(4) confusion

420. Patients with pneumoconiosis, e.g., asbestosis or silicosis, often require surgery on other organs. In the preoperative assessment, one must recognize that

(1) the lung volumes will be increased
(2) the x-ray abnormality may not reflect the functional changes
(3) early airway closure is the hallmark
(4) fibrosis usually is present

Questions 421 and 422

A 67-year-old female has been at bed rest over the weekend after a fracture of her tibia. She undergoes open reduction and internal fixation. During the procedure, she suffers a pulmonary embolus from a venous thrombus.

421. If this patient were intubated and mechanically ventilated, one would expect

(1) a sudden decrease in $ETCO_2$
(2) a stable $PaCO_2$
(3) a normal $ETCO_2$ after 1 hour
(4) a sudden increase in ETN_2

422. If this patient were under epidural anesthesia, with an oxygen mask and $ETCO_2$ catheter placed, one would expect

(1) a sudden decrease in $ETCO_2$
(2) a stable $PaCO_2$
(3) a normal $ETCO_2$ after 1 hour
(4) a sudden increase in minute ventilation

423. Molar concentration of a gas in the gaseous phase is

(1) inversely proportional to temperature
(2) dependent on the molecular properties of the gas
(3) proportional to partial pressure
(4) proportional to molecular weight

424. Molar concentration of a gas in the aqueous phase is

(1) inversely proportional to temperature
(2) dependent on the molecular properties of the gas and liquid
(3) proportional to partial pressure
(4) proportional to molecular weight

Answers and Explanations

263. **(B)** As an air-fluid mixture, the lung tends to sag with gravity, causing a gravity-induced pleural pressure gradient. Pleural pressure increases 0.25 cm H_2O every centimeter down the lung. Thus, it is lowest at the apex and highest at the base. Pleural pressure is related to body position in that, in different positions, different areas will be dependent. Pleural pressure variation is caused by the hydrostatic pressure exerted by gravity. Because the lungs are composed of air and fluid, the pressure gradient is less than the 1 cm H_2O per centimeter elevation found in blood vessels. *(1:794; 4:1740, 1755; 5:681)*

264. **(D)** The Laplace law, $P = 2T/R$, states that the pressure within an elastic sphere is directly proportional to the tension of the wall and inversely proportional to the radius of the curvature. In this case, the sphere is the alveolus. Alveoli are lined with a film of surfactant which lends stability to the alveoli by decreasing the surface tension as the radius of the alveolus becomes smaller. Without this ability to vary surface tension, small alveoli, which have a smaller radius of curvature and thus a higher pressure, would empty into large alveoli and alveolar stability would be lost. *(1:794; 5:690)*

265. **(D)** Surfactant is a substance containing dipalmitoyl lecithin produced by the Type II alveolar epithelial cells of the lung. The substance is important in adults as well as newborns, providing alveolar stability. It is 90% lipid and 10% protein. *(1:793; 5:690)*

Vital capacity	Inspiratory reserve volume	Inspiratory capacity	Total lung capacity
	Tidal volume		
	Expiratory reserve volume	FRC	
	Residual volume		

266. **(D)** The FRC is composed of expiratory reserve volume and residual volume. Tidal volume plus inspiratory reserve volume comprise the inspiratory capacity. The other options are not designated capacities. *(1:804–5; 4:862; 5:694)*

267. **(C)** The maximum volume of air inhaled from after a tidal inspiration is referred to as the inspiratory reserve volume. Vital capacity includes all of inspiration and expiration. Inspiratory capacity includes the tidal volume. *(1:804–5; 4:862; 5:694)*

268. **(B)** The metabolic functions of the lung include the inactivation of many substances but not epinephrine. *(1:877; 4:1738; 5:633, 685)*

269. **(B)** Postoperative pulmonary compromise is primarily restrictive, a decreased FRC. It is most profound after nonlaparoscopic upper abdominal surgery. Carboxyhemoglobin levels return to normal after 24 hours of smoking cessation, but mucociliary clearance and sputum production are impaired for several weeks. *(1:809; 5:2294, 2714)*

270. **(A)** Muscle paralysis will decrease chest and abdominal wall tone, improving compliance. The decrease in FRC and the supine position will move the patient to a less compliant region of the lung volume relationship. The endotracheal tube increases airway resistance, worsening dynamic compliance. Controlled ventilation changes the pressure gradient entirely; the airway pressure must be supra-atmospheric instead of subatmospheric. *(1:794; 4:689; 5:689)*

271. **(D)** Humidification of inhaled gases is at its optimum with the patient breathing through the nose. Atropine will dry out secretions. An endotracheal tube and a tracheostomy both bypass the humidification process of the nose. A closed system will keep humidity within the system. *(1:581, 1391; 5:1585, 2547)*

272. **(A)** The principal muscle of respiration is the diaphragm. The other muscles, although involved, do not exert nearly the influence of the diaphragm. The diaphragm can account for the entire amount of inspired air during quiet respiration, and 60% of air movement in maximal ventilation. *(1:791; 4:1379; 5:1464)*

273. **(B)** The dependent areas are better ventilated, since the alveoli in the dependent areas are smaller and more compliant. *(1:800; 4:1755; 5:680)*

274. **(A)** Inspiratory force is a measurement of reserve of ventilatory effort. It is a simple method that does not depend on patient cooperation. Therefore, it is particularly useful in the unconscious or anesthetized patient. A pressure in excess of 20 cm H_2O below atmospheric pressure is considered the minimum to show adequate reserve. Expiratory force correlates with coughing ability. *(1:614, 1388; 5:1470, 2715)*

275. **(C)** Patients with pickwickian syndrome have a decreased expiratory reserve volume and higher intra-abdominal pressure when supine. These patients have alveolar hypoventilation and increased work of breathing, and one would expect this to improve after weight loss. *(1:1035; 4:508; 5:1031)*

276. **(D)** Cartilage extends only through the bronchi. Its absence defines the first bronchiolar generation. Large airways (>2 mm) cause 90% of the airway resistance. There are several generations of airways that participate in gas exchange. Tracheal rigs are U shaped and incomplete posteriorly. *(1:792; 4:859; 5:1518)*

277. **(E)** Nonperfused alveoli contribute to dead space. All of the other options contribute to physiologic shunt by providing a path for blood to bypass oxygenation. *(1:802; 4:877; 5:697)*

278. **(D)** CPAP with oxygen to the unventilated lung will decrease the shunt and attenuate the desaturation. This implies that some alveoli are in continuity with the proximal airways. Continued blood flow to the atelectatic unventilated lung causes a shunt. CO_2 removal is usually not a problem. *(1:829; 4:1767; 5:1899)*

279. **(D)** After bilateral carotid endarterectomy, a patient will be more susceptible to hypoxemia because of denervation of the carotid bodies. The patient will not respond to hypoxemia with hyperventilation. In addition, the resting PCO_2 is elevated, and the response to small doses of opioids may be accentuated. Hypertension is not a constant finding in these patients. *(1:798; 5:177, 179)*

280. **(E)** Adequacy of alveolar ventilation is obtained by measuring the carbon dioxide tension. Oxygen values may be affected by ventilation, but the measure of ventilation is carbon dioxide. *(1:821; 5:697, 1648)*

281. **(B)** The measure of oxygen tension uses the Clark electrode (the most common technique). This measurement is affected by temperature. The Severinghaus electrode is used for the measurement of carbon dioxide. Hemoglobin concentration is related to the oxygen content of blood, but not the partial pressure. *(1:758; 4:864; 5:1444)*

282. **(C)** Anesthesia usually causes a decrease in FRC, which leads to hypoxemia. All of the other options can lead to hypoxemia: hypoventilation by decreased FRC and increased shunt, hyperventilation by shift of the oxyhemoglobin dissociation curve to the left and decreased cardiac output, supine position by decreased ventilation and decreased FRC, and increased airway pressure by change in the ventilation-perfusion relationships. *(1:805; 5:707)*

283. (A) Since oxygen consumption does not decrease as much as cardiac output, the mixed venous blood will have a lower oxygen content. With a fixed shunt, the more desaturated venous blood that is shunted past the lungs will lower the resulting arterial oxygenation. CO_2 may rise slightly. Urinary output usually decreases with decreased cardiac output. Dead space will increase. Lower overall perfusion will increase the proportion of the lungs in Zone. *(1:803; 4:876; 5:1442)*

284. (C) Patients who are hyperventilated for long periods of time have their carbon dioxide stores depleted. Postoperatively, these patients hypoventilate in an effort to restore their carbon dioxide and, in so doing, may become hypoxemic if not given supplemental oxygen. *(1:799; 4:110; 5:718, 2133)*

285. (D) The carbon dioxide retention is due primarily to impaired pulmonary mechanics causing inefficient ventilation. The minute ventilation is normal, but the rapid shallow breathing increases V_D/V_T. Loss of connective tissue in the lung causes expiratory obstruction. Oxygen should be administered for hypoxia, not hypercarbia, since in some patients hypercarbia may be exacerbated. *(1:802; 4:238; 5:2264, 6:177–87)*

286. (A) Tracheal mucus flow and, necessarily, ciliary activity are impeded by deviation from ambient oxygen tension. Tracheal mucus flow is increased in patients with obstructive lung disease. Tracheal mucus flow is an important cleansing mechanism. Mucus flow and ciliary activity are depressed by cigarette smoke. *(5:711)*

287. (C) Closing volume is the lung volume at which small airways collapse, isolating the alveoli beyond. Closing volume is related to lung elasticity, thus is greater in the young and old. Factors that change FRC (general anesthesia, obesity, and position) can cause bronchiole closing with each breath, but do not change closing volume, per se. CC is measured by a single breath from maximal exhalation of a tracer gas, or by nitrogen washout after using 100% O_2 as the tracer gas. Measurement of closing volume will detect changes in the small airways before they would become apparent by spirometry. *(1:1035; 4:510; 5:696)*

288. (E) The CO_2 response curve is left shifted by norepinephrine, acidosis and hypoxia. Naloxone will reverse opioid depression. Peripheral chemoreceptors contribute 15% of the control. *(1:799; 5:178)*

289. (A) Leukotrienes, one branch of arachidonic acid metabolism, are potent pulmonary vascular constrictors. *(1:1300; 4:1729, 1738; 5:685)*

290. (C) The peripheral chemoreceptors are primarily responsible for the hypoxic drive to respiration. They are shut off at very high oxygen tensions. The peripheral chemoreceptors respond to changes in oxygen tension, not content, and are located in the carotid arteries. *(1:799)*

291. (B) Diffusion is rarely the limiting component of oxygen transport. High altitude lowers the PaO_2, the other components contribute to shunt. *(1:800; 5:1440)*

292. (D) DLCO is a measure of the diffusing capacity of the lung. Increased hemoglobin and pulmonary blood volume will affect it. It is not a lung capacity in the same sense as FRC. *(1:806; 5:1011)*

293. (D) General anesthesia leads to an alteration of chest wall mechanics, resulting in decreased FRC. Breathing at a lower lung volume may also affect the lung mechanics, leading to increased elastic recoil and decreased compliance. *(1:795, 805; 4:688; 5:174,708)*

294. (B) The curve is shifted to the right with acidosis. The other options are correct. *(1:202, 670; 5:699)*

295. (B) The patient with restrictive disease will have a decrease in FVC. The FEV_1 may be decreased as well, but the FEV_1/FVC ratio usually is normal. *(1:807; 4:232; 5:1000)*

296. **(B)** This patient represents a common situation. The clinical scenario shows acute respiratory acidosis from increasing impairment of ventilation and mild hypoxemia. In view of his cervical cord injury and his other sites of trauma, treatment of the respiratory failure is of utmost importance. Given this patient's history, his deterioration, and the concurrent injuries, the treatment of choice is to intubate and ventilate him. *(1:1266; 4:572; 5:2474)*

297. **(D)** This patient is not functioning on a hypoxic drive; therefore, 40% oxygen should not contribute to his respiratory depression. The other options are all possible. *(1:1266; 4:572; 5:2474)*

298. **(A)** This patient lacks any ability to clear his airway because he has a high cervical lesion. This inability to cough will be the major contributor to respiratory failure, unless vigorous chest physical therapy is instituted. While alert, aspiration should not be a risk, since his lesion is not high enough to impair his gag reflex. Respiratory compromise from shock, massive transfusion, or prolonged high FIO_2 is possible, but a lesser risk. Supine position actually improves diaphragm movement in patients with no other respiratory muscles. *(1:1266; 4:572; 5:2474)*

299. **(E)** This patient has an elevated dead space, a normal inspiratory capacity, a high PCO_2, a normal vital capacity, and abnormal O_2 saturation. *(1:804; 4:232; 5:694)*

300. **(A)** To correct the patient's high PCO_2 level, either the tidal volume could be increased or the dead space should be decreased. Since the tidal volume is normal, it is more rational to decrease the dead space. *(1:804; 4:232; 5:694)*

301. **(B)** Hysteresis is present when the path of deformation on application of a force differs from the path taken when the force is withdrawn. This is seen in the inflation-volume curves and is due to the presence of alveolar stability, which is possible because of surfactant and its effect on surface tension. *(1:794; 5:690)*

302. **(B)** The curve for saline is also shifted to the left, showing that less pressure is needed to inflate and deflate the lung with saline. The force opposing inflation of the lung is the surface tension. When air is involved and surfactant is present, air spaces will remain open, and the curve differs as shown. *(1:794; 5:690)*

303. **(C)** All of the factors cited are presumptive evidence of ventilation. The only sure measure of demonstrating effective gas exchange is the analysis of blood gases, specifically carbon dioxide. *(1:820, 1489; 4:1094; 5:1648)*

304. **(D)** Pulmonary vascular resistance is the result of cardiac output, the state of vasomotor tone, and pressure. A change in cardiac output will normally be followed by changes in the radius of the vessels to allow maintenance of normal pressure. The resistance is equal to pressure divided by flow. *(1:794, 801; 4:1729; 5:165–6)*

305. **(E)** Hypoxic pulmonary vasoconstriction causes diversion of blood flow from hypoxic to nonhypoxic lung tissue. In the usual circumstance, blood is diverted to areas where flow is decreased, e.g., to the apices. This causes a decrease in V_D/V_T and physiologic shunt. *(1:794; 4:1765; 5:687)*

306. **(A)** During anesthesia, exhalation becomes more active, resulting in more work. The work of breathing involves both elastic and resistance work. The optimum rate is about 15 breaths per minute in normal adults. In the patient with restrictive disease, short shallow breaths decrease the effort. In general, humans adjust their breathing pattern to minimize the work of breathing while maintaining adequate ventilation. *(1:794,808; 5:177,692)*

307. **(D)** Anatomic dead space increases with increased lung volume. The normal dead space is 1 mL/lb of body weight or 2 mL/kg. Equipment dead space may greatly increase the amount of dead space, which can be disproportionately large in infants. *(1:802; 4:876; 5:697)*

308. **(B)** P_{50} is the PO_2 level on the oxyhemoglobin dissociation curve at which hemoglobin is 50% saturated. The normal value is 27 mmHg and may change with the other factors that cause a shift in the oxyhemoglobin dissociation curve. *(1:202; 5:700)*

309. **(C)** Carbon monoxide binds hemoglobin avidly, forming carboxyhemoglobin. Because oxyhemoglobin and carboxyhemoglobin absorb light at the same wavelength, it is not possible to detect carbon monoxide poisoning with pulse oximetry. *(1:1390; 5:2671)*

310. **(C)** The chemoreceptors of the carotid body sense PaO_2 (less than 65 mmHg) and send afferents via the glossopharyngeal nerve *(1:798; 5:179)*

311. **(D)** The Bohr effect describes the effect of carbon dioxide on the uptake of oxygen. Option B describes the Haldane effect. Carbonic anhydrase, hydrogen ion, and catalytic enzymes are not involved. *(5:703)*

312. **(D)** Hypercapnia may result from rebreathing of CO_2 while using an Ayre's T-piece at low fresh gas flow. The other options all cause decreased carbon dioxide tension. *(1:580; 4:109; 5:293)*

313. **(D)** In Zone 1, the alveolar pressure is higher than the arterial or venous pressure; therefore, these areas act as dead space units. *(1:800; 4:1754; 5:680)*

314. **(E)** In Zone 2, the blood flow is dependent on the arterial pressure and the alveolar pressure. These change with the status of ventilation. Venous pressure is still not a determinant of blood flow in this zone. *(1:800; 4:1754; 5:680)*

315. **(A)** In Zone 3, blood flow is determined by the arteriovenous pressure difference. Dead space is low, since the units are being perfused, alveolar pressure is not high, and the venous pressure is lower than arterial but higher than alveolar pressure. *(1:800; 4:1754; 5:680)*

316. **(D)** Gravity is the cause of the hydrostatic pressure gradient in the arterial and venous circulations. If the density of the alveolar gas were closer to blood, the pressure gradient in the alveoli would be closer to the vascular gradient. *(1:800; 4:1754; 5:680)*

317. **(A)** The diaphragm is innervated by the phrenic nerve. *(1:791; 5:1660)*

318. **(E)** MMV is measured over 15 seconds. It is a measure of respiratory muscle and chest wall compliance, and is most affected by chronic obstructive pulmonary disease (COPD). *(1:806; 4:1013; 5:1002)*

319. **(B)** The main buffer in the body is the bicarbonate system. All the other listed systems are quantitatively less important. *(1:166; 4:961; 5:1605)*

320. **(A)** Hemoglobin is the primary noncarbonic buffer. Bicarbonate provides 50% of the body's buffering and hemoglobin provides 35%. Hemoglobin is a weaker acid than carbonic acid (p*K* 6.8 vs. 6.1). Hemoglobin exists within the red cell as a potassium salt. *(1:166; 4:961; 5:1605)*

321. **(D)** Analysis of arterial blood gases is the most reliable method of documenting hypoxemia, but pulse oximetry gives accurate, fast, and continuous results and so is best as a monitor. Detection of cyanosis and circulatory responses are all subjective and may not be accurate. Residual anesthetics may blunt normal responses to hypoxia, and hypoxemia may be present without apnea. The pulse oximeter permits the detection of hypoxemia in less time than transcutaneous oxygen readings. *(1:670; 4:2319; 5:1448)*

322. **(D)** Residual anesthetic, both inhaled and intravenous, inhibit hypoxic drive. It may or may not respond to opioid reversal, depending on the anesthesia technique employed. The patient with hypoventilation may be hypotensive. The best way to identify the hypoxia is a continuous convenient monitor: pulse oximetry. *(1:1385; 4:2319; 5:179)*

323. **(B)** The time for nitrogen elimination is dependent on the volume of the breaths and the underlying lung condition. Patients with lung disease will eliminate the nitrogen more slowly, as will smokers and the elderly. A normal patient will eliminate most nitrogen after 20 L total expired volume, approximately 4 to 5 vital capacity breaths. *(1:598; 5:1456)*

324. **(A)** Chemoreceptor stimulation leads to an increased ventilatory effort. Cyanosis requires at least 5 g/dL deoxyhemoglobin in blood to be apparent. Severe hypoxemia can cause bradycardia, in addition to a rapid fall in blood pressure and circulatory collapse. *(1:798; 4:2319; 5:177,716)*

325. **(A)** Type I alveolar cells form the majority of the alveolar gas exchange surface. When injured, Type II cells create new Type I cells. Type II cells produce surfactant. Type III cells are alveolar macrophages. *(1:793; 5:704)*

326. **(C)** Diffusion nonequilibrium is the theoretical state where diffusion rates limit the exchange of gases across the alveolar membrane. In practice this seems to be unimportant in human physiology for CO_2 and probably O_2. Factors that decrease the contact time of the blood and alveolar gas would increase the gradient. Both obstructing airflow or slowing blood flow would have the opposite effect. *(1:800; 4:860; 5:1440)*

327. **(C)** The normal value for A-aDO_2 is about 8 mmHg in a young individual breathing room air and increases linearly with age to about 25 mmHg in the eighth decade of life. The increase is due to changes in arterial oxygen tension and not the alveolar oxygen tension. It is not a good tool for measuring *V/Q* changes, since it is difficult to measure alveolar oxygen tension. Increasing dead space has little effect either on arterial or alveolar oxygen tensions. *(1:804; 5:700, 1010)*

328. **(A)** Oxygen is transported bound to hemoglobin and in the dissolved state. Carbon dioxide changes hemoglobin affinity, but glucose does not. *(1:202; 5:700)*

329. **(B)** The Haldane effect refers to the shift in the CO_2 dissociation curve caused by altered levels of oxygen. Low PO_2 shifts the curve so that the blood is able to pick up more CO_2. *(5:702, 1256, 1439–40)*

330. **(C)** Ninety-five percent of CO_2 enters the red cell; 5% remains as dissolved CO_2. Thirty percent binds to amino acids as the carbamino compounds, and the other 65% is hydrolyzed by carbonic anhydrase to carbonic acid, which dissociates to bicarbonate. Carboxyhemoglobin is CO bound to hemoglobin. *(1:800; 5:703)*

331. **(D)** Shunting refers to blood that goes through the lung without being oxygenated. Ventilation in excess of perfusion is dead space or wasted ventilation. *(1:803; 4:877; 5:690)*

332. **(B)** The cuff of an endotracheal tube should be inflated enough to prevent aspiration and still allow capillary flow in the underlying trachea. A pressure of 20 mmHg will allow this in a high-volume, low-pressure cuff. The normal perfusion pressure of the tracheal mucosa is 25 to 35 mmHg. *(4:1081; 5:1650, 2546)*

333. **(D)** When patients breathe an increased FIO_2, there is an increased amount of shunt present due to absorption atelectasis. This occurs in all air spaces. The lung units with low *V/Q* ratios have a greater tendency to collapse. *(1:803; 5:690, 711)*

334. **(A)** The Bohr equation is used to calculate dead space. It requires measurement of expired carbon dioxide and alveolar carbon dioxide. It does not require mixed venous CO_2 or O_2 measurement. *(1:803; 4:876; 5:697)*

335. **(A)** The ciliated epithelium extends only to the terminal bronchioles. *(1:792; 5:161)*

336. **(C)** The internal laryngeal branch is almost entirely sensory. The other nerves are almost entirely motor. *(1:722; 4:1066; 5:2737)*

337. **(B)** The musculature of the bronchioles encircles them. The musculature becomes thinner distally, but it is relatively thicker in relation to

the diameter of the bronchiole. The musculature is important in maintaining the size of the bronchiolar lumen. The elastic fibers are embedded in the musculature. *(1:793; 4:859)*

338. **(B)** The recurrent laryngeal nerve innervates the vocal cord. In its absence, the preserved activity of the cricothyroid muscle places the cord in a paramedian position. The airway is patent, but the cords cannot properly oppose to create a normal voice. *(1:595; 5:2538)*

339. **(A)** Carbon monoxide does not lead to cyanosis but to hemoglobin that is cherry red. The other hemoglobins are associated with cyanosis and stagnant blood may appear cyanotic. Desaturated blood also appears dark. *(1:1390; 4:1640; 5:2671)*

340. **(B)** Energy sources are added to the stored blood to support erythrocyte ion pumps and membrane integrity. This anaerobic metabolism causes the pH to fall. Potassium ions leave the red cells in exchange for hydrogen ions, and thus the plasma potassium levels rise. Seventy percent viability of red cells 24 hours after transfusion was the criterion used for determining the maximum shelf life of banked blood. 2,3-DPG phosphatase is more active in the acid environment, causing a rapid decrease in 2,3-DPG, which will shift the oxygen dissociation curve to the left. *(1:206; 4:993; 5:1806–7)*

341. **(A)** The pulmonary problem in scleroderma is a restrictive disease. The chest wall and the lung itself contribute to the restriction. There is fibrosis and a decreased vital capacity leading to decreased arterial saturation. The V_D/V_T ratio increases. *(1:510–1; 6:511)*

342. **(A)** The Hering-Breuer reflex is now thought to have little importance in the adult human. It is a stimulus prompted by stretching of small airways. The reflex is depressed by barbiturates. *(1:798; 4:1069; 5:173)*

343. **(E)** The slope of the response curve is an index of the patient's sensitivity to carbon dioxide. The slope and position may change. Increased work leads to a flatter slope. Anesthetics differ in their effect on the curve. *(1:801; 4:1069; 5:178)*

344. **(E)** Anaphylaxis releases bronchoconstrictive mediators (like leukotrienes). Pulmonary fibrosis and surfactant have their effect on pulmonary compliance. *(1:800; 5:692)*

345. **(E)** PaO_2 above 100 mmHg does not depress the response to CO_2. The other options all have an effect on the response. *(1:800; 4:1069; 5:178)*

346. **(D)** Type III alveolar cells are macrophages, which are not stationary elements of the alveolar lining. *(1:793; 4:860; 5:703)*

347. **(D)** The circulation to the lung is of two systems, but the bronchial system is expendable, as is seen in lung transplantation. The pressure in the pulmonary circuit is low relative to the left side. The arterial branches follow the airway branching. *(1:794; 4:859; 5:684)*

348. **(B)** A flow/volume loop has flow on the y-axis and volume on the x-axis. An air leak would be manifest by an open curve. *(1:806; 4:232; 5:1008)*

349. **(C)** Dynamic tracheal stenosis would be seen with inspiratory or expiratory flattening, depending on whether the affected area was extrathoracic or intrathoracic, respectively. Point 2 is the point of peak inspiratory flow, while maximum expiration is at point 1. Point 3 is at maximum inspiration and point 4 at peak expiratory flow. Since a flow/volume loop is made from maximum expiration to maximum inspiration and back, the point at rest, FRC, cannot be determined. Without knowing the residual volume, FRC, or tidal volume, none of the classical lung volumes can be determined. Vital capacity, the sum of inspiratory reserve, expiratory reserve, and tidal volumes can be determined, but it is a capacity, not a volume. *(1:806; 4:232; 5:1008)*

350. **(E)** This curve is characteristic of obstructive lung disease, with a higher residual volume and scooping of the expiratory limb from small airway closure. *(1:806; 4:232; 5:1008)*

351. **(B)** This curve is characteristic of restrictive lung disease, with smaller volumes and well preserved flows relative to volume. *(1:806; 4:232; 5:1008)*

352. **(C)** The FRC measurement is possible through the use of body plethysmography or a trace gas (such as helium) dilution. The residual volume component of the FRC cannot, by definition, be measured with spirometry. An esophageal balloon is used to estimate pleural pressure, and is not relevant to this measurement. *(1:805; 4:862; 5:1006)*

353. **(B)** Dead space effects are primarily on ventilation (carbon dioxide). Unventilated alveoli dilute the end-tidal concentration of CO_2. $PaCO_2$ thus rises. *(1:802; 5:697)*

354. **(D)** In the lateral position, lung relationships change. The perfusion is now greater in the dependent lung. Compliance differs between the lungs because the dependent lung is at a lower FRC. *(1:821; 4:1750; 5:1864)*

355. **(C)** Although the fraction of O_2 remains constant, the total pressure is lower, so the partial pressure of O_2 drops. $PaCO_2$ falls as a response to hypoxia and increased respiratory drive. High-altitude pulmonary edema is seen in some people. The best treatment is to return to lower altitudes, although calcium channel blockers, steroids, and barotherapy can be effective. *(1:798; 5:2687)*

356. **(C)** PEEP does increase FRC and *V/Q* ratios. It may not improve oxygenation and it may increase lung compliance. *(1:830; 4:139, 520; 5:698)*

357. **(E)** All the options are correct. Dead space is increased in the lung apex which is an area of high *V/Q* ratio. It is known to be increased after administration of atropine. *(1:801; 4:478, 876; 5:680)*

358. **(E)** All of the agents listed have been shown to cause bronchodilation. *(1:400; 4:1143; 5:155)*

359. **(B)** The work of breathing involves both elastic and resistive work. If breathing is deep and slow, more effort is expended in overcoming elastic work. Most of the breathing effort is expended in inspiration. *(1:794; 4:1739; 5:693)*

360. **(E)** All the options are correct. Carbaminohemoglobin is the form that CO_2 takes when bound to hemoglobin. *(1:800; 4:963; 5:703)*

361. **(A)** The first three options are true. The reason for the rocking-boat ventilation pattern is that the diaphragm is the only muscle not inhibited with deep anesthesia (unlike the intercostals). *(1:399; 4:1143; 5:171)*

362. **(A)** A decrease in FRC results from the supine position, with the resultant increased intra-abdominal pressure and cephalad shift of the diaphragm. Bronchial dilation would not decrease FRC. *(4:688; 5:175)*

363. **(E)** All of the options are true. Anything that increases pressure in the pulmonary circulation will decrease hypoxic pulmonary vasoconstriction. The first three options all increase intravascular pressure. It is also known that anesthetic agents reduce hypoxic pulmonary vasoconstriction. Hypoxic pulmonary vasoconstriction is important for *V/Q* matching and oxygenation under one-lung anesthesia. *(1:831; 4:1765; 5:1891)*

364. **(C)** Decreased cardiac output will result in little change in O_2 consumption, so mixed venous O_2 content will fall, some of which, will be shunted through the lung and worsen oxygenation. Increased oxygen consumption will decrease mixed venous O_2, with the same effect. *(1:803; 4:597, 877; 5:683)*

365. **(C)** The patient with hypercapnia has increased catecholamines, respiratory acidosis, and increased plasma potassium. The potassium comes from the liver and other tissues. *(1:166; 4:961; 5:718)*

366. **(D)** Hypocapnia leads to decreased cerebral blood flow, decreased ionized calcium, and decreased oxygen delivery to the tissues. There is an inhibition of hypoxic pulmonary

vasoconstriction leading to *V/Q* mismatch. *(1:168; 4:998; 5:718)*

367. (C) Pulse oximetry is affected by perfusion and light interference. The standard oximeter does not distinguish alternative hemoglobins. *(1:670–1; 4:92; 5:1448–9)*

368. (A) The anesthetized patient has decreased ability to respond to increased CO_2, low oxygen, and metabolic acidemia. The ability to respond to added airway resistance is not lost. *(1:398–400; 4:1143; 5:159–61,178–80)*

369. (E) All of these factors may lead to pulmonary edema. Surgery can cause capillary damage from rough handling and can interrupt lymphatic drainage. Injudicious fluid administration can add more fluid than the pulmonary circulation can handle. *(1:167–7, 991, 1468; 4:986,2357; 5:2713)*

370. (B) Respiratory quotient is the rate of CO_2 output divided by the rate of O_2 uptake. It may vary with the metabolic substrate and, therefore, will not always be 0.8. *(1:804; 5:2906–7)*

371. (B) PEEP increases FRC and decreases the work of breathing. Compliance is increased and the lung volume is increased. *(1:764; 4:1468–9; 5:2820)*

372. (E) All of the options are true, if positioning is done properly. This requires careful placement of rolls under the body in the prone position so that no pressure is placed on the abdomen. *(1:641; 4:688; 5:1152–3)*

373. (A) Supplemental oxygen and maneuvers to improve ventilation and prevent atelectasis are important. The use of carbon dioxide is not a current treatment. *(1:1385–92; 4:137; 5:2711–5)*

374. (A) The adult trachea is a midline structure of 10 to 15 cm in length. Half its length is intrathoracic. It bifurcates at the level of the fifth thoracic vertebra. *(1:595; 4:860; 5:1618)*

375. (B) Airway resistance is a dynamic measurement obtained by dividing pressure by flow. It is dependent on the caliber of the airway. The pattern of flow, laminar or turbulent, will change the measured resistance. *(1:795–6; 4:860; 5:690–1)*

376. (D) The FRC/CC relationship is important in determining if the airway will remain open. If CC is below FRC, lung areas will stay open. If CC is above FRC, lung areas will collapse with each breath. If CC is above FRC plus tidal volume, lung areas will stay collapsed. In obesity, FRC is decreased, while CC stays constant. *(1:1035; 4:510; 5:694–7; 6:446)*

377. (E) All of the options are correct. V_D/V_T ratio is a measure of wasted ventilation. If the percentage of tidal volume that is wasted goes up, one must compensate by increasing minute ventilation. *(1:802–3; 4:876; 5:697–8)*

378. (A) When paralyzed, the diaphragm is pushed cephalad by abdominal contents in the supine and lateral positions. Under spontaneous, anesthetized breathing, active expiration with increased abdominal tone also pushes the diaphragm cephalad. Retraction and packing in the abdomen can also decrease the diaphragm's descent. The last option, second gas effect, is incorrect. The second gas effect is the concentrating effect of the initial uptake of nitrous oxide on the other alveolar gases. *(1:384; 4:688; 5:1153)*

379. (D) The forced vital capacity (FVC) is the volume of gas that can be exhaled from a maximal inhalation. The results vary with patient cooperation. The FEV_1 is that portion of the procedure done in the first second. The FEV_1 is greatly reduced in obstructive disease. Inspiratory reserve volume is only a portion of the FVC. *(1:805; 4:233; 5:1000–1)*

380. (D) The primary stimulus is pH. Carbon dioxide reacts with water to form hydrogen ion; thus, CO_2 is involved indirectly. Hypoxia acts peripherally at the carotid bodies *(1:798; 5:177)*

381. (B) Each gas contributes to the total pressure in an amount that is proportional to its molecular fraction. Dalton's law states that each gas exerts the same pressure that it would if it alone occupied the container. Solubility plays a role in dissolved gases. *(1:379; 4:1020)*

382. **(A)** The composition of alveolar gas differs by the first three options since there is active CO_2 production, O_2 consumption, and humidification. Nitrogen undergoes no metabolism and insignificant excretion, so at equilibrium, no net transfer occurs across the alveolar membrane. *(4:1133; 5:697–9, 1439)*

383. **(C)** The patient has a respiratory alkalosis. His oxygen tension is low. Calculation of his expected PO_2 using the alveolar air equation shows his oxygen tension should be over 100 mmHg. Although the patient's oxygenation seems to be stable, it is important to follow this type of patient because there may be increasing difficulty in the subsequent hours. It is also important to consider that this patient may have compromised his oxygenation as a result of carbon monoxide poisoning. If there were thermal injury to the airway, prophylactic intubation would be indicated to maintain patency in the face of the subsequent edema. *(1:168–9; 4:964; 5:1603–5, 2495)*

384. **(A)** Patients who have flame burns are at greater risk from smoke and flame damage to the lungs than are those who have another type of burn, e.g., a chemical burn or scald. Injury to the upper airway is from thermal injury. The lower airway can be damaged by smoke and toxic fumes. Early damage is seen by 2 days, the effects of ARDS (adult respiratory distress syndrome) occur in the 2- to 5-day period, and pulmonary emboli are a continuing risk for the duration of the acute recovery. Tracheostomy is rarely needed until late in the process (weeks) unless there is another injury to the face or mouth. *(1:1275; 4:2166; 5:2495)*

385. **(E)** His oxygenation will be affected by pulmonary edema (should it occur), sickle cell disease (should it be present), carbon monoxide poisoning, and alkalosis. The alkalosis will affect oxygenation by shifting the oxyhemoglobin dissociation curve to the left, causing an increased affinity of hemoglobin for oxygen. *(1:506–7, 1274–5; 4:2166; 6:725–34)*

386. **(D)** Rapid, shallow respirations magnify the effect of dead space ventilation on total ventilation. As the tidal volume decreases, dead space, which is fixed, becomes a larger percentage, and there is a decrease in the actual alveolar ventilation. Respiratory rate increases with decreased compliance. *(1:802–3; 5:697–8)*

387. **(C)** Carbon monoxide binds hemoglobin avidly, and causes hemoglobin to appear cherry red, so cyanosis is less likely. Pulse oximeters that do not add another wavelength of light may read carboxyhemoglobin as oxyhemoglobin, falsely raising the O_2 saturation measured. Carbon monoxide is best displaced by a high FIO_2 (1.0 or even hyperbaric oxygen therapy), so merely adjusting for O_2 saturation is insufficient. Carboxyhemoglobin levels above 15% are significant and above 50% are lethal. *(1:1275; 4:2166; 5:1450–2, 2671–2)*

388. **(A)** The circumstances cited may be due to poor ventilation because the patient has been placed in an unphysiologic position. Cardiac output may drop as a result of compression of the vena cava, and isoflurane anesthesia may cause cardiac depression and systemic vasodilation. Blood loss should not be a problem at this stage of the procedure. *(1:394,650–3; 4:1750; 5:2197)*

389. **(A)** After positioning a patient, it is important to confirm that good ventilation is present. If the blood pressure is low, decreasing the amount of volatile agent being delivered is prudent. The kidney rest should be lowered to rule out caval compression. Inserting an arterial catheter is not curative, and will distract from more urgent therapy. *(1:650–3; 4:1750; 5:2197)*

390. **(A)** The surgical approach to the kidney sometimes opens the diaphragm, causing a pneumothorax. This can cause hypoventilation, and the resulting hypercapnia could account for the hypertension and ectopic beats. In addition, any procedure on the adrenal gland may involve a hormone-releasing tumor, such as a pheochromocytoma. *(1:1027–8; 4:313; 5:2197–8)*

391. **(B)** DLCO depends on diffusion of CO across perfused alveoli to be avidly bound to hemoglobin. Exercise and increased hemoglobin promote CO uptake, while fibrosis impedes diffusion and pulmonary embolism decreases effective gas transfer surface. *(1:806; 4:1742; 5:1010–1)*

392. **(A)** Basilar alveoli expand more than the apical ones, but end up about the same size. Residual volume is independent of the state of inspiration. Pleural pressure decreases, balancing the increased elastic recoil of the lung. Pulmonary vascular resistance is lowest at FRC. *(1:801–2, 804–7; 4:1752; 5:165–6, 693–4)*

393. **(B)** The LMA sits nonocclusively above the vocal cords, so it cannot prevent aspiration or laryngospasm. Positive pressure ventilation can also safely be accomplished in patients who are not at increased risk for aspiration. *(1:601–4; 4:1076; 5:1625–7)*

394. **(B)** The superior laryngeal nerve innervates the epiglottis and superior cords, while the recurrent laryngeal nerve innervates the trachea below the cords. The glossopharyngeal nerve innervates the posterior tongue and inferior soft palate. The anterior ethmoid nerve innervates the nose. *(1:617–9; 4:1066; 5:2537–9)*

395. **(D)** Equal ventilation of both sides is consistent with proper tube placement, although it does not guarantee it. Unequal ventilation obviously demonstrates poor tube placement or lung pathology. The distance from the lips to the carina in a female is about 25 cm. Therefore, the tube distance of 30 cm from the upper teeth is excessive. The carina lies at the level of the fourth to fifth thoracic vertebra. Therefore, the tube placement by chest x-ray also is incorrect. *(1:595,608–9; 4:109, 1068; 5:1633–6)*

396. **(C)** Laminar airflow through a tube is described by the Poiseuille's equation and is dependent on the radius and length of the tube, the radius being more important. Viscosity is more important than density in resistance to flow. *(1:795; 4:452; 5:690–1)*

397. **(B)** Patients breathing 100% oxygen are at risk of developing pulmonary toxicity. They can also develop absorption atelectasis secondary to the oxygen being breathed in the absence of an inert gas. PO_2 values are not present in excess of atmospheric pressure, and retrolental fibroplasia is not a concern in an adult. *(5:711, 2676–8)*

398. **(E)** The oxygen concentration in the blood is determined by the concentration in the alveoli, the efficiency of the lungs, and oxygen consumption. The alveolar oxygen concentration is determined by the inspired partial pressure ($FIO_2 \times$ barometric pressure), dead space, other gas movement, and ventilation. Lung efficiency decreases with age and some positions. *(1:804; 4:2319; 5:1010)*

399. **(E)** Patients with scoliosis present many intraoperative problems. Congenital scoliosis may be associated with cardiac, pulmonary, and neurologic abnormalities. Congenital and acquired scoliosis can cause a restrictive pulmonary pattern, resulting in hypoxemia, pulmonary vascular hypertension, and cor pulmonale. Blood conservation strategies are an important component of the anesthetic plan. *(1:1110–1; 4:280; 5:2418–9; 6:211)*

400. **(B)** Increased body temperature and increased hydrogen ion concentration both shift the oxyhemoglobin dissociation curve to the right, thereby increasing the P_{50}. Teleologically, during exercise, muscles generate heat and produce lactic acid, so this response helps unload oxygen. Methemoglobin is a hemoglobin molecule in which the iron is in the ferric state and is unable to combine with oxygen. *(1:202–3; 4:998; 5:699–701)*

401. **(E)** The rate of pulmonary complications correlates strongly with all these factors. In addition, the existence of chronic or acute pulmonary disease and the specific location of the incision can influence the level of postoperative pulmonary dysfunction. *(1:809–10, 1385–6; 4:241; 5:1085–9)*

402. (C) Mixed venous oxygen is usually about 40 mmHg or a saturation of 75%. It is best sampled from the pulmonary artery, so adequate mixing of blood from the inferior and superior vena cavae has occurred. Since mixed venous oxygen reflects oxygen usage by the body, it can be an indicator of tissue hypoxia. *(1:678; 4:818; 5:1331–2)*

403. (B) Blue nail polish and methylene blue both have absorbance in the 660-nm wavelength region, thereby falsely depressing the calculated saturation. Hyperbilirubinemia has no effect in actual experiments, though it does reduce nonpulsatile saturation readings. High altitude may change the PaO_2, but not the accuracy of oximetry. *(1:670–1; 4:92; 5:1448–53)*

404. (C) The pulmonary vascular system is well endowed with smooth muscle, though it lacks the muscular arterioles of the systemic circulation. There are rich connections to the autonomic nervous system and many α-adrenergic and β-adrenergic receptors. Injection of norepinephrine will lead to vasoconstriction. *(1:877–8; 4:722; 5:165–6)*

405. (B) Neurogenic pulmonary edema may follow head injury or other causes of increased intracranial pressure and can cause hypoxemia. It is caused by massive sympathetic discharge, and is best treated by reduction of intracranial pressure and support of oxygenation and ventilation. Diuretics are not indicated unless hypervolemia is also present. It is known that it will not develop in a denervated lung. *(4:1640; 6:208)*

406. (E) All of the options are true of CC. It is the sum of closing volume and residual volume. It is the phase in which intrapleural pressure closes, or at least constricts, some small bronchioles in dependent lung areas, slowing their contribution to ventilation. *(4:689; 5:693–7, 2437–8,2842)*

407. (C) Transpulmonary pressure is the pressure gradient across the lung measured as the pressure difference between the airway opening and the pleural surface. *(1:794,801; 4:1753; 5:693–7)*

408. (C) Dead space will increase if cardiac output is not maintained. Shunt will be increased in normal patients, but not in those with COPD. Nitroprusside is more potent than isoflurane and as potent as nicardipine at increasing pulmonary shunt when titrated to the same end point. *(1:773–7)*

409. (A) Vital capacity is the full exhalation from total lung capacity without time limit for the maneuver. CC is the earliest test of small airway disease, but requires gas analysis. Maximal breathing capacity is very effort dependent. FEV_1 is altered in obstructive disease. *(1:804–7; 4:232; 5:695–6)*

410. (A) Using nomograms, or actual calculations, these gas tensions are consistent with a pure acute respiratory acidosis. This degree of alveolar hypoventilation would cause a PO_2 in the 30's without supplemental O_2. Since O_2 is not an acid and requires no buffering, PO_2 history cannot be inferred from a single blood gas measurement. *(1:168–9; 4:964, 1070; 5:1439,1609)*

411. (A) Acute respiratory acidosis is best treated by improved ventilation. If the patient is unstable or unconscious, controlled ventilation with endotracheal intubation would be wise. Oxygen should be administered to improve PO_2 to supply tissues. Administering bicarbonate for a pure respiratory problem is not indicated, and will increase PCO_2. Finding and addressing the underlying cause of the hypoventilation is important. Possible causes might be pulmonary infection or bronchospasm. *(1:168–9; 4:964, 1070; 5:1603–5)*

412. (A) Asthma is a reversible constriction of bronchial smooth muscle. Bronchial tone is increased by vagal stimuli and decreased by β-adrenergic stimulation. Corticosteroids, often administered topically (by inhalation), also block bronchial constriction. Cromolyn has no place in the immediate preparation, since it does not relieve bronchospasm. It would be helpful in long-term prophylaxis of bronchoconstriction. *(1:478–9,817; 4:236; 5:1859, 2852–3; 6:198–9)*

413. **(D)** Even in the recovery room, the patient can be started on a stir-up regimen to help him or her handle secretions. Opioids should be used carefully in an amount that will treat the pain but still allow the patient to breathe deeply and mobilize secretions. *(1:1378–80; 4:2306; 5:2718–20)*

414. **(C)** This describes a method of emergency ventilation that can be used to oxygenate a patient adequately. If there is no obstruction above, pneumothorax is unlikely but one of the possible complications of the procedure. Gastric dilatation does not occur, since the air will be vented to the atmosphere. Pulmonary aspiration is unlikely because the flow is sufficient to blow the secretions out of the larynx but may nevertheless occur. Ventilation may be inadequate to prevent hypercapnia. *(1:630–1; 4:650; 5:2459)*

415. **(E)** In the patient with blunt anterior chest injury, respiratory distress, and hypotension, all of these diagnoses must be considered. *(1:1270–2; 5:2479–82)*

416. **(A)** Ventilating a patient by means of a bag and mask unit may result in gastric distention, thereby decreasing effective ventilation. The patient's condition will deteriorate, and with a shift of the mediastinum, cardiac arrest may occur. *(1:1272; 5:1623–4)*

417. **(E)** Bullae are an end-stage manifestation of emphysema with findings of CO_2 retention, increased residual volume, and decreased DLCO. The chest x-ray and CT will show a honeycombing pattern with relatively normal lung parenchyma compressed by the large air sacs. *(1:840–1; 4:1782; 5:1916–7; 6:179)*

418. **(D)** The bullae can expand or rupture with nitrous oxide or large tidal volumes. Blind chest tube insertion can also rupture a bulla, causing a persistent fistula. Ventilating the nonbullous side is the safest course *(1:840–1; 4:1782; 5:1917; 6:181)*

419. **(E)** Fat emboli can cause microcirculatory injury to the lungs, skin, and brain. Fever is almost always present. *(1:1114–5; 4:295; 5:2425; 6:175–6)*

420. **(C)** Pneumoconiosis is associated with fibrosis. Since a restrictive component is present, the diaphragm may move very little. The x-ray may look much worse than the functional state. *(1:806; 4:234; 6:209–10)*

421. **(B)** A pulmonary embolus causes a sudden increase in dead space. $ETCO_2$ will fall because nonperfused alveoli are still being ventilated. With no change in ventilatory pattern, the $PaCO_2$ will rise. Once $PaCO_2$ rises to steady state, $ETCO_2$ will return to normal. N_2 will only be seen if an air embolus occurs. *(1:668–9; 4:1635; 5:697–8)*

422. **(E)** A pulmonary embolus causes a sudden increase in dead space. $ETCO_2$ will fall because nonperfused alveoli are still being ventilated. Since the patient will attempt to maintain a stable $PaCO_2$, minute ventilation will increase. $ETCO_2$ will remain low until the embolus resolves. When no oxygen is administered, blood gases after a pulmonary embolus usually show hypoxemia and hypocarbia. *(1:668–9,802; 5:714–5; 6:172–5)*

423. **(B)** For an ideal gas, which most physiological situations approximate well, the molar concentration is P/RT. R is a universal constant, and does not vary with molecular properties such as molecular weight. *(1:379–80; 4:1012, 1128)*

424. **(A)** The molar concentration in the liquid phase is proportional to the molar concentration in the gas phase and a solubility factor which in turn depends on the molecular properties of the components. Molecular weight is not important, however. *(1:379–80; 4:1128, 2087)*

CHAPTER 5

Nervous System

Questions

DIRECTIONS (Questions 425 through 465): Each of the numbered items or incomplete statements in this section is followed by answers or by completions of the statement. Select the ONE lettered answer or completion that is BEST in each case.

425. Body temperature may change intraoperatively because of all the following EXCEPT

(A) exposure of body surfaces
(B) intravenous fluids
(C) muscle relaxants
(D) humidity of inspired gases
(E) oxygen concentration

426. Heat production in the body is decreased by

(A) shivering
(B) brown fat
(C) activity
(D) circulating norepinephrine
(E) neuromuscular diseases

427. The temperature-regulating mechanism is in the

(A) forebrain
(B) pons
(C) thalamus
(D) hypothalamus
(E) cerebellum

428. The respiratory control center

(A) contains no expiratory neurons
(B) receives no mechanical sensory input
(C) receives input from airway oxygen receptors
(D) is located in the medulla and pons
(E) is activated by peripheral CO_2 tension

429. The total volume of cerebrospinal fluid (CSF) in the adult is

(A) 10 to 20 mL
(B) 20 to 40 mL
(C) 40 to 60 mL
(D) 60 to 100 mL
(E) 120 to 150 mL

430. The normal CSF pressure in a recumbent patient is

(A) 20 cm H_2O
(B) 20 mm H_2O
(C) 110 mm H_2O
(D) 110 cm H_2O
(E) 50 mm H_2O

431. Absorption of CSF takes place through

(A) ependymal cells
(B) arachnoid villi
(C) the pia mater
(D) the foramen of Monro
(E) the foramen of Magendie

432. CSF flows through all of the following EXCEPT the

(A) arachnoid villi
(B) cerebral ventricle
(C) lateral ventricle
(D) subarachnoid space
(E) epidural space

433. The specific gravity of normal CSF is

(A) 1.000 to 1.003
(B) 1.100 to 1.200
(C) 1.003 to 1.009
(D) 1.200 to 1.300
(E) under 1.000

434. The A-beta fiber is

(A) an efferent to muscle spindles
(B) a large nerve fiber associated with transmission of deep touch and proprioception
(C) an afferent sensory nerve conducting pain
(D) a preganglionic sympathetic
(E) not myelinated

435. Preganglionic autonomic nerve fibers are

(A) A-alpha fibers
(B) A-beta fibers
(C) A-gamma fibers
(D) B fibers
(E) C fibers

436. The smallest nerve fiber, a postganglionic fiber associated with slow conduction, is the

(A) A-alpha fiber
(B) A-beta fiber
(C) A-gamma fiber
(D) B fiber
(E) C fiber

437. The largest nerve fibers associated with motor function are

(A) A-alpha and A-beta fibers
(B) A-gamma and A-delta fiber
(C) not myelinated
(D) B fibers
(E) C fibers

438. Integrity of all of the following structures can be monitored by sensory-evoked potentials EXCEPT the

(A) dorsal columns
(B) cerebellum
(C) thalamus
(D) vestibular-cochlear nerve
(E) inferior colliculus

439. The blood supply to the motor tracts of the spinal cord is primarily from the

(A) posterior spinal artery
(B) anterior spinal artery
(C) penetrating branches of the radicular arteries
(D) artery of Adamkiewicz
(E) basilar artery

440. Parkinson's disease is associated with dysfunction of the

(A) cerebellum
(B) lateral ventricle
(C) pons
(D) basal ganglia
(E) aqueduct of Sylvius

441. A nuclear group in the brain involved in transmission of sensory information to the cortex is the

(A) cerebellum
(B) caudate
(C) hypothalamus
(D) thalamus
(E) hippocampus

442. All of the following cranial nerves contain parasympathetic efferent fibers EXCEPT the

(A) oculomotor nerve (III)
(B) trigeminal nerve (V)
(C) facial nerve (VII)
(D) glossopharyngeal nerve (IX)
(E) vagus nerve (X)

443. The electroencephalographic (EEG) waveform with a frequency range of 8 to 13 Hz is associated with

(A) alpha activity
(B) beta activity
(C) theta activity
(D) delta activity
(E) gamma activity

444. The compressed spectral array (CSA) recording of the EEG

(A) is more accurate than a standard tracing
(B) presents data in a format of amplitude, time, and frequency
(C) is more accurate for determination of sudden events
(D) requires no training for interpretation
(E) should be in use for all neurosurgical procedures

445. Somatosensory-evoked potentials (SSEP)

(A) are disrupted by small doses of thiopental
(B) are not affected by deep levels of halothane anesthesia
(C) are evaluated from the standpoint of amplitude only
(D) are rendered unreadable by neuromuscular blocking agents
(E) are little affected by nitrous oxide

446. In the face of a declining supply of nutrients

(A) neuronal function deteriorates in an all-or-none fashion
(B) there is no reserve below the normal level of cerebral blood flow (CBF)
(C) irreversible neuronal damage occurs with EEG evidence of ischemia
(D) the actual CBF level at which function deteriorates varies with anesthetics
(E) an isoelectric EEG indicates irreversible neuronal damage

447. The intermediolateral gray column

(A) extends from C7 to S1
(B) is the origin of the preganglionic fibers of the sympathetic nervous system
(C) conducts only sensory impulses
(D) contains the anterior horn cells
(E) includes the substantia gelatinosa

448. The substantia gelatinosa

(A) has the highest concentration of opioid receptors in the spinal cord
(B) is located in the lateral columns
(C) is part of the dorsal column
(D) is in the motor area of the brain
(E) is in the ventral column

449. Which of the following is NOT involved in sensory transmission?

(A) dorsal root ganglion
(B) spinothalamic tract
(C) parietal cortex
(D) ventral, posterior, and lateral nuclei of the thalamus
(E) precentral gyrus

450. In the adult, the spinal cord ends

(A) at the lower border of the second sacral vertebra
(B) at the lower border of the first lumbar vertebra
(C) at a segment in the lower lumbar region depending on the patient's height
(D) midway between the third and fourth lumbar vertebra
(E) at the lumbosacral junction

451. The oculocardiac reflex involves all of the following structures EXCEPT

(A) vagus nerve
(B) trigeminal ganglion
(C) ophthalmic division of V
(D) oculomotor nerve
(E) brain stem

452. Hormonal products of the anterior pituitary include all of the following EXCEPT

(A) growth hormone
(B) luteinizing hormone
(C) antidiuretic hormone (ADH)
(D) follicle-stimulating hormone
(E) thyrotropin

453. ICP can be monitored with all of the following EXCEPT

(A) a catheter implanted directly into brain parenchyma
(B) by a pressure transducing bolt in the subarachnoid space
(C) by a pressure transducer in the epidural space
(D) a ventriculostomy catheter
(E) a catheter in the subdural space

454. The Glasgow Coma Score is based on

(A) eye opening, verbal response, and motor response
(B) assessment of knee jerk and other motor reflexes
(C) assessment of pupil size and brain stem reflexes
(D) assessment of respiration and autonomic brain stem functions
(E) assessment of EEG

455. All of the following are used to minimize damage caused by cerebral ischemia EXCEPT

(A) thiopental
(B) nimodipine
(C) isoflurane
(D) halothane
(E) hypothermia

456. Intracerebral steal

(A) is a term used to describe focal flow decrease caused by cerebral vasodilatation
(B) is a phenomenon caused by alkalosis
(C) is a condition caused only by brain tumors
(D) always occurs during anesthesia
(E) is always helpful to ischemic tissue

457. An increase in $PaCO_2$ from 40 to 50 mmHg will increase CBF

(A) 1 to 2 mL/100 g/min
(B) 5 to 10 mL/100 g/min
(C) 10 to 20 mL/100 g/min
(D) 25 to 50 mL/100 g/min
(E) none at all

Questions 458 and 459 refer to Figure 5-1:

458. On the relatively flat part of the curve

(A) there is a small increase in pressure with increased volume
(B) there is a small pressure increase showing poor compensatory effect
(C) pressure increases are compensated for by increased blood flow
(D) there is a large pressure increase
(E) compensatory mechanisms are not functional

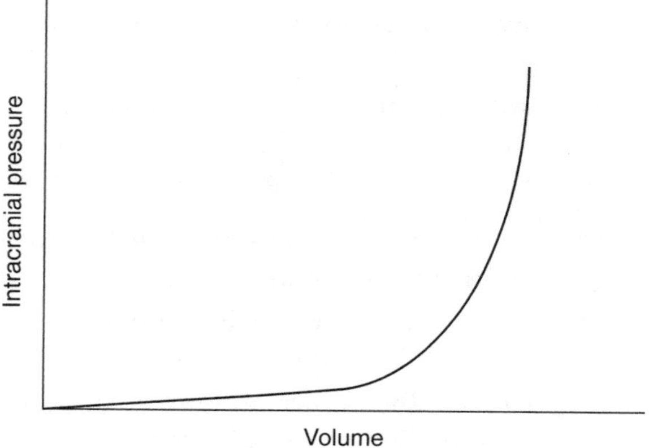

FIG. 5-1

459. ICP measurements similar to those shown in the graph are obtained from a

(A) needle in the caudal canal
(B) spinal needle
(C) transducer in the extradural space within the cranium
(D) carotid artery transducer
(E) jugular bulb transducer

460. Which is the most profound chemical stimulus for regulation of CBF?

(A) metabolic alkalosis
(B) hypothermia
(C) hyperthermia
(D) carbon dioxide
(E) hypercalcemia

461. The oxygen reserves of the brain are

(A) infinite
(B) capable of maintaining function for 25 minutes
(C) greater under anesthesia
(D) very low
(E) carried primarily in the cerebral hemispheres

462. Loss of cerebral autoregulation

(A) will cause the cerebral blood pressure to fall with systemic hypertension
(B) always affects the entire brain at the same time
(C) may be focal
(D) only affects the lower end of the blood pressure range
(E) has no clinical significance

463. CSF pressure may be increased by

(A) coughing
(B) long expiratory time
(C) short expiratory time
(D) low expiratory resistance
(E) negative expiratory phase

464. Cerebral perfusion pressure is determined as

(A) MAP + ICP
(B) SBP – ICP
(C) MAP – ICP
(D) MAP + CVP – ICP
(E) SBP – CVP

MAP: mean arterial blood pressure
ICP: intracranial pressure
SBP: systolic arterial blood pressure
CVP: mean central venous pressure

465. Propofol does all of the following EXCEPT

(A) decreases CBF
(B) crosses the blood-brain barrier
(C) reduces ICP
(D) decreases cerebral metabolic rate
(E) decreases the latency of somatosensory-evoked potentials

DIRECTIONS (Questions 466 through 496): For each of the items in this section, ONE or MORE of the numbered options is correct. Choose answer

(A) if only 1, 2, and 3 are correct
(B) if only 1 and 3 are correct
(C) if only 2 and 4 are correct
(D) if only 4 is correct
(E) if all are correct

466. The cranial nerve concerned with auditory sensation is

(1) located in the posterior fossa
(2) near the brain stem
(3) close to cranial nerves V and VII
(4) adjacent to nerves conducting vestibular information

467. The celiac plexus

(1) contains visceral afferent and efferent fibers
(2) receives afferents from greater, lesser, and least splanchnic nerves
(3) innervates abdominal viscera
(4) has no somatic fibers

468. Intracranial contents that contribute to ICP include

(1) brain parenchyma
(2) arterial blood
(3) CSF
(4) venous blood

469. CSF is

(1) formed in the choroid plexus
(2) functions as a cushion for the brain
(3) is replaced three times daily
(4) produced in the lumbar area

470. Neurotransmitters in the central nervous system that exert an inhibitory action on postsynaptic neurons include

(1) dopamine
(2) glycine
(3) serotonin
(4) GABA

471. The reticular formation

(1) is located in the cerebellum
(2) has inhibitory functions
(3) has only sensory input
(4) is concerned with sleep

472. A typical nerve cell has

(1) numerous branching dendrites
(2) three or four axons to conduct impulses
(3) a dendritic zone rich in receptors
(4) a myelin sheath that covers the entire cell

473. CSF is formed

(1) as a secretion from the choroid plexus
(2) according to need, increasing greatly if needed
(3) by an energy requiring process
(4) in an amount of about 12 mL per day under normal conditions

474. The brain stem is supplied with blood from the

(1) vertebral artery
(2) cingulate artery
(3) basilar artery
(4) posterior cerebral artery

475. CSF has

(1) a lower pH than plasma
(2) a buffer capacity equal to blood
(3) a pH which changes quickly with alteration of blood PCO_2
(4) a normal bicarbonate level of 55 mg/dL

476. A typical reflex arc includes a(n)

(1) sense organ
(2) afferent neuron
(3) efferent neuron
(4) synapse on a peripheral effector

477. The EEG

(1) alone is sufficient for the diagnosis of brain death, because a flat line is synonymous with death
(2) shows increased cortical activity with hypercapnia
(3) is not useful in the diagnosis of convulsive disorder
(4) will not be affected by hysterical seizures

478. The reticular activating system (RAS)

(1) is a specific area with a specific pathway system
(2) is located in the cerebral cortex
(3) has input from the spinal cord only
(4) is necessary for the conscious, alert state

479. When the cranium is open, the ICP is

(1) unaltered by anesthetic agents
(2) dependent on the position of the head relative to the heart
(3) equal to ambient pressure
(4) cannot be measured by a ventricular catheter

480. The act of vomiting is an integrated activity that includes

(1) activation of the vomiting center in the cerebral cortex
(2) the closing of the glottis
(3) the closure of esophageal and gastric cardiac sphincters
(4) activation of the chemoreceptor trigger zone in the area postrema

481. Sweating may occur as a

(1) thermal-regulating mechanism
(2) response to hypoglycemia
(3) response to emotional stimuli
(4) general reaction to sympathetic stimuli

482. Body temperature in the patient under general anesthesia

(1) is closely regulated by the hypothalamus
(2) is known to be elevated if the patient is sweating
(3) is best assessed by skin temperature
(4) will tend to drift to reach ambient temperature

483. Body temperature regulation involves

(1) heat production by the muscle
(2) heat production by the liver
(3) heat dissipation by the skin
(4) heat dissipation by the lungs

484. In the patient with an acute injury at the C4–5 level, one would expect

(1) a major loss of diaphragmatic power
(2) no breathing
(3) grossly impaired alveolar ventilation
(4) disturbance of heart rate

485. The normal brain

(1) has a constant metabolic rate
(2) has autoregulation of flow over the blood pressure range of 50 to 150 mmHg
(3) has a constant blood flow
(4) requires fructose for energy

486. Factors important in regulation of ICP include the

(1) CSF
(2) intracranial tissue
(3) blood
(4) dura

487. The blood-brain barrier

(1) permits free passage of bicarbonate ion
(2) does not include the brainstem
(3) is impermeable to carbon dioxide
(4) is composed of tight junctions in the vascular endothelium

488. Mannitol decreases cerebral edema because it

(1) is an osmotic diuretic
(2) pulls water across an intact blood-brain barrier
(3) rapidly dehydrates the brain
(4) is a peripheral vasodilator

489. Hyperventilation may lead to

(1) cerebral vasodilatation
(2) tetany
(3) shift of the oxyhemoglobin dissociation curve to the right
(4) decreased cardiac output

490. In administration of anesthesia to a patient with Arnold-Chiari malformation

(1) coughing on the endotracheal tube should be prevented
(2) extreme flexion-extension of the neck should be avoided
(3) postoperative respiratory depression may be encountered
(4) myelomeningocele is common

491. In the patient with amyotrophic lateral sclerosis, anesthetic techniques should include

(1) avoidance of succinylcholine
(2) avoidance of respiratory depressants
(3) minimal use of relaxants
(4) avoidance of halothane

492. Autonomic hyperreflexia

(1) occurs with a lesion above T5
(2) is initiated only by bladder distention
(3) is accompanied by severe hypertension
(4) is accompanied by tachycardia

493. In the anesthesia of a patient with myotonic dystrophy, it is important to consider that

(1) halothane may cause shivering postoperatively and precipitate a myotonic response
(2) myotonia can be relieved by nondepolarizing relaxants
(3) myotonia can be precipitated by succinylcholine
(4) neostigmine prevents myotonia

494. Considerations in the care of a patient with arteriovenous malformation include

(1) critical control of blood pressure during induction
(2) barbiturate-induced coma for control of CBF during resection
(3) hypertension to treat vasospasm
(4) malignant brain swelling

495. In a patient with a subarachnoid hemorrhage, it is not uncommon to find

(1) hyponatremia
(2) dysrhythmias
(3) cerebral vasospasm
(4) ECG abnormalities consistent with ischemia

496. A patient exhibiting Cushing's reflex would have

(1) intracranial hypertension
(2) arterial hypertension
(3) bradycardia
(4) ICP and CBF passively related to blood pressure

Answers and Explanations

425. (E) Oxygen concentration does not influence body temperature. Exposure to low ambient temperature may cause hypothermia. This is especially a problem in small children because of their large body surface to mass ratio. Cold intravenous fluids and dry inspired gases are sources of heat loss and a fall in body temperature. Muscle relaxants prevent shivering, a mechanism by which the body maintains a normal temperature. *(1:683–8; 2:613–6; 4:2442–6; 5:1572–7)*

426. (E) Heat production in the body is decreased by neuromuscular diseases which inhibit shivering, a major contributor to thermogenesis. Shivering, brown fat, and physical activity increase heat production by the body. Circulating norepinephrine regulates nonshivering thermogenesis in skeletal muscle and brown fat. *(1:683–4; 2:612; 4:2442–6; 5:1572–4)*

427. (D) The hypothalamus is responsible for controlling the temperature set point. *(1:261, 683–4; 2:611; 4:2442; 5:1572–3)*

428. (D) The respiratory control center, located in the pons and medulla, integrates information from peripheral mechanical receptors and peripheral and central chemosensors. The apneustic center contains both inspiratory and expiratory neurons. Peripheral oxygen sensors are located in the carotid body (not the airway). Receptors for hydrogen ion are centrally located; receptors for carbon dioxide do not affect respiratory drive. *(1:796–7; 5:170–1)*

429. (E) The total volume of CSF is 120 to 150 mL. There is a constant turnover of CSF with a production rate of 0.35 to 0.40 mL/min. *(1:747; 4:244, 1615; 5:832)*

430. (C) The usual CSF pressure is 110 mm H_2O (about 8 mm Hg). *(1:747; 2:495; 5:813–4)*

431. (B) The absorption of CSF takes place through the arachnoid villi. The ependymal cells and the pia mater are not involved. The foramina of Monro and Magendie are conduits for CSF flow. *(1:747; 2:493–4; 4:244, 1366–7; 5: 832, 2162–3)*

432. (E) CSF flows through the cerebral ventricles and the subarachnoid space before absorption in the arachnoid villi. There is no CSF in the epidural space. *(1:747; 2:493–4; 4:244, 1366–7; 5:2162-3)*

433. (C) The specific gravity of CSF ranges from 1.003 to 1.009 at 37°C; the density ranges from 0.9998 to 1.0008 at 37°C. Specific gravity is defined as the ratio of the density of substance at a specific temperature to that of water at the same temperature. *(1:698; 4:1367)*

434. (B) The A-beta fibers are large, myelinated fibers transmitting sensation of deep touch and proprioception. *(1:263–4; 4:1297; 5:577)*

435. (D) B fibers are small fibers of preganglionic autonomic nerves. *(1:263–4; 4:1297; 5:577)*

436. (E) A very small postganglionic fiber with slow conduction is the C fiber. *(1:263–4; 4:1297; 5:577)*

437. (A) The largest fibers are A-alpha and A-beta. They are associated with motor function. Both are myelinated and exhibit fast conduction. *(1:263–4; 4:1297; 5:577)*

438. (B) The cerebellum is not monitored by sensory-evoked potentials. The dorsal columns of the spinal cord transmit proprioception and are assessed by SSEPs. The vestibular-cochlear nerves (cranial nerve VIII) and the inferior colliculus are assessed by brainstem auditory-evoked potentials. Peaks corresponding to thalamic nuclei are detected with both. *(1:683, 754–6; 4:890–6; 5:1527–9)*

439. (B) The descending motor tracts are located in the ventral white matter in the spinal cord. The primary blood supply is from the anterior spinal artery. *(1:856–7; 4:2119–20; 5:2086–8; 6:146–7)*

440. (D) Parkinsonism is due to pathology in the basal ganglia of the brain. A nonintention tremor may be a symptom. *(1:276, 501–2; 5:1095; 6:258)*

441. (D) The thalamus is located at the base of the brain and contains many distinct nuclear groups. Sensory information is processed here before conduction to the cortex. Distinct peaks corresponding to the thalamus are found in somatosensory and brain stem auditory-evoked potentials (BAEPs). *(1:123; 4:889; 5:1527–9)*

442. (B) The trigeminal nerve contains somatic afferent fibers from the face and supplies motor innervation to the muscle of mastication. The oculomotor nerve contains parasympathetic efferents that control pupillary constriction; the facial nerve contains parasympathetic fibers supplying the submandibular gland; the glossopharyngeal nerve supplies the parotid gland; and the vagus nerve supplies the heart, respiratory system, and gut. *(1:263; 4:724; 5:621–3)*

443. (A) Alpha activity has a frequency range of 8 to 13 Hz. Beta activity has a frequency of greater than 13 Hz. Theta rhythm has an activity of 4 to 7 Hz. Delta rhythm has an activity of less than 4 Hz. *(1:752; 4:884; 5:1513–4)*

444. (B) The CSA presents data from the EEG in a format of amplitude, time, and frequency. Although this format is more convenient, some specificity and accuracy is lost. The CSA is not as accurate for sudden events. In spite of its convenience, a certain amount of training and familiarity is needed to interpret the tracing accurately. CSA is not necessary for every anesthetic. *(1:752–3; 4:885; 5:1515–6)*

445. (E) SSEPs are little affected by nitrous oxide. They are not disrupted by small doses of thiopental. Deep levels of halothane will disrupt the tracings. The tracings evaluate latency and amplitude. *(1:753–6; 4:893; 5:1532–7)*

446. (D) In humans undergoing carotid endarterectomy, volatile anesthetics have been shown to alter the cortical blood flow level at which EEG evidence of ischemia first develops; halothane had the least effect, isoflurane reduced the threshold most, and enflurane was intermediate between the two. Although neuronal function deteriorates progressively with decreasing blood flow, not until CBF is less than half (22 mL/100 g/min) the normal level is EEG evidence of ischemia manifest. The EEG is isoelectric at 15 mL/100 g/min, but irreversible damage does not occur until flow falls below 6 mL/100 g/min. *(1:942; 2:492; 4:1854–5; 5:1864–5)*

447. (B) The intermediolateral gray column of the spinal cord contains the origin of the preganglionic fibers of the sympathetic nervous system. It extends from T1 to S3. Sensory transmissions are conducted by white matter. Anterior horn cells are in the ventral horn of the gray matter; the substantia gelatinosa is in the dorsal horn. *(1:264–5, 691, 4:723–6; 5:621–3)*

448. (A) The substantia gelatinosa has the highest concentration of opioids in the cord. This area is in the dorsal horn of the spinal cord. *(1:1406, 1437; 4:1512–6; 5:385)*

449. (E) The precentral gyrus is primarily involved with motor functions. Sensation involves the dorsal root ganglion cell, the lateral spinothalamic tract, the thalamic nuclei, and the postcentral gyrus. *(1:1403–4; 5:2729–30, 2764–5)*

450. (B) The spinal cord ends at the lower border of the first lumbar vertebra in the adult. The subarachnoid space continues further caudally and ends at S2. Although in children the cord ends in the lower lumbar region and moves rostral as the child reaches adulthood, the position of the end of the cord is not dependent on patient height. *(1:691–2; 4:1365; 5:1654–5)*

451. (D) The oculocardiac reflex is characterized by bradycardia (even to the point of asystole) produced by pressure or traction on the eye. The afferent limb of the reflex is composed of fibers from the ophthalmic division of the trigeminal nerve which synapse on brain stem neurons and ultimately increase vagal efferent activity. Activation of the reflex can occur from manipulation of any of the structures which carry afferent impulses, including the trigeminal ganglion and the long ciliary nerves which innervate the globe. The oculomotor nerve is not involved in the reflex. *(1:973–4; 4:2184; 5:739, 2530)*

452. (C) ADH and oxytocin are secreted by the posterior pituitary. Growth hormone, luteinizing hormone, follicle-stimulating hormone, thyrotropin, prolactin, and adrenocorticotropin are all secreted by the anterior pituitary. *(1:767–9; 4:315; 5:2158–9; 6:435–6)*

453. (A) ICP is routinely monitored by transducing a small catheter introduced into the cerebral ventricle or subdural space. Similarly, a hollow bolt with pressure transducer attached or a bolt-mounted transducer can be placed into the subdural space. Alternatively, ICP can be measured by placing a pressure transducer into the epidural space. A catheter implanted directly into brain parenchyma is not used. *(1:757–8; 2:499; 4:246)*

454. (A) The Glasgow Coma Score is based on the ability to open the eyes and verbal and motor responses. Responses to knee jerk, pupil size, respiration, and EEG are assessed as part of a general neurologic examination but are not part of the Glasgow Coma Score. *(1:777–8; 4:246; 5:2152)*

455. (D) Hypothermia can protect the brain from periods of ischemia of moderate duration. Thiopental, isoflurane, and the calcium channel blocker nimodipine, also may provide some degree of cerebral protection. Halothane has no cerebral protective benefits. *(1:750–2; 4:2372–3; 5:837–42, 2150–1)*

456. (A) Intracerebral steal describes a decrease in blood flow to ischemic tissue due to vasodilatation in tissues surrounding the ischemic area. Because blood vessels in the ischemic area are already maximally dilated, dilation of vessels in adjacent tissue shunts blood away from deprived areas. The phenomenon is not helpful to compromised tissue. *(1:746; 5:840–1)*

457. (C) CBF increases 1 to 2 mL/100 g/min with each 1 torr rise in $PaCO_2$ in the normal physiologic range of $PaCO_2$. For this reason it is important to have good airway control at all times when dealing with patients with increased ICP. *(1:746; 2:497; 4:1610; 5:816–7)*

458. (A) On the flat part of the compliance curve, there is a slow rise in pressure for increases in volume. In this area of the curve, there is good compensation. As one moves to the right, there is a sudden increase in pressure with small changes in volume. *(1:747–8; 2:494–5 4:1614–6; 5:2127–30)*

459. (C) These types of tracings are obtained with transducers located in the extradural space within the cranium. Small increments of volume are added, and the pressure changes are noted. *(1:747–8; 4:1614–6; 5:2127–30)*

460. (D) The most profound chemical stimulus affecting CBF is carbon dioxide. Changes in electrolytes and acid-base balance may have an effect, but the predominant effect is from $PaCO_2$. Hypothermia decreases CBF by decreasing cerebral metabolic rate. *(1:746; 4:1617–20; 5:816–7)*

461. (D) The oxygen reserves are very low. The reserves are not changed under anesthesia, but oxygen use is decreased. Since oxygen reserves are so low, the brain is subject to hypoxia with any bout of ischemia. *(1:745; 4:2365; 5:833–4)*

462. **(C)** Loss of autoregulation may be focal and only affect part of the brain. For this reason, changes in $PaCO_2$ may cause diversion of blood from the areas that are ischemic to areas that are still regulated. Loss of autoregulation may have great significance, but it is difficult to know how to manipulate the CO_2 to get the best effect. *(1:746; 2:496; 4:1620–1; 5:817)*

463. **(A)** CSF pressure is increased with coughing. Coughing increases intrathoracic and intra-abdominal pressure which is transmitted to the cranial vault. Maneuvers that decrease intrathoracic pressure will decrease ICP. *(1:762–5; 4:718–9; 5:2144–5)*

464. **(C)** Cerebral perfusion pressure is the fundamental concept underlying the autoregulation curve. It represents the blood pressure available for global perfusion of the brain. Because the brain is enclosed in the cranium, which functions as a noncompliant container, CPP is defined as the pressure of the blood entering the cranium (MAP) minus the pressure exerted within the cranium (ICP). This assumes that central venous pressure is less than ICP, as is the case when the head is elevated and there is no resistance to venous outflow from the cranium. *(1:747; 2:496; 4:244; 5:817)*

465. **(E)** Propofol is a potent metabolic suppressant. Since flow-metabolism coupling is maintained, there is a significant reduction in CBF, blood volume, and ICP. Propofol, as do most anesthetic agents, increases the latency of SSEPs. *(1:334, 754–6; 2:503; 4:1623; 5:821–2, 1525–36)*

466. **(E)** The vestibular-cochlear nerve (cranial nerve VIII) conducts both auditory sensation and the output of the vestibular apparatus concerned with motion sensation. The fibers traverse the posterior fossa before entering the brain stem and are adjacent to cranial nerves V and VII. BAEPs are often monitored intraoperatively during surgery on or near these structures. *(1:754, 766; 4:896–7; 5:1530–2)*

467. **(E)** The celiac plexus is a major abdominal plexus composed of a number of ganglia and innervates most of the abdominal contents. Nerve fibers are derived from T5 through T12 and convey only visceral afferent and efferent information with no somatic inputs. *(1:263–5; 4:724; 5:621–3)*

468. **(E)** All intracranial contents contribute to determination of ICP in a closed cranium. These include brain parenchyma, CSF, and intracranial blood volume in all compartments (arterial, capillary, and venous). *(1:747–8; 2:493–4; 4:1614–6; 5:2127–30)*

469. **(A)** The majority of CSF is formed in the choroid plexus; transependymal diffusion from the brain interstitium also contributes to total volume. Continuous production results in a complete turnover three times daily. The brain and spinal cord essentially float in this fluid. *(1:746–7; 2:493–4; 4:1366–7; 5:832–3)*

470. **(E)** The major inhibitory transmitter in the central nervous system is GABA, but other transmitters may have inhibitory effects as well. All of the options have an inhibitory effect. *(1:744; 4:1213; 5:109–14)*

471. **(C)** The RAS has both inhibitory and excitatory functions. The RAS is located in the brain stem. It has sensory and motor inputs. *(1:122, 1406–7; 4:1104; 5:109)*

472. **(B)** The typical nerve cell has numerous dendrites that branch out and a dendritic zone that is rich in receptors. The nerve cell has only one axon. The myelin sheath covers the axon only. *(1:1403–4; 4:1332–4)*

473. **(B)** CSF is formed by filtration and by active transport, the latter responsible for about two-thirds of the total amount. The active transport process requires energy. The amount formed is 0.35 to 0.40 mL/min. *(1:746–7; 2:493–4; 4:1366–7; 5:832)*

474. **(B)** The two vertebral arteries join to form the basilar artery. This unpaired midline structure gives rise to many branches and perforating arteries that supply the brain stem and the cerebellum. Transient ischemic attacks or strokes of the basilar arterial system result in unconsciousness from loss of RAS function. In addition, motor function can be lost from ischemia of the cerebral pyramids which contain long motor tracts to the brain stem nuclei and spinal cord. *(1:854–7; 6:246–7)*

475. **(B)** The pH of CSF is less than plasma. Since the blood-brain barrier is more permeable to CO_2 than to bicarbonate and CSF has minimal buffering capacity, the CSF pH will change quickly with alterations of PCO_2 in the blood. *(1:746–7; 2:493–7; 4:1367)*

476. **(E)** All of the options are present in a reflex arc. *(1:261, 265)*

477. **(C)** The EEG is not sufficient for the diagnosis of brain death. One must get a careful history, and the patient must be assessed carefully. Patients who are cold or under the influence of barbiturates may display a flat line. Cortical activity will increase with hypercapnia. Hysterical seizures are not seen on the EEG. *(1:752–3; 4:1986–7; 5:1525)*

478. **(D)** The RAS is a nonspecific area in the brainstem responsible for maintaining the conscious, alert state. It receives input from higher and lower centers. *(1:122; 4:1104; 5:109)*

479. **(B)** ICP is maintained by a closed cranium. Opening of the cranium at surgery relieves ICP and it equals ambient pressure or zero. While positioning of the head and anesthetic agents will impact on brain relaxation, they cannot alter ICP. *(1:762–3; 2:493–4; 4:1614; 5:2127–30)*

480. **(C)** The act of vomiting requires the activation of the vomiting center, which is located in the reticular formation of the medulla. It requires the opening of the esophageal and gastric cardiac sphincters. *(1:994; 4:134–6)*

481. **(E)** Sweating is an important temperature-regulating mechanism. It occurs in response to sympathetic nervous system-mediated stimulation of the sweat glands and can result from a number of physiologic perturbations, including hypoxia, or as a result of emotional stress. *(1:683–8; 2:611; 4:2442–6; 5:1572–4)*

482. **(D)** In the patient under general anesthesia, core temperature will tend to drift toward ambient temperature. Thus, it is important to take steps to both prevent heat loss and to actively treat hypothermia. The problem is especially common during surgery involving open body cavities. *(1:683–8; 2:613–6; 4:2442–6; 5:1576–7)*

483. **(E)** Regulation of body temperature is a complex mechanism involving heat production by muscles and liver and heat loss from skin and lungs. The conduct of anesthesia impacts considerably on all of these organ systems. *(1:683–8; 2:610–2; 4:2442–6; 5:1572–4)*

484. **(B)** The phrenic nerve arises from cervical segments 3–5 and supplies motor innervation to the diaphragm. The intercostal muscles receive motor innervation from nerves originating in the thoracic cord. Although disruption of the cervical cord between C4 and C5 would, therefore, severely compromise ventilation, some diaphragmatic function would remain intact. In the absence of hypoxia, no disturbance in heart rate is expected. *(1:1268; 2:509; 4:1828; 6:279–80)*

485. **(A)** In the normal brain, metabolic rate and blood flow are relatively constant. The autoregulatory range in which constant blood flow is maintained occurs between a MAP of 50 to 150 mmHg. Fructose is not a required energy source of the brain. *(1:745–6; 2:496; 4:1608, 1620; 5:814–7)*

486. **(A)** Factors of importance in regulating ICP are CSF, intracranial tissue, and blood flow. The dura is not a regulating mechanism. *(1:747–8; 2:494–7; 4:1614–6; 5:2127–34)*

487. **(D)** The blood-brain barrier is present in all regions of the brain and spinal cord and is formed by tight junctions between vascular endothelial cells and foot processes of the glia. Carbon dioxide freely crosses the blood-brain barrier; bicarbonate is charged and crosses the blood-brain barrier more slowly. *(1:746–7; 4:1618–9; 5:832–3, 816–7)*

488. **(A)** Mannitol does all of these. While rapid administration can produce transient peripheral vasodilation, this does not contribute to relief of cerebral edema. *(1:762; 2:505; 4:1632; 6:237–8)*

489. **(C)** Hyperventilation may lead to tetany and decreased cardiac output. The cerebral vasculature is constricted, and the oxyhemoglobin curve is shifted to the left. *(1:746; 5:2132; 6:238–9)*

490. **(E)** The Arnold-Chiari malformation is characterized by brain stem compression from herniation of the cerebellar tonsils through the foramen magnum and caudal displacement of the brain stem. Hydrocephalus is common from obliteration of the foramina of Lushka and Magendie which drain the fourth ventricle. Therefore, increases in ICP may be encountered. Extreme flexion-extension of the neck may increase brain stem compression. Postoperative respiratory and hemodynamic patterns should be monitored closely because of possible brain stem compression. The syndrome is commonly associated with myelomeningocele. *(1:1189; 5:2164; 6:256)*

491. **(A)** The patient with amyotrophic lateral sclerosis is chronically weak and has muscle wasting. Succinylcholine should be avoided. Muscle relaxants, if required, should be used in low doses. Respiratory depressants are avoided. If the patient has any respiratory compromise postoperatively, artificial ventilation is in order. Although these patients may exhibit exaggerated changes in hemodynamics in response to inhalational anesthetics, there is no specific recommendation to avoid these agents. *(1:434, 503; 5:534; 6:257)*

492. **(B)** Autonomic hyperreflexia occurs with spinal lesions above T5. Any noxious stimulation, e.g., urinary catheter insertion, may lead to hypertension accompanied by sweating and bradycardia. The anesthesiologist must be prepared to treat the hypertension immediately. *(1:1109–10; 2:509; 4:258; 5:1044; 6:280)*

493. **(B)** The patient with myotonic dystrophy presents an anesthetic problem with regard to muscle relaxation. While succinylcholine may cause myotonia, myotonic episodes are not necessarily reversed by nondepolarizing muscle relaxants. Furthermore, neostigmine may aggravate myotonia. Postoperative shivering in the recovery room may induce myotonia. At the end of the procedure, mechanical ventilation is preferable to reversal of muscle relaxants. *(1:433, 493–4; 2:210; 5:536–7; 6:519–21)*

494. **(D)** Arteriovenous malformations are high-flow, low-pressure lesions, and hypertension-induced hemorrhage is rare, if not nonexistent. Vasospasm is also rare. Intraoperative hemorrhage can be severe if rupture occurs. In addition, reperfusion breakthrough (development of severe brain swelling as a result of the sudden alteration in blood flow and perfusion pressure in the brain parenchyma surrounding the malformation) during resection can result in severe intraoperative problems. Treatment includes induced hypotension and high-dose barbiturates. *(1:772–3; 4:253–4, 1646; 5:2151–2)*

495. **(E)** After subarachnoid hemorrhage, rhythm disturbances and ischemic ECG changes are common. Spasm of the cerebral arteries can result in decreased CBF and stroke. Hyponatremia can occur from SIADH (syndrome of inappropriate secretion of ADH). *(1:769–72; 4:250, 1641–2; 5:2146–51)*

496. **(E)** Cushing's triad of intracranial hypertension, arterial hypertension, and reflex bradycardia is often the terminal stage of brain ischemia. During development of cerebrovasomotor paralysis, increases in blood pressure increase cerebral blood volume, enhancing cerebral edema and further increasing ICP. Ultimately, cerebral perfusion pressure decreases. *(1:770; 4:245; 5:739)*

CHAPTER 6

Renal, Hepatic, Endocrine, Hematologic, and Metabolic Systems

Questions

DIRECTIONS (Questions 497 through 533): Each of the numbered items or incomplete statements in this section is followed by answers or by completions of the statement. Select the ONE lettered answer or completion that is BEST in each case.

497. Total body water in an adult constitutes

(A) 10% of body weight
(B) 20% of body weight
(C) 30% of body weight
(D) 40% of body weight
(E) 60% of body weight

498. The principal intracellular ion is

(A) Na^+
(B) K^+
(C) Cl^-
(D) HCO_3^-
(E) Mg^{2+}

499. Magnesium

(A) concentration is related to irritability of nervous tissue
(B) concentration is low intracellularly
(C) is not excreted in the urine
(D) is not present in bile
(E) is not present in gastric juice

500. The circulation to the kidney

(A) is autoregulated over a mean arterial pressure range of about 80 to 160 mmHg
(B) is not regulated by neural factors
(C) is innervated by sympathetic nerves originating in T2-T3
(D) is not affected by epinephrine
(E) is constricted by prostaglandin E_2

501. The glomerulus contains

(A) an artery, a vein, and a tubule
(B) only a tuft of capillaries
(C) a nephron
(D) an ascending limb and a descending limb
(E) the macula densa

502. The countercurrent mechanism of the kidney involves

(A) the nephron and the glomerulus
(B) the arteries and veins
(C) a collecting duct and a distal tubule
(D) the loops of Henle and the vasa recta
(E) the proximal tubule and the distal tubule

503. Liver biotransformation reactions involve all of the following EXCEPT

(A) oxidation
(B) auto-oxidation
(C) reduction
(D) conjugation
(E) hydrolysis

504. The enterohepatic circulation

(A) describes the storage of bile in the gallbladder
(B) is most important for the reabsorption of bile salts
(C) is unimportant in the liver metabolism of drugs
(D) allows all the reabsorbed substances to avoid the systemic circulation
(E) promotes excretion of drug conjugates via the gastrointestinal tract

505. Jaundice may result from all of the following EXCEPT

(A) excessive production of bilirubin
(B) increased uptake of bilirubin into hepatic cells
(C) intrahepatic obstruction of ducts
(D) defects in bilirubin conjugation
(E) Gilbert's disease

506. The liver receives its blood supply from

(A) the hepatic artery only
(B) the portal vein only
(C) both the hepatic artery and the portal vein
(D) vessels that run in the center of the lobules
(E) the superior mesenteric artery

507. Hepatic blood flow

(A) is closely regulated during surgery and anesthesia
(B) increases with arterial hypoxemia
(C) is decreased with sympathetic stimulation
(D) is closely regulated by dopamine
(E) responds slowly to bodily needs

508. The liver affects glucose metabolism by all of the following mechanisms EXCEPT

(A) glycogen storage
(B) gluconeogenesis
(C) glycogenolysis
(D) insulin production
(E) conversion of galactose to glucose

509. The shock wave during extracorporeal shock wave lithotripsy is timed to coincide with a particular point in the electrocardiogram (ECG), and occurs

(A) at the peak of the P wave.
(B) 100 msec after the peak of the P wave
(C) 20 msec after the peak of the R wave
(D) 200 msec after the peak of the R wave
(E) at the peak of the T wave

510. Unconjugated bilirubin

(A) is nontoxic
(B) is secreted into the intestinal tract
(C) is the product of white cell breakdown
(D) is conjugated with glucuronic acid
(E) breaks down to biliverdin

511. Ascites

(A) follows chronic decreased portal vein pressure
(B) follows periods of hyperalbuminemia
(C) is usually accompanied by hypernatremia
(D) may have an adverse cardiopulmonary effect
(E) should be removed rapidly to avoid reaccumulation

512. If a patient has prerenal failure, the urine will

(A) be dilute
(B) be concentrated

(C) have a specific gravity of approximately 1.010
(D) be excreted in large amounts
(E) have a reddish tinge due to presence of red blood cells

513. A patient with myxedema is admitted for emergency abdominal surgery. A finding consistent with the myxedematous state is

(A) fine, soft hair
(B) moist skin
(C) bradycardia
(D) heat intolerance
(E) pitting edema of the eyelids

514. Fluid and blood replacement in the patient with myxedema

(A) should be guided by blood pressure
(B) should be guided by electrocardiographic voltage
(C) does not differ from that in normal patients
(D) should be guided by invasive arterial and central venous pressure monitoring
(E) should be accompanied by rapid restoration of the euthyroid state

515. A patient with a recent onset (3 weeks) of clinical hyperthyroidism is admitted for repair of a tendon laceration and has not taken propylthiouracil for 4 days. At this time the

(A) patient should be essentially euthyroid
(B) patient should be considered as a high-risk hyperthyroid patient
(C) therapy with propylthiouracil may have rendered the patient hypothyroid
(D) patient would probably have bradycardia
(E) patient should have general anesthesia if at all possible

516. A patient should be considered at risk of pituitary-adrenal axis suppression after daily doses of glucocorticoids for

(A) 2 days
(B) 1 week
(C) 1 month
(D) 2 months
(E) 6 months

517. The patient with a diagnosis of pheochromocytoma

(A) requires immediate surgery
(B) should be treated for 10 to 14 days with an α-adrenergic antagonist
(C) can be anesthetized regardless of the level of blood pressure readings
(D) is usually hypervolemic
(E) should have a Swan-Ganz catheter in place preoperatively

518. The most important goal in the treatment of the diabetic patient undergoing anesthesia is to

(A) keep blood sugar in the normal range
(B) prevent glycosuria
(C) prevent hypoglycemia
(D) prevent ketoacidosis
(E) prevent acetonuria

519. Hypoglycemia in the awake patient

(A) is identical to that in the anesthetized patient
(B) is characterized by bradycardia
(C) is characterized by hypertension
(D) is due to ketoacidosis
(E) is characterized by a marked parasympathetic response

520. A male diabetic patient is anesthetized for the emergent drainage of a large abscess. His usual dose of NPH insulin had been given in the morning, and the procedure lasts only 15 minutes. After surgery his insulin requirement is expected to

(A) decrease
(B) be unaffected
(C) increase greatly; therefore, he should have frequent injections empirically
(D) remain high for a week
(E) increase

521. Hyperosmolar coma in a diabetic

(A) usually occurs in young people
(B) occurs at osmolar levels of 150 to 175 mOsm/L
(C) occurs in the absence of ketonemia
(D) is usually accompanied by oliguria
(E) requires treatment with large doses of insulin

522. In a patient who receives 4 units of warmed blood in 1 hour

(A) electrocardiographic changes are expected
(B) serum calcium will fall to very low levels
(C) signs of tetany will be seen even with normal calcium values
(D) calcium chloride, 10 g/h, should be administered
(E) no treatment is necessary

523. A patient is to be operated on for a tumor of the small bowel. In the preoperative interview, a history of flushing, diarrhea, and joint pain is elicited. There are also symptoms compatible with congestive heart failure. A likely diagnosis is

(A) Zollinger-Ellison syndrome
(B) carcinoid syndrome
(C) pheochromocytoma
(D) Peutz-Jeghers syndrome
(E) adrenal tumor with metastasis

524. A patient with Sipple's syndrome (multiple endocrine neoplasia type IIa) is scheduled for a thyroidectomy. Soon after induction, hypertension of 210/130 is recorded. A likely cause for this is

(A) light anesthesia
(B) pheochromocytoma
(C) inadvertent injection of a pressor agent
(D) hypercarbia
(E) allergic response to an anesthetic agent

525. In the patient with cirrhosis

(A) the serum albumin level will be elevated
(B) excessive sodium is lost in the urine
(C) pancuronium is more effective
(D) serum gamma globulin level will be low
(E) less thiopental is required for induction

526. The optimal anesthetic regimen for a patient undergoing liver transplantation

(A) will avoid fentanyl
(B) will avoid nondepolarizing relaxants
(C) is a balanced technique
(D) will avoid halogenated hydrocarbons
(E) will depend on the cause of the liver failure and the patient's status

527. The patient with acute viral hepatitis

(A) is not affected by surgical procedures
(B) is an acceptable candidate for general anesthesia for elective surgery if the degree of liver enzyme elevation is mild
(C) is at high risk for perioperative mortality
(D) should have an inhalational induction to avoid thiopental
(E) should never have a general anesthetic

528. A patient who has had a transsphenoidal hypophysectomy for acromegaly several years ago presents for cholecystectomy. Which of the following hormones should be given in the perioperative period?

(A) adrenocorticotropic hormone (ACTH)
(B) thyroid-stimulating hormone (TSH)
(C) vasopressin
(D) cortisol
(E) insulin

529. Which of the following statements is TRUE regarding the effects of medications on blood sugar?

(A) Propranolol potentiates the hyperglycemic response to stress.
(B) Captopril increases insulin requirements in the diabetic.
(C) Hydrochlorothiazide potentiates the hypoglycemia produced by glyburide.

(D) Atenolol decreases insulin release from the pancreas.
(E) Prednisone increases the blood glucose concentration.

530. Cryoprecipitate contains all of the following clotting factors EXCEPT

(A) factor VIII
(B) factor IX
(C) factor XIII
(D) von Willebrand factor
(E) fibrinogen

531. The treatment of a hemolytic transfusion reaction may involve the immediate administration of all of the following EXCEPT

(A) crystalloid intravenous fluids
(B) furosemide
(C) hydrocortisone
(D) sodium bicarbonate
(E) mannitol

532. Which of the following statements is FALSE regarding disseminated intravascular coagulation (DIC)?

(A) DIC is usually due to the abnormal consumption of clotting factors.
(B) DIC may be treated with heparin.
(C) Gram-negative endotoxemia is a common cause of DIC.
(D) Abnormalities in laboratory tests include a prolonged prothrombin time and decreased values for platelet count and plasma fibrinogen.
(E) Regardless of the etiology, therapy of DIC is directed toward replacement of clotting factors and inhibition of the clotting cascade.

533. A patient has had a subtotal parathyroidectomy for parathyroid hyperplasia consisting of the removal of $3^1/_2$ of the four glands. If an excessive amount of parathyroid tissue was removed, significant hypocalcemia might first occur how long after the removal of the last gland?

(A) 1 hour
(B) 6 hours
(C) 18 hours
(D) 60 hours
(E) 96 hours

DIRECTIONS (Questions 534 through 581): For each of the items in this section, ONE or MORE of the numbered options is correct. Choose answer

(A) if only 1, 2, and 3 are correct
(B) if only 1 and 3 are correct
(C) if only 2 and 4 are correct
(D) if only 4 is correct
(E) if all are correct

534. Albumin

(1) levels are lower in the neonate
(2) is the major plasma protein
(3) is necessary for maintenance of oncotic pressure
(4) has a half-life of approximately 3 weeks

535. Facts affecting interpretation of liver function tests include

(1) albumin is a good indicator of hepatocyte function but has a long half-life
(2) coagulation factors have a short half-life, therefore reflect recent changes
(3) dye-removal tests are not specific
(4) bilirubin tests are not specific for liver function

536. Hematocrit values in patients with cirrhosis may reflect

(1) coagulation disorders
(2) increased plasma volume
(3) bleeding
(4) megaloblastic anemia

537. Perioperative management of the patient with carcinoid syndrome should include

(1) infusion of epinephrine to treat bronchospasm
(2) the use of vasopressin to treat severe intraoperative hypotension
(3) the preferential use of muscle relaxants in the benzylisoquinoline class as opposed to those in the steroid class
(4) administration of octreotide preoperatively

538. Antidiuretic hormone (ADH)

(1) release is under control of osmoreceptors in the hypothalamus
(2) release is from the posterior pituitary
(3) release is inhibited by increased stretch of the atrial baroreceptors
(4) acts on the proximal convoluted tubule

539. The patient with long-standing obstructive sleep apnea

(1) benefits from a short-acting benzodiazepine given at bedtime
(2) should be kept intubated and mechanically ventilated for the night following surgery
(3) usually requires large doses of opioids to treat postoperative pain
(4) is at high risk for developing postoperative hypoxemia

540. The perioperative management of the patient with gout should include the avoidance of

(1) cyclooxygenase (COX) inhibitors
(2) succinylcholine
(3) β-adrenergic antagonists
(4) hypovolemia

541. Five percent dextrose solutions should not be used in patients

(1) for transurethral resection (TUR) of the prostate
(2) for appendectomy
(3) for intracranial procedures
(4) with diabetes

542. Persons with which of the following abnormalities are considered to be hypercoagulable?

(1) Factor V Leiden
(2) Protein S deficiency
(3) Protein C deficiency
(4) Antithrombin III deficiency

543. Treatment of hyperkalemia includes

(1) elimination of exogenous sources
(2) correction of cause of endogenous sources
(3) administration of glucose with insulin
(4) administration of acidifying solutions

544. A young adult with Duchenne's muscular dystrophy who is to undergo general anesthesia

(1) often has delayed gastric emptying
(2) has an increased risk of malignant hyperthermia
(3) can safely receive nondepolarizing muscle relaxants
(4) is usually a difficult intubation

545. Relatively rare diseases that are much more common in patients with the acquired immunodeficiency syndrome include

(1) pneumonia due to *Pneumocystis carinii*
(2) Kaposi sarcoma
(3) infection due to *Mycobacterium avium* complex
(4) B-cell lymphoma

546. Effect(s) of vasopressin include(s)

(1) production of dilute urine
(2) decreased urine flow
(3) decreased urine osmolality
(4) vasoconstriction

547. The syndrome of inappropriate secretion of antidiuretic hormone (SIADH) includes

(1) hyponatremia
(2) low serum osmolality
(3) excessive renal secretion of sodium
(4) normal renal function

548. Magnesium

(1) is an intracellular cation
(2) is mobilized from stores in bone
(3) is needed for protein synthesis
(4) is bound to plasma proteins

549. Parenteral nutrition

(1) with lipids can be administered by peripheral vein
(2) has a low incidence of phlebitis if the dextrose concentration is kept below 5%
(3) uses lipids as an important source of calories
(4) uses protein as a calorie source

550. The blood supply to the liver is by two vessels, the hepatic artery and the hepatic portal vein. These vessels differ in that

(1) 60% of the blood supply comes from the hepatic artery
(2) the portal vein provides 50% of the oxygen supply
(3) the portal vein blood is more fully saturated than the hepatic artery
(4) the portal vein supplies the bulk of the nutrients to the liver

551. Clotting factor(s) that is (are) vitamin K dependent is (are)

(1) factor I
(2) factor VII
(3) factor V
(4) factor X

552. Autoregulation of the hepatic blood flow

(1) involves the hepatic artery
(2) involves the portal vein
(3) is via the sympathetic nervous system
(4) is via the parasympathetic nervous system

553. The perioperative management of the patient with sickle cell anemia should include the avoidance of

(1) hypoxemia
(2) hypothermia
(3) hypovolemia
(4) a low hematocrit

554. A cirrhotic patient

(1) cannot produce normal amounts of protein as a result of hepatocyte dysfunction
(2) will always have jaundice, since bilirubin will accumulate
(3) may not metabolize succinylcholine normally
(4) will not be able to metabolize lidocaine as a result of a lack of pseudocholinesterase

555. The liver disease that results from alcohol ingestion

(1) can be prevented by a nutritious diet
(2) is usually present without hematologic involvement
(3) does not involve any gastrointestinal disorders
(4) is usually accompanied by vitamin deficiency

556. The teenage girl with anorexia nervosa typically has

(1) osteoporosis
(2) increased risk of intraoperative dysrhythmias
(3) delayed gastric emptying
(4) hypertension due to increased sympathetic activity

557. The mechanism for active transfer of sodium out of the cell and potassium into the cell

(1) is referred to as the sodium-potassium pump
(2) requires no energy, being a passive movement
(3) is located in the cell membrane
(4) functions independently of the sodium concentration

558. The pH

(1) of a solution is the logarithm to the base 10 of the reciprocal of the hydrogen ion concentration
(2) of a solution is the negative logarithm of the hydrogen ion concentration
(3) of water in which hydrogen ion and hydroxyl ion are in equal concentration is 7.0
(4) will decrease 1 unit for each 10-fold decrease in hydrogen ion concentration

559. When evaluating renal function, one must consider that

(1) proteinuria is always abnormal
(2) a specific gravity of 1.023 or greater demonstrates good concentrating function
(3) blood urea nitrogen (BUN) elevation is indicative of renal dysfunction
(4) more creatinine is produced by muscular persons

560. Extracellular fluid loss may occur with

(1) small bowel obstruction
(2) rapidly developing ascites
(3) burns
(4) fever and hyperventilation

561. Mechanisms by which blood glucose can increase during surgery include

(1) intravenous administration
(2) secretion from the liver secondary to a decreased insulin secretion
(3) secretion from the liver secondary to an increased catecholamine output
(4) uptake of glucose into tissue

562. Insulin should be administered to a diabetic patient during surgery

(1) by adding it to the container of intravenous fluids
(2) only when monitoring urine and blood for sugar
(3) in the usual daily dosage
(4) only if glucose is being infused

563. The juvenile-onset diabetic is noted for

(1) heavy habitus
(2) good response to oral hypoglycemics
(3) blood sugar in the range of 50 to 100 mg/dL
(4) very low levels of circulating insulin

564. Hormones secreted by the adrenal cortex include

(1) mineralocorticoids
(2) glucocorticoids
(3) sex hormones
(4) norepinephrine

565. Glucocorticoids

(1) are produced in the adrenal cortex under influence of ACTH
(2) are active in carbohydrate metabolism
(3) are active in protein metabolism
(4) include aldosterone

566. In diabetes insipidus

(1) the serum sodium is high
(2) the osmolality of the serum is high
(3) the urine is dilute
(4) thirst need not be present

567. Clinical manifestations of hyperthyroidism include

(1) sweating
(2) vasoconstriction

(3) tachycardia
(4) shivering

568. The patient with hyperaldosteronism usually has

(1) excess secretion of hormone from the adrenal medulla
(2) hypertension
(3) acidosis
(4) hypokalemia

569. von Willebrand's disease

(1) is a congenital bleeding disorder
(2) usually affects only females
(3) is usually associated with a normal platelet count
(4) is caused by factor VIII deficiency

570. Anesthetic agents acceptable for renal transplantation surgery include

(1) isoflurane
(2) propofol
(3) thiopental
(4) metocurine

571. The patient having a thyroidectomy for hyperthyroidism

(1) has an increased minimum alveolar concentration (MAC)
(2) is given iodide for 1 to 2 weeks prior to surgery
(3) should be heavily sedated in the preoperative period
(4) may require propranolol intravenously

572. Atropine is usually omitted from premedication in the patient with hyperthyroidism because it

(1) interferes with interpretation of pulse rate
(2) interferes with pupillary constriction
(3) interferes with sweating mechanisms
(4) slows gastric emptying

573. Laboratory abnormalities common in patients with end-stage renal disease include

(1) anemia
(2) thrombocytopenia
(3) hyperkalemia
(4) hypomagnesemia

574. Important maneuver(s) in the perioperative period in a patient with Gilbert's syndrome include(s)

(1) avoidance of fasting
(2) preoperative transfusion of fresh frozen plasma
(3) avoidance of stress
(4) recognition of the etiology of the laboratory abnormality

575. Hereditary causes for cirrhosis include

(1) hemochromatosis
(2) α_1-antitrypsin deficiency
(3) Wilson's disease
(4) antithrombin III deficiency

576. The management of ascites secondary to cirrhosis may include

(1) administration of spironolactone
(2) administration of indomethacin
(3) sodium restriction
(4) administration of captopril

577. The renal failure accompanying the hepatorenal syndrome

(1) is associated with kidneys that are normal on biopsy
(2) may be improved by the administration of dopamine by infusion
(3) is oliguric
(4) may be readily diagnosed by the abnormalities in urinary sediment analysis

578. The sickling of red blood cells in a patient with sickle cell anemia

(1) is a reversible process
(2) occurs when deoxygenated hemoglobin molecules precipitate
(3) impairs the clotting cascade
(4) may cause infarction in tissues with a high oxygen extraction ratio

579. The patient with hemophilia B

(1) should receive vitamin K preoperatively
(2) is male
(3) has a decreased plasma concentration of factor VIII
(4) has a prolongation in the partial thromboplastin time

580. In the preoperative pharmacologic management of pheochromocytoma

(1) α-adrenergic blockade is instituted before β-adrenergic blockade
(2) diuretic antihypertensives are contraindicated
(3) 1 to 2 weeks of therapy are recommended
(4) dosage is adjusted according to the levels of urinary catecholamine metabolites

581. True statement(s) involving hemolytic transfusion reactions include

(1) ABO incompatibility is the most likely cause
(2) repeating the serologic testing of the blood using additional antibodies usually identifies incompatibility with a rare antigen
(3) under general anesthesia, the usual presenting sign is hemoglobinuria
(4) the severity of the hemolytic reaction is unrelated to the volume of transfused blood

Answers and Explanations

497. **(E)** The total body water in the adult constitutes 60% of the body weight. *(1:170; 4:943; 5:1763; 6:374)*

498. **(B)** The principal intracellular ion is potassium. The second most important is magnesium. Sodium is the most important extracellular ion. *(1:185; 4:946; 5:1768; 6:375)*

499. **(A)** Magnesium concentration is related to the irritability of nervous tissue. There is a high intracellular magnesium concentration. It is excreted in the urine even in the face of a low concentration in the blood, and is present in the bile and the gastric juice. *(1:194; 4:972–4; 5:1772; 6:387–8)*

500. **(A)** The renal circulation is autoregulated over a wide range of blood pressures. There are many neural factors which control the vessel diameter and the renal blood flow, including the renin-angiotensin system. Innervation comes from T4 to T12, the vagus, and the splanchnic nerves. Epinephrine also has an influence. Prostaglandin E_2 causes vasodilatation. *(1:1007; 4:330; 5:2178)*

501. **(B)** The glomerulus is a tuft of capillaries. The veins, Bowman's capsule, and the tubules are important structures but are not part of the glomerulus. *(1:1005; 5:777)*

502. **(D)** The countercurrent mechanism involves the loops of Henle and the vasa recta. By this mechanism, the kidney is able to maintain concentration gradients. All of the renal structures, the nephron, the glomerulus, and the tubules, are involved in the process of excretion, but the parallel loops and the vasa recta are components of the countercurrent mechanism. *(1:1008; 5:784, 2178)*

503. **(B)** Biotransformation of drugs in the liver includes all of the mechanisms listed except auto-oxidation. *(1:1073, 1316; 3:14; 5:70)*

504. **(B)** The enterohepatic circulation is the mechanism whereby some substances are reabsorbed from the gut into the portal blood and then excreted from the liver back into the gut. The mechanism is most important for bile salts. It is of importance for drugs that undergo this process. The mechanism can function whether or not the substance is conjugated. *(1:1071–2; 5:748)*

505. **(B)** Jaundice can result from all of the options except increased uptake of bilirubin into hepatic cells. The uptake is necessary to prevent jaundice. *(1:1074–5; 4:339–40, 344–5; 5 2219–20; 6:316–7)*

506. **(C)** The liver receives blood from the hepatic artery and the hepatic portal vein. The hepatic artery is a branch of the aorta. The vessels, except for the central vein, run in the interlobular spaces. *(1:1067; 4:1905; 5:743–4; 6:313–4, 319)*

507. **(C)** Hepatic blood flow is decreased with sympathetic stimulation. Hepatic blood flow is difficult to regulate under anesthesia, and surgical retraction and packing can decrease this value. Hypoxemia leads to increased vascular resistance and thus also decreases hepatic blood flow. Hepatic blood flow responds quickly to body needs, and dopamine does not appear to

play a role in its regulation. *(1:1071; 4:1905; 5:745–7; 6:313–4)*

508. **(D)** Insulin production occurs in the pancreas and not in the liver. Glycogen storage, gluconeogenesis, glycogenolysis, and the conversion of galactose to glucose all take place in the liver and are important to glucose metabolism. *(1:1072; 4:1908–12; 5:747–8, 2243; 6:311)*

509. **(C)** The shock wave is timed to occur 20 milliseconds after the peak of the R wave which corresponds to a point during the absolute refractory period. Shock-wave-induced dysrhythmias are therefore minimized. *(1:1022; 4:1969; 5:2196; 6:364)*

510. **(D)** Unconjugated bilirubin is conjugated with glucuronic acid to form a water-soluble compound that can be excreted. Unconjugated bilirubin is toxic in high concentrations and can lead to kernicterus. It is not secreted into the intestinal tract. Bilirubin results from the breakdown of hemoglobin from red cells, not white cells. Biliverdin also results from the breakdown of hemoglobin and is converted to bilirubin. *(1:1071; 4:336; 5:2215; 6:305, 316–7)*

511. **(D)** Ascites may be associated with severe cardiorespiratory symptoms. Ascites follows chronic increased portal vein pressure. It is usually accompanied by hyponatremia. Ascitic fluid should be removed slowly to avoid hypotension. *(1:1085; 4:352; 5:1791–2, 2223; 6:310)*

512. **(B)** When trying to diagnose the cause of renal failure, one should obtain a urine specimen before treatment is begun. The urine will be concentrated in the patient with prerenal failure and will have a higher specific gravity. The urine need not be blood-tinged. Oliguria is present. *(1:179, 1472–3; 4:2155; 5:789; 6:357–8)*

513. **(C)** Patients with myxedema have coarse hair and dry skin. Cold intolerance and bradycardia are present. The edema is nonpitting. *(1:1122; 4:306–7; 5:1047; 6:417–8)*

514. **(D)** These patients should be well monitored. Too rapid replacement of thyroid hormone may cause myocardial ischemia. The ECG and blood pressure are not sufficient to monitor fluid status. *(1:1123; 4:306–8; 6:419–20)*

515. **(B)** The mean time for onset of action for propylthiouracil is 8 days, and it takes about 6 weeks before the euthyroid state is reached. This patient would probably still have tachycardia. If a regional anesthetic is feasible, it might be safer for the inadequately treated hyperthyroid patient. *(1:1121; 4:306; 5:1046; 6:413, 415, 417)*

516. **(B)** There are many estimates of time to development of adrenal suppression, but the definite time is not known. In general, additional glucocorticoid therapy is recommended on the day of surgery in any patient who has been on glucocorticoids for a week for the past 12 months. *(1:1129; 4:311; 5:1040)*

517. **(B)** The patient with pheochromocytoma should be well prepared preoperatively. This may take 2 weeks of α-adrenergic blockade. These patients are usually hypovolemic. While the level of monitoring varies with the amount of concurrent illness in the patient, a pulmonary artery catheter is not mandatory in all cases. *(1:1131; 4:313–4; 5:1043; 6:431–3)*

518. **(C)** The most important goal is to prevent hypoglycemia. One can monitor glucose levels and try to keep them in the normal range, but this should not be done at the risk of hypoglycemia. Many regimens can be used for perioperative glucose control. *(1:1135; 4:316; 5:1024, 1781; 6:407–8)*

519. **(C)** Hypoglycemia may be manifest as sympathetic discharge, including hypertension and tachycardia. The skin is cold and clammy. Ketoacidosis is more likely to occur with insulin deficiency. *(1:1136; 5:1026; 6:403–4)*

520. **(E)** After surgery, the insulin requirement will be increased for a short time. For the first 72 hours after surgery, there should be close monitoring of the sugar. *(1:1134; 4:316–7; 5:1778–9)*

521. (C) Hyperosmolar coma may occur in the absence of ketosis. The osmolarity is over 350 mOsm (normal is about 285 mOsm). The condition is seen more often in the elderly and is accompanied by polyuria. Treatment is with rehydration and small doses of insulin. *(1:1135–6; 6:410)*

522. (E) If blood is warmed as it is given, there should be no ill effects, and no treatment is necessary. The serum calcium will not fall if blood is given slowly, and there will be no signs of tetany. *(1:206; 4:2422; 5:1812)*

523. (B) The symptoms are compatible with carcinoid syndrome. The Zollinger-Ellison syndrome is associated with hypersecretion of gastric acid. Pheochromocytoma is associated with hypertension and a catecholamine-secreting tumor. Peutz-Jeghers syndrome is associated with multiple polyps of the gastrointestinal tract. *(1:1046; 4:389; 5:1108–9; 6:332–3)*

524. (B) The patient with Sipple's syndrome who develops hypertension must be suspected of having a pheochromocytoma. Light anesthesia and hypercarbia should be quickly ruled out. Allergic reactions are not usually associated with hypertension. The inadvertent injection of a pressor agent may occur but should be easy to rule out. *(1:1130–1; 4:1894; 5:1042; 6:430–1)*

525. (E) Decreased plasma albumin levels decrease the bound fraction of thiopental and result in a greater fraction of free thiopental. Serum gamma globulin is higher in cirrhosis, and pancuronium has a larger volume of distribution; therefore, it is less effective for a given dose. Patients with cirrhosis excrete sodium-poor or sodium-free urine. *(1:332, 1094; 4:347–8,352–3, 1914–5; 5:2184, 2217–9; 6:312)*

526. (E) There is no ideal anesthetic technique for liver transplantation, although no class of anesthetic drugs is contraindicated. The effects of the agents should be titrated against the patient responses. *(1:1355–7; 4:1937–9; 5:2249; 6:319)*

527. (C) The patient with acute viral hepatitis should not be a surgical candidate unless the benefits clearly outweigh the risks. A perioperative mortality rate of approximately 10% has been reported. *(1:1067, 1083–4; 4:346; 5:2216; 6:318)*

528. (D) A patient who has had a transsphenoidal hypophysectomy for acromegaly is likely to be lacking all of the hormones of the anterior pituitary (growth hormone [GH], ACTH, TSH, follicle-stimulating hormone [FSH], luteinizing hormone [LH], prolactin). The patient probably has been maintained on a glucocorticoid and thyroxine, both of which should be given in the perioperative period. The posterior pituitary is usually not removed during a transsphenoidal hypophysectomy; therefore, the patient is unlikely to require vasopressin therapy. *(1:1137; 4:315, 1647–8; 5:1051; 6:436–7)*

529. (E) Glucocorticoids, such as prednisone, have a hyperglycemic effect. Nonspecific β-adrenergic antagonists, such as propranolol, antagonize the hyperglycemic response to stress. Hydrochlorothiazide also causes hyperglycemia, and antagonizes the hypoglycemic effect of glyburide. Insulin release from the pancreas is stimulated by β_2-adrenergic agonists, and would not be affected by atenolol, which is specific for the β_1-adrenergic receptor. Inhibitors of angiotensin-converting enzyme, such as captopril, have no effect on glucose homeostasis. *(1:1133; 3:120, 252–3, 256, 776, 820–1, 1659; 5:2262)*

530. (B) Factor IX is not present in cryoprecipitate. Treatment of factor IX deficiency (hemophilia B) involves transfusion of prothrombin complex which contains factors II, VII, IX, and X. *(1:231; 4:2417; 5:1824; 6:348, 493)*

531. (C) In a hemolytic transfusion reaction, the immediate goals include maintenance of intravascular volume and increasing urine flow with diuretics to prevent acute tubular necrosis from hemoglobinemia. Alkalinizing the urine may decrease the likelihood of hemoglobin precipitation in the renal tubules. There is no indication for the immediate use of cortisol. *(1:207; 4:2418–9; 5:1816–7)*

532. **(E)** DIC has many causes, and the most effective therapy is to treat the underlying cause, if possible. Anticoagulants (e.g., heparin, although controversial) and inhibitors of fibrinolysis (e.g., ε-aminocaproic acid) may be helpful. *(1:230–1; 4:936–7; 5:2867–8; 6:496–7)*

533. **(D)** After parathyroidectomy, the serum calcium concentration falls for several days, reaching a nadir on the second or third postoperative day. The acute onset of hypocalcemia (i.e., within a few hours) is very unlikely. *(1:1125; 4:1955; 6:422).*

534. **(E)** All of the options are true. Albumin levels are lower in the neonate than in the adult. Oncotic pressure is largely determined by the concentration of albumin, which is the primary plasma protein. Since albumin has a long half-life, it is not a good test for acute changes in liver function. If the albumin level is low, the disease has been present for some time. *(1:1072; 4:335–6,1912–3; 5:747, 753–6)*

535. **(E)** All of the options are true. Albumin is a good test, but it has a long half-life, and recent changes in liver function may not be reflected. Coagulation factors, on the other hand, have a shorter half-life and will reflect recent changes. Dye-removal tests are not specific because abnormalities may reflect changes in blood flow, uptake, binding, and excretion. *(1:1073–5; 4:340–2; 5:753–6)*

536. **(E)** A decreased hematocrit in the patient with cirrhosis may reflect many factors: a coagulation defect leading to bleeding, increased plasma volume leading to dilution of red cells, megaloblastic anemia, and increased hemolysis. Many of these factors are interrelated. *(1:1089; 4:349; 5:760–3; 6:309–11)*

537. **(C)** Measures should be taken to decrease the likelihood of a carcinoid crisis. These include giving octreotide to prevent or decrease the release of factors by the carcinoid tumor, and avoiding medications that increase release of factors such as catecholamines and benzylisoquinoline muscle relaxants such as mivacurium and cisatracurium. Vasopressin is the preferred drug for treating hypotension. *(1:1046–7; 5:1108–10; 6:332–4)*

538. **(A)** ADH release is in response to decreased plasma osmolarity detected by osmoreceptors in the hypothalamus. It is also released in response to decreased tension in the stretch receptors in the atrium. ADH is secreted from the posterior pituitary. The principal effect is on the collecting tubules. *(1:1137; 4:315,960; 5:1487)*

539. **(D)** The patient with obstructive sleep apnea is at high risk for developing postoperative hypoxemia and for this reason usually warrants a longer period of postoperative observation. This is not, however, an indication for routine postoperative intubation and ventilation. Apneic episodes are more likely in the patient who uses nighttime sedatives or alcohol. Patients with obstructive sleep apnea are also at increased risk for opioid-induced hypoventilation or apnea. *(1:1038, 1385–6; 5:1030–1; 6:444–5)*

540. **(D)** Hypovolemia may decrease renal excretion of uric acid and increase the risk of an acute attack. Although aspirin increases uric acid levels, other COX inhibitors, especially indomethacin, are effective in preventing or treating an attack. There is no reason to avoid succinylcholine or β-adrenergic antagonists. *(6:460)*

541. **(B)** Five percent dextrose solutions should not be used in the patient undergoing TUR or intracranial procedures. The patient for TUR may absorb hypotonic fluids, and the intravenous administration of 5% dextrose will only make the symptoms worse. In the patient for neurosurgery, cerebral edema may be aggravated by the administration of dextrose solutions. In addition, hypoxic damage is worsened by hyperglycemia. Although 5% dextrose solution may not be the fluid of first choice in the patient undergoing appendectomy, it would probably not cause harm, either. Diabetic patients are usually managed with both insulin and 5% dextrose in the perioperative period. *(1:763, 1018–20; 4:957–8, 1972–3; 5:2191–2, 2915)*

542. **(E)** Protein C, protein S, and antithrombin III are endogenous anticoagulants. Their deficiencies make patients hypercoagulable, as does the presence of the abnormal form of factor V designated as factor V Leiden. *(1:217, 226; 5:1839, 2868; 6:495)*

543. **(A)** Hyperkalemia may be treated by the administration of insulin and glucose solutions, eliminating exogenous sources (one of the major causes), and correcting causes of endogenous sources. Treatment includes alkalinizing the blood, not acidifying it. *(1:188–9; 4:976–7; 5:1770; 6:383)*

544. **(A)** Duchenne's muscular dystrophy is often accompanied by delayed gastric emptying and susceptibility to malignant hyperthermia. Nondepolarizing muscle relaxants may be used if needed. There is no definite association with airway difficulties. *(1:491–3; 5:535–6; 6:517–8)*

545. **(E)** The two listed infections and the two listed malignancies are all much more common in patients with AIDS. *(6:567)*

546. **(C)** Vasopressin causes a concentrated urine and antidiuresis. In addition, the urine has a high osmolality. Vasoconstriction also occurs. *(1:1137; 4:315–6; 5:1052–3; 6:373)*

547. **(E)** Inappropriate secretion of ADH is often seen after surgery. The syndrome includes hyponatremia and low serum osmolality. The kidneys excrete an excessive amount of sodium under the influence of ADH. Underlying renal function is normal. *(1:181–3; 4:315–6; 5:1052–3; 6:374–6)*

548. **(E)** Magnesium is an important intracellular ion. It may be mobilized from large stores in bone when needed for enzyme and protein synthesis. It is bound to plasma proteins. *(1:193–4; 4:972–3; 5:1772–4; 6:387–8)*

549. **(E)** Parenteral nutrition is very important in the pre and postoperative care of patients, however, there may be problems associated with its use. It can be administered through a peripheral vein, and the problem with phlebitis is less if the glucose concentration is limited. Both lipids and proteins are used as sources of calories. *(1:1478; 5:1034–5, 2908; 6:453)*

550. **(C)** The liver has a dual blood supply. Only 25% of the blood is supplied by the hepatic artery. The oxygen supply is evenly divided by the two vessels, even though the portal vein blood is more unsaturated. Most of the nutrients come from the portal vein. *(1:1067; 4:1905–7; 5:743)*

551. **(C)** The vitamin K-dependent factors are II, VII, IX, and X. *(1:228; 4:373; 5:749; 6:503)*

552. **(B)** Autoregulation of hepatic blood flow is predominantly in the hepatic artery and via the sympathetic nervous system. The portal vein is not primarily involved in the autoregulation nor is the parasympathetic nervous system. *(1:1071; 4:1905–6; 5:745–6)*

553. **(E)** A low hematocrit decreases oxygen delivery to peripheral tissues. Hypoxemia, hypovolemia, and hypothermia increase the risk of red cell sickling and sludging. *(1:506–7; 5:1113; 6:480–2)*

554. **(B)** A cirrhotic patient does not have normal hepatocytes and cannot normally produce protein, although jaundice may not be present. Succinylcholine may not be handled normally due to decreases in plasma pseudocholinesterase which is made in the liver. Lidocaine metabolism by cytochrome P450 may be decreased in cirrhosis, but lidocaine is not metabolized by pseudocholinesterase. *(1:1072, 1075, 1094; 4:335–8, 1908–14; 5:2218, 2225; 6:307–16)*

555. **(D)** Alcohol ingestion in large amounts will lead to liver disease in spite of good nutrition. Hematologic disorders are due to clotting deficiencies and to the alcohol itself. Gastrointestinal problems, including pancreatitis, are common. *(1:1084, 1089; 4:353–4; 5:760; 6:307–11)*

556. **(A)** Such patients may have profound electrolyte disturbances increasing the risk of dysrhythmias. Osteoporosis may result from malnutrition. Orthostatic hypotension and delayed gastric emptying are common. *(5:1034; 6:451–2)*

557. (B) Sodium transport via the sodium-potassium pump is a process that requires energy. The pump is in the cell membrane. The process is related to the sodium concentration, since sodium is the rate-limiting step. *(1:185; 4:1303–4; 5:1769)*

558. (A) The first three statements are correct. The pH will increase 1 unit for each 10-fold decrease in hydrogen ion concentration. *(1:165; 5:1601; 6:388–91)*

559. (C) Proteinuria may be normal with exercise or stress, and BUN elevation may occur with a high-protein diet or when there is blood present in the gastrointestinal tract. *(1:1015; 5:1493, 2179; 6:342–3)*

560. (E) Extracellular fluid may be lost by all of the mechanisms listed. The loss may not be apparent on examination and may require a careful history and examination to determine. For example, in the case of ascites, it is important to determine how fast the accumulation has occurred. In many cases, the accumulation is slow, and there is little cardiovascular instability. If the accumulation is fast, there may be great instability. *(1:1044–5, 1087–8, 1276; 4:952–3; 5:1790–4; 6:310, 727–8)*

561. (A) Blood sugar rises during surgery by many mechanisms, including administration of intravenous fluids, secretion from the liver due to catecholamines, and insulin suppression. Uptake into the tissues leads to a decrease in blood sugar. *(1:1072, 1394; 4:316–7; 5:1776, 1779; 6:409–10)*

562. (C) Administration of insulin should be carefully monitored during surgery and blood sugar should be followed closely. The intravenous line containing the insulin should not be the primary line but should be piggy-backed into another line. There is also concern that some of the insulin will be adsorbed to the intravenous tubing. The dosage should be titrated to needs. *(1:1134–5; 4:1957–8; 5:1779–80; 6:408–9)*

563. (D) The juvenile-type diabetic is usually brittle, with blood sugars in the higher ranges. These patients are usually thin and have very low levels of circulating insulin. They have a poor response to oral hypoglycemics. *(1:1133; 5:1776–7; 6:395–7)*

564. (A) Mineralocorticoids, glucocorticoids, and sex hormones are secreted by the adrenal cortex. Norepinephrine is secreted by the adrenal medulla. *(1:1125; 4:310–1, 1582; 5:1035; 6:425)*

565. (A) Glucocorticoids are produced under stimulation by ACTH and are active in carbohydrate, protein, and fatty acid metabolism. Aldosterone is a mineralocorticoid. *(1:1126; 4:310–1; 5:1035–6; 6:425)*

566. (E) In diabetes insipidus, the serum sodium and the serum osmolality are high. The urine is dilute. In spite of polyuria, thirst need not be present. *(1:1137–8; 5:1767–8; 6:437–8)*

567. (B) Sweating and tachycardia are compensatory mechanisms to rid the body of the increased heat being produced in the patient with hyperthyroidism. Vasoconstriction is not seen, but vasodilatation is. Shivering is not a heat-losing mechanism. *(1:1120–1; 4:305–6; 5:1046; 6:412–3)*

568. (C) The patient with hyperaldosteronism has hypertension and hypokalemia. The excess secretion comes from the adrenal cortex. Alkalosis is present. *(1:1127; 4:1593–4; 5:1038; 6:429)*

569. (B) von Willebrand's disease is a congenital bleeding disorder. The defect is in von Willebrand factor and has an autosomal-dominant transmission in most cases, and thus not seen only in females. von Willebrand factor participates in the process of platelet adhesiveness. Thus, affected patients usually have a prolonged bleeding time and a normal platelet count. *(1:224–6; 4:372–3; 5:2868; 6:492–3)*

570. **(A)** Most drugs used in anesthesia may be employed in the regimen for renal transplantation surgery, although there may need to be alterations from the usual doses. Metocurine is not metabolized by the liver and is cleared exclusively by renal mechanisms. Since the transplanted kidney may not begin to function immediately, metocurine should be avoided. *(1:1353; 4:772, 1979–80; 5:2241; 6:366)*

571. **(C)** The patient with hyperthyroidism does not have an increased MAC. Iodide is given for 1 to 2 weeks preoperatively to decrease thyroid hormone release and to decrease the vascularity of the gland. Propranolol may also be needed pre- or intraoperatively to manage tachycardia. In contrast to the usual practice of many years ago, heavy preoperative sedation is no longer necessary. In fact, many patients scheduled for thyroidectomy are admitted on the day of surgery. *(1:1121; 4:305–6, 1952–4; 5:1046; 6:414–5)*

572. **(B)** Anticholinergics interfere with interpretation of the pulse rate and with the sweating mechanisms. Pupillary activity and gastric emptying are not concerns in these patients. *(1:287–8; 4:305–6, 1952–4; 5:523; 6:415)*

573. **(B)** Patients with end-stage renal disease are anemic because the kidney is the primary source of erythropoietin. They have an increased bleeding time due to platelet dysfunction, but the platelet count is not necessarily decreased. Electrolyte abnormalities include hyperkalemia and hypermagnesemia. *(1:1010–2; 4:324–6; 5:1101, 1486; 6:346–8)*

574. **(D)** Gilbert's syndrome is a common and entirely benign cause of unconjugated hyperbilirubinemia. Although serum bilirubin levels may rise in response to fasting or stress, they are not elevated to hazardous levels. The most important measure in the perioperative period is to elicit the history of Gilbert's syndrome from the patient so that the isolated abnormality of hyperbilirubinemia is not investigated further. *(1:1096–7; 4:340; 5:755; 6:316)*

575. **(A)** Hemochromatosis and Wilson's disease result in cirrhosis due to the accumulation of excessive amounts of iron and copper, respectively, in the liver. α_1-Antitrypsin deficiency causes bullous emphysema as well as cirrhosis. Antithrombin III deficiency is a disorder of the clotting cascade which results in excessive thrombosis. *(1:1084–5, 1365; 4:348; 5:2215–7; 6:308–9)*

576. **(B)** The most effective class of diuretics in the cirrhotic patient with ascites are the aldosterone antagonists, such as spironolactone. Since cirrhotics have difficulty excreting sodium, restriction of dietary sodium acts to decrease the formation of ascites and edema fluid. Inhibitors of prostaglandin synthesis, such as aspirin and indomethacin, are contraindicated in ascites because they decrease renal sodium excretion. An angiotensin-converting enzyme inhibitor such as captopril has no specific therapeutic effect in a patient with ascites, and may interact with spironolactone to cause hyperkalemia. *(1:1087; 4:352–3, 1918–9; 5:1791–2, 2223; 6:310)*

577. **(A)** The oliguric renal failure accompanying the hepatorenal syndrome is thought to be due to decreased renal perfusion, and thus may respond to a dopamine infusion which may increase renal blood flow. Urinary sediment and renal biopsy are normal. *(1:1088–9; 4:351–2, 1922–3; 5:751, 2248; 6:311)*

578. **(C)** In sickle cell anemia, the abnormal hemoglobin molecule precipitates when deoxygenated. When a significant fraction of the hemoglobin molecules in a red cell precipitates, the cell irreversibly assumes sickle shape. This alteration in red cell shape increases blood viscosity, decreases microcirculatory flow, and causes platelet aggregation and fibrin deposition. Tissues with a high oxygen extraction ratio are at increased risk for infarction because hemoglobin molecules are more likely to become deoxygenated. *(1:506; 4:364–7; 5:1113; 6:478–80)*

579. **(C)** The patient with hemophilia B has a decreased plasma concentration of factor IX and will have a prolonged partial thromboplastin time (PTT). Vitamin K will not increase the synthesis of factor IX in this genetic deficiency. Hemophilia B is a sex-linked trait affecting males. *(1:226; 4:370–2; 5:1116–7; 6:492)*

580. **(A)** Phenoxybenzamine, a long-acting, nonselective α-adrenergic antagonist is begun 1 to 2 weeks prior to surgery. Once blood pressure control is obtained, if tachycardia persists, a β-adrenergic antagonist is added. Administration of a β-adrenergic antagonist prior to adequate α-adrenergic blockade may result in worsening of the hypertension. Because patients with pheochromocytoma are hypovolemic, diuretics are contraindicated in the control of the hypertension. The α- and β-adrenergic antagonists are titrated to blood pressure and heart rate control, and do not decrease the urinary excretion of catecholamine metabolites. *(1:1131–3; 4:314–5; 5:1042–4; 6:431–2)*

581. **(B)** Most hemolytic transfusion reactions are caused by clerical errors in which ABO-incompatible red cells are transfused. Many of the common signs of a hemolytic transfusion reaction (fever, chills, chest pain, hypotension) may be masked by general anesthesia, and hemoglobinuria and coagulopathy are the signs most likely to be noted. The severity of the reaction is directly related to the volume of transfused cells, so immediate discontinuation of the transfusion is mandatory if a hemolytic reaction is suspected. Only a small percentage of hemolytic transfusion reactions are caused by the failure to detect incompatibility during serologic testing. *(1:207; 4:2418–9; 5:1815–6)*

CHAPTER 7

Pharmacology

Questions

DIRECTIONS (Questions 582 through 762): Each of the numbered items or incomplete statements in this section is followed by answers or by completions of the statement. Select the ONE lettered answer or completion that is BEST in each case.

582. The most important site of drug transformation is usually the

(A) liver
(B) spleen
(C) kidney
(D) lungs
(E) bloodstream

583. The first-pass effect refers to

(A) the biotransformation of a drug in its vehicle of administration
(B) the change of a drug by enzymes in muscle
(C) biotransformation of a drug as it passes through the intestinal mucosa and liver
(D) the drug lost by urinary excretion
(E) the drug lost by fecal excretion

584. Bioavailability of a drug refers to the amount of drug that

(A) is administered intramuscularly
(B) is administered orally
(C) reaches the liver
(D) is excreted by the kidney
(E) reaches its site of action

585. Renal clearance of a drug

(A) is usually of little importance
(B) has no relationship to creatinine clearance
(C) is constant for a given drug
(D) varies with pH, urine flow rate, and renal blood flow
(E) may exceed renal blood flow

586. Halogenation of a hydrocarbon

(A) usually increases its flammability
(B) increases its anesthetic potency
(C) has no effect
(D) makes it more of a cardiac irritant
(E) makes it explosive

587. Halothane structurally is

(A) an ether
(B) a methane derivative
(C) an ethane derivative
(D) an ethane derivative substituted with fluorine, chlorine, and iodine
(E) an ethane derivative substituted with fluorine, bromine, and iodine

588. Local anesthetics have their effect at the

(A) presynaptic nerve terminal
(B) postsynaptic nerve terminal
(C) GABA receptor
(D) membrane
(E) calcium channel

589. Halothane is stored

(A) in clear bottles with a preservative
(B) in tinted bottles with the preservative butylated hydroxytoluene
(C) without a preservative
(D) with provision to prevent photochemical decomposition
(E) in plastic bottles

590. Isoflurane

(A) is a poor muscle relaxant compared with halothane
(B) has a vapor pressure of 175 mmHg at 20°C
(C) stimulates ventilation
(D) has a MAC of approximately 1.2%
(E) should be delivered by spontaneous ventilation

591. The most potent of the inhalation anesthetics is

(A) halothane
(B) desflurane
(C) nitrous oxide
(D) sevoflurane
(E) isoflurane

592. The cardiovascular effects of isoflurane are characterized by

(A) stable cardiac rhythm
(B) sensitization of the heart to epinephrine
(C) a slowing of the heart rate
(D) increased stroke volume
(E) decreased cardiac output

593. A 62-year-old male with a history of diabetes mellitus, hypertension, and coronary artery disease is admitted to the hospital for unstable angina. He was recently diagnosed with heparin-induced thrombocytopenia type II and is now scheduled to undergo off-pump coronary artery bypass grafting (CABG) for severe three-vessel coronary artery disease. A suitable agent for intraoperative anticoagulation would be

(A) heparin
(B) warfarin
(C) argatroban
(D) danaparoid
(E) clopidogrel

594. Each of the following is an ether EXCEPT

(A) halothane
(B) desflurane
(C) enflurane
(D) isoflurane
(E) sevoflurane

595. Desflurane

(A) is an ether derivative
(B) has a blood-gas coefficient of 0.9
(C) has a vapor pressure of 450 mmHg at 20°C
(D) is unstable in the presence of soda lime
(E) contains no fluorine atoms

596. The action of thiopental after injection is terminated by

(A) its elimination unchanged by the kidneys
(B) its biotransformation by the liver
(C) its being bound to proteins
(D) its redistribution
(E) being taken up in fatty tissues

597. Propofol as compared to thiopental

(A) is less likely to provoke bronchospasm
(B) if administered in equipotent doses for induction of anesthesia causes less reduction in systemic blood pressure
(C) causes adrenal suppression after prolonged infusion
(D) has no effect on cerebral metabolic rate
(E) does not cause excitatory motor activity

598. The pH of a 2.5% solution of thiopental is

(A) 2
(B) 4

(C) 6
(D) 8
(E) 10

599. Thiopental administration in humans is followed by

(A) increased myocardial contractility, decreased stroke volume
(B) decreased myocardial contractility, increased stroke volume
(C) decreased myocardial contractility, decreased blood pressure
(D) decreased myocardial contractility, increased blood pressure
(E) decreased stroke volume, increased blood pressure

600. If a patient has a need for a high cardiac output and high systemic blood pressure, the muscle relaxant of choice would be

(A) rocuronium
(B) cisatracurium
(C) pancuronium
(D) vecuronium
(E) succinylcholine

601. Thiopental administration in hypnotic doses will

(A) increase cerebral blood flow
(B) have a depressant effect on uterine contractions
(C) paralyze skeletal muscle
(D) increase EEG activity
(E) decrease cerebral oxygen consumption

602. After injection of thiopental, 4 mg/kg, there would be

(A) decreased heart rate
(B) arrhythmias
(C) baroreceptor-mediated decrease in sympathetic activity
(D) more pronounced changes in patients with heart disease
(E) no change in cardiovascular parameters

603. A patient has been given an injection of ketamine in a dose calculated to be sufficient for anesthesia. His eyes remain open, and there is slight nystagmus and occasional purposeless movements. This is an indication that

(A) the dose is inadequate
(B) more ketamine should be given to stop the movements
(C) the dose is excessive
(D) the dose is adequate for anesthesia
(E) the patient is having a seizure

604. Which of the following agents should be avoided in a patient with heparin-induced thrombocytopenia type II?

(A) danaparoid
(B) lepirudin
(C) argatroban
(D) warfarin
(E) benzodiazepines

605. Droperidol is

(A) an opioid of intermediate potency
(B) an acetylcholinesterase inhibitor
(C) a butyrophenone derivative
(D) a short-acting (1 to 2 hours) sedative
(E) eliminated totally by the kidney

Questions 606 through 608

A 61-year-old male with a past medical history of diverticulosis is admitted to the intensive care unit after undergoing an emergent laparotomy with sigmoid colectomy and creation of Hartman's pouch for perforated diverticulitis. According to the surgeon, there was frank contamination of the peritoneal cavity. Despite broad-spectrum antibiotic coverage, the patient develops symptoms consistent with septic shock on postoperative day one.

606. An infusion of recombinant human activated protein c (APC) is started in addition to aggressive fluid resuscitation and vasopressor therapy. Which of the following statements about APC is true?

(A) It has gram-negative antibiotic activity.
(B) It has procoagulant activity.
(C) It is recommended for use in patients with severe sepsis.
(D) It is used as a continuous infusion to support systolic blood pressure.
(E) APC infusion should be limited to 48 hours duration.

607. Despite aggressive fluid resuscitation and blood pressure support with norepinephrine, the patient remains hypotensive. A two-dimensional echocardiogram is obtained which shows no wall motion abnormalities, a hyperdynamic left ventricle, and a calculated cardiac index of 4.8 L/min/m^2. Which of the following agents would be most useful in addition to the norepinephrine for the treatment of this patient's hypotension?

(A) milrinone
(B) epinephrine
(C) phenylephrine
(D) dobutamine
(E) vasopressin

608. On the second postoperative day, the patient continues to require high-dose vasopressor support and a laboratory workup reveals relative adrenal insufficiency. Low-dose hydrocortisone therapy is started. Which of the following statements is true?

(A) In patients with septic shock and relative adrenal insufficiency, corticosteroids enhance the vascular reactivity to other vasoactive substances.
(B) Corticosteroids in the setting of septic shock will enhance the bactericidal effect of broad-spectrum antibiotics.
(C) Addition of corticosteroids to the therapeutic regimen will improve glycemic control.
(D) The major action of hydrocortisone in this setting is sodium retention with subsequent improvement of volume status.
(E) Administration of corticosteroids will most likely cause leukopenia in this patient.

609. Local anesthetics in the injectable form are

(A) salts
(B) acids
(C) bases
(D) proteins
(E) lipids

610. Rocuronium

(A) is a long-acting muscle relaxant
(B) can cause bronchoconstriction
(C) has a mean onset time at three times the ED95 of about 45 seconds
(D) causes significant histamine release when used in a dose adequate for rapid-sequence induction
(E) is metabolized by plasma esterases

611. A 9-year-old boy sustained 40% burns to the anterior portion of his body. On the 20th day after the burn, he is brought to the operating room for a skin graft. Anesthesia is induced with thiopental, and succinylcholine is injected for relaxation. The ECG shows a bizarre tracing followed by asystole. The most likely cause of the arrhythmia is

(A) an overdose of thiopental
(B) hypoxia
(C) hyperkalemia

(D) electrocution
(E) administration of the wrong drug

612. Antibiotics that possess neuromuscular blocking properties include all of the following EXCEPT

(A) neomycin
(B) kanamycin
(C) penicillin
(D) gentamicin
(E) amikacin

613. Expected cardiovascular changes after ketamine administration include

(A) elevated diastolic pressure, normal systolic pressure
(B) elevated diastolic and systolic pressures
(C) decreased diastolic and systolic pressures
(D) decreased diastolic pressure, increased systolic pressure
(E) no change in blood pressure

614. Echothiophate iodide eyedrops, chronically administered, will have an effect similar to

(A) β-adrenergic antagonists
(B) a monoamine oxidase (MAO) inhibitor
(C) atropine
(D) organophosphate insecticides
(E) thiopental

615. A muscle relaxant with a steroid structure is

(A) mivacurium
(B) atracurium
(C) succinylcholine
(D) pancuronium
(E) doxacurium

616. A small dose of a nondepolarizing muscle relaxant given 3 minutes before an intubating dose of succinylcholine

(A) increases the dose of succinylcholine required
(B) will not prevent the rise in intracranial pressure
(C) is useful in preventing arrhythmias
(D) doubles the time to recovery from neuromuscular blockade
(E) permits faster intubation

617. Thiopental is contraindicated in

(A) porphyria congenita
(B) porphyria cutanea tarda
(C) acute intermittent porphyria
(D) myotonia
(E) chorea

618. When succinylcholine is administered in a dose of 1 mg/kg, one may expect all of the following EXCEPT

(A) an intraocular pressure increase
(B) an increase in intragastric pressure
(C) bradycardia in infants and children
(D) tachycardia after multiple doses in adults
(E) hyperkalemia in paraplegic patients

619. A 58-year-old woman with a diagnosis of myasthenia gravis underwent a laparotomy. Her anesthesia consisted of thiopental, nitrous oxide, fentanyl, and oxygen, with pancuronium as a muscle relaxant in a total dose of 1 mg. During the procedure, she received an IV dose of gentamicin and remained apneic at the end of the procedure. The cause of the apnea could be ascribed to all of the following EXCEPT

(A) thiopental
(B) myasthenia gravis
(C) pancuronium
(D) fentanyl
(E) gentamicin

620. A drug is administered by IV injection. Approximately what percentage of the drug remains after four half-lives have elapsed?

(A) 1%
(B) 4%
(C) 6%
(D) 10%
(E) 12%

621. Vecuronium

(A) has a duration of about 2 minutes
(B) is a depolarizing muscle relaxant
(C) is difficult to reverse
(D) has very little cumulative effect
(E) depends on the kidney for elimination

622. The administration of anticholinesterase drugs will

(A) prolong all neuromuscular blockade
(B) always reverse nondepolarizing agents
(C) shorten the block of a depolarizing agent
(D) reverse nondepolarizing agents if the plasma concentration of the drug is low enough
(E) reverse the action of a depolarizing agent if only partial paralysis is present

623. Which of the following effects is NOT mediated by the α_1-adrenoceptor?

(A) contraction of the bladder sphincter
(B) contraction of the radial muscle of the iris
(C) vasoconstriction in the skin
(D) relaxation of the uterus
(E) inhibition of renin secretion

624. A patient is admitted for emergency orthopedic surgery. Preliminary data show blood urea nitrogen (BUN) value of 85 mg/dL and a serum potassium of 6.0 mEq/L. The least desirable drug for intubation would be

(A) rocuronium
(B) vecuronium
(C) cisatracurium
(D) mivacurium
(E) pancuronium

625. Propofol

(A) increases cerebral blood flow
(B) is very soluble in aqueous solutions
(C) causes catecholamine release
(D) has little analgesic property
(E) has a longer sleep time than thiopental

626. The most sensitive test to determine adequate recovery from neuromuscular blockade is

(A) 5-second head lift
(B) 5-second hand grip
(C) inspiratory force
(D) tactile response to double-burst stimulation (DBS)
(E) tactile response to train-of-four

627. The physical properties of isoflurane include

(A) flammability in oxygen
(B) red color
(C) pungent odor
(D) chemical breakdown when exposed to sunlight
(E) addition of thymol as a preservative

628. Factors that will potentiate nondepolarizing neuromuscular blockade include all of the following EXCEPT

(A) respiratory acidosis
(B) large body surface area burn
(C) administration of a volatile anesthetic
(D) hypothermia
(E) hypermagnesemia

629. Propranolol hydrochloride

(A) is an α-adrenergic antagonist
(B) is a β-adrenergic agonist
(C) causes increased heart rate
(D) causes decreased cardiac output
(E) causes increased cardiac contractility

Questions 630 through 632 refer to Figure 7-1.

630. The figure shows the evoked twitch response following an unknown drug. The segment (1) is the baseline before drug administration, and (2) is the response 3 minutes after drug administration. From these two segments, you can say that the drug

(A) is not a muscle relaxant
(B) is a depolarizing muscle relaxant
(C) is a nondepolarizing muscle relaxant

(D) caused complete paralysis
(E) caused 90% depression

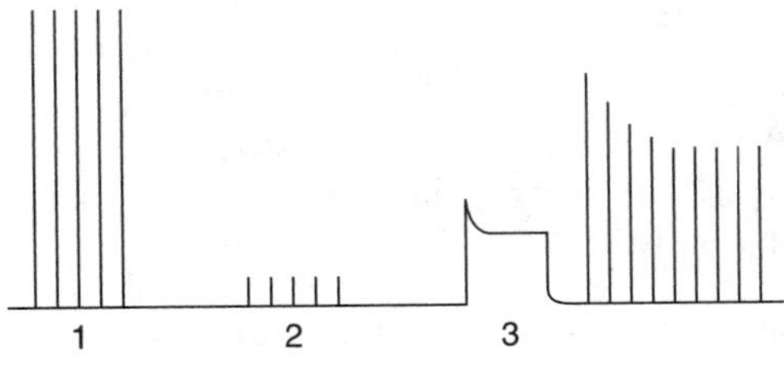

FIG. 7-1

631. At time point (3), a tetanic stimulation was applied with the depicted result. Following this, one could state that

(A) the drug is not a muscle relaxant
(B) the drug is a depolarizing muscle relaxant
(C) the drug is a nondepolarizing muscle relaxant
(D) either the drug is a nondepolarizing muscle relaxant, or a dual block is present
(E) the tracing shown is due to artifact

632. The inability to sustain contraction (fade) during tetanic contraction is due to

(A) inability to transmit rapid electrical impulses
(B) cellular loss of potassium
(C) depletion of cellular DNA
(D) inability of recording apparatus to show all contractions
(E) the inability of the end plate to release sufficient acetylcholine

633. Transdermal scopolamine

(A) is effective in the prevention of postoperative nausea and vomiting (PONV) if administered preoperatively
(B) is effective in the treatment of PONV when administered postoperatively
(C) patch should be applied to the deltoid muscle
(D) is devoid of any side effects
(E) in a multilayered adhesive unit has a duration of action of about 16 hours

634. The atypical antipsychotics like clozapine and olanzapine differ from antipsychotic agents like haloperidol and chlorpromazine in that they

(A) cause less extrapyramidal effects
(B) do not produce hypotension
(C) do not cause weight gain and increased appetite
(D) have less anticholinergic effects
(E) have not been associated with a risk of new-onset type 2 diabetes

635. A patient who had been given vecuronium received neostigmine, 3 mg, at the termination of the surgical procedure. Six minutes later, a large amount of white, frothy secretions was noted in the endotracheal tube. Vigorous suctioning was required to remove these secretions in order to ventilate the patient. The treatment of choice for such secretions is

(A) atropine
(B) digoxin
(C) more neostigmine
(D) readministration of vecuronium
(E) use of a ventilator

636. The patient with myasthenia gravis

(A) has normal reactions to muscle relaxants
(B) reacts abnormally to relaxants only when the condition is not well controlled
(C) has decreased sensitivity to nondepolarizing relaxants
(D) has an increased sensitivity to nondepolarizing relaxants
(E) has an increased sensitivity to depolarizing relaxants

637. If succinylcholine is given to facilitate intubation, and the patient is allowed to recover fully from the depolarizing block, and then is given pancuronium, one would expect

(A) a lower than normal amount of pancuronium is needed to reestablish neuromuscular blockade
(B) the pancuronium to have a shorter duration of action as a result of the succinylcholine
(C) no change in the duration of action of pancuronium
(D) no change in the duration of action of pancuronium but less intensity of relaxation
(E) the development of a phase II block

638. The formation of acetylcholine involves all of the following EXCEPT

(A) nerve terminal
(B) choline
(C) acetyl-CoA
(D) choline acetyltransferase
(E) cholinesterase

639. Botulinus toxin interferes with neuromuscular transmission by

(A) preventing synthesis of acetylcholine
(B) preventing breakdown of acetylcholine
(C) preventing storage of acetylcholine
(D) preventing release of acetylcholine
(E) blocking receptors for acetylcholine

640. The administration of atropine may be followed by all of the following EXCEPT

(A) no change in systemic vascular resistance
(B) no change in heart rate in old age
(C) angina pectoris
(D) decreased AV conduction time
(E) decreased oxygen utilization

641. Depolarizing neuromuscular block is characterized by all of the following EXCEPT

(A) muscle fasciculation
(B) initial absence of fade
(C) posttetanic facilitation
(D) antagonism of block by vecuronium
(E) potentiation of block by anti-cholinesterase

642. A patient scheduled for surgery is taking propranolol 40 mg daily for hypertension. This drug

(A) should be stopped immediately, and the procedure can be done the next day
(B) should be stopped and surgery post-poned for 2 weeks
(C) should be continued during surgery and then stopped
(D) will not affect the anesthetic regimen in any way
(E) may cause rebound hypertension if abruptly stopped

643. Which of the following has only α-adrenergic agonist activity?

(A) epinephrine
(B) norepinephrine
(C) methoxamine
(D) isoproterenol
(E) ephedrine

Questions 644 through 660

Many patients are unaware of the reasons for which they are taking their medications. When provided with a list of medications by a patient, an anesthesiologist must be able to recognize the medications and know the likely pathologic states for which the medications are indicated. For questions 644 through 660, a medication is followed by five diseases or pathologic states. Choose the ONE disease for which the medication may be indicated.

644. Lisinopril

(A) panic disorder
(B) hypertension
(C) paroxysmal supraventricular tachycardia
(D) premature ventricular contractions (PVCs)
(E) hyperthyroidism

645. Cyclosporine

(A) bone pain due to cancer
(B) glaucoma
(C) intestinal roundworms
(D) tuberculosis
(E) graft versus host disease

646. Omeprazole

(A) peptic ulcer disease
(B) diabetic gastroparesis
(C) vertigo
(D) urinary tract infection
(E) type II diabetes mellitus

647. Tamoxifen

(A) reflux esophagitis
(B) anxiety
(C) Parkinson's disease
(D) breast cancer
(E) hypertension

648. Acetazolamide

(A) glaucoma
(B) asthma
(C) peptic ulcer disease
(D) deep venous thrombosis
(E) atrial fibrillation

649. Selegiline

(A) schizophrenia
(B) insomnia
(C) grand mal epilepsy
(D) Parkinson's disease
(E) migraine headaches

650. Pyrazinamide

(A) urinary tract infection
(B) anemia
(C) transient ischemic attacks
(D) tuberculosis
(E) hypertension

651. Glyburide

(A) myasthenia gravis
(B) type II diabetes mellitus
(C) angina pectoris
(D) glaucoma
(E) hypercalcemia

652. Amiodarone

(A) Wolff-Parkinson-White (WPW) syndrome
(B) hyperthyroidism
(C) angina pectoris
(D) pulseless electrical activity
(E) ventricular tachycardia

653. Fenoxfenadine

(A) hay fever
(B) insomnia
(C) schizophrenia
(D) chronic pain
(E) inflammatory bowel disease

654. Cromolyn sodium

(A) deep venous thrombosis
(B) asthma
(C) claudication
(D) angina pectoris
(E) atrial fibrillation

655. Hydroxychloroquine

(A) rheumatoid arthritis
(B) pinworms
(C) vaginal candidiasis
(D) trichomoniasis
(E) giardiasis

656. Dofetilide

(A) muscle spasticity
(B) trigeminal neuralgia
(C) atrial fibrillation
(D) peripheral edema
(E) insomnia

657. Allopurinol

(A) systemic lupus erythematosus
(B) gout
(C) osteoarthritis
(D) rheumatoid arthritis
(E) psoriatic arthritis

658. Leuprolide

(A) acromegaly
(B) prostate cancer
(C) hyperthyroidism
(D) type II diabetes mellitus
(E) Addison's disease

659. Fluphenazine

(A) schizophrenia
(B) insomnia
(C) attention deficit disorder
(D) reflex sympathetic dystrophy
(E) diabetic neuropathy

660. Methysergide

(A) dementia
(B) dysmenorrhea
(C) carcinoid syndrome
(D) diabetes insipidus
(E) asthma

Questions 661 and 662 refer to Figure 7-2.

661. The train-of-four stimulus is depicted in the figure. Stimulation is at 2 Hz. The response in A is normal. By looking at B you know that

(A) a depolarizing block is present
(B) the patient is partially paralyzed
(C) the patient is partially paralyzed, but one would need a baseline recording to know how much
(D) the train-of-four ratio is 50%
(E) the patient could sustain a head lift

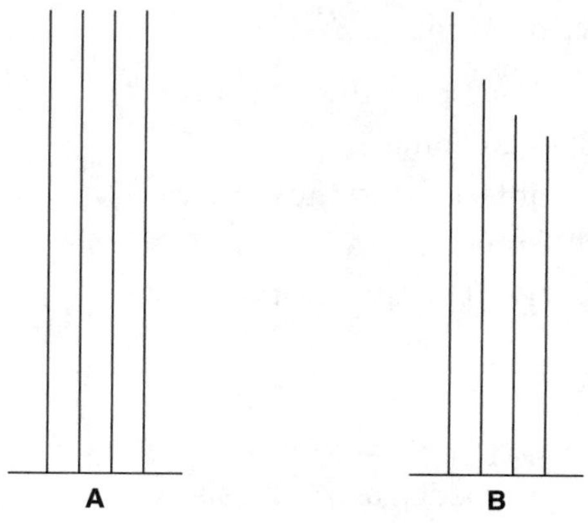

FIG. 7-2

662. The train-of-four ratio for a depolarizing block is

(A) variable
(B) 50%
(C) 60%
(D) 75%
(E) 100%

663. Phenytoin

(A) is the drug of choice for complete heart block
(B) is broken down chiefly by cholinesterase
(C) is useful in treatment of arrhythmias associated with digoxin
(D) increases diastolic depolarization
(E) increases automaticity

664. The calcium channel blocking drugs

(A) increase myocardial contractility
(B) increase anesthetic requirements
(C) are poorly bound to plasma protein
(D) interfere with the flow of calcium ion through cellular membranes
(E) are all about equally effective

665. Lepirudin and argatroban exert their anticoagulant effect by

(A) vitamin K antagonism
(B) inhibition of platelet aggregation
(C) glycoprotein IIb/IIIa inhibition

(D) increasing the rate of the thrombin-antithrombin reaction
(E) direct thrombin inhibition

666. Lorazepam may produce all of the following effects EXCEPT

(A) extrapyramidal effects
(B) decreased anxiety
(C) sedation
(D) respiratory depression
(E) amnesia

667. A patient who is taking lithium for treatment of manic-depressive attacks is scheduled to have general anesthesia. The lithium

(A) need not be considered in the anesthetic regimen
(B) may affect both depolarizing and nondepolarizing muscle relaxants
(C) decreases the duration of nondepolarizing muscle relaxants
(D) should be stopped 2 weeks before surgery
(E) may increase anesthetic requirements

668. If a patient is taking a MAO inhibitor, which of the following should be avoided?

(A) local anesthetics
(B) halothane
(C) vecuronium
(D) meperidine
(E) aspirin

669. All of the following are adverse effects of furosemide EXCEPT

(A) renin release
(B) ototoxicity
(C) hypokalemia
(D) acidosis
(E) hyperglycemia

670. The most common electrolyte alteration caused by the thiazide diuretics is

(A) hypokalemia
(B) hypoglycemia
(C) hyperchloremia
(D) hypernatremia
(E) hyperuricemia

671. Nifedipine

(A) is quite effective in supraventricular tachycardia
(B) is used for the treatment of ischemic heart disease
(C) has a half-life of 30 minutes
(D) is a peripheral vasoconstrictor
(E) is an effective drug for ventricular tachycardia

Questions 672 and 673

A 25-year-old man is admitted after abdominal trauma. After induction of anesthesia, a Foley catheter is inserted. There is little urine output. Blood pressure is 90/60. The urine/plasma osmolality ratio is found to be 2.5.

672. The treatment of choice is

(A) ethacrynic acid
(B) mannitol
(C) furosemide
(D) 500 mL of crystalloid solution
(E) fluid restriction

673. Later, the patient's CVP is found to be 22 mmHg. The treatment of choice should be

(A) furosemide
(B) mannitol
(C) 50% glucose
(D) 500 mL of crystalloid solution
(E) 500 mL of 5% albumin

674. The interaction of protamine and heparin to terminate anticoagulation is of

(A) competition for binding sites
(B) a chemical interaction leading to an inactive compound
(C) pH change
(D) a conformational change
(E) platelet stimulation

675. Echothiophate causes prolongation of succinylcholine by

(A) chemical interaction
(B) interaction at site of absorption
(C) altered protein binding
(D) competition for binding sites
(E) altered metabolism

676. Epinephrine causes a prolongation of activity of local anesthetics by

(A) chemical interaction
(B) decreasing absorption
(C) altered protein binding
(D) competition for binding sites
(E) altered metabolism

677. The interaction of phenobarbital and phenytoin can be described as one of

(A) chemical interaction
(B) interaction at site of absorption
(C) altered protein binding
(D) competition for binding sites
(E) altered metabolism

678. The combination of nitrous oxide at 0.5 MAC plus isoflurane 0.5 MAC is one of

(A) antagonism
(B) potentiation
(C) additive effect
(D) synergism
(E) no effect

679. A 42-year-old female with a body mass index (BMI) of 32 and a history of severe PONV is scheduled for a laparoscopic cholecystectomy for symptomatic cholelithiasis. Your anesthetic plan includes the administration of a 5-HT_3 receptor antagonist and droperidol for the prevention of PONV. If you were to add a third agent, the most useful drug would be

(A) metoclopramide
(B) prochlorperazine
(C) famotidine
(D) dexamethasone
(E) ephedrine

680. Quinidine is

(A) useful only when given intravenously
(B) about 90% bound to plasma protein
(C) useful only for atrial arrhythmias
(D) known to counteract neuromuscular blockers
(E) contraindicated in patients with renal impairment

681. Lidocaine

(A) is eliminated chiefly by the liver
(B) is effective orally
(C) is toxic at levels over 1 mcg/mL
(D) toxicity is noted by the appearance of hematuria
(E) is useful in supraventricular tachycardia

682. An IV agent that contains polyethylene glycol in its formulation is

(A) thiopental
(B) methohexital
(C) lorazepam
(D) midazolam
(E) propofol

Questions 683 and 684 refer to Figure 7-3.

683. If A and B are two different medications producing the same effect, then the figure shows that

(A) A has higher efficacy than B
(B) A has lower efficacy than B

(C) A and B must act via different receptors to produce their effects
(D) A is more potent than B
(E) A has a higher ED_{50} than B

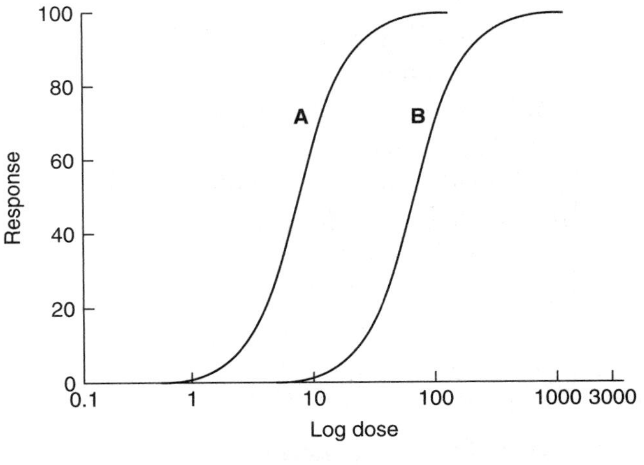

FIG. 7-3

684. If A depicts the dose-response relationship of a drug X acting alone, and B depicts the dose-response relationship of drug X in the presence of drug Y, then it can be said that drug Y

(A) is a competitive antagonist of drug X
(B) is a noncompetitive antagonist of drug X
(C) must act via a different receptor than drug X
(D) decreases the efficacy of drug X
(E) increases the potency of drug X

685. A drug that is associated with pulmonary toxicity is

(A) doxorubicin
(B) bleomycin
(C) vincristine
(D) methotrexate
(E) L-asparaginase

686. Methohexital

(A) is metabolized to a greater extent than thiopental
(B) is converted to active metabolites
(C) has a longer terminal half-life than thiopental
(D) causes histamine release from mast cells
(E) is contraindicated in asthma

687. Midazolam

(A) is contraindicated in the child
(B) is shorter acting than thiopental
(C) is associated with less frequent venous irritation than diazepam
(D) suppresses adrenal cortical function
(E) has a high incidence of histamine release

688. Lidocaine, in a 1% solution,

(A) is highly protein-bound after IV injection
(B) has a duration of 60 to 120 minutes when used for peripheral nerve blockade
(C) is not effective topically
(D) is frequently associated with allergic reactions
(E) frequently causes tachycardia

689. The addition of epinephrine to a local anesthetic

(A) prolongs the duration of the drug's activity
(B) causes a higher peak level of the drug
(C) decreases intensity of the block
(D) increases surgical bleeding
(E) causes increased toxicity

690. Sufentanil

(A) is less potent than fentanyl
(B) is more fat soluble than fentanyl
(C) is redistributed to terminate its effect when given in high doses for cardiac surgery
(D) causes more histamine release than fentanyl
(E) produces burst suppression on EEG at high doses

691. Procaine is

(A) metabolized to a great extent in the lung
(B) metabolized in the cerebrospinal fluid (CSF)
(C) excreted unchanged in the urine
(D) broken down into active metabolites
(E) more slowly metabolized in patients with liver disease

692. Flecainide is

(A) an antiarrhythmic drug in Class IC
(B) a lidocaine analog
(C) a muscle relaxant
(D) the drug of choice for local anesthesia reactions
(E) administered only orally

693. Alfentanil

(A) has a larger volume of distribution when compared to fentanyl
(B) is one-third more potent than fentanyl
(C) has a longer terminal half-life than fentanyl
(D) has its effect terminated by redistribution
(E) does not affect the MAC of enflurane

694. Dexmedetomidine

(A) is a hypnotic agent used for the induction of general anesthesia
(B) can markedly reduce the MAC of inhalational anesthetics
(C) when bolused intravenously at 2 mcg/kg will result in an initial decrease in blood pressure and increase in heart rate
(D) will result in adrenal suppression when infused continuously for postoperative sedation
(E) causes more respiratory depression than opioids and benzodiazepines

695. Prostaglandin E_1 (PGE_1)

(A) is a potent vasoconstrictor
(B) should be given by bolus
(C) should always be stopped before inducing anesthesia
(D) is useful in closing a patent ductus arteriosus
(E) may cause hypotension

696. Physostigmine

(A) may cause central anticholinergic syndrome
(B) does not cross the blood-brain barrier
(C) is less effective than neostigmine in treating emergence delirium
(D) may reverse the sedative effect of benzodiazepines
(E) often produces an uncomfortably dry mouth

697. The patient on a tricyclic antidepressant

(A) may have an increased number of arrhythmias
(B) should have halothane and pancuronium as drugs of choice
(C) should be cautioned to stop the medication before surgery
(D) may become hypotensive with ketamine
(E) may have short emergence with thiopental

698. A 66-year-old male with a history of diabetes mellitus, hypertension, and coronary artery disease is to undergo urgent surgery for an expanding 6.4-cm infrarenal abdominal aortic aneurysm. He had a myocardial infarction 2 months ago at which time he underwent percutaneous coronary angioplasty with placement of drug-eluting stents to the left anterior descending and right coronary arteries. Current medications include clopidogrel and aspirin, which were started after the stent placement. With respect to the combination of these two drugs and the intraoperative risk of bleeding it is true that

(A) both drugs exert their antiplatelet effect through the same mechanism
(B) clopidogrel does not add significantly to the risk of bleeding posed by aspirin
(C) if these drugs were stopped one day prior to surgery, the risk of bleeding would be minimized

(D) both drugs can be antagonized
(E) clopidogrel should be discontinued 7 days prior to surgery in elective cases where surgical hemostasis is needed

699. A patient is brought to the operating room with a solution of vancomycin infusing. You should

(A) stop the infusion
(B) run the infusion as fast as possible to get the drug in before induction
(C) disconnect the IV line and reconnect it as a piggyback infusion to give more control
(D) disregard the fact that vancomycin is present
(E) insist that the surgeon use an alternative antibiotic

700. Terbutaline

(A) is an α-adrenergic agonist
(B) is a selective β_2-adrenergic agonist
(C) causes more tachycardia than isoproterenol
(D) should be avoided in patients with heart disease
(E) causes hyperkalemia

701. The patient who has recently abused cocaine

(A) will be calm and sedated
(B) may be treated with propranolol
(C) exhibits signs of sympathetic blockade
(D) will have bradycardia
(E) will have hypotension

702. Meperidine

(A) is a naturally occurring opioid
(B) is more potent than morphine
(C) is excreted unchanged in the urine
(D) is primarily metabolized in the liver
(E) is not useful orally

703. Ranitidine

(A) is a histamine H_1 receptor antagonist
(B) will decrease the pH of gastric fluid
(C) has a half-life of about a day
(D) must be given parenterally
(E) has a low incidence of CNS toxicity

704. Verapamil

(A) is very useful in the treatment of supraventricular tachycardia
(B) is contraindicated in patients with asthma
(C) is especially useful when combined with propranolol
(D) is a potent vasoconstrictor
(E) has no effect on the pacemaker cells

705. Chemical substances released from mast cells and basophil storage granules at the time of an anaphylactic reaction include all of the following EXCEPT

(A) histamine
(B) bradykinin
(C) leukotrienes
(D) calcium
(E) prostaglandins

706. Glycopyrrolate

(A) is a quaternary amine
(B) crosses the blood-brain barrier with ease
(C) is associated with the central cholinergic syndrome
(D) is a cholinergic agonist
(E) is a naturally occurring belladonna alkaloid

707. Milrinone

(A) is a catecholamine used for treatment of congestive heart failure
(B) is an antiarrhythmic drug
(C) causes peripheral vasodilation
(D) inhibits Na,K-ATPase
(E) has a longer elimination half-life than inamrinone

708. The methylxanthine group of drugs

(A) includes caffeine
(B) have a strong β_1-adrenergic mimetic effect
(C) stimulates production of phosphodiesterase
(D) causes bronchoconstriction
(E) leads to a decrease in cyclic adenosine monophosphate (cAMP)

709. Hydralazine

(A) is a vasodilator due to its catecholamine inhibitor properties
(B) may cause bradycardia
(C) may lead to a lupus-like syndrome
(D) leads to diuresis and sodium loss
(E) has beneficial effects in patients with angina

710. Gabapentin

(A) is approved for the monotherapy of partial seizures
(B) is a benzodiazepine
(C) is metabolized by the liver
(D) is useful for the treatment of migraine, chronic pain, and bipolar disorder
(E) increases the plasma levels of carbamazepine and phenobarbital when used concomitantly

711. A major side effect of valproic acid is

(A) decreased renal function
(B) elevation in liver enzymes
(C) potentiation of muscle relaxants
(D) arrhythmias
(E) anemia

712. Sublingual drug administration

(A) leads to lower drug levels compared to oral administration
(B) is more effective for ionized drugs
(C) circumvents the first-pass effect
(D) leads to rapid liver breakdown of the drug
(E) requires a much larger dose for effectiveness

713. If a patient is to receive calcium for hypocalcemia, the calcium

(A) should be mixed with sodium bicarbonate
(B) can be given as the chloride or the gluconate
(C) may lead to hypertension
(D) may be given intravenously or intramuscularly
(E) gluconate salt is preferred

714. A patient is a 45-year-old male with a history of hypertension controlled with a diuretic. He is scheduled for a hernia repair. He is completely asymptomatic, but his potassium is 3.0 mEq/L. The appropriate management is to

(A) cancel the procedure
(B) start a potassium infusion and proceed
(C) proceed with the procedure but not administer potassium
(D) give 40 mEq of potassium by mouth before starting
(E) proceed with the procedure only if it can be done under regional anesthesia

715. Metoclopramide

(A) is a dopaminergic agonist
(B) decreases gastric acid secretion
(C) stimulates motility of the upper gastrointestinal tract
(D) may lead to vomiting
(E) leads to an ileus and increased small intestinal transit time

716. Perioperative therapy with β-adrenoceptor antagonists

(A) should be restricted to patients undergoing cardiac surgery procedures
(B) has only been shown to improve outcome in patients undergoing vascular surgery procedures
(C) reduces cardiac morbidity and mortality in patients with or at risk for coronary artery disease who undergo noncardiac surgery
(D) has its effect on the renin-angiotensin system
(E) leads to tachycardia

717. A patient is scheduled for surgery who is a Jehovah's Witness. She adamantly refuses to receive blood products. The procedure may require volume replacement. A product that may be used is

(A) 5% albumin
(B) washed red cells
(C) autologous blood
(D) hetastarch
(E) platelets to decrease bleeding, thus making transfusion unnecessary

718. A 21-year-old female is emergently taken to the operating room for exploratory laparotomy after sustaining multiple injuries including a grade 4 splenic rupture in a high-speed motor vehicle accident. On transfer to the operating room table, the patient is noted to have pulseless ventricular fibrillation, and cardiopulmonary resuscitation according to ACLS guidelines is initiated. Despite resuscitation according to protocol including the administration of epinephrine IV, the patient remains in ventricular fibrillation. Which of the following drugs is recommended as an alternative to epinephrine in this setting?

(A) phenylephrine 80 mcg IV push
(B) adenosine 6 to 12 mg IV push
(C) vasopressin 40 U IV push
(D) sodium bicarbonate 50 mEq IV push
(E) overdrive pacing with isoproterenol

719. Dobutamine

(A) is primarily an α-adrenergic agonist
(B) has primarily β_1-adrenergic effects
(C) causes decreased renal blood flow
(D) is associated with severe increases in heart rate
(E) is a naturally occurring catecholamine

720. Edrophonium, in a dose of 1 mg/kg

(A) has a slower onset time than neostigmine
(B) has a much shorter duration than neostigmine
(C) has greater muscarinic side effects than neostigmine
(D) has a faster onset and decreased duration than neostigmine
(E) should be preceded by atropine

721. A patient with hypertension is being treated with clonidine. This drug

(A) has its primary effect in the brain stem
(B) should be quickly tapered before surgery
(C) should be given by infusion during surgery
(D) is a central analgesic that is reversed by naloxone
(E) increases the MAC of halothane

722. A patient, who has a known history of heart disease and asthma, is seen in the recovery room with hypertension. An infusion of labetalol is started. All of the following statements about this drug are true EXCEPT

(A) it may precipitate wheezing in this patient
(B) it is a β-adrenergic antagonist
(C) it is an α-adrenergic antagonist
(D) it is conjugated in the liver
(E) the IV and oral doses are the same

723. All of the following statements are true of protamine when used for heparin antagonism EXCEPT

(A) patients on protamine-containing insulin are at high risk for adverse effects to IV protamine
(B) it is a synthetic salt
(C) it can cause anaphylactic reactions
(D) it should be administered slowly
(E) it should be administered into a peripheral vein

724. Esmolol

(A) is a β_1-adrenergic agonist
(B) has a half-life of 4 hours
(C) is contraindicated in the patient with AV block
(D) is more likely than propranolol to cause bronchospasm
(E) is metabolized in the liver

725. A 45-year-old male is to have a gallbladder procedure. He has had heart disease and is on amiodarone. This drug

(A) should be stopped before surgery
(B) has a half-life of 4 hours
(C) is used for ventricular arrhythmias
(D) is eliminated by the kidneys
(E) has no effects on β-adrenergic receptors

726. Sodium nitroprusside is to be used for treatment of intraoperative hypertension. This drug

(A) causes venous dilatation only
(B) will be needed in increased doses if the patient has been previously treated with propranolol
(C) may cause cyanide toxicity in high doses, evidenced by alkalosis and increasing drug dosage needed to achieve the same result
(D) may cause acidosis as a sign of toxicity
(E) may cause a toxicity that is evidenced by an acidosis that is responsive to sodium bicarbonate

727. A local anesthetic that inhibits the reuptake of norepinephrine is

(A) procaine
(B) cocaine
(C) bupivacaine
(D) mepivacaine
(E) lidocaine with epinephrine

728. Hydroxyzine

(A) is an antihistamine
(B) is a phenothiazine derivative
(C) is a potent analgesic
(D) produces amnesia
(E) should always be given intravenously

729. Postoperative pain control with methadone

(A) is limited by its short half-life
(B) is more effective with oral administration
(C) is used on an every 2-hour regimen
(D) may take 48 hours to obtain a stable effect
(E) does not depress respiration

730. Dopamine

(A) is a transmitter confined to the CNS
(B) stimulates dopaminergic receptors only at an infusion rate of 10 mcg/kg/min
(C) decreases renal blood flow
(D) increases cardiac output by stimulating β_1-adrenergic receptors
(E) decreases pulmonary artery pressure

731. A patient has undergone a laparotomy for a bowel obstruction. A rapid-sequence induction was performed using succinylcholine for muscle relaxation, and pancuronium was subsequently given. Neuromuscular blockade was reversed with neostigmine. A few minutes after extubation, the surgical dressing is stained with blood and the surgeon decides that the wound must be reexplored. If another rapid-sequence induction is to be performed

(A) succinylcholine will be ineffective
(B) succinylcholine will lead to severe hyperkalemia

(C) succinylcholine will be effective, however, the time of onset will be delayed
(D) succinylcholine will be effective, however, its duration will be prolonged
(E) succinylcholine will behave as it did during the first rapid-sequence induction in this patient

732. Vecuronium

(A) has a longer duration in children as compared to adults
(B) has an onset time independent of dose
(C) is totally eliminated by the kidneys
(D) activity is not prolonged by hepatic disease
(E) inhibits histamine catabolism

733. Cimetidine may impair the metabolism of all of the following EXCEPT

(A) lidocaine
(B) propranolol
(C) diazepam
(D) atracurium
(E) theophylline

734. A patient is scheduled for glaucoma surgery. Timolol eyedrops have been in use for 2 months. This drug

(A) is a β-adrenergic agonist
(B) may be beneficial in the patient with asthma
(C) may lead to tachycardia under anesthesia
(D) is not absorbed systemically
(E) does not affect the size of the pupil

735. Drug elimination may be via all of the following organs EXCEPT

(A) breast
(B) kidney
(C) lung
(D) sweat glands
(E) muscle

736. Scopolamine

(A) increases heart rate better than atropine
(B) is a better antisialogogue than atropine
(C) has poorer amnestic properties than glycopyrrolate
(D) has poorer sedative properties than glycopyrrolate
(E) is better for decreasing gastric acid than glycopyrrolate

737. Dantrolene

(A) has a half-life of about 36 hours
(B) reduces levels of intracellular calcium
(C) causes marked cardiac depression
(D) should be administered daily for 3 days after an episode of malignant hyperthermia
(E) causes nephrotoxicity

738. Glucagon has all of the following effects EXCEPT

(A) decreased insulin secretion
(B) inotropic cardiac effects
(C) relaxation of gastrointestinal smooth muscle
(D) increased hepatic gluconeogenesis
(E) increased lipolysis in adipose tissue

739. Atracurium

(A) has both prejunctional and postjunctional effects
(B) liberates histamine at all dose levels
(C) liberates histamine at all infusion rates
(D) breaks down into laudanosine, which has caused seizures in humans
(E) causes increased blood pressure

Questions 740 through 742 refer to Figure 7-4.

740. The figure shows the blood concentration of a drug as a function of time after an IV bolus of 400 mg. The volume of distribution of this medication in the central compartment is

(A) 5 L
(B) 8 L
(C) 10 L
(D) 12 L
(E) 15 L

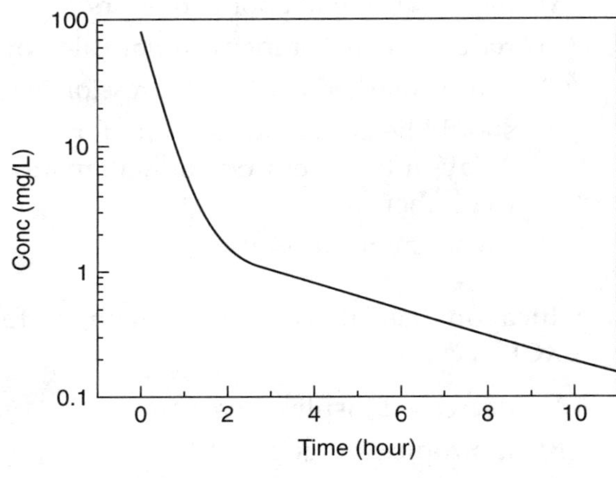

FIG. 7-4

741. The redistribution half-life for this drug is

(A) 5 minutes
(B) 10 minutes
(C) 15 minutes
(D) 20 minutes
(E) 30 minutes

742. The terminal half-life for this drug is

(A) 1 hour
(B) 2 hours
(C) 3 hours
(D) 4 hours
(E) 5 hours

743. A 72-year-old male with Parkinson's disease on levodopa is scheduled for surgery. The anesthetic plan should include

(A) stopping levodopa for 24 hours before induction
(B) avoidance of phenothiazines
(C) use of neuroleptanesthesia as the technique
(D) anticipated need for larger than usual doses of pressors
(E) use of high concentrations of a volatile agent in lieu of a nondepolarizing muscle relaxant

744. A patient who has had surgical ablation of the pituitary gland is scheduled for surgery. She is on desmopressin twice daily. The drug

(A) may be given intravenously during the procedure
(B) may lead to hypotension
(C) has a half-life of a few minutes in the circulation
(D) is equally effective in nephrogenic diabetes insipidus
(E) may increase blood loss at the time of surgery

745. Dextran

(A) may lead to severe allergic reactions during its initial administration to a patient
(B) may lead to clot formation by accelerating coagulation
(C) interferes with crossmatching of blood due to the hemolysis it causes
(D) stays in the intravascular space for 3 hours
(E) blocks release of histamine

746. Mivacurium

(A) is a steroidal compound
(B) is eliminated unchanged by the kidneys
(C) has a faster onset than succinylcholine
(D) has a longer duration than vecuronium
(E) is suitable for use by infusion

747. Edrophonium

(A) inhibits acetylcholinesterase by the same mechanism as neostigmine
(B) inhibits acetylcholinesterase by the same mechanism as pyridostigmine
(C) has a longer duration than neostigmine
(D) forms a chemical bond with acetylcholinesterase
(E) has a shorter time of inhibition than neostigmine or pyridostigmine

748. A patient with hemiplegia is undergoing surgery during which pancuronium is administered for muscle relaxation. Twitch monitors have been placed on both the affected and unaffected sides. The patient may respond with

(A) hyperkalemia
(B) neuromuscular block on the normal side with decreased block on the affected side
(C) neuromuscular block on the affected side with decreased block on the normal side
(D) an exaggerated response on both sides
(E) a normal response on both sides

749. Age-related differences in drug response in the elderly include

(A) increase in MAC
(B) decreased rate of hepatic glucuronidation of morphine
(C) lower required induction dose of thiopental
(D) shorter recovery time to normal ventilatory response after fentanyl
(E) lower intubating dose of vecuronium

750. Nonparticulate antacid administration

(A) should be given 3 hours before surgery
(B) decreases gastric volume
(C) may lead to pulmonary distress if aspiration occurs
(D) has a lag time of 1 hour for effectiveness
(E) is aimed at raising the pH to at least 2.5

751. Histamine H_2 antagonists administered in the immediate preoperative period

(A) facilitate gastric emptying
(B) should be used in all patients
(C) increase the pH of fluid in the stomach as well as fluid secreted after administration
(D) protect against aspiration of gastric juice
(E) do not change the need for a cuffed tube

752. All of the following statements are true of the use of diazepam used as a preanesthetic medication EXCEPT that it

(A) will decrease anxiety
(B) will decrease recall
(C) will decrease the likelihood of local anesthetic toxicity
(D) should be given intramuscularly
(E) is a better amnestic if scopolamine is given with it

753. A drug that is a mixed opioid agonist-antagonist is

(A) butorphanol
(B) naloxone
(C) bleomycin
(D) methohexital
(E) midazolam

754. Which of the following induction agents causes postoperative adrenal suppression?

(A) thiopental
(B) midazolam
(C) ketamine
(D) etomidate
(E) propofol

755. Patients with alcohol abuse

(A) will have increased anesthetic requirements in the acute state of intoxication
(B) will have reduced anesthetic requirements in the chronic abuse state
(C) will develop tolerance to its CNS effects with chronic usage
(D) will develop tolerance to its respiratory effects with chronic usage
(E) are more resistant to the toxic effects of local anesthetics

756. Which of the following is a β-adrenergic antagonist?

(A) isoproterenol
(B) dobutamine
(C) nadolol
(D) albuterol
(E) ritodrine

757. A drug that is a tricyclic antidepressant is

(A) pargyline
(B) phenelzine
(C) isocarboxazid
(D) haloperidol
(E) doxepin

758. The local anesthetic with the LEAST toxicity after accidental intravascular injection is

(A) procaine
(B) chloroprocaine
(C) tetracaine
(D) lidocaine
(E) mepivacaine

759. The correct order for the lipid solubilities of opioids is

(A) alfentanil > fentanyl > sufentanil > morphine
(B) alfentanil > sufentanil > fentanyl > morphine
(C) morphine > fentanyl > sufentanil > alfentanil
(D) sufentanil > alfentanil > fentanyl > morphine
(E) sufentanil > fentanyl > alfentanil > morphine

760. Which of the following statements is TRUE of heparin?

(A) It inhibits several steps in the intrinsic pathway of blood clotting.
(B) Its dosage may be adjusted by estimating the patient's clotting ability via the bleeding time.
(C) It should not be injected subcutaneously.
(D) It interacts with drugs which inhibit the liver microsomal enzyme system.
(E) It should never be combined in therapy with an oral anticoagulant.

761. Which of the following classes of agents may be used as a nighttime sedative with the LEAST disturbance of physiologic sleep?

(A) phenothiazines
(B) barbiturates
(C) butyrophenones
(D) benzodiazepines
(E) tricyclic antidepressants

762. The p*K*a of mepivacaine is 7.6. At physiologic pH (7.4), what percentage of mepivacaine molecules are in the uncharged form?

(A) 3.9%
(B) 6.1%
(C) 39%
(D) 61%
(E) 100%

DIRECTIONS (Questions 763 through 921): For each of the items in this section, ONE or MORE of the numbered options is correct. Choose answer

(A) if only 1, 2, and 3 are correct
(B) if only 1 and 3 are correct
(C) if only 2 and 4 are correct
(D) if only 4 is correct
(E) if all are correct

763. A patient who has recently self-administered cocaine is to be anesthetized for emergency surgery. Which of the following recommendations should be followed in this patient?

(1) avoid epinephrine
(2) avoid β-adrenergic antagonists
(3) avoid halothane
(4) avoid morphine

764. (An) agent(s) known to increase cerebral blood flow is (are)

(1) halothane
(2) isoflurane
(3) sevoflurane
(4) thiopental

765. If a patient has increased intracranial pressure and a volatile agent is to be used, methods that can be used to mitigate problems include

(1) use of isoflurane as compared to halothane
(2) hyperventilating the patient to produce hypocarbia
(3) induction and maintenance of anesthesia with thiopental
(4) administration of a vasopressor to decrease cerebral blood flow

766. Ephedrine

(1) is a catecholamine
(2) is a direct-acting pure α-adrenergic agonist
(3) can be used for long periods without tachyphylaxis
(4) does not decrease uterine blood flow

767. Conduction blockade by local anesthetics is by blockade of flow of

(1) calcium ions
(2) chloride ions
(3) magnesium ions
(4) sodium ions

768. Calcium channel blockers

(1) should be continued through the perioperative period
(2) have a profound negative inotropic effect when administered to a patient who has been given a high-dose opioid anesthetic
(3) act with inhalational anesthetics to prolong atrioventricular conduction time
(4) may antagonize the effect of muscle relaxants

769. In the distribution phase of an IV drug

(1) the delivery of the drug to tissues is independent of blood flow
(2) highly charged, lipid-insoluble drugs distribute to the vessel-rich group of tissues because of poor uptake by fat
(3) distribution to the vessel-poor group of tissues is facilitated by binding to plasma proteins
(4) the interstitial concentration of the drug is affected by the pH of the interstitial fluid

770. Drug clearance

(1) may be due to metabolism to an inactive product
(2) may be due to metabolism to less active products
(3) may be due to elimination of an unchanged drug
(4) is independent of drug concentration

771. Noncatecholamine sympathomimetics include

(1) ephedrine
(2) dopamine
(3) methoxamine
(4) isoproterenol

772. The loading dose of a drug

(1) helps reach a tissue concentration in a shorter time
(2) causes a higher level than needed for a short time
(3) decreases the time to a steady-state plasma concentration
(4) may lead to undesirable effects

773. Which of the following statements about the pharmacologic action(s) of nesiritide is (are) true?

(1) It causes arterial and venous dilation.
(2) It is derived from brain-type natriuretic peptide.
(3) It will increase renal blood flow and glomerular filtration rate.
(4) It suppresses the action of norepinephrine, angiotensin, and endothelin.

774. Anesthetic agents for which specific antagonists exist include

(1) fentanyl
(2) butorphanol
(3) midazolam
(4) propofol

775. The elevation of intracranial pressure seen with succinylcholine is

(1) prevented by pretreatment with a nondepolarizing muscle relaxant
(2) associated with transient hypercalcemia
(3) not seen in patients who are hyperventilated and given thiopental
(4) due to fasciculation of the sacrospinalis muscles

776. Epinephrine

(1) stimulates α-adrenergic and β-adrenergic receptors
(2) has strong chronotropic effects
(3) may lead to cardiac arrhythmias
(4) may decrease systemic vascular resistance

777. Isoproterenol may be used

(1) in treating heart failure with pulmonary hypertension
(2) as a temporary pacemaker
(3) as a bronchodilator
(4) to treat overzealous β-adrenergic blockade

778. Droperidol, used as a preanesthetic medication,

(1) may produce dysphoria
(2) increases intraoperative rhythm disturbances
(3) may produce extrapyramidal symptoms
(4) increases PONV

779. Which of the following statements about amiodarone is (are) true?

(1) It is the drug of choice for the treatment of WPW syndrome.
(2) It is considered the most effective of the antiarrhythmic drugs for the prevention of recurrences of atrial fibrillation.
(3) Hypotension does not occur with IV administration.
(4) It is a therapeutic option in the treatment of refractory ventricular fibrillation and tachycardia.

780. Nifedipine

(1) reduces arteriolar resistance and blood pressure
(2) may cause tachycardia as a compensatory effect
(3) is not suited for treatment of arrhythmias
(4) is a coronary vasoconstrictor

781. Systemic signs of local anesthetic toxicity include

(1) dizziness
(2) tinnitus
(3) difficulty in focusing
(4) drowsiness

782. The cardiac effects of local anesthetics

(1) occur by interference with sodium conductance
(2) may be manifest as ventricular fibrillation
(3) are worse with bupivacaine because the effect is of longer duration
(4) are easily treated with diazepam

783. The use of the priming principle to shorten the onset time for neuromuscular blocking agents

(1) uses divided doses of the drug
(2) is a reliable method for obtaining adequate blockade in an emergency
(3) may lead to the patient having difficulty in swallowing and protecting the airway in the interval period
(4) requires an interval period of about 120 seconds

784. Drug elimination from the body

(1) is faster than distribution
(2) may occur by metabolism
(3) is usually complete after two half-lives
(4) may occur by excretion

785. As compared with atracurium, cisatracurium

(1) has the same ED50
(2) has a shorter duration
(3) has the same potency
(4) releases less histamine

786. When using an infusion of a muscle relaxant

(1) monitoring is not necessary once the steady state is reached
(2) the first twitch should be maintained
(3) volatile agents will not affect relaxant requirements
(4) atracurium should not be mixed in Ringer's lactate solution

787. Administration of phenylephrine

(1) is useful in treating hypotension accompanying spinal anesthesia
(2) leads to an increased cardiac output
(3) causes increases in pulmonary artery occlusion pressure
(4) is given in doses of 10 to 25 mg/70 kg

788. Theophylline is

(1) a catecholamine
(2) an adenosine antagonist
(3) an adrenergic antagonist
(4) a phosphodiesterase inhibitor

789. Halothane sensitizes the heart to catecholamines. This sensitization is

(1) potentiated by hypoxemia
(2) potentiated by hypercarbia
(3) diminished by β-adrenergic antagonists
(4) diminished by calcium channel blockers

790. Doxorubicin may produce a cardiomyopathy that

(1) is present only during therapy
(2) is dose dependent
(3) is not affected by radiation therapy
(4) can be evaluated with echocardiography

791. Cardiovascular changes seen with volatile anesthetics include

(1) blood pressure decrease due to depression of cardiac output with halothane
(2) blood pressure decrease due to vasodilatation and depression of contractility with sevoflurane
(3) blood pressure decrease due to decreased peripheral vascular resistance with isoflurane
(4) more pronounced heart rate changes with halothane than with isoflurane

792. Nitrous oxide

(1) decreases pulmonary vascular resistance
(2) increases cardiac output when given with opioids
(3) decreases blood pressure when administered with isoflurane
(4) increases blood pressure when given with halothane

793. Drug(s) considered safe for the patient with a history of malignant hyperthermia is (are)

(1) thiopental
(2) lidocaine
(3) droperidol
(4) pancuronium

794. Increasing the dose of pancuronium from its ED95 to twice the ED95

(1) shortens the onset of block to less than 2 minutes
(2) increases the intensity of the block
(3) lengthens the time of recovery
(4) has no effect on heart rate

795. Physostigmine has been shown to reverse the CNS effects of

(1) diphenhydramine
(2) diazepam
(3) scopolamine
(4) meperidine

796. IV benzodiazepines

(1) are equally effective amnestic agents
(2) do not lower blood pressure when used as the sole induction agent
(3) are not associated with respiratory depression
(4) are metabolized by the liver

797. Digoxin

(1) increases myocardial contractile force
(2) slows heart rate
(3) inhibits Na,K-ATPase
(4) is potentiated by calcium

798. α-adrenoceptors in the adrenergic nervous system

(1) are stimulated by norepinephrine
(2) are blocked by labetalol
(3) cause vasoconstriction
(4) cause bronchial dilatation

799. The histamine release seen with administration of muscle relaxants may cause

(1) bradycardia
(2) erythema
(3) increases in peripheral vascular resistance
(4) hypotension

800. The decrease in cerebral metabolic rate caused by inhaled anesthetics is

(1) most profound with isoflurane
(2) independent of dose
(3) associated with a decrease in cerebral electrical activity
(4) present during seizure activity

801. Esmolol

(1) is an ester
(2) metabolism is antagonized by succinylcholine
(3) can blunt the response to intubation
(4) has a duration longer than that of propranolol

802. Drugs effective in the treatment of depression include

(1) trazodone
(2) buspirone
(3) fluoxetine
(4) benztropine

803. A child is brought to the emergency room for evaluation of his disoriented state. This began shortly after he ate some berries in the garden. He is noted to be very warm, flushed, and to have dilated pupils. Appropriate actions include administration of

(1) atropine
(2) aspirin
(3) diphenhydramine
(4) physostigmine

804. A patient has been on oral steroid therapy for a dermatologic problem for 2 years. He or she is to undergo a cholecystectomy. The appropriate approach to steroid coverage may include administration of hydrocortisone sodium succinate

(1) 100 mg PO at the time of oral benzodiazepine premedication
(2) 100 mg IV at the time of induction
(3) 100 mg IV the night before surgery
(4) 100 mg IV 12 hours after induction

805. Cefazolin is being administered when there is a sudden increase in airway pressure, wheezing, tachycardia, and decreased blood pressure. The appropriate treatment should include

(1) stopping the cefazolin infusion
(2) 100% oxygen
(3) a bolus of IV fluid
(4) 100 mcg epinephrine IV

806. Tachyphylaxis

(1) can occur with administration of ephedrine
(2) occurs with continuous administration of norepinephrine
(3) describes the rapidly decreasing effectiveness of certain drugs with repeat administration
(4) is also known as ceiling effect

807. High doses of fentanyl may cause hypotension as a result of

(1) histamine release
(2) a direct vasodilatory effect
(3) myocardial depression
(4) bradycardia

808. Digoxin toxicity may be manifest as

(1) prolonged P-R interval
(2) nausea and vomiting
(3) multifocal PVC's
(4) tachycardia

809. β_2-Adrenergic receptors in the autonomic nervous system

(1) are stimulated by isoproterenol
(2) are antagonized by esmolol
(3) cause glycogenolysis
(4) cause vasoconstriction

810. Magnesium deficiency may be

(1) seen in the chronic alcoholic
(2) manifest as cardiac arrhythmias
(3) associated with skeletal muscle spasm
(4) associated with CNS irritability

811. Opioid agonists produce

(1) dilated pupils
(2) nausea and vomiting mediated through the gastrointestinal tract
(3) good amnesia
(4) unconsciousness at high doses

812. A patient scheduled for an emergency procedure gives a history of heroin abuse. Problems that must be anticipated include

(1) full stomach
(2) exposure to AIDS
(3) exposure to hepatitis
(4) decreased need for anesthetics

813. Toxicity from administration of sodium nitroprusside

(1) may be due to cyanide or thiocyanate ions
(2) is aggravated by the previous treatment with captopril
(3) is seen more often in patients who are nutritionally deprived
(4) will not occur if the dose is kept under 2 mg/kg/h

814. Drugs that completely or partly depend on renal elimination and, therefore, must be avoided or used in reduced dosage in the patient with renal failure include

(1) penicillin
(2) sodium nitroprusside
(3) digoxin
(4) pancuronium

815. Drugs suitable for the patient with hyperthyroidism include

(1) thiopental
(2) propranolol
(3) succinylcholine
(4) aspirin

816. Isoflurane-mediated coronary steal

(1) involves the diversion of blood away from areas of heart muscle supplied by collaterals
(2) results in a higher incidence of perioperative ischemia when isoflurane is used in patients with coronary artery disease
(3) results from the vasodilatory effect of isoflurane on the coronary circulation
(4) is associated with increased postoperative mortality in patients undergoing CABG under isoflurane anesthesia as compared to nitrous oxide/opioid anesthesia

817. Corticosteroids given for the treatment of asthma

(1) have an immediate onset of action
(2) are best given by the inhalation route to decrease onset time
(3) have no use in the anesthetized patient
(4) should involve only those drugs with little mineralocorticoid effect

818. Drug therapy appropriate for ventricular arrhythmias includes

(1) lidocaine
(2) procainamide
(3) amiodarone
(4) adenosine

819. Mechanism(s) that accurately describe the effect of the drugs used to produce deliberate hypotension is (are)

(1) sodium nitroprusside only dilates resistance vessels
(2) nicardipine causes coronary and peripheral vasodilatation
(3) isoflurane causes decreased cardiac output and peripheral vascular resistance
(4) nitroglycerin dilates capacitance vessels

820. Anesthetic management of the patient in renal failure may include

(1) propofol for induction
(2) fentanyl for analgesia
(3) maintenance of anesthesia with isoflurane
(4) pancuronium for intubation

821. Anesthetic drugs that are appropriate for the patient with asthma include

(1) ketamine
(2) atropine
(3) thiopental
(4) halothane

822. Ionization of a drug

(1) is a function of pH of the fluid in which it is dissolved
(2) may render a drug inactive
(3) is a function of the p*K*a of the drug
(4) may affect the ability of the drug to cross membranes

823. Pharmacologic differences in the obese patient as compared to the lean patient include

(1) a decreased degree of metabolism of isoflurane
(2) an increased terminal half-life of thiopental
(3) a prolonged duration of action of succinylcholine
(4) an increased volume of distribution of midazolam

824. Which of the following agents have antiemetic effects?

(1) droperidol
(2) prochlorperazine
(3) ondansetron
(4) ranitidine

825. The pharmacologic effects of ketamine include

(1) depression of the medial thalamic nuclei
(2) no effects on porphyrin metabolism
(3) maintenance of respiratory rate and tidal volume at near normal levels
(4) loss of gag reflex

826. The principal disadvantage(s) of methohexital is (are)

(1) poor water solubility
(2) pain on injection
(3) low pH of solution
(4) involuntary muscle movements

827. A patient with acute intermittent porphyria

(1) voids urine-containing uroporphyrins, which is pathognomonic for the disease
(2) will likely suffer a cardiac arrest if administered thiopental for induction of anesthesia
(3) will have urine negative for porphobilinogen
(4) may be given morphine safely

828. Systemic toxicity to a local anesthetic is related to the

(1) dose of drug
(2) rapidity of injection
(3) vascularity of the site injected
(4) inherent toxicity of the drug injected

829. After IV injection, the distribution of thiopental will show

(1) rapidly falling concentration in blood
(2) increasing concentration in the viscera
(3) a rising concentration in muscle that follows the concentration in the viscera
(4) a slowly rising concentration in the fatty tissue

830. A patient who has been treated in the past with high doses of cisplatin may have which of the following permanent problems?

(1) renal insufficiency
(2) chronic pancreatitis
(3) peripheral neuropathy
(4) congestive heart failure

831. A comparison of dissociative and inhalational anesthesia shows that

(1) the relaxed, nonresponding state is present in each
(2) there is decreased muscle tone in each
(3) the presence of movement is a sign of insufficient dosage in each
(4) the eyes may remain open during dissociative anesthesia

832. The signs of nondepolarizing block include

(1) fade with train-of-four stimulation
(2) fade with tetanic stimulation
(3) posttetanic facilitation
(4) no change after administration of anticholinesterase drug

833. Healing of a duodenal ulcer may be facilitated by

(1) antacids
(2) sucralfate
(3) cimetidine
(4) omeprazole

834. Local anesthetic agents block nerve conduction by

(1) blocking conduction through the myelin sheath
(2) acting from within the nerve cell
(3) depolarizing the membrane
(4) blocking ionic movement through membrane pores

835. The short duration of methohexital is due to

(1) its low pH
(2) its inactivation by metabolism
(3) low fat solubility
(4) its redistribution

836. If a patient has an allergic reaction to an injection of lidocaine obtained from a multiple-dose vial, the patient

(1) will probably not have a reaction if a second injection is made with lidocaine from a fresh, single-dose vial
(2) will probably have no reaction to procaine
(3) may be exhibiting a reaction to methylparaben
(4) is probably reacting to a bacterial contaminant previously introduced into the vial

Questions 837 through 839

A 46-year-old woman enters the hospital for correction of strabismus. She has been using echothiophate iodide drops for 2 years. She undergoes induction of anesthesia with thiopental, and succinylcholine is given to facilitate intubation. Anesthesia is maintained with 70% N_2O in O_2 and fentanyl. During the procedure she is mechanically ventilated, and no untoward effects occur during the case. At the end of the surgery, the patient does not awaken and does not move or breathe.

837. Possible causes of the apparent prolonged anesthesia include

(1) fentanyl overdosage
(2) prior use of echothiophate iodide
(3) pseudocholinesterase deficiency
(4) nitrous oxide

838. Maneuvers that can be done to ascertain the cause of the prolongation include

(1) use of a nerve stimulator
(2) inhalation of carbon dioxide
(3) administration of naloxone
(4) administration of another dose of succinylcholine

839. Application of a nerve stimulator reveals that the patient has no twitch or tetanic response to nerve stimulation. There is no response to opioid antagonism. The proper treatment includes

(1) administration of pralidoxime
(2) administration of a stimulant, e.g., epinephrine
(3) administration of neostigmine
(4) continued ventilation

840. The membrane potential of a muscle cell is

(1) about –90 mV at rest
(2) primarily dependent on the relative concentrations of intracellular and extracellular potassium
(3) dependent on the selective permeability of the cell membrane
(4) changed rapidly with depolarization

841. Intramuscular injection

(1) permits a more rapid onset than after subcutaneous injection
(2) into the gluteus maximus permits a more rapid onset than if the drug were injected into the quadriceps
(3) is preferred over subcutaneous injection for irritating substances
(4) cannot be used for nonaqueous solutions

842. Succinylcholine may cause a prolonged effect in patients

(1) whose pseudocholinesterase is not inhibited by dibucaine
(2) with severe liver disease
(3) with a low fluoride number for pseudocholinesterase
(4) who are pregnant

843. Cardiovascular changes noted after ketamine administration include

(1) decreased systolic blood pressure
(2) increased heart rate
(3) decreased diastolic blood pressure
(4) increased cardiac output

844. A patient has been receiving lithium for manic-depressive illness. At the time of surgery, which of the following drugs would be potentiated?

(1) pancuronium
(2) succinylcholine
(3) thiopental
(4) neostigmine

845. Signs of histamine release include

(1) erythema
(2) hypertension
(3) bronchospasm
(4) bradycardia

846. Inhibitor(s) of acetylcholinesterase that act by forming a carbamyl ester at the esteratic site of the enzyme is (are)

(1) neostigmine
(2) physostigmine
(3) pyridostigmine
(4) edrophonium

847. Drug antagonism may occur when two drugs

(1) chemically combine in the body
(2) both affect a physiologic system in a similar way
(3) compete for the same receptor site
(4) displace the dose-response curve in the same direction

848. When reversing nondepolarizing neuromuscular blockade

(1) atropine should be given at the same time as neostigmine
(2) glycopyrrolate should be given at the same time as edrophonium
(3) neostigmine should be titrated until reversal is complete, up to a maximum dose of 15 mg in the 70-kg patient
(4) lack of fade during DBS is a good indicator of reversal

849. Vomiting may be induced by the administration of

(1) apomorphine
(2) chlorpromazine
(3) ipecac
(4) dimenhydrinate

850. The myasthenia gravis patient

(1) has weakness of muscles innervated by cranial nerves
(2) usually has diaphragmatic weakness
(3) is very sensitive to nondepolarizing agents
(4) has focal sensory deficits

851. The purpose of digoxin therapy is to

(1) increase depressed myocardial contractility
(2) achieve higher stroke volume at lower filling pressure
(3) move the patient onto a better Frank-Starling cardiac function curve
(4) control severe tachyarrhythmias

852. In a patient with myotonia, one would expect muscle spasm that

(1) occurs after succinylcholine
(2) does not relax with pancuronium
(3) is not prevented by spinal anesthesia
(4) is alleviated by quinine

853. Acetylcholine effects that are potentiated by neostigmine include

(1) bradycardia
(2) sweating
(3) skeletal muscle contraction
(4) urinary retention

854. Which of the following circumstances increase the chance of clinically significant carbon monoxide poisoning during general anesthesia?

(1) first case on Monday morning
(2) use of desflurane
(3) use of high fresh gas flows over prolonged periods
(4) use of Baralyme as the CO_2 absorbent

855. Figure 7-5 shows the ratio of the concentrations of sevoflurane in alveolar gas to inspired gas as a function of time. Factor(s) which will increase the initial slope is (are):

(1) increased inspired sevoflurane concentration
(2) presence of nitrous oxide in inspired gas
(3) increased minute ventilation
(4) increased cardiac output

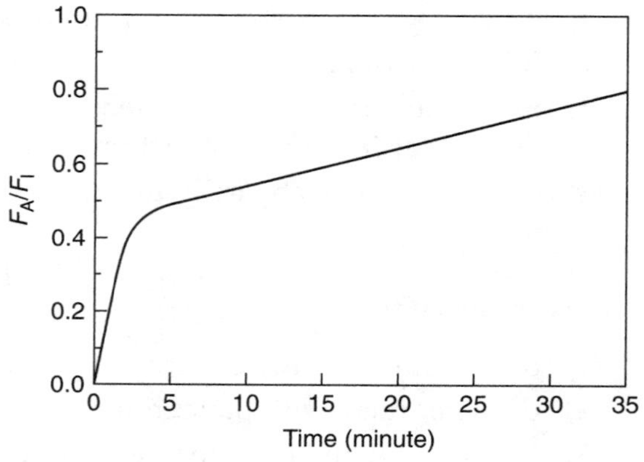

FIG. 7-5

856. Succinylcholine is contraindicated in

(1) amyotrophic lateral sclerosis
(2) massive tissue trauma
(3) paraplegia
(4) burns

857. Two drugs are said to have an additive effect when

(1) one drug will accelerate the speed of onset of the other
(2) one drug will prolong the action of the other
(3) both drugs are of the same chemical family
(4) the combined effect of the two drugs is the algebraic sum of the individual actions

Questions 858 and 859 refer to Figure 7-6.

858. Drugs A and B were given by IV bolus every 24 hours. From the graph, it can be concluded that

(1) drug A does not maintain persistent therapeutic levels when administered every 24 hours
(2) drug A exhibits cumulative properties
(3) drug B is cumulative when administered every 24 hours
(4) drug B is completely destroyed in less than 24 hours

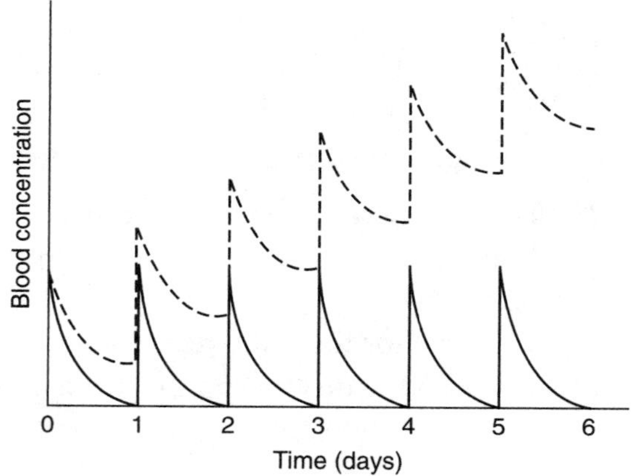

FIG. 7-6

859. In addition, the graph shows that

(1) drug A has a long half-life in the body
(2) a large initial dose of both drugs was given
(3) different doses are given for initial doses and maintenance doses
(4) in the case of drug B, the body does not remove one dose before another is administered

860. Pyridostigmine, when compared with neostigmine

(1) has a longer duration of action
(2) acts at a different site on acetylcholinesterase
(3) is more dependent on the kidneys for elimination
(4) has a more rapid onset of action

861. β-Adrenoceptors prepare the body for fight or flight by

(1) increasing the cardiac output
(2) bronchoconstriction
(3) increasing the circulating glucose
(4) contracting the ciliary muscle

862. All of the following agents increase hypoxic pulmonary vasoconstriction EXCEPT

(1) halothane
(2) fentanyl
(3) thiopental
(4) sodium nitroprusside

863. Metoprolol

(1) is a selective β_1-adrenergic antagonist
(2) inhibits the vasodilator effect of isoproterenol
(3) inhibits the chronotropic effect of isoproterenol
(4) is synergistic with the inotropic effect of isoproterenol

864. A diabetic patient is to undergo anesthesia for vein stripping. On the morning of surgery, he receives one-half dose of his insulin requirement and his usual morning dose of 40 mg of propranolol. Anesthesia is maintained with isoflurane. In the recovery room 4 hours later, his blood sugar is 52 mg/dL. The cause of this could be

(1) stress of surgery
(2) propranolol
(3) isoflurane
(4) insulin

865. Adverse effects of digoxin include

(1) altered color vision
(2) hunger
(3) dizziness
(4) alopecia

866. The Eaton-Lambert syndrome is characterized by

(1) increased sensitivity to nondepolarizing muscle relaxants
(2) nondepolarizing neuromuscular blockade which is readily reversed by neostigmine
(3) decreased release of acetylcholine at the neuromuscular junction
(4) decreased sensitivity to depolarizing muscle relaxants

867. The instillation of phenylephrine eyedrops used before eye muscle surgery may

(1) be beneficial because of its miotic effect
(2) lead to increased conjunctival bleeding
(3) be ineffective if used in 2.5% solution
(4) cause hypertension

868. Drugs to be avoided in persons deficient in glucose-6-phosphate dehydrogenase include

(1) acetaminophen
(2) primaquine
(3) cefazolin
(4) nitrofurantoin

869. The neuroleptic state is characterized by

(1) amnesia
(2) reduced motor activity
(3) analgesia
(4) indifference to the environment

870. Large doses of IV morphine for anesthetic use

(1) are always accompanied by loss of consciousness
(2) cause increased liberation of catecholamines
(3) cause cardiac instability
(4) cause decreased cardiac output if nitrous oxide is added

871. The anesthetic state achieved with ketamine is characterized by

(1) analgesia to somatic pain
(2) inhibition of reflexes, e.g., gag reflex
(3) cardiovascular stimulation
(4) marked respiratory depression

872. A patient is infected with *Clostridium perfringens* and has gas gangrene. Which of the following agents is likely to be effective against this organism?

(1) penicillin G
(2) cefotetan
(3) clindamycin
(4) gentamicin

873. Certain drugs may affect the enzyme cytochrome P450. Which of the following statements is (are) TRUE?

(1) Cimetidine inhibits cytochrome P450.
(2) After treatment with phenobarbital, cytochrome P450 activity is decreased.
(3) Chronic ingestion of ethanol increases cytochrome P450 levels.
(4) Nitrous oxide binds to the iron atom of cytochrome P450 and prevents oxygen binding.

874. The metabolism of nitrous oxide in humans

(1) yields nitric oxide as the main product
(2) occurs in the gut
(3) is increased by phenobarbital pretreatment
(4) is a reductive process

875. Triamterene

(1) is used in combination with other diuretics to prevent potassium loss
(2) is a competitive antagonist of aldosterone
(3) acts primarily on the distal tubule
(4) is less effective in the presence of Conn's syndrome (primary hyperaldosteronism)

876. Figure 7-7 depicts the relative effect of thiopental in three physiologic states. From this we can say that

(1) in the normal patient, there is an initial high concentration followed by rapidly decreasing concentration
(2) a patient in shock will have decreased effect from thiopental
(3) the anxious patient will require a larger dose
(4) the bleeding patient has a faster decrease in brain concentration as a result of the thiopental being lost with the blood

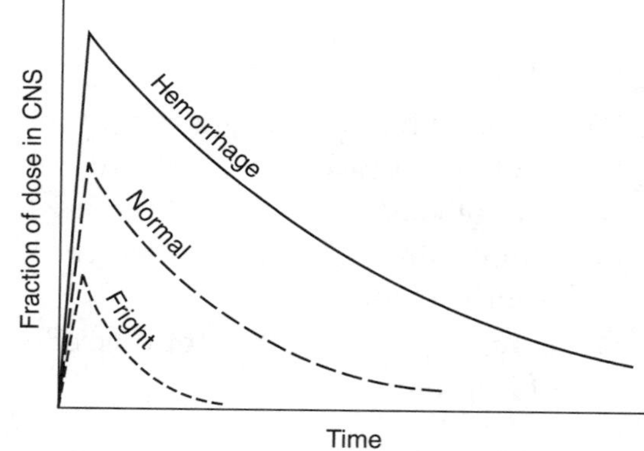

FIG. 7-7

877. Amyl nitrite is used in the emergency treatment of cyanide poisoning because it

(1) irreversibly binds cyanide ion
(2) causes vasodilation
(3) competes with cyanide ion for binding to cytochromes
(4) oxidizes hemoglobin

878. When methyl methacrylate cement is used to secure the femoral component of a hip prosthesis in the femoral canal

(1) hypertension is a common occurrence
(2) hypoxemia is a common occurrence
(3) the patient should have a right atrial catheter for the removal of an air embolus
(4) the polymerization reaction of the cement is an exothermic reaction capable of burning the patient

879. Tiotropium bromide

(1) is a β-adrenergic agonist
(2) is a muscarinic receptor antagonist
(3) is shorter acting than ipratropium bromide
(4) should be administered on the day of surgery

880. Benzodiazepines with active metabolites include

(1) lorazepam
(2) midazolam
(3) oxazepam
(4) diazepam

881. Ondansetron

(1) facilitates gastric emptying
(2) decreases gastroesophageal reflux
(3) is sedating at higher doses
(4) has antiemetic effects

882. Flumazenil

(1) is more highly bound to plasma protein than midazolam
(2) has active metabolites
(3) is longer acting than the injectable benzodiazepines used in anesthesia
(4) may antagonize the sedative effects of midazolam

883. Ropivacaine

(1) is supplied as an optically active compound
(2) produces motor blockade of longer duration than sensory blockade
(3) is less likely than bupivacaine to cause ventricular arrhythmias after accidental IV injection
(4) has a low degree of toxicity by virtue of its rapid intravascular metabolism

884. In comparison to thiopental, which of the following effects occur to a greater extent after an equianesthetic dose of propofol?

(1) Myocardial depression
(2) Binding to plasma proteins
(3) Vasodilation
(4) Decreased cerebral perfusion pressure

885. Opioids which have a shorter time to peak effect than morphine include

(1) alfentanil
(2) sufentanil
(3) fentanyl
(4) remifentanil

886. Effective single-agent prophylaxis against a wound infection caused by *Staphylococcus aureus* may be achieved by the administration of

(1) penicillin G
(2) vancomycin
(3) ampicillin
(4) cephalothin

887. A withdrawal syndrome consisting of rebound hypertension can occur after abrupt cessation of therapy with

(1) propranolol
(2) hydrochlorothiazide
(3) clonidine
(4) hydralazine

888. Which of the following agents can be used to treat hyperlipidemia?

(1) clofibrate
(2) lovastatin
(3) niacin
(4) cholestyramine

889. Vitamin D

(1) requires metabolic activation by both the liver and kidney for activity
(2) is a precursor to parathyroid hormone
(3) facilitates calcium absorption in the intestine
(4) decreases bone resorption and promotes bone deposition

890. Adverse effects to thiazide diuretics include

(1) hypokalemia
(2) glucose intolerance
(3) hyperuricemia
(4) duodenal ulceration

891. Which of the following actions is (are) involved in the antidepressant effect of amitriptyline?

(1) competitive antagonism at the 5-HT receptor
(2) inhibition of MAO
(3) depletion of norepinephrine stores
(4) inhibition of norepinephrine reuptake

892. An IV injection of PGE_1 will

(1) decrease blood pressure
(2) decrease gastric acid secretion
(3) cause bronchodilation
(4) result in watery diarrhea

893. Which of the following agents may be administered to treat postpartum hemorrhage?

(1) ergonovine
(2) $PGF_{2\alpha}$
(3) oxytocin
(4) progesterone

894. Fenoldopam

(1) activates α- and β-adrenoceptors
(2) causes a dose-related decrease in renal blood flow and glomerular filtration rate
(3) is a selective D_2 receptor agonist approved for the short-term parenteral infusion to prevent contrast-induced nephropathy
(4) is a selective D_1 receptor agonist approved for the short-term parenteral treatment of severe hypertension

895. Carbon monoxide

(1) has a higher binding affinity for hemoglobin than oxygen
(2) does not produce toxicity in the presence of an increased FIO_2
(3) causes oxygen to be released less efficiently from hemoglobin in tissues
(4) combines and dissociates from hemoglobin at a rate equal to that of oxygen

896. Which of the following effects may occur with chronic use of the combination oral contraceptive pill?

(1) increased risk of thromboembolic disorders
(2) increased risk of endometrial carcinoma
(3) impairment of glucose tolerance
(4) peptic ulcer disease

897. Adverse effects due to phenytoin include

(1) decreased seizure threshold
(2) hirsutism
(3) gingival hyperplasia
(4) megaloblastic anemia

898. Mannitol is

(1) filtered by the glomerulus
(2) a weak acid
(3) pharmacologically inert
(4) almost completely reabsorbed from the renal tubule

899. Transmembrane transport mechanisms without saturability or specificity include

(1) filtration
(2) facilitated diffusion
(3) passive nonionic diffusion
(4) active transport

900. Based on the neurotransmitter aberrations which are thought to be responsible for the symptoms of Parkinson's disease, which of the following drug types would be therapeutic in a patient with Parkinson's disease?

(1) a MAO inhibitor
(2) a muscarinic antagonist
(3) a dopaminergic agonist
(4) a 5-HT (serotonin) antagonist

901. The withdrawal syndrome in a person physically dependent on which of the following drugs can be fatal?

(1) amphetamine
(2) secobarbital
(3) heroin
(4) diazepam

902. Spironolactone

(1) is contraindicated in the presence of hyperkalemia
(2) depresses pituitary adrenocorticotropic hormone (ACTH) secretion
(3) can produce androgenic effects
(4) interferes with aldosterone biosynthesis

903. Which of the following are NOT metabolized by catechol- *O*-methyltransferase?

(1) amphetamine
(2) isoproterenol
(3) ephedrine
(4) dopamine

904. Active metabolites of heroin include

(1) morphine
(2) apomorphine
(3) acetylmorphine
(4) codeine

905. Method(s) of drug administration which circumvent the first-pass effect include

(1) sublingual administration
(2) rectal administration
(3) topical administration
(4) enteric coated tablets

906. Low-molecular-weight heparins

(1) have a more predictable pharmacokinetic profile than unfractionated heparin
(2) are administered intravenously
(3) cause a lower incidence of heparin-induced thrombocytopenia than unfractionated heparin
(4) all have equivalent antithrombotic effects

907. The majority of first-generation histamine H_1-receptor antagonists are also competitive antagonists at which of these other receptors?

(1) nicotinic
(2) histamine H_2
(3) dopaminergic
(4) muscarinic

908. A patient is admitted to the hospital approximately an hour after supposedly ingesting 10 g of aspirin. The patient has not vomited. Which of the following effects are likely to be seen?

(1) increased carbon dioxide production
(2) miosis
(3) hyperventilation
(4) hypothermia

909. Which of the following agents are utilized in the combination chemotherapy of tuberculosis?

(1) Rifampin
(2) Pyrimethamine
(3) Streptomycin
(4) Tetracycline

910. Cytochrome P450 performs which of the following metabolic reactions?

(1) acetylation of hydralazine
(2) demethylation of ketamine
(3) conjugation of morphine to glucuronide
(4) oxidation of thiopental to pentobarbital

911. Aspirin and acetaminophen both

(1) inhibit prostaglandin biosynthesis equieffectively
(2) have a high incidence of causing abdominal discomfort
(3) cause methemoglobinemia
(4) are effective antipyretic agents

912. Which of the following medications are controlled substances?

(1) Cocaine
(2) Midazolam
(3) Morphine
(4) Thiopental

913. Drug metabolism usually results in which of the following effects?

(1) alteration of the drug to yield an inactive metabolite
(2) increasing the water solubility of the drug
(3) increasing the rate of renal excretion of the drug
(4) increasing the duration of action of the drug

914. In a patient who is physically dependent on morphine, which of the following are symptoms of the abstinence syndrome that begin within 12 hours of the last dose of morphine?

(1) sweating
(2) seizures
(3) lacrimation
(4) pilomotor activity ("gooseflesh")

915. Which of the following enzymes are required for norepinephrine synthesis?

(1) dopamine β-hydroxylase
(2) catechol-*O*-methyltransferase
(3) tyrosine hydroxylase
(4) phenylethanolamine-*N*-methyltransferase

916. Adverse effects to captopril include

(1) cough
(2) renal failure in patients with renal artery stenosis
(3) proteinuria
(4) loss of sense of taste

917. Hypotension associated with morphine administration results from

(1) vasodilatation due to histamine release
(2) centrally mediated vasodilation
(3) increased vagal stimulation
(4) myocardial depression

918. The rapidly perfused (vessel-rich) tissues include

(1) heart
(2) kidney
(3) brain
(4) muscle

919. Ketorolac 60 mg and morphine 10 mg produce approximately equivalent magnitudes of which of the following effects after intramuscular injection?

(1) nausea
(2) sedation
(3) ventilatory depression
(4) analgesia

920. The administration of digoxin

(1) causes increased myocardial contractility
(2) increases myocardial oxygen consumption
(3) increases the P-R interval
(4) increases heart rate

921. The CNS response to an induction dose of an IV anesthetic agent may be altered by

(1) hypovolemia
(2) hypoalbuminemia
(3) age
(4) renal failure

Answers and Explanations

582. (A) The liver is the most important site of drug transformation. The kidney is responsible for the removal of many drugs, and the metabolism of a few. There are some drug reactions that occur in the lung and a few that occur in the spleen. *(3:12)*

583. (C) The first-pass effect is the biotransformation of a drug as it passes through the intestinal mucosa and liver. In some cases, much of the pharmacologic activity is lost at this point, leaving little of the drug to have its desired effect. *(3:5–7,12)*

584. (E) Bioavailability is the amount of drug that reaches its site of action in active form. *(3:5–8)*

585. (D) Renal clearance of a drug in an individual with normal renal function may vary from patient to patient and from time to time, depending on such variables as urine pH and flow rate and renal blood flow. Excretion is correlated with renal blood flow. Renal elimination is correlated with creatinine clearance, and one may use changes in creatinine clearance to gauge how doses must be altered. *(3:18–20)*

586. (B) Halogenation decreases flammability and increases anesthetic potency. Cardiac irritability is not increased. *(1:377)*

587. (C) Halothane is a halogenated hydrocarbon in the ethane series. The halogen substitutions are fluorine, chlorine, and bromine. *(3:378)*

588. (D) Local anesthetics act at the nerve membrane. The local anesthetics block sodium channels so that the membrane cannot conduct nerve impulses. *(3:450–1)*

589. (D) Halothane is stored in tinted bottles, and the preservative is thymol. The bottles are glass, not plastic. *(4:1125)*

590. (D) Isoflurane is a better relaxant when compared to halothane. The vapor pressure is 238 mmHg at 20°C. Isoflurane depresses ventilation and should be delivered by assisted or controlled ventilation. The value for MAC is correct. *(3:352; 4:1144–6)*

591. (A) The MAC for halothane in O_2 is 0.75 at 1 atm, making it the most potent of the inhalation anesthetics listed. Of the agents listed the order from most to least potent is halothane > isoflurane > sevoflurane > desflurane > N_2O. *(1:378)*

592. (A) Isoflurane is characterized by a stable cardiac rhythm that is maintained in the face of epinephrine injection in conservative doses. A tachycardia is noted that leads to a decrease in stroke volume. Cardiac output is maintained. *(3:352)*

593. (C) Use of heparin in a patient with heparin-induced thrombocytopenia type II can lead to life-threatening thrombotic complications and is therefore not indicated. Warfarin is an oral anticoagulant. Danaparoid is a mixture of nonheparin glycosaminoglycans and is approved for prophylaxis of deep vein thrombosis. It can be administered SC or IV, however, it has a long half-life of about 24 hours. Clopidogrel is an antiplatelet agent for the secondary prevention of stroke and the reduction of cardiac events after percutaneous coronary intervention. Argatroban is a direct thrombin inhibitor with a relatively short half-life of 1 to 2 hours that can be continuously infused during CABG surgery. *(3:1525, 1530–1, 1534–5; 5:1116)*

594. (A) All of the options have an ether structure with the exception of halothane, which is a straight chain hydrocarbon. *(1:378; 3:348; 4:1125)*

595. (A) Desflurane is an ether that contains only fluorine as the halogen. It is stable in the presence of soda lime. Its blood-gas partition coefficient is 0.42 and its vapor pressure is 664 mmHg at 20°C. *(3:348, 353–4; 4:1054–5)*

596. (D) The action of thiopental is terminated by redistribution. The drug is bound to protein, and various concomitant drugs may interfere with binding. It is taken up in fatty tissues, but that is not the mechanism of terminating its effect. *(3:343)*

597. (A) Excitatory phenomena after propofol induction occur at about the same frequency as with thiopental, however, bronchospasm appears to occur less frequently. In equipotent doses, propofol produces a dose-dependent decrease in blood pressure that is more pronounced than with thiopental administration. Propofol and thiopental cause a similar decrease in cerebral metabolic rate. Adrenal suppression occurs after the administration of etomidate. *(1: 333–4; 3:344–6; 5:323–4)*

598. (E) Thiopental is a very alkaline solution. When inadvertently mixed with acidic solutions, e.g., opioids or muscle relaxants, a precipitate will occur. *(4:1211)*

599. (C) The administration of thiopental will cause vasodilation and a mild decrease in myocardial contractility. *(3:344)*

600. (C) Pancuronium is likely to increase cardiac output and blood pressure, while vecuronium and cisatracurium are devoid of any cardiovascular side effects. Succinylcholine may cause bradycardia that may affect blood pressure. Rocuronium may lead to an increase in heart rate at doses greater than 0.6 mg/kg. *(4:781–2)*

601. (E) Thiopental will decrease cerebral oxygen consumption. Cerebral blood flow and EEG activity will decrease. Uterine contractions are not diminished, and there is little effect on skeletal muscle tone. *(3:344)*

602. (D) Injection of a sleep dose of thiopental leads to an increased heart rate, probably brought about by baroreceptor-mediated changes in sympathetic activity. These changes are more pronounced in the patient with heart disease. *(3:344)*

603. (D) Traditional signs of the anesthetic state are not seen with ketamine administration. Purposeless movements are seen, and these do not indicate the need for more anesthetic. *(1:336–7; 3:346–7; 5:347)*

604. (D) Warfarin should not be used until the thrombocytopenia has resolved because it can cause venous limb gangrene or multicentric skin necrosis. Danaparoid, lepirudin, and argatroban are anticoagulants that are approved for patients with heparin-induced thrombocytopenia. Benzodiazepines have no clinical effect on the coagulation system. *(3:1525–6, 1536; 5:1116)*

605. (C) Droperidol is a butyrophenone derivative. It is a short-acting neuroleptic that is used in anesthesia as an antiemetic. It is eliminated from the body by hepatic mechanisms. *(1:555; 3:490; 5:2598)*

606. (C) Recombinant human activated protein c (APC) has anticoagulant activity and has been shown to decrease mortality in patients with severe sepsis. The drug is infused at 24 mcg/kg/h for 96 hours and does not increase systolic blood pressure. Caution is advised in patients with thrombocytopenia and coagulopathy because of an increased risk of bleeding. *(5:2796)*

607. (E) Arginine vasopressin (AVP) levels in patients with septic shock are inappropriately low, and when administered exogenously, vasopressin can dramatically increase systolic blood pressure in this patient population. The proposed mechanism is twofold: first, endogenous AVP depletion may be the result of excessive baroreceptor-mediated release from sustained hypotension, and second, AVP appears to restore vascular sensitivity to norepinephrine. Milrinone, epinephrine, and dobutamine are not indicated because this patient has good myocardial contractility. Phenylephrine is a selective α_1-adrenoceptor agonist and does not provide additional benefit. *(3:146, 222–3, 228–9, 805, 927; 5:806–7)*

608. (A) Administration of low-dose corticosteroids to patients in septic shock has been shown to decrease the time to discontinuation of vasopressor therapy. One of the major actions of corticosteroids on the cardiovascular system is to enhance the vascular reactivity to other vasoactive substances, for example, norepinephrine or phenylephrine. Corticosteroids do not improve the bactericidal activity of broad-spectrum antibiotics, they will increase blood glucose levels, and cause a leukocytosis due to increased release of polymorphonuclear leucocytes from the bone marrow. Sodium retention is much more pronounced with administration of mineralocorticoids; glucocorticoids play a permissive role in the renal excretion of free water. *(3:1659–60; 5:2795–6)*

609. (A) Local anesthetics in the free base form are poorly soluble in water and therefore local anesthetic solutions are supplied as the hydrochloride salts. The acidic pH increases the stability of epinephrine, since basic solutions promote the breakdown of catecholamines. *(1:455; 3:371–2; 4:1308–10)*

610. (B) Rocuronium is a nondepolarizing aminosteroid muscle relaxant with an intermediate duration of action. Nondepolarizing muscle relaxants including rocuronium can block muscarinic receptors present in the airways thereby causing bronchoconstriction. The ED_{95} of rocuronium is 0.3 mg/kg; administration of three times the ED_{95} will result in a mean onset time of about 75 seconds without histamine release. Because of its fast onset of action, it is considered an alternative to succinylcholine to facilitate endotracheal intubation. It undergoes hepatic metabolism and elimination. *(1:430–1; 3:196–9; 5:513–4, 2214, 2379)*

611. (C) Succinylcholine in the susceptible patient causes the release of potassium from inside the cell. The efflux of potassium, which is normal at the motor end plate, is seen over a much wider area of the membrane in the burned patient. This may result in a massive outpouring of potassium and cardiac arrest. *(3:203; 4:2176)*

612. (C) Penicillin does not have neuromuscular blocking properties. The aminoglycoside antibiotics are associated with decreased muscle function or potentiation of muscle relaxants in a dose-related fashion. This is probably due to a prejunctional inhibition of acetylcholine release. *(3:1230)*

613. (B) Ketamine administration is associated with increases in both systolic and diastolic blood pressures. The mechanism for the increases is direct CNS stimulation as well as the liberation of catecholamines. *(1:337; 3:347; 4:1218; 5:348)*

614. (D) The organophosphate insecticides act by causing an inhibition of cholinesterase. After the patient has been on the drops for a few weeks, the level of cholinesterase will drop to about 20% of normal. Echothiophate eyedrops, chronically administered, may cause a prolonged effect in patients who are then given succinylcholine. In most cases, this presents no problem, but it must be borne in mind when succinylcholine is being administered for short cases. *(3:181–2; 4:2187; 5:2537)*

615. (D) Of the drugs listed, only pancuronium is a steroid. *(3:198)*

616. (A) The small blocking dose of a nondepolarizing muscle relaxant given before an intubating dose of succinylcholine increases both the dose of succinylcholine required as well as the time to the onset of muscle paralysis. The rise in intracranial pressure is attenuated, but there is no effect on succinylcholine-induced arrhythmias. The time to recovery is not affected. *(3:201; 4:769–70)*

617. (C) Thiopental is contraindicated in patients with acute intermittent porphyria. The other porphyrias listed as options are not associated with enzyme induction. *(3:344; 4:557–8, 1219; 6:456–7)*

618. (D) Succinylcholine causes an increase in intraocular pressure and intragastric pressure. It causes bradycardia in infants and children and in adults after the second dose. Severe hyperkalemia may occur in patients with loss of nervous input to large areas of muscle. *(4:767–70)*

619. (A) In the case cited, any of the options except thiopental may be involved in the apnea. Thiopental, in usual doses, should not cause apnea after a procedure as long as a laparotomy. Any of the other options or a combination of the options can be the cause. *(4:1220; 1597)*

620. (C) After one half-life has elapsed, 50% of the drug remains; after two half-lives, 25%; after three half-lives, 12.5%; after four half-lives, 6.25%; and after five half-lives, 3.125%. *(3:22–3)*

621. (D) Vecuronium has little cumulative effect. It is a nondepolarizing relaxant with a duration of about 30 minutes. It is not difficult to reverse and it is eliminated chiefly in the bile. *(4:776–7; 5:493–4, 506)*

622. (D) Administration of anticholinesterase drugs will reverse the block of a nondepolarizing agent only if the latter drug is in low enough concentration. Administration of an anticholinesterase will prolong the block of succinylcholine by inhibiting its metabolism. *(4:750, 768; 5:492)*

623. (D) Uterine relaxation is mediated by β_2-adrenoceptors. All of the other effects listed are mediated by α_1-adrenoceptors. *(3:119–20)*

624. (E) Of the nondepolarizing muscle relaxants mentioned, the least desirable agent would be pancuronium, due to its prolonged terminal half-life in renal failure. The intermediate-acting muscle relaxants have a shorter duration of action even in the presence of renal failure. *(1:1013; 4:769, 775, 778–9; 5:2185–86)*

625. (D) Propofol is an IV agent with a fast onset and short time to awakening. It is essentially insoluble in water and there is very little analgesic effect. When compared to thiopental, the sleep time is shorter. *(4:1212–6; 5:318–21)*

626. (D) While all of the options are indicators of recovery, an objective measurement of recovery from neuromuscular block is better than a clinical assessment. DBS with mechano-, electro-, or acceleromyography is the method of choice. If this is unavailable, tactile response to DBS is preferable to tactile estimation of the train-of-four ratio, because it is easier to appreciate fade with DBS. Normal tidal volume is not sufficient to protect the airway. There must be enough force to generate a cough, and about 15 mL/kg are needed. Hand grip is a clinical assessment and must be sustained to correlate with recovery. Inspiratory force is a good measurement and should be at least 25 cm H_2O negative pressure. The advantage of inspiratory force is that it can be assessed in the patient who is still asleep and unable to respond. *(4:919; 5:486, 1565–66)*

627. (C) Isoflurane is a clear, colorless, stable, volatile liquid with a pungent odor. It contains no preservative and undergoes little biotransformation. *(3:352; 4:1144–6)*

628. (B) Patients with burns over a large portion of their body surface area are resistant to nondepolarizing neuromuscular blockade. The other factors all potentiate such blockade. *(4:2176; 5:872–3; 6:732)*

629. (D) Propranolol, a nonselective β-adrenergic antagonist, causes decreased cardiac output, heart rate, and contractility. *(3:253–4)*

630. (E) From the data presented in the two segments, the only statement that can be made is that there is a 90% block. The block could represent either type of blockade. Complete relaxation is not present. *(4:915–6; 5:484–6)*

631. **(D)** The drug is a muscle relaxant, since tetanic stimulation produces fade. The drug may be either a depolarizing or nondepolarizing relaxant. If the cumulative dose of a depolarizing relaxant is sufficient to cause a phase II block, fade may be demonstrated. *(4:915–8; 5:1563)*

632. **(E)** When a nondepolarizing muscle relaxant is present, acetylcholine cannot be mobilized in sufficient quantities to sustain contraction. It is not a result of recording inabilities or potassium. Cellular DNA is not a factor. *(4:760)*

633. **(A)** The transdermal therapeutic system containing scopolamine should be applied to the postauricular mastoid region where drug absorption is especially efficient. The drug is much more effective in prevention of PONV than in treatment of nausea once it has occurred. The patch has duration of action of about 72 hours; side effects include dry mouth, drowsiness, and blurred vision. *(3:171; 5:2720–1)*

634. **(A)** The atypical antipsychotics have less risk of extrapyramidal effects. They can cause hypotension and clozapine, olanzapine and quetiapine have been associated with an increased risk of new-onset type 2 diabetes as well as increased appetite and weight gain. Anticholinergic effects are least frequently caused by potent antipsychotics like haloperidol. *(3:485–6, 490–1)*

635. **(A)** Anticholinesterase drugs have potent muscarinic stimulating effects. These may be exhibited in the heart, bronchioles, or gut. It is important to consider each of these areas in the particular patient whose relaxants are being reversed. Atropine or glycopyrrolate should be used with the reversal drug. Although the muscarinic manifestations are not common, their occurrence may be quite troublesome, such as the production of copious respiratory secretions. *(3:183)*

636. **(D)** The patient with myasthenia gravis has an altered sensitivity to relaxants. These patients are more sensitive to nondepolarizing agents. This sensitivity is present with or without control of the disease. These patients may also be resistant to depolarizing relaxants, particularly those treated with pyridostigmine. *(4:1810–1; 6:525–6)*

637. **(A)** After a patient has had and recovered from a dose of succinylcholine, it takes a smaller amount of nondepolarizing drug than normal to establish a neuromuscular block. This may be due to continued desensitization of the end plate. A longer duration of the nondepolarization block is also expected. *(4:761)*

638. **(E)** Cholinesterase is involved in the breakdown rather than the formation of acetylcholine. In the nerve terminal, choline is acetylated under the control of the enzyme choline acetyltransferase. The acetyl moiety is derived from acetyl-CoA. *(3:126)*

639. **(D)** Botulinus toxin leads to paralysis by preventing the release of acetylcholine at nerve terminals. *(3:127,143; 6:557)*

640. **(E)** Atropine is not always predictable, especially in the elderly. Its administration will lead to increased heart rate, increased oxygen utilization, and possibly angina pectoris. It causes decreased conduction time and has no effect on systemic vascular resistance. *(3:164–6)*

641. **(C)** A depolarization block does not manifest posttetanic facilitation. The block is characterized by fasciculation and is antagonized by nondepolarizing muscle relaxants and potentiated by anticholinesterase agents. There is no fade initially, but fade may appear with the onset of a phase II block. *(3:200; 4:767–70; 5:484–6,1563)*

642. **(E)** Propranolol should not be terminated suddenly, since a withdrawal syndrome has been reported with sudden termination. In general, most antihypertensive agents should be continued through the perioperative period making heart rate and blood pressure easier to control. *(5:1120–1)*

643. (C) Methoxamine is the drug with pure α-adrenergic effects. Epinephrine, norepinephrine, and ephedrine have mixed effects. Isoproterenol has only β-adrenergic effects. *(3:232, 220–6, 237, 228)*

644. (B) Lisinopril is an inhibitor of angiotensin-converting enzyme used in the therapy of hypertension. *(3:823)*

645. (E) Cyclosporine is an immunosuppressive agent used to prevent rejection of transplanted organs and in the therapy of graft versus host disease in bone marrow transplant recipients. *(3:1466–7)*

646. (A) Omeprazole is an inhibitor of the proton pump in the gastric mucosa and decreases gastric acid secretion. It is used to treat peptic ulcer disease. *(3:1007)*

647. (D) Tamoxifen is an estrogen antagonist used to prevent recurrence of breast cancer. *(3:1440)*

648. (A) Acetazolamide is an inhibitor of carbonic anhydrase. It decreases the formation of aqueous humor and lowers intraocular pressure in glaucoma. *(3:766)*

649. (D) Selegiline is an inhibitor of MAO type B. It is used in Parkinson's disease. *(3:559)*

650. (D) Pyrazinamide is one of many medications used in combination to treat tuberculosis. *(3:1281)*

651. (B) Glyburide is an oral hypoglycemic agent used to treat non-insulin-dependent (type II) diabetes mellitus. *(3:1701)*

652. (E) Amiodarone is the currently recommended pharmacologic treatment for ventricular tachycardia. The nonpharmacologic treatment of choice would be synchronized cardioversion. *(5:1404)*

653. (A) Fexofenadine is a second-generation histamine H_1-receptor antagonist used to treat allergic conditions such as hay fever. It does not appreciably cross the blood-brain barrier and therefore lacks sedative effects. *(3:653)*

654. (B) Cromolyn sodium is used in the prophylaxis of asthma. It decreases histamine release from mast cells in response to an allergen. *(3:742)*

655. (A) Hydroxychloroquine is an antimalarial agent that also has antiinflammatory actions in persons with rheumatoid arthritis and other connective tissue disorders. *(3:1077)*

656. (C) Dofetilide is a class III antiarrhythmic agent that is an action potential-prolonging drug. It is effective in maintaining sinus rhythm in patients with atrial fibrillation. *(3:947, 960)*

657. (B) Allopurinol is an inhibitor of xanthine oxidase that decreases uric acid formation. It is used in the therapy of gout. *(3:721)*

658. (B) Leuprolide is an agonist at the receptor for gonadotropin-releasing hormone. It is used in the palliation of prostate cancer. *(3:1441)*

659. (A) Fluphenazine is an antipsychotic agent used to treat schizophrenia and other psychotic disorders. *(3:486)*

660. (C) Methysergide is an antagonist of serotonin (5-hydroxytryptamine). It is used to treat some of the symptoms associated with carcinoid syndrome. *(3:284)*

661. (B) The patient is partially paralyzed. The block that is present is not a depolarizing block. The train-of-four ratio is 75% and therefore the patient could not sustain a head lift. The advantage of the train-of-four is that it is not necessary to have a control tracing, since it serves as its own control. *(4:916, 919; 5:1552–3)*

662. (E) The train-of-four ratio for a depolarizing block is 100%, since there is no fade. The fourth twitch will be as high as the first. *(4:768; 5:1552–3)*

663. (C) Phenytoin is useful in treatment of arrhythmias associated with digoxin toxicity. It is eliminated through the kidneys after being conjugated with glucuronic acid. It decreases diastolic depolarization and automaticity. *(3:963)*

664. (D) The calcium channel blockers interfere with the flow of calcium through membranes. All are highly bound to protein. Calcium channel blockers increase coronary blood flow, decrease myocardial contractility, and may decrease anesthetic requirements up to 25%. *(3:853–6; 5:1122)*

665. (E) Lepirudin and argatroban are direct thrombin inhibitors. *(3:1526; 5:1116)*

666. (A) Lorazepam is a benzodiazepine and therefore does not cause extrapyramidal effects as are seen with the phenothiazines. It usually produces a state of sedation accompanied by decreased anxiety and anterograde amnesia. Respiratory depression may also occur. *(3:409–10)*

667. (B) Lithium may prolong the effect of both depolarizing and nondepolarizing relaxants. It is not necessary to stop the drug before induction, but it is necessary to consider the drug when planning the anesthesia. In addition to the effect on relaxants, the drug may lead to a reduction of anesthetic requirements. *(4:264; 5:517, 2654)*

668. (D) When meperidine is given to a patient on a MAO inhibitor, hyperpyrexia and CNS excitation, which may be fatal, may result. Local anesthetics, halothane, vecuronium, and aspirin, should not cause problems. *(3:468, 594)*

669. (D) Use of furosemide leads to alkalosis. Renin release follows decreased intravascular volume. Hypokalemia is one of the most common problems in the patient on diuretics. Hyperglycemia occurs but is not as common as with thiazides. Ototoxicity usually is seen with chronic treatment with high doses, and is probably related to changes in the endolymph. *(3:769–72)*

670. (A) Hypokalemic, hypochloremic alkalosis is one of the most common problems in the patient being treated with thiazides. Hyperuricemia may occur, but it is not as common as hypokalemia. Hyponatremia and hyperglycemia also may occur. *(3:776)*

671. (B) Nifedipine is effective for coronary vasospasm due to its vasodilating activity. It has no antiarrhythmic effect. Its half-life is about 4 hours. *(3:856–8)*

672. (D) In this patient with oliguria, the first priority is to replace volume. A fluid challenge is appropriate. *(4:142–3)*

673. (A) This patient has normal renal function as indicated by the urine/plasma osmolality ratio, and his initial oliguria (due to hypovolemia) is treated with fluid. If this same patient should later develop a high CVP, the treatment of choice would be a loop diuretic. If mannitol or an osmotic-type diuretic is administered in this situation, a further increase in CVP would be expected initially. *(3:768; 4:1548–9)*

674. (B) The binding of protamine, a basic protein, and heparin, an acidic protein, results in a compound with no anticoagulation effect. There is no other mechanism involved. *(3:1525)*

675. (E) Echothiophate treatment leads to a prolongation of succinylcholine effect by inhibiting succinylcholine metabolism. Echothiophate is a cholinesterase inhibitor with a prolonged effect. Since pseudocholinesterase breaks down succinylcholine, patients who are taking these eyedrops may experience prolonged paralysis after succinylcholine. *(4:750; 5:2537)*

676. (B) Epinephrine causes vasoconstriction, which leads to decreased absorption of the local anesthetic, thus prolonging its effect. This action may also delay metabolism. Protein binding, chemical interaction, and binding site competition are not involved. *(3:372–3)*

677. (E) The interaction of phenytoin and phenobarbital is one of altered metabolism due to enzyme induction. If a patient has a stable drug level, this can be changed by starting another drug that may induce the enzyme system that metabolizes the first drug. *(3:468)*

678. (C) The combination of nitrous oxide in 0.5 MAC concentration and isoflurane in 0.5 MAC concentration will lead to an anesthetic effect of 1 MAC. For all practical purposes, nitrous oxide and a halogenated agent are the only agents that are administered together, but if two halogenated agents could be used together, their effect would be additive. *(3:337; 4:1136–8; 5:1239–43)*

679. (D) The most useful drug to add to this regimen is dexamethasone, which has been shown to be effective either as a first-line agent, or in combination with 5-HT_3 receptor antagonists and/or droperidol. Metoclopramide is a short-acting prokinetic agent mainly used to ameliorate nausea and vomiting that accompany gastrointestinal dysmotility syndromes. Prochlorperazine should not be administered since the patient has already received a D_2 receptor blocker (droperidol). Famotidine is an H_2 receptor antagonist and not indicated for the prevention or treatment of PONV. Ephedrine is a noncatecholamine α- and β-adrenergic agonist and is a potent CNS stimulant, it is not recommended for the prevention or treatment of PONV. *(3:237, 651, 1026, 1031; 5:2720–1)*

680. (B) Quinidine is orally effective and highly bound to protein. It is useful against both atrial and ventricular arrhythmias. Quinidine potentiates neuromuscular blockers and may be the Class I antiarrhythmic of choice for people with renal disease. *(3:965–6)*

681. (A) Although well absorbed from the gastrointestinal tract, lidocaine is extensively metabolized on its first pass through the liver. The symptoms of toxicity are usually CNS stimulation, and usually occur at blood levels greater than 5 mcg/mL. Lidocaine is useful in ventricular arrhythmias. *(3:961–2)*

682. (C) Lorazepam is formulated with polyethylene glycol and propylene glycol. All of the other agents are water soluble, with the exception of propofol, which is essentially insoluble in water. *(4:1213; 5:318, 335)*

683. (D) A and B have the same efficacy because they produce the same maximum response. Because A is located to the left of B on the x-axis, A is said to be more potent than B; A will produce a particular magnitude of effect at a lower dose than will B. The dose at which the half-maximal response occurs, the ED_{50}, is lower for A than for B. It is likely that A and B act via the same receptor. *(3:50)*

684. (A) The effect of drug Y is to shift the dose-response relationship of X to the right with no effect on the maximum response to X. This is the definition of a competitive antagonist. It is likely that X and Y act via the same receptors. *(3:39–42)*

685. (B) Bleomycin is associated with pulmonary fibrosis, sometimes long after the drug has been stopped. Doxorubicin has cardiac toxicity. Vincristine is associated with myelosuppression. Methotrexate toxicity is manifest as immunosuppression. L-Asparaginase is associated with hepatic toxicity and allergic reactions. *(3:1430)*

686. (A) Methohexital is metabolized to a greater extent than thiopental. The drug is metabolized to hydroxymethohexital, which is inactive. The terminal half-life is shorter than that of thiopental. Methohexital does not release histamine from mast cells and is not contraindicated in asthma. *(3:343–4; 4:1219–21)*

687. (C) Midazolam is associated with fewer cases of venous irritation than diazepam. The drug is not contraindicated in children; in fact, it is very useful in children. The duration is longer than that of thiopental. There is no adrenal cortical depression and no histamine release. *(3:406–408; 5:337–340)*

688. (B) Lidocaine (1%) has a duration of action of about 60 to 120 minutes. It is about 70% protein bound compared to bupivacaine, which is 95% protein bound. Lidocaine is effective topically and there are few allergic reactions. The drug is not associated with tachycardia. *(3:374–5, 379,961–2)*

689. **(A)** Addition of epinephrine prolongs the action of a local anesthetic by decreasing its absorption. This leads to decreased plasma concentrations and decreased toxicity. The vasoconstrictor may also lead to decreased surgical bleeding. It does not decrease the intensity of the block. *(3:372–3)*

690. **(B)** Sufentanil is more potent and more fat soluble than fentanyl. At high doses of sufentanil sometimes used for cardiac surgery, its effects are terminated more by metabolism and elimination than by redistribution. Neither fentanyl nor sufentanil release histamine. The EEG effects of a large dose of an opioid agonist are a decrease in frequency and an increase in amplitude. *(4:1243, 1246)*

691. **(E)** The ester compounds are more slowly metabolized in patients with liver disease, since the metabolism is via cholinesterase. The metabolism takes place in the plasma and the liver. No metabolism occurs in the CSF, since there is no cholinesterase in the CSF. The drugs are broken down to inactive metabolites and excreted by the kidneys. *(4:1348)*

692. **(A)** Flecainide is an antiarrhythmic drug in Class IC. The drug is not a muscle relaxant and can be administered orally or parenterally. It is not the drug of choice for local anesthetic reactions. *(3:960–1)*

693. **(D)** The effect of alfentanil is primarily terminated by redistribution after bolus administration. The volume of distribution is smaller than that of fentanyl, the terminal half-life is shorter, and it is about 7.5 times less potent. Like other opioid analgesics, alfentanil acts to reduce the MAC of volatile anesthetics. *(4:1245–6)*

694. **(B)** Dexmedetomidine is an α_2-adrenergic agonist with 1600-fold greater selectivity for the α_2-receptor than for the α_1-receptor. It is approved for short-term sedation (≤24 hours) but can be used as an adjunct for general anesthesia as well as where it can markedly reduce the MAC of inhalational anesthetics. It causes less respiratory depression than opioids and benzodiazepines, and has not been associated with adrenal suppression. When administered as an IV bolus, it will cause an initial increase in blood pressure and decrease in heart rate. Dexmedetomidine is not indicated for the induction of general anesthesia. *(3:358; 5:355–9)*

695. **(E)** PGE_1 may cause hypotension due to vasodilation. The drug should not be stopped before induction, since it may be required to keep a ductus arteriosus open. The drug is administered by infusion, since it has a short duration. *(3:674, 679)*

696. **(D)** Physostigmine may be effective in reversing the sedative effects of benzodiazepines. The drug crosses the blood-brain barrier well since it is a tertiary amine. It is used in the treatment of central anticholinergic syndrome and can cause hypersalivation. *(4:136; 5:662, 2718)*

697. **(A)** Arrhythmias are common in the patient on a tricyclic antidepressant and the combination of halothane, pancuronium, and a tricyclic has been reported to be associated with severe intraoperative arrhythmias. The drug should not be stopped prior to surgery, but the anesthesiologist must be aware of its interactions with anesthetic agents. Ketamine and catecholamines may cause an exaggerated hypertensive response in these patients. *(4:263)*

698. **(E)** Aspirin irreversibly blocks the production of thromboxane A_2 in platelets, whereas clopidogrel exerts its antiplatelet effect by blocking the platelet ADP receptor. Because of their discrete mechanisms, the effect of these drugs is additive or even synergistic, which results in a higher risk for excessive bleeding during surgical procedures. There are no antagonists available for these medications. Because of their long duration of action, stopping the drugs one day prior to surgery would not decrease the risk of bleeding. Current recommendations are to stop clopidogrel 7 days prior to surgery in elective cases where surgical hemostasis is needed. *(3:1534; 5:1126, 1982–5)*

699. **(C)** Vancomycin is associated with histamine release when given rapidly. The drug should not be given in the primary IV line but should be given by piggyback so that the infusion rate can be accurately controlled. *(3:1264)*

700. (B) Terbutaline is a selective β_2-adrenergic agonist. It causes less tachycardia than isoproterenol. It is preferred in the patient with heart disease compared to nonselective β-adrenergic agonists. *(3:229–30)*

701. (B) The patient showing signs of cocaine toxicity may be excited, anxious, and restless, and have hypertension and tachycardia. *(3:634–5; 6:642–3)*

702. (D) Meperidine is primarily metabolized in the liver. About 5% is excreted unchanged. The drug is a synthetic opioid and may be used orally. The potency is about one-tenth that of morphine. *(3:592–4)*

703. (E) Ranitidine has less CNS toxicity than cimetidine. It is a histamine H_2 antagonist that increases the pH of the gastric fluid. It has a half-life of about 3 hours. It is effective orally as well as parenterally. *(3:1010–11)*

704. (A) Verapamil is a calcium channel blocker that is useful in the treatment of supraventricular tachycardia. It is useful in the patient with asthma but should not be combined with propranolol, since it may cause profound bradycardia. The drug causes vasodilatation. It slows the rate in pacemaker cells. *(3:856–7, 947–9, 952)*

705. (D) The release of mediators from mast cells and basophils in anaphylaxis is a calcium- and energy-dependent process. Histamine is released from mast-cell storage granules. The other three options, bradykinin, leukotrienes, and prostaglandins, are rapidly synthesized and are all mediators of the reaction. *(3:71; 5:1092)*

706. (A) Glycopyrrolate is a quaternary amine that is an anticholinergic agent. It is a synthetic alkaloid, and not naturally occurring like atropine and scopolamine. Since it is a quaternary amine, it does not cross the blood-brain barrier and is not associated with the central cholinergic syndrome. *(3:168–9)*

707. (C) Milrinone and inamrinone are phosphodiesterase inhibitors used for the short-term parenteral inotropic support in severe heart failure. They exhibit their therapeutic effect through inhibiting the breakdown of cyclic AMP and have no antiarrhythmic activity. Because milrinone has a shorter half-life and fewer side effects, it is a better choice than inamrinone for this indication. *(3:927)*

708. (A) The methylxanthine group of drugs includes caffeine. This group has strong β_2-adrenergic-like effects such as bronchodilation but does not act via β_2-adrenergic receptors. The mechanism of action is by phosphodiesterase inhibition, which leads to an increase in cyclic AMP. *(3:743–4)*

709. (C) Hydralazine may be associated with a lupus-like syndrome. It is a direct vasodilator and may cause reflex tachycardia, which limits its usefulness in the patient with angina. There may be retention of sodium and water that may require diuretics. *(3:885–6)*

710. (D) Gabapentin is approved for the treatment of partial seizures in adults when used in addition to other antiseizure drugs. It is not metabolized, and excreted unchanged in the urine. It is also used for the treatment of migraine, chronic pain, and bipolar disorder and concomitant use does not increase the plasma levels of carbamazepine or phenobarbital. *(3:538–9)*

711. (B) Valproic acid commonly causes elevation of hepatic enzymes in plasma. *(3:537)*

712. (C) Sublingual administration avoids the first-pass effect and bypasses the liver. This route is not useful with all drugs, only those that are unionized, e.g., nitroglycerin. This route leads to higher blood levels compared to oral administration. *(3:6–7)*

713. (B) Either formulation of calcium can be used and there is usually no preference. They can only be given by the IV route. The combination of calcium and sodium bicarbonate will form a precipitate. Calcium infusions usually cause vasodilation and hypotension, except when there is existing hypotension due to severe hypocalcemia. *(3:1719)*

714. (C) There is a growing consensus that the concern over hypokalemia is unfounded. If the likely cause of the hypokalemia is known and the patient is asymptomatic, there is no need to cancel the procedure in an otherwise healthy patient. Intraoperative infusions may cause hyperkalemia, and rapid administration of potassium may result in asystole. It is not possible to replenish the intracellular potassium over a short period of time. *(4:979; 5:1106–7; 6:381–2)*

715. (C) Metoclopramide stimulates upper gastrointestinal motility. It is an antiemetic and has no effect on acid secretion in the stomach. Transit time is shorter and there is no ileus, since motility is stimulated. *(3:1025–6)*

716. (C) Perioperative β-blockade in patients with established coronary artery disease, or at risk for perioperative myocardial injury, is a Level 1 recommendation by the American Heart Association/American College of Cardiology, and should not be restricted to patients undergoing vascular surgery. Prospective, randomized controlled clinical trials have demonstrated marked reductions in the incidence of cardiac death and nonfatal myocardial infarctions with these agents, which resulted in decreased 2-year mortality rates in this patient population as compared to the control group. *(5:905, 2068; 6:14)*

717. (D) Hetastarch is a synthetic colloid. The other options all include administration of blood products. Some people of this belief will accept blood that is autologous; others will require that the blood be in circuit continuously, e.g., in a cell saver. It is important to understand the particular patient's requirements before undertaking the procedure. *(4:990–1)*

718. (C) Vasopressin is recommended in the ACLS guidelines as an alternative agent to epinephrine in cases of refractory ventricular fibrillation and pulseless ventricular tachycardia. Because of its potential adverse effects, sodium bicarbonate should not be routinely administered during CPR. Phenylephrine, adenosine, and isoproterenol are not recommended for the treatment of refractory ventricular fibrillation. *(3:805; 5:2935–9)*

719. (B) Dobutamine is a synthetic catecholamine and contains a chiral center. The preparation used in clinical practice contains a racemic mixture of (+) and (–) enantiomers in equal amounts and the apparent effect of these two isomers is β_1-adrenergic stimulation. Dobutamine increases cardiac output and contractility. At higher dosage levels, α-adrenergic agonist effects may be seen. It causes increased renal blood flow. *(3:228–9)*

720. (E) Edrophonium should be given with, or preceded by, atropine. The drug has a fast onset but, in the dose cited, a duration as long as neostigmine. The muscarinic effects are fewer than with neostigmine. *(5:521–3)*

721. (A) Clonidine has its primary effect in the brain stem. The drug should not be tapered quickly because that can precipitate a withdrawal syndrome manifested as hypertension. The central analgesia is not reversed by naloxone and the MAC of halothane is decreased. *(3:233–4,358; 5:1121)*

722. (E) Labetalol is both an α-adrenergic and a β-adrenergic antagonist and is subject to the first-pass effect, so that the oral dose must be higher to obtain a therapeutic level. The drug may precipitate wheezing because of its effect on bronchial β-adrenergic receptors. Labetalol is conjugated in the liver. *(3:255)*

723. (B) Protamine is a naturally occurring protein salt. Patients who have been sensitized to protamine are more susceptible to reactions. The drug causes fewer reactions if injected slowly and into a peripheral vein. *(3:1525; 4:2398–2400)*

724. (C) Esmolol is a β_1-adrenergic antagonist and is contraindicated in the patient with AV block. The drug has a half-life of under 15 minutes. It is less likely than propranolol to cause bronchospasm, since it blocks β_1-adrenergic receptors, not β_2-adrenergic receptors. *(3:256)*

725. (C) Amiodarone is useful for both atrial and ventricular arrhythmias, but it is usually reserved for ventricular arrhythmias that are refractory to other therapy, since there are many side effects. Amiodarone has a half-life of many weeks, therefore stopping it before surgery would have little effect. It is eliminated through the liver. Autonomic effects of amiodarone include a noncompetitive β-adrenergic receptor blockade. *(3:948,953,956)*

726. (D) Sodium nitroprusside may lead to cyanide toxicity, which may be manifested as acidosis and a tolerance to the drug's effects. This drug affects both venous and arterial vessels. The acidosis that is seen is treated by stopping the drug and should not be treated by sodium bicarbonate. The hypotensive effect of sodium nitroprusside is potentiated by prior treatment with β-adrenergic antagonists. *(3:889–90; 4:1691–2; 5:2187)*

727. (B) Cocaine is the only local anesthetic that blocks the reuptake of norepinephrine. The addition of epinephrine to a local anesthetic solution does not affect norepinephrine reuptake. *(3:374)*

728. (A) Hydroxyzine is an antihistamine and a sedative. It has limited analgesic properties and is not an amnestic agent. It is not approved for IV use. *(3:656–7)*

729. (D) Postoperative pain control with methadone requires some time to obtain stable levels. The drug has a long half-life and is usually given in twice daily dosage. It does depress respiration. The oral dose is about 50% as effective as the same parenteral dose. *(3:596–7)*

730. (D) Dopamine has different actions depending on the dosage: at infusion rates greater than 10 mcg/kg/min, there is stimulation of α-adrenergic receptors. At lower infusion rates (approximately 1 to 2 mcg/kg/min), dopaminergic receptors are stimulated which leads to an increase in renal blood flow. Dopamine causes an increase in pulmonary artery pressure, making it a poor choice in the patient with right heart failure. Dopamine is a transmitter in both the central and peripheral nervous systems. *(3:226–7)*

731. (D) The sequence that is proposed may result in a prolonged duration of succinylcholine action because the prior administration of neostigmine will result in pseudocholinesterase inhibition and decreased succinylcholine metabolism. *(4:768; 5:492)*

732. (E) Vecuronium inhibits histamine-*N*-methyltransferase, the enzyme responsible for degrading histamine. It has a shorter duration of action in children. The onset time will be faster if the dose is increased. About 10% is eliminated through the kidneys and the activity is prolonged by hepatic disease. *(4:776; 5:514)*

733. (D) Cimetidine may decrease the metabolism of all of the drugs listed except atracurium via inhibition of cytochrome P450. Atracurium is not metabolized by cytochrome P450. *(3:205,258, 408, 1011)*

734. (E) Timolol does not affect the size of the pupil. It is absorbed systemically and may lead to bradycardia, since it is a β-adrenergic antagonist. Asthma may be precipitated. *(3:254, 1834–6)*

735. (E) Drug elimination is principally by the liver, kidneys, and the lung. Other avenues are sweat, saliva, and breast milk, although these are not of quantitative significance in most cases. While some drugs (e.g., remifentanil) are metabolized in muscle, there is no elimination via muscle. *(3:18–9)*

736. (B) Scopolamine is a potent antisialogogue. It also has potent amnestic and sedative properties. It is not as effective as atropine in increasing heart rate, nor is it as effective as glycopyrrolate in decreasing gastric acid secretion. *(3:162–6)*

737. (B) Dantrolene decreases intracellular calcium and causes generalized muscle weakness. The half-life is about 9 hours. Hepatotoxicity is common after prolonged therapy. Cardiac toxicity is not usually seen. While continued intensive care for a few days is indicated after an episode of malignant hyperthermia, continued administration of dantrolene is not indicated after the initial episode is terminated. *(3:204, 206; 4:2427, 2438–40; 6:716–21)*

738. (A) Glucagon increases insulin secretion and increases plasma glucose levels. *(3:1707–8)*

739. (A) Atracurium has both pre and postjunctional effects. Histamine release is more likely only with higher doses (>0.5 mg/kg) injected rapidly (in less than a minute). Laudanosine has caused seizures at high plasma levels in experimental animals, but this problem has not been observed in humans. The effect of atracurium on blood pressure is usually minor, but hypotension may occur if a large dose is given rapidly. *(4:760, 778–9)*

740. (A) At zero time, the blood concentration is 80 mg/L. Volume of distribution is the dose of the drug divided by the concentration. Thus, the volume of distribution in the central compartment immediately after the injection of the drug is 400 mg/80 mg/L or 5 L. *(3:20–2)*

741. (C) To obtain the redistribution half-life, the initial straight portion of the curve is extrapolated. At zero time, the blood concentration is 80 mg/L. At 1 hour, the extrapolated line gives a blood concentration of 5 mg/L. The time required for the blood concentration to drop from 80 to 5 mg/L is four half-lives: after one half-life, the blood concentration will be 40 mg/L; after two half-lives, 20 mg/L; after three half-lives, 10 mg/L; and after four half-lives, 5 mg/L. Since four half-lives occurred in 1 hour, then the redistribution half-life must be 0.25 hour or 15 minutes. *(3:21)*

742. (C) To obtain the terminal half-life, the final straight portion of the curve is extrapolated. At 10 hours, the blood concentration is 1 mg/L. The blood concentration is 2 mg/L at 7 hours, and 4 mg/L at 4 hours. It is apparent that the blood concentration is declining by half every 3 hours, thus the terminal half-life must be 3 hours. *(3:21)*

743. (B) The patient with Parkinson's disease has a deficiency in central dopaminergic activity. Thus, phenothiazines and neuroleptanesthesia (which employs a dopaminergic antagonist) are relatively contraindicated. There is no contraindication to muscle relaxants and the response to pressors is not abnormal. Since levodopa has a short duration of action, it should be continued up until the time of surgery. *(5:1095–6; 6:258–9)*

744. (A) Desmopressin is usually taken twice daily by intranasal administration. In contrast to vasopressin that is rapidly inactivated, the duration of the effect of desmopressin is several hours. Desmopressin may increase blood pressure, but it does so to a much lesser degree than vasopressin. Desmopressin increases the circulating concentrations of factor VIII and von Willebrand factor, and has no effect in patients with nephrogenic diabetes insipidus. While there is not likely that a need for intraoperative administration of desmopressin would arise if the patient was given a dose by inhalation preoperatively; it may be given by IV infusion to decrease serum osmolality and increase water retention. *(3:803–4; 4:315–6; 5:1117, 2861)*

745. (A) Since dextran occurs naturally in table sugar and is synthesized by gut bacteria, an allergic reaction may occur even in a patient who has never received IV dextran. Dextran decreases coagulation by impairing platelet function. Crossmatching of blood is affected by rouleaux formation, which interferes with the crossmatch. The solution stays in the intravascular space for about 12 hours. Histamine is released during dextran infusion. *(4:990; 5:1787, 1826)*

746. (E) Mivacurium is short acting and may be administered by infusion. It is a benzylisoquinoline derivative and is broken down by plasma cholinesterase. The drug has a slower onset than succinylcholine and a shorter duration than vecuronium. *(4:779–81; 5:508–9)*

747. (E) Edrophonium does not form a chemical bond with acetylcholinesterase. The inhibition is shorter than neostigmine or pyridostigmine. *(3:179)*

748. **(B)** Patients with hemiplegia may have a normal block on the normal side but decreased block on the affected side. If one is monitoring the degree of block on the affected side, there may be a false interpretation of the results and more relaxant administered when none is needed. *(4:1628)*

749. **(C)** The elderly generally require a lower dose of thiopental to induce anesthesia. With increasing age, MAC for inhalational anesthetics decreases, and recovery of normal ventilatory drive after fentanyl administration is delayed. In contrast, with increasing age there is no alteration in the intubating dose of vecuronium. The rate of hepatic synthetic reactions, such as the glucuronidation of morphine, is also unchanged in the elderly, while the rate of hepatic oxidative and reductive reactions declines with increasing age. *(4:481–2; 6:742–3)*

750. **(E)** The aim of antacid administration is to raise the pH. It should be given 15 to 30 minutes before surgery. Antacids will not decrease gastric volume but may actually increase it. Nonparticulate antacids should not cause any problem if aspirated. There is no lag time. *(4:2007, 2245–7; 5:2600)*

751. **(E)** Pharmacologic preparation of the patient does not preclude good airway management. Histamine H_2-receptor antagonists do not protect against aspiration, do not facilitate gastric emptying, nor do they have any effect on fluid that is already present in the stomach. If administered 2 to 3 hours before induction of anesthesia, they will increase gastric pH and by decreasing gastric acid secretion lead to lower residual gastric volumes. There is no need to use histamine H_2-receptor antagonists in all patients. *(4:2247–8; 5:2599–2601)*

752. **(D)** Diazepam used for premedication should be administered orally or intravenously. Intramuscular injections are erratically absorbed and painful. Local anesthetic toxicity is decreased because the seizure threshold is raised. Diazepam will decrease anxiety and recall. The latter effect is increased if scopolamine is also administered. *(4:54–6, 1212)*

753. **(A)** Butorphanol is a mixed opioid agonist-antagonist. Naloxone is a pure opioid antagonist, methohexital a short-acting barbiturate, midazolam a benzodiazepine, and bleomycin a drug used in cancer chemotherapy. *(3:599–601)*

754. **(D)** Etomidate is the only induction agent which causes decreased steroid production. Etomidate inhibits the activity of the 17α-hydroxylase and 11β-hydroxylase enzymes that participate in the conversion of cholesterol to cortisol and other steroids. *(3:345–6; 4:1219)*

755. **(C)** Patients who abuse alcohol manifest tolerance to its CNS effects but not to its ventilatory effects. The acutely intoxicated person requires less anesthesia, whereas the chronic alcoholic who is not acutely intoxicated requires more. Because alcoholic liver disease may lead to a decreased capacity to metabolize local anesthetics, local anesthetic toxicity may be more likely in the alcoholic. *(3:430; 4:549)*

756. **(C)** Of the drugs listed, only nadolol is a β-adrenergic antagonist. Isoproterenol is a β-adrenergic agonist and dobutamine is a β_1-adrenergic agonist. Albuterol and ritodrine are β_2-adrenergic agonists. *(3:254)*

757. **(E)** Doxepin is a tricyclic antidepressant. Pargyline, phenelzine, and isocarboxazid are MAO inhibitors. Haloperidol is a butyrophenone antipsychotic agent. *(3:451, 656)*

758. **(B)** Chloroprocaine is the least toxic local anesthetic by virtue of its rapid hydrolysis after accidental IV injection. *(1:458; 4:1348)*

759. **(E)** The correct order for the lipid solubilities of opioids is sufentanil > fentanyl > alfentanil > morphine. *(4:1243)*

760. **(A)** Heparin increases the activity of antithrombin III which neutralizes the activated forms of factors II, IX, X, XI, XII, and XIII and of kallikrein. The effect of heparin is estimated by the activated partial thromboplastin time. It may be given subcutaneously and it may be given while waiting for the effect of an oral anticoagulant to occur. *(3:1522–6)*

761. (D) Benzodiazepines produce the least depression of rapid eye movement (REM) sleep and a pattern of sleep most nearly physiologic. *(3:403)*

762. (C) The Henderson-Hasselbalch equation states that pH = p*K* + log[proton acceptor/proton donor]. With a weak base like mepivacaine, the uncharged form is the proton acceptor, and 39% is present in this form at physiologic pH. Even without a calculator, an approximation is possible. If the pH were equal to the p*K* of the drug, equivalent amounts of the charged and uncharged moieties would be present. In this case, the pH is slightly lower than the p*K*, favoring slightly less than half of the drug to be in the uncharged form. *(3:4)*

763. (B) Cocaine causes the release of catecholamines and prevents their reuptake. Exogenously administered catecholamines given intraoperatively may be expected to have an exaggerated effect. Because halothane may sensitize the heart to catecholamine-induced arrhythmias, it should be avoided. β-Adrenergic antagonists are given for cocaine-induced arrhythmias. There is no reason to withhold opioids for analgesia. *(1:1321; 4:551; 5:633,1124)*

764. (A) The halogenated hydrocarbons increase cerebral blood flow to various degrees, halothane having the most profound effect. Thiopental decreases cerebral blood flow. *(3:344, 351–5; 4:1143, 1147, 1150, 1215)*

765. (A) Although all halogenated inhalational anesthetics increase cerebral blood flow, isoflurane does so to a lesser degree than halothane. Such an increase in cerebral blood flow may be prevented or decreased by hypocarbia and barbiturates. The administration of vasopressors does not decrease cerebral blood flow. *(4:1147, 1215, 1619)*

766. (D) Ephedrine is useful in obstetrics, since it does not decrease uterine blood flow. Ephedrine is a direct α-adrenergic and β-adrenergic agonist as well as having an indirect action via the release of norepinephrine. Tachyphylaxis develops readily to the indirect action. *(3:237; 4:490)*

767. (D) Conduction blockade by local anesthetics is due to blockade of sodium conductance. Calcium, magnesium, and chloride conductances are not involved. *(3:368–9, 4:1299–1303)*

768. (B) Calcium channel antagonists should not be stopped in the preoperative period because of the possibility of worsening ischemia or hypertension. In patients undergoing inhalational anesthesia, they prolong atrioventricular conduction time, while in patients undergoing a high-dose opioid anesthetic, they have little effect on cardiac output. Calcium channel blockers also potentiate neuromuscular blocking agents. *(4:1559–62)*

769. (D) A weak base is less ionized at higher pH values, thus rendering it more lipid soluble and more likely to distribute to the interstitial fluid. The converse is true of weak acids. Increased blood flow to a tissue increases the distribution of drugs to that tissue. Distribution is decreased by binding to plasma protein and by decreased lipid solubility. *(3:9–11)*

770. (E) Drug clearance may occur by metabolism to inactive or less active products or by elimination of the unchanged drug. Clearance is independent of drug concentration. *(3:12–8)*

771. (B) All sympathomimetic drugs are not catecholamines. Ephedrine and methoxamine are examples. Both dopamine and isoproterenol are catecholamines. *(3:226–8, 232, 237)*

772. (E) The loading dose of a drug is the dose given to achieve a faster blood concentration. This decreases the time needed to achieve a steady state level. With some drugs, the larger dose can have detrimental effects, e.g., a larger dose of relaxant may lead to a longer time of paralysis or cardiovascular effects. *(3:27)*

773. (E) Nesiritide is the recombinant form of endogenous brain natriuretic peptide (BNP).

It is approved for the treatment of patients with acutely decompensated congestive heart failure. *(5:795–6; 6:109)*

774. (A) Naloxone may reverse the effects of opioid agonists such as fentanyl and butorphanol, and flumazenil may reverse the effects of a benzodiazepine such as midazolam. There is no specific antagonist for the hypnotic effects of propofol. *(4:1250–1, 1253, 1287)*

775. (B) The exact mechanism by which succinylcholine causes an increase in intracranial pressure is unknown. Induction with a generous dose of thiopental, hyperventilation, and pretreatment with a nondepolarizing muscle relaxant all act to mitigate this effect. *(4:770; 5:491)*

776. (E) Epinephrine is the most widely used catecholamine. It has many uses because of its effects on α-adrenergic and β-adrenergic receptors. Arrhythmias have been caused by epinephrine. There are strong chronotropic effects that may limit its use as an inotrope. At very low doses (0.1 mcg/kg), there may be β_2-adrenergic-mediated vasodilation. *(3:220–5)*

777. (E) Isoproterenol is useful in the treatment of heart failure with associated pulmonary hypertension. The drug can be used as a temporary pacemaker, since it has good chronotropic effects. The β-adrenergic effects lead to bronchodilatation. Since the drug is a β-adrenergic agonist, it can be used for treatment of overzealous β-adrenergic blockade. *(3:228)*

778. (B) Droperidol is known to cause dysphoria that limits its usefulness as a premedicant. Extrapyramidal symptoms may occur, and this must be considered in the patient with parkinsonism. It has been shown to prevent PONV. *(5:2598)*

779. (C) Amiodarone is a structural analogue of thyroid hormone and exerts a variety of pharmacologic effects. Hypotension after IV administration of the drug is due to vasodilation and depression of myocardial performance. Adenosine, verapamil, and procainamide are some of the drugs used for the treatment of cardiac dysrhythmias associated with the WPW syndrome, amiodarone is not indicated. *(3:953, 956; 5:2935–9; 6:83, 87, 123)*

780. (A) Nifedipine does not affect conduction through the AV node at routine clinical doses and is used as a coronary vasodilator and antihypertensive. The arterial dilatation caused by nifedipine may cause a reflex tachycardia in some patients. *(3:856)*

781. (E) Systemic signs of local anesthetic toxicity involve all of the options. When a block is being administered, it is important to look for these signs. *(3:373; 4:1403–4; 5:592–4)*

782. (A) The cardiac effects of local anesthetic toxicity occur by interference with sodium conductance. Ventricular tachycardia and fibrillation may occur. The toxicity with bupivacaine is of longer duration and, therefore, harder to treat. Diazepam has no place in the treatment of the cardiac effects of local anesthetic toxicity but is helpful in the treatment of CNS toxicity. *(3:373, 537–8; 4:1351–2)*

783. (B) The use of the priming principle to hasten the onset of action of muscle relaxants uses divided doses with a time interval that is thought to be optimal at about 4 minutes. The method is not consistent in giving good intubating conditions. Another problem is that one must wait for the time to elapse. During this time, the sensitive patient may have trouble with swallowing and airway control. *(4:783; 5:504)*

784. (C) Drug elimination may occur by metabolism or excretion and is usually complete after 5 half-lives. Elimination is slower than distribution. *(3:9–18, 20–22)*

785. (D) As compared with atracurium, cisatracurium has a lower ED50 and is therefore more potent. It is less likely to release histamine, even when the relative dose is much larger. Its duration is slightly longer than that of atracurium after an equivalent dose. *(1:427–9; 4:778–9; 5:494)*

786. **(C)** When using an infusion of a muscle relaxant, monitoring is important during the entire period of the infusion because of the possibility of accumulation of the relaxant. It is also important to maintain the first twitch of the train-of-four to ensure there is not an overdose. Volatile anesthetics decrease the amount of relaxant needed. Mixing atracurium in Ringer's lactate solution may hasten the rate of degradation because Ringer's lactate has a higher pH than saline or dextrose solutions. *(4: 783–4, 981; 5:455, 486, 494)*

787. **(B)** Phenylephrine may be used to treat the hypotension that accompanies spinal anesthesia. It also increases the pulmonary artery occlusion pressure. Cardiac output may be decreased by the reflex bradycardia that is often seen with phenylephrine. The usual dose is about 50 to 150 mcg/min in an adult. *(4:738–40, 1382–3)*

788. **(C)** Theophylline is a competitive adenosine antagonist and a phosphodiesterase inhibitor. The former effect is thought to be the mechanism of its pharmacologic action. The latter effect occurs only at higher concentrations than those achieved during therapy. *(3:743–4)*

789. **(E)** Halothane sensitization of the heart is increased with hypoxemia and hypercarbia. This effect is diminished by both α-adrenergic and β-adrenergic antagonists and by calcium channel blockers. *(1:298; 4:1142–3)*

790. **(C)** The cardiomyopathy seen with doxorubicin may be seen long after the chemotherapy has been discontinued. This side effect is dependent on the cumulative dose of doxorubicin, and may be evaluated by echocardiography. Patients who have received radiation therapy to the mediastinum may have a greater likelihood of this effect. *(3:1428)*

791. **(B)** Halothane decreases blood pressure primarily by decreasing cardiac output, isoflurane and sevoflurane primarily by decreasing peripheral vascular resistance. Halothane does not have as great an effect on heart rate as does isoflurane. *(4:1142, 1145; 5:201–2)*

792. **(D)** Nitrous oxide increases systemic and pulmonary vascular resistances. When it is given with halothane or isoflurane, blood pressure and cardiac output are increased compared to that with the volatile agent alone, probably because of an increase in sympathetic nervous system activity. When it is given with opioids, it causes a decrease in cardiac output, probably as a result of its effect on systemic vascular resistance. *(4:1140)*

793. **(E)** All of the drugs cited are considered safe for the patient with malignant hyperthermia. Drugs that increase the heart rate may make diagnosis difficult, since one of the first signs of malignant hyperthermia is tachycardia. *(4:2440–1; 5:1184–5)*

794. **(A)** Increasing the dose of pancuronium to twice its ED95 will shorten the onset of the block and may be used to achieve more rapid intubating conditions. If this is done, the consequences include a prolonged effect, more profound block, and a dose-related increase in heart rate. *(4:775; 5:505)*

795. **(E)** Physostigmine is an interesting drug that has many uses in reversing drug effects. Since it can penetrate the blood-brain barrier, it acts to increase the activity of central cholinergic neurons. This action specifically antagonizes the CNS effects of scopolamine and diphenhydramine that have muscarinic antagonist activity. In addition, there is a nonspecific action to antagonize the effects of opioids and benzodiazepines, although specific antagonists (such as naloxone and flumazenil) are preferable. *(3:189; 4:136; 5:662, 2717–8)*

796. **(D)** Benzodiazepines are not equally effective in producing amnesia. They are all metabolized in the liver. Hypotension usually occurs on induction, but to a lesser degree than with thiopental or propofol. Respiratory depression and respiratory arrest may occur. This is a particular risk in the elderly. *(4:1217–8, 1220–3; 5:338, 340)*

797. **(E)** Digoxin is an inotrope and may cause bradycardia. The mechanism involves inhibition of Na,K-ATPase. Calcium will potentiate the effects of digoxin. *(3:916–8)*

798. (A) α-Adrenergic receptors are stimulated by norepinephrine, causing vasoconstriction, and are blocked by labetalol. Bronchioles are not affected by β-adrenergic receptor agonists. *(3:119, 226, 255)*

799. (C) Histamine release causes tachycardia and hypotension. Erythema is common, and there is decreased peripheral vascular resistance. In addition, there may be wheezing and increased airway pressures as a result of bronchospasm. *(3:649–51)*

800. (B) The decrease in cerebral metabolic rate is most profound with isoflurane and it is associated with a decrease in cerebral electrical activity. The decrease in metabolic rate is dose dependent. Cerebral metabolic rate rises dramatically during seizure activity. *(4:886, 1144, 1146–7; 5:825–31)*

801. (B) Esmolol is an ester that is metabolized by esterases that are distinct from those that metabolize succinylcholine. Therefore, there should be no prolongation of the effect of esmolol when succinylcholine is also administered. Esmolol has a very short duration, and can be used to blunt the tachycardia and hypertension that may accompany intubation. *(3:256; 4:748; 5:657, 1648)*

802. (B) Both trazodone and fluoxetine are atypical (nontricyclic) antidepressants. Buspirone is an atypical antianxiety agent. Benztropine is a centrally acting anticholinergic used to treat Parkinson's disease and the extrapyramidal effects of antipsychotic agents. *(3:278–80, 472–3, 560)*

803. (D) The patient has all the symptoms of central anticholinergic syndrome that may follow the ingestion of jimson weed that contains atropine. The typical symptoms of "dry as a bone, blind as a bat, red as a beet, hot as a stove, and mad as a hatter" are due to the central and peripheral antimuscarinic effects of atropine. Physostigmine can cross the blood-brain barrier and will counteract both the central and peripheral effects. *(3:181–3, 189; 5:662)*

804. (C) Coverage of steroid therapy at the time of surgery is important to avoid stress reactions and acute adrenal insufficiency. Replacement doses are based on the anticipated stress of the operative procedure. A dose at induction, followed by an infusion or parenteral doses at regular intervals, is recommended until the patient is able to return to his or her preoperative oral regimen. Hydrocortisone sodium succinate is not for oral administration. *(3:1664; 4:1592; 5:1039–41)*

805. (E) A patient with the presented symptoms may be having an anaphylactic reaction. The first action should be to stop the suspected offender. The treatment is to ensure oxygenation, to improve the relative volume status, and to support the blood pressure with vasopressors. *(4:2391–2; 5:1092–3)*

806. (B) Tachyphylaxis, or acute tolerance to a drug that develops rapidly, is seen with ephedrine because of depletion of norepinephrine stores. Repeated or continuous administration of norepinephrine does not reduce its effectiveness. Ceiling does occur when a patient does not exhibit a drug effect even at infinitely high doses, for example with administration of opioids. *(1:254; 3:134; 5:447, 650)*

807. (D) The hypotension seen with fentanyl administration is due to bradycardia. There is no direct effect on the vessels or on the myocardium, and there is no histamine release. *(4:1240–1; 5:394–6)*

808. (A) Digoxin toxicity is manifest as nausea and vomiting, visual disturbances, prolonged P-R interval, and arrhythmias, especially multifocal PVCs. Bradycardia is common. *(3:958; 6:113–4)*

809. (B) β_2-Adrenergic receptors are stimulated by isoproterenol that results in peripheral vasodilation and increased glycogenolysis. Esmolol is a selective β_1-adrenergic antagonist. *(3:137, 256)*

810. **(E)** Magnesium deficiency may by manifested as muscle spasm, arrhythmia, or CNS irritability. It should be suspected in the chronic alcoholic and in anyone who is nutritionally deprived. *(4:972–4; 5:1772)*

811. **(D)** The opioids usually produce unconsciousness at high dosage, but it is important to recognize that unconsciousness is not absolute. The pupils are constricted. Opioids do not produce reliable amnesia, and the nausea and vomiting seen are mediated through the CNS in the chemoreceptor trigger zone. *(3:579–87)*

812. **(A)** The known heroin abuser may have a full stomach due to the slower emptying time of the gastrointestinal tract. This population has a higher incidence of AIDS and hepatitis. These persons may be highly tolerant to opioids and require huge doses for analgesia. *(3:632–3; 4:532–5; 5:938)*

813. **(B)** Captopril given before the nitroprusside is started may decrease the amount of nitroprusside needed to give the desired hypotension. Patients more at risk are those who are nutritionally deprived because of a deficiency of sulfur-containing substrates that detoxify cyanide to thiocyanate. Cyanide does not accumulate if the infusion rate of nitroprusside is kept below 2 mcg/kg/min. *(3:889–90, 1892–3)*

814. **(E)** Thiocyanate is a metabolite of sodium nitroprusside, and its half-life is prolonged in patients with renal failure, which increases the risk of cyanide toxicity. Penicillin and digoxin are primarily eliminated by renal excretion, and this must be taken into consideration when choosing the dose. Pancuronium is also predominantly eliminated by the kidney and has a decreased clearance and prolonged duration of action in renal disease. *(3:918, 1198; 4:772; 5:527, 2185, 2187)*

815. **(A)** The patient with hyperthyroidism should be adequately prepared preoperatively. This usually involves the preoperative administration of β-adrenergic antagonists to decrease the heart rate. Aspirin is usually avoided because it may displace the thyroid hormone from its plasma protein-binding sites. There is no reason to avoid thiopental or succinylcholine during anesthesia. *(4:305–6; 5:1046–7; 6:411–7)*

816. **(B)** For coronary *steal* to occur, there must be a stenosed vessel supplying an area of myocardium that is dependent on flow through collateral vessels. Isoflurane may then cause vasodilation that diverts blood away from these collaterals, leading to ischemia in the collateral-dependent area of myocardium. In spite of this theory and the fact that many persons have "steal-prone" coronary anatomy, there is no evidence that the use of isoflurane is associated with an increase in perioperative ischemia or mortality. *(3:352; 4:1145; 5:205–6,914–5)*

817. **(D)** Corticorticoids used for the treatment of asthma should be of the glucocorticoid type, since no mineralocorticoid effect is desired. The onset time is measured in hours. Giving the drug by inhalation does not decrease the onset time, but it may decrease the systemic toxicity because lower doses may be used. *(3:738–40,1672)*

818. **(A)** All of the drugs listed are antiarrhythmics, however, adenosine is indicated only for the treatment of supraventricular tachycardias. *(3:953, 957, 961–2, 963–4; 4:1566–9, 1572–5)*

819. **(C)** Sodium nitroprusside dilatates both the resistance and capacitance vessels, whereas nitroglycerin primarily dilatates capacitance vessels. Nicardipine is a calcium channel blocker that causes coronary and peripheral vasodilatation. Isoflurane in clinically meaningful doses dilates the peripheral vasculature without any significant effect on cardiac output. *(1:774; 3:352, 845, 889; 4:1144–5, 1553, 1691)*

820. **(A)** The patient with renal failure usually does not need an altered dose of medications the effects of which are terminated by redistribution. Pancuronium is primarily eliminated unchanged by the kidney. *(1:1012–14; 5:2182–7)*

821. **(E)** Ketamine is unique among IV induction agents in that it has a bronchodilatory effect. Because ketamine causes copious salivation, atropine is usually administered with it.

Halothane also has bronchodilating activity and does not irritate the airway. Thiopental has little effect on pulmonary resistance. *(4:236–7, 1217–8; 5:159–61, 333, 347–8)*

822. **(E)** Ionization of a drug is important in the drug's function, since charged particles do not cross membranes properly. The degree of ionization is a function of the pH of the solution and the p*K*a of the drug as given by the Henderson-Hasselbalch equation. *(3:3–4)*

823. **(C)** Obese patients have an increased volume of distribution and an increased terminal half-life for highly lipid-soluble agents such as midazolam and thiopental. The degree of isoflurane metabolism is not significantly changed. Obese subjects have higher levels of pseudocholinesterase activity; when a larger dose of succinylcholine is given to an obese patient, the duration of action is similar to that seen when a similar dose (in mg/kg) is given to a lean patient. *(4:514–7, 522; 5:336–7, 1033–4; 6:449)*

824. **(A)** Droperidol and prochlorperazine have antiemetic effects by virtue of their blockade of dopaminergic receptors in the chemoreceptor trigger zone, and ondansetron blocks 5-HT_3 receptors, also in the chemoreceptor trigger zone. Ranitidine is a histamine H_2-receptor antagonist that decreases gastric acid secretion but has no direct antiemetic activity. *(4:2325–6; 5:2599–2600, 2720–1)*

825. **(A)** Ketamine causes a depression of certain thalamic and cortical areas. There is maintenance of respiratory rate and tidal volume, and the gag reflex is preserved in most cases. Ketamine may be given safely to patients with porphyria; however, postoperative psychosis related to administration of the drug may be difficult to distinguish from those attributable to the disease. *(3:346–7; 4:279, 1216–8; 5:346–8, 1098)*

826. **(C)** The principal disadvantages of methohexital are pain on injection and involuntary muscle movements. The pH of the solution is high and it is water soluble. *(4:1211, 1216; 5:328, 2655)*

827. **(D)** Acute intermittent porphyria is an inborn error of metabolism in which porphobilinogen cannot be converted to uroporphyrin. Therefore, the urinalysis is positive for porphobilinogen, which is pathognomonic, and negative for uroporphyrins. Thiopental induces the enzyme δ-aminolevulinic acid synthetase that results in increased synthesis of porphobilinogen, and therefore will worsen acute intermittent porphyria. This disease is associated with a constellation of symptoms and signs, the worst of which are neurologic, however, cardiac arrest with thiopental is not among them. Morphine may be safely administered. *(4:279; 5:1098)*

828. **(E)** The toxicity of the local anesthetic drugs can be altered by adjusting the options cited. Limiting the dose, injecting over a longer period of time, choosing a less toxic local anesthetic, and avoiding sites of high vascularity when possible will lessen the risk of adverse effects. *(4:1350–2; 5:592–4)*

829. **(E)** Injection of thiopental is followed by a rapid decrease in concentration in the blood with a subsequent rise in the viscera. Following the rise in the viscera is the concentration rise in the lean tissues, such as muscle. The concentration gradually builds up in adipose tissue. *(4:1220–1; 5:329–30)*

830. **(B)** Both renal insufficiency and peripheral neuropathy may persist after courses of therapy with cisplatin. Congestive heart failure has been reported after high doses of doxorubicin. *(3:1433–4)*

831. **(D)** Dissociative anesthesia is noted by the presence of open eyes and occasional muscle movements that are not signs of inadequate anesthesia. Muscle tone may be increased in dissociative anesthesia. *(1:336–7; 5:346–8)*

832. **(A)** A nondepolarizing block will be reversed by an anticholinesterase agent. Signs of a nondepolarizing block include fade with train-of-four stimulation, fade with tetanic stimulation, and posttetanic facilitation. *(4:916–8; 5:484–6)*

833. **(E)** All of these agents are effective in facilitating the healing of a duodenal ulcer. *(3:1009–11, 1012–14, 1017)*

834. **(C)** Local anesthetics block conduction by blocking ionic movement through pores in the cell membrane. These pores, or ionic channels, are blocked from the inside of the nerve cell. Local anesthetics do not depolarize the membrane. Conduction does not occur through the myelin sheath. *(3:368–73; 4:1306–9; 5:579–80)*

835. **(D)** The principal mechanism behind the termination of the drug effect of methohexital is redistribution. This drug is metabolized to a greater extent than thiopental, but increased metabolism is associated with the shorter terminal half-life of methohexital, and not its more rapid rate of redistribution. The pH of a solution of methohexital is high, and so is its fat solubility. *(4:1220–1; 5:328–30)*

836. **(B)** This patient is most likely having a reaction to the methylparaben preservative present in multiple-dose vials of lidocaine. Persons allergic to methylparaben are also likely to be allergic to local anesthetics of the ester type, such as procaine. Since true allergy to lidocaine is extremely rare, a second dose of preservative-free lidocaine is very unlikely to cause a reaction. Bacterial contamination of a preservative-containing solution is also unlikely. *(4:1353–4; 5:596–7)*

837. **(A)** Succinylcholine will have a prolonged effect in a person deficient in pseudocholinesterase, either on the basis of a genetic defect or because they have taken a cholinesterase inhibitor such as echothiophate, which is readily absorbed from the eye. Opioid overdosage may also cause apnea and unresponsiveness. Since 70% N_2O does not cause apnea, it is not likely to be associated with this patient's current signs. *(4:750, 768; 5:1563, 2717–18, 2536–7)*

838. **(B)** In trying to determine the cause, a nerve stimulator can be used to assess the level of muscle relaxation. If no twitches are elicited, a prolonged effect of succinylcholine is demonstrated. An opioid antagonist such as naloxone may be administered to determine if excessive opioid effect is causing apnea and unconsciousness. Administration of carbon dioxide by inhalation is technically difficult in most settings and will not provide a definitive answer. There is no reason to administer another dose of succinylcholine. *(4:768, 1250–1; 5:420–1, 1563)*

839. **(D)** Once it has been established that there is no twitch, the only course to take is to ventilate the patient as long as is necessary. Even with no measurable pseudocholinesterase activity, paralysis due to an intubating dose of succinylcholine should wear off within an hour or two due to renal elimination of unmetabolized succinylcholine. Neostigmine will not reverse the paralysis if pseudocholinesterase deficiency is the cause. Pralidoxime is unlikely to be effective in reactivating cholinesterase that is chronically inhibited by echothiophate. A stimulant will not be effective because the patient is not asleep. *(3:185, 189; 4:768; 5:420–1, 492, 2507–8)*

840. **(E)** The resting membrane potential is about –60 to –90 mV and is primarily due to the difference in the intra- and extracellular potassium concentrations. The membrane potential is dependent on selective permeability of the membrane to various ions. The membrane potential changes rapidly with depolarization. *(4:758–9)*

841. **(B)** The intramuscular route may be used for aqueous or nonaqueous solutions and permits more rapid onset than after subcutaneous injection. Irritating substances that may cause pain if injected subcutaneously may be tolerated by the intramuscular route. Absorption from the deltoid and quadriceps muscles is more rapid than from the gluteus maximus. *(3:7)*

842. **(E)** There are several genes that code for abnormal forms of plasma cholinesterase. Atypical cholinesterase, which metabolizes succinylcholine at a reduced rate, is not inhibited by dibucaine. Similarly, fluoride-resistant cholinesterase is not inhibited by fluoride ion, and is designated as having a low fluoride number that denotes the percentage of inhibition produced by fluoride. Both severe liver disease and pregnancy are associated with decreased synthesis of plasma cholinesterase molecules. *(4:768–9; 5:487–8)*

843. (C) Ketamine stimulates the sympathetic nervous system by both central and peripheral mechanisms. There is therefore an increase in systolic and diastolic blood pressures, heart rate, and cardiac output. *(3:347–7; 4:1218; 5:347–8)*

844. (A) Lithium potentiates the effects of barbiturates and therefore will increase the sleep time of thiopental. In addition, lithium will potentiate the neuromuscular blockade produced by succinylcholine and pancuronium. Lithium will make reversal of neuromuscular blockade by neostigmine more difficult. *(4:264; 5:517, 2654)*

845. (B) Signs of histamine release include erythema, hypotension, bronchospasm, and tachycardia. *(3:649–51)*

846. (A) Neostigmine, pyridostigmine, and physostigmine form a chemical complex at the esteratic site of the acetylcholinesterase molecule. In contrast, edrophonium forms an electrostatic bond at the anionic site and has hydrogen bonding at the esteratic site. *(3:179; 5:519)*

847. (B) Drug antagonism occurs when two drugs have effects that tend to counteract one another. This may occur by combination, as in the case of heparin-protamine interaction, or they may compete for the same receptor site, as with muscle relaxants and acetylcholine. *(3:41–2, 73)*

848. (D) Because of their rapid onset times, atropine should be given at the same time as edrophonium in order to minimize side effects. Similarly, glycopyrrolate should be given at the same time as neostigmine. In the 70 kg-patient, a 5-mg dose of neostigmine will produce maximal inhibition of acetylcholinesterase, and a larger dose may actually cause neuromuscular blockade. The disappearance of fade during DBS indicates adequate recovery. *(4:919–20; 5:519–24)*

849. (B) Both apomorphine and ipecac may be used to induce vomiting in a patient who has ingested a toxic substance. Chlorpromazine and dimenhydrinate both have antiemetic properties. *(3:75–6)*

850. (B) The patient with myasthenia gravis has weakness in the muscles innervated by cranial nerves and often experiences ophthalmoplegia and ptosis. Although the accessory muscles of respiration may also be weak, diaphragmatic weakness is uncommon. These patients are very sensitive to nondepolarizing muscle relaxants. Myasthenia gravis is exclusively a motor disease. *(4:1808–11; 5:1098–9; 6:522–7)*

851. (E) The aim of therapy with digoxin is to increase contractility and to control tachyarrhythmias. By moving the patient's hemodynamic profile to a better curve, higher stroke volume is possible with lower filling pressure. *(3:916–20)*

852. (E) Succinylcholine will lead to spasm in patients with myotonia. Neither nondepolarizing neuromuscular blocking agents nor the administration of spinal anesthesia will prevent or treat the muscle spasms in myotonia because the pathologic lesion is at the muscle membrane level. Quinine will relieve the spasm. *(4:277, 281–2; 5:543–6; 6:519–20)*

853. (A) The effects of neostigmine include bradycardia, sweating, and increased skeletal muscle contraction. The bladder is stimulated to contract. *(3:121, 181–3)*

854. (E) All inhalational anesthetics can produce CO due to their interaction with strong bases in carbon dioxide absorbents, particularly when the absorbent is dry (such as when it has been flushed with dry gas during an entire weekend). The relative propensity for producing carbon monoxide is desflurane > isoflurane. Carbon monoxide production with the use of sevoflurane and halothane is negligible. Recommendations to decrease the risk of carbon monoxide production include low fresh gas flows, use of fresh absorbent, and the use of soda lime instead of baralyme. *(5:255–6)*

855. **(A)** An increased initial slope of the curve indicates an increased rate of sevoflurane uptake. An increased rate of uptake will occur if the inspired concentration of sevoflurane is increased, if nitrous oxide is added to the inspired gas mixture (the second gas effect), or if minute ventilation is increased. An increased cardiac output will cause a decreased rate of anesthetic uptake. *(4:1130–4; 5:134–5)*

856. **(E)** Succinylcholine is relatively contraindicated in all of these cases, but factors that must be taken into consideration are the extent of the burn or trauma, the elapsed time since the burn or trauma, and the amount of muscle wasting. *(4:769; 5:530, 533–4, 2457–8)*

857. **(D)** Additive effect refers to a combined effect that is the algebraic sum of the individual actions. *(3:72)*

858. **(B)** In the graph, the blood level of drug A decreases to nearly zero between doses. Drug B shows increasing levels over a period of time, i.e., it accumulates. *(3:22–4)*

859. **(D)** The graph shows that drug B is not completely eliminated before the next dose is given. The relative size of the dose cannot be determined from the diagram, but each of the doses is the same size. The half-life of drug A is not long. *(3:22–4)*

860. **(B)** Pyridostigmine has a slower onset and longer duration of action as compared to neostigmine. The kidneys are responsible for about 75% of the elimination of pyridostigmine and about 50% of the elimination of neostigmine. Both drugs bind to the identical site on acetylcholinesterase. *(3:179, 183; 5:2168)*

861. **(B)** β-Adrenergic receptors prepare the body for emergencies by increasing the cardiac output, dilating the bronchioles, increasing glucose, and relaxing the ciliary muscle to facilitate distant vision. *(3:121)*

862. **(E)** None of the agents increase hypoxic pulmonary vasoconstriction. Halothane and sodium nitroprusside inhibit hypoxic pulmonary vasoconstriction, while thiopental and fentanyl have no effect. *(4:1766)*

863. **(B)** Metoprolol is a selective β_1-adrenergic antagonist. It inhibits the inotropic and chronotropic effects of isoproterenol and does not inhibit the vasodilator effects of that drug. *(3:255–7)*

864. **(C)** Propranolol causes hypoglycemia (by inhibiting insulin release), as does insulin. Isoflurane has little effect on blood sugar, although the stress of surgery may cause an increase. *(3:120–1, 258, 1699; 4:316–7; 6:409)*

865. **(B)** One of the side effects of digoxin is altered color vision often manifested as halos surrounding objects. Another side effect is dizziness. Digoxin is not associated with hunger but may cause anorexia or nausea. Alopecia is not a side effect. *(3:920–1)*

866. **(B)** Patients with the Eaton-Lambert syndrome have increased sensitivity to both depolarizing and nondepolarizing muscle relaxants because of decreased release of acetylcholine at the neuromuscular junction. The syndrome does not respond to acetylcholinesterase inhibitors and nondepolarizing neuromuscular blockade is difficult to reverse. *(4:23–4,387; 5:863; 6:527–8)*

867. **(D)** Phenylephrine eyedrops may be absorbed systemically and lead to hypertension. It is important to know the concentration being used. The effect that is needed by the surgeon, i.e., mydriasis or vasoconstriction, is usually achieved with the 2.5% concentration. *(3:1836–7; 4:2186)*

868. **(C)** Primaquine and nitrofurantoin can cause hemolysis in persons deficient in glucose-6-phosphate dehydrogenase. This effect does not occur with acetaminophen or cefazolin. *(3:704, 1080, 1085, 1161, 1212)*

869. **(C)** The neuroleptic state is characterized by reduced motor activity and indifference to the surroundings. Neither amnesia nor analgesia is produced when a neuroleptic agent is administered by itself. *(3:357–8)*

870. **(D)** IV morphine in large doses does not cause cardiac instability; however, the addition of nitrous oxide may cause myocardial depression. Even with large doses, the patient may be awake. Catecholamine release is inhibited. *(4:1249–50; 5:394)*

871. **(B)** Ketamine anesthesia is characterized by analgesia and cardiovascular stimulation. Airway reflexes may be preserved, and ventilation is not depressed. *(3:346–7; 1:327–8; 5:346–8)*

872. **(A)** *C. perfrigens* is a gram-positive rod. The drug of choice for gas gangrene is penicillin G. In the unlikely event that a particular strain is resistant, or if the patient is allergic to penicillin, clindamycin or a second-generation cephalosporin such as cefotetan may also be used. *(3:1151)*

873. **(B)** Phenobarbital and ethanol are both enzyme inducers, and increase the synthesis of cytochrome P450 molecules. Cimetidine inhibits metabolism of many drugs by inhibiting cytochrome P450. There is no inhibitory effect of nitrous oxide on cytochrome P450. *(3:16–7, 356, 430)*

874. **(C)** Human enzymes are not thought to be able to metabolize nitrous oxide. The reduction of nitrous oxide to nitrogen occurs to a small degree in the gut and is catalyzed by bacterial enzymes. *(5:236)*

875. **(B)** Triamterene inhibits potassium secretion in the distal tubule and does not interact with aldosterone nor is its action dependent on the aldosterone level. In combination with potassium-losing diuretics, it can minimize potassium loss. *(3:777–9)*

876. **(B)** The same dose of thiopental is not appropriate for all patients. An understanding of the diagram is important for the proper use of IV medications. A patient in shock will have an increased effect of thiopental because the fraction of the cardiac output reaching the brain is higher in shock. Furthermore, the rate of redistribution from the brain is slower because of decreased perfusion of other tissues. The patient who is anxious and has an increased cardiac output will have faster redistribution, and the effect will be shorter. *(4:1221–2; 5:332–3)*

877. **(D)** Amyl nitrite oxidizes hemoglobin to methemoglobin that binds cyanide tightly and keeps it in the peripheral circulation. Although amyl nitrite is a vasodilator, this effect is unimportant in the management of cyanide toxicity. *(3:1893)*

878. **(C)** Methyl methacrylate is a vasodilator associated with hypotension. During the insertion of a prosthesis into the femoral canal, emboli of fat and marrow are common, resulting in hypoxemia. An air embolus of hemodynamic consequence is much less likely. The polymerization of methyl methacrylate is an exothermic reaction and tissue can be burned if allowed to contact curing cement. *(4:296–7; 5:2413–6)*

879. **(C)** Tiotropium bromide is a quaternary ammonium muscarinic antagonist for the treatment of COPD. When inhaled, the action of the drug is confined almost exclusively to the mouth and airways, it does not cross the blood-brain barrier and is considered a bronchoselective bronchodilator with a longer duration of action than ipratropium bromide. Bronchodilators should be continued on the day of surgery. *(3:168)*

880. **(C)** Oxazepam, an active metabolite of diazepam, and lorazepam are not metabolized to active compounds, but are conjugated to glucuronide and excreted. Midazolam is hydroxylated to a compound with about one-tenth the activity as a benzodiazepine agonist. *(3:406–8)*

881. **(D)** Ondansetron is a 5-HT_3 receptor antagonist. It is an antiemetic and acts at the chemoreceptor trigger zone. It produces no sedation and has few, if any, adverse effects. It has no effect on gastrointestinal motility. *(3:333, 1029–31; 4:2325; 5:2720–1)*

882. **(D)** Flumazenil is a benzodiazepine antagonist and therefore may reverse midazolam-induced sedation. It has no active metabolites and is shorter acting than the benzodiazepines it may be used to antagonize. It has a rapid onset of action due to its limited protein binding (approximately 50%). *(3:412; 4:1287; 5:344)*

883. **(B)** Only the S-isomer of ropivacaine is present in solutions for clinical use, in contrast to other optically active local anesthetics that are supplied as racemic mixtures. Ropivacaine is much less toxic than bupivacaine and produces motor blockade that is of shorter duration than sensory blockade. Like other amide local anesthetics, ropivacaine is metabolized in the liver. *(3:376; 4:1354; 5:595–6)*

884. **(E)** All of these effects occur with greater magnitude after an equivalent anesthetic dose of propofol. *(4:1215, 1218–20; 5:320–5)*

885. **(E)** The time to the peak effect of fentanyl and its derivatives is shorter than that for morphine. *(4:1237)*

886. **(C)** Most strains of *S. aureus* produce beta-lactamase that will render penicillin G and ampicillin ineffective. Cephalothin and other first-generation cephalosporins and vancomycins are both effective drugs for single-agent prophylaxis of such a wound infection. *(3:1144–5, 1147)*

887. **(B)** Rebound hypertension can occur after abrupt cessation of β-adrenergic antagonists such as propranolol and centrally acting antihypertensives such as clonidine. *(3:880, 884)*

888. **(E)** All of these agents lower plasma lipid levels. *(3:984, 989, 991, 993)*

889. **(B)** Vitamin D is structurally unrelated to parathyroid hormone and requires hydroxylation by both hepatic and renal enzymes for biologic activity. It promotes both bone decalcification and intestinal calcium absorption. *(3:1725–8)*

890. **(A)** Thiazide diuretics may cause potassium depletion and increased serum uric acid levels, sometimes precipitating gout. Hyperglycemia and exacerbation of diabetes mellitus may also occur. *(3:776)*

891. **(D)** Decreased activity of certain aminergic neurons is associated with depression; amitriptyline inhibits the reuptake of norepinephrine and 5-HT (serotonin). *(3:451)*

892. **(E)** PGE_1 causes vasodilation, resulting in decreased blood pressure, and contraction of gastrointestinal longitudinal smooth muscle that decreases the transit time of the contents of the GI tract. Additionally, PGE_1 increases mucus, water, and electrolyte secretion into the intestine; thus there is a watery diarrhea. PGE_1 also directly relaxes bronchial smooth muscle and decreases the volume and acid concentration of gastric secretions. *(3:674–7)*

893. **(A)** Ergonovine, $PGF_{2\alpha}$, and oxytocin increase uterine contractions and may decrease postpartum hemorrhage. Progesterone usually relaxes the uterus. *(3:284, 675, 1558–9, 1619)*

894. **(D)** Fenoldopam is a selective D_1 receptor agonist that is indicated for short term, rapid reduction of blood pressure in severe hypertension. The drug is devoid of D_2-, α- and β-adrenoceptor activity and causes a dose-related increase in renal blood flow and glomerular filtration rate. Fenoldopam does not mitigate the severity of contrast-induced nephropathy. *(3:227; 5:799)*

895. **(B)** Carbon monoxide shifts the oxygen-hemoglobin dissociation curve to the left; thus oxygen is bound more tightly and is less easily released in tissues. Carbon monoxide combines with hemoglobin at one-tenth the rate of oxygen, but dissociates from hemoglobin at only 1/2200 the rate of oxygen. The affinity of hemoglobin for carbon monoxide is therefore 220 times that of oxygen, and an increased FIO_2 provides no protection from carbon monoxide. *(3:1880–1)*

896. **(B)** The combination oral contraceptive pill increases the risk of thromboembolic disease and may exacerbate latent diabetes mellitus. No carcinogenic effect on the endometrium has been demonstrated, nor has an adverse effect on the gastrointestinal mucosa been noted. *(3:1625–7)*

897. **(E)** All of these effects may occur during phenytoin therapy. *(3:530)*

898. **(B)** Mannitol is filtered by the glomerulus and is negligibly reabsorbed. It is a nonelectrolyte that has almost no pharmacologic effects aside from its osmotic activity after IV injection. *(3:767–8)*

899. **(B)** Filtration occurs when drugs diffuse through aqueous pores in membranes. Passive nonionic diffusion is the process by which lipid-soluble compounds diffuse through the lipoidal portions of membranes. Facilitated diffusion and active transport use carrier molecules for certain ligands and are thus subject to saturability. *(3:3–4)*

900. **(A)** The basal ganglia are thought to contain inhibitory dopaminergic neurons and excitatory cholinergic neurons. In Parkinson's disease, there is a loss of dopaminergic neurons and a relative excess in cholinergic neuronal activity. Therefore, both dopamine agonists and acetylcholine antagonists are therapeutic. In addition, a MAO inhibitor will decrease dopamine degradation and increase dopaminergic activity. *(3:555–60)*

901. **(C)** Withdrawal from barbiturates and benzodiazepines can result in grand mal seizures, which may be fatal. Opioid and amphetamine withdrawal, while uncomfortable, are not considered to be life-threatening. *(3:628–36)*

902. **(B)** Spironolactone is a competitive antagonist of aldosterone binding to its renal receptor. Adverse effects include exacerbation of preexisting hyperkalemia and the production of androgenic effects as a result of its ability to bind to the testosterone receptor. *(3:779–81)*

903. **(B)** Isoproterenol and dopamine are catecholamines and are substrates for catechol-O-methyl transferase. Amphetamine and ephedrine do not contain the necessary catechol nucleus. *(3:135, 216–9)*

904. **(B)** Heroin is diacetyl morphine. One or both of the acetyl groups may be removed via hydrolysis to yield acetylmorphine and morphine, respectively. *(3:590, 632)*

905. **(A)** The first-pass effect is the biotransformation of a drug as it passes through the intestinal mucosa and liver. In some cases, much of the pharmacologic activity is lost at this point, leaving little of the drug to have its desired effect. Since the rectum is drained by veins that empty into the inferior vena cava, and not the hepatic portal vein, rectal administration, as well as sublingual and topical administration, circumvent the first-pass effect. *(3:5–8)*

906. **(B)** Low-molecular-weight heparins have a more predictable pharmacokinetic profile than standard heparins, which allows weight-adjusted subcutaneous administration without routine laboratory monitoring. If monitoring is deemed necessary, antifactor Xa activity can be determined. The incidence of heparin-induced thrombocytopenia is lower as compared to standard heparin. Even if similar antifactor Xa activity is achieved with any of these agents, it cannot be assumed that they produce equivalent antithrombotic effects. *(3:1523–4)*

907. **(D)** Most first-generation histamine H_1-receptor antagonists to some degree also block the muscarinic receptor. The second-generation H_1-receptor antagonists have no such effect. *(3:653)*

908. **(B)** Aspirin causes the uncoupling of oxidative phosphorylation and therefore there will be an increase in carbon dioxide production with resulting hyperventilation. Body temperature will be increased because of the increased metabolic rate. Miosis is characteristic of intoxication with opioids. *(3:699)*

909. **(B)** Rifampin and streptomycin are commonly used in the combination chemotherapy of tuberculosis. Pyrimethamine is an antimalarial agent, and tetracycline has no activity against tuberculosis. *(3:1080, 1240–1, 1277–8, 1280–1)*

910. **(C)** Cytochrome P450 performs many oxidative reactions, including the oxidative demethylation of ketamine to norketamine and the desulfuration of thiopental to pentobarbital. The acetylation of hydralazine is catalyzed by acetyltransferase, and the conjugation of morphine is catalyzed by glucuronyl transferase. *(3:12–3)*

911. **(D)** Aspirin and acetaminophen are both effective inhibitors of prostaglandin biosynthesis in the CNS, and are thus both effective antipyretic agents. Only aspirin inhibits prostaglandin biosynthesis in peripheral tissues, and only aspirin is associated with epigastric distress. *(3:697–700, 703–5)*

912. **(E)** All of these medications are controlled substances. Cocaine and morphine are listed in Schedule II, thiopental in Schedule III, and midazolam in Schedule IV. *(3:1908)*

913. **(A)** Drug metabolism usually results in a decreased duration of action because the drug is rendered inactive, more water soluble, and more readily excreted by the kidney. *(3:11–2)*

914. **(B)** Sweating and lacrimation are two of the early signs of the abstinence syndrome. Chills and pilomotor activity are prominent effects a day or two after the last dose of morphine. Seizures are not a part of the opioid withdrawal syndrome, but do occur with abrupt withdrawal of barbiturates and benzodiazepines. *(5:1094)*

915. **(B)** Tyrosine hydroxylase converts tyrosine to dopa. Dopamine β-hydroxylase converts dopamine to norepinephrine. Phenylethanolamine-*N*-methyltransferase converts norepinephrine to epinephrine. Catechol-*O*-methyltransferase catalyzes the degradation of norepinephrine to normetanephrine. *(3:131)*

916. **(E)** All of these effects occur with captopril. Cough and loss of sense of taste are common. Renal failure is likely in persons with renal artery stenosis. Proteinuria is less common, but is more likely to occur in persons with underlying renal impairment. *(3:828–9)*

917. **(A)** The hypotension that is associated with morphine administration is not due to cardiac depression. Morphine-induced hypotension is due to centrally mediated vasodilatation, histamine-induced vasodilation, and increased vagal activity. *(4:1240–1; 5:394–6)*

918. **(A)** The vessel-rich group includes those organs that receive the bulk of circulation. It includes the heart, brain, kidneys, and liver. Muscle is more poorly perfused. *(3:8–9)*

919. **(D)** Ketorolac is a nonsteroidal antiinflammatory agent that has high efficacy as an analgesic. Its major adverse effect is epigastric distress. It is much less likely than morphine to cause sedation, ventilatory depression, or nausea. *(3:709; 5:2734–5)*

920. **(B)** Administration of digoxin leads to increased contractility and a longer P-R interval. The heart rate decreases. Oxygen consumption will decrease if the failing heart is made to pump more efficiently. *(3:916–8)*

921. **(E)** IV anesthetics used for induction are highly lipid soluble. The CNS response may be affected by factors that alter the amount of drug reaching the brain in the time immediately following injection. Thus, the hypovolemic patient has a greater fraction of the cardiac output reaching the brain, and redistribution is less rapid due to the decreased blood flow to peripheral tissues. The patient with renal failure if often hypoalbuminemic; such patients need a decreased dose of induction agent due to less protein binding that causes a higher unbound fraction to be available. The elderly require lower induction doses of IV anesthetics, although the mechanism is unclear. *(1:1283; 4:481–2, 514–7, 522, 1218; 5:100, 1330–2, 2183–4, 2475; 6:742–3)*

PART II

Clinical Sciences

CHAPTER 8

General Anesthesia

Questions

DIRECTIONS (Questions 922 through 947): Each of the numbered items or incomplete statements in this section is followed by answers or by completions of the statement. Select the ONE lettered answer or completion that is BEST in each case.

922. The first person to hold an academic position in anesthesiology at an American university was

(A) Ivan Magill
(B) Arthur Guedel
(C) Ralph M. Waters
(D) Henry K. Beecher
(E) William T.G. Morton

923. A patient with obstructive lung disease has an altered anesthetic induction with an insoluble agent because of

(A) decreased cardiac output
(B) increased perfusion
(C) increased PCO_2
(D) uneven ventilation
(E) decreased minute volume

924. A patient involved in a motor vehicle accident required a prolonged extraction time from the vehicle at the scene. After transport by helicopter to the hospital, the patient is brought to the operating room for removal of a ruptured spleen and fixation of multiple lower extremity fractures. After induction of general anesthesia and endotracheal intubation, an esophageal temperature probe is inserted. The patient's temperature is 32°C. All of the following problems may be expected EXCEPT

(A) metabolic acidosis
(B) impairment of the intrinsic clotting cascade
(C) cardiac dysrhythmias
(D) platelet dysfunction
(E) profound peripheral vasodilation

925. The term MAC refers to

(A) the median anesthetic concentration
(B) an anesthetic dosage that prevents movement after skin incision in 50% of patients
(C) a measurement not affected by age
(D) a measurement that is pertinent only to volatile anesthetics
(E) the mean alveolar concentration

926. Signs of inadequate general anesthesia include all of the following EXCEPT

(A) eyelid movement
(B) pupillary constriction
(C) hyperventilation
(D) sweating
(E) limb movement

927. During a surgical procedure to reattach the retina, the surgeon injects a bubble of sulfur hexafluoride into the vitreal cavity. Prior to the injection of sulfur hexafluoride, the anesthesiologist should

(A) hyperventilate the patient to produce hypocarbia
(B) discontinue all fluorinated anesthetic gases
(C) ensure that all extraocular muscles are maximally paralyzed via the administration of a muscle relaxant
(D) temporarily cease mechanical ventilation
(E) discontinue nitrous oxide

928. A patient was brought to the operating room for the emergency repair of a liver laceration and removal of the spleen following blunt abdominal trauma in a motor vehicle accident. Postoperatively in the postanesthesia care unit (PACU) the patient has a worsening metabolic acidosis, poor urine output, and the abdomen is tense and distended. The best method for evaluating the patient for abdominal compartment syndrome would be to measure

(A) cardiac output via echocardiography
(B) cardiac output via a thermodilution pulmonary artery catheter
(C) pressure inside the bladder
(D) simultaneous PCWP and CVP values
(E) abdominal girth every hour

929. A patient has chronic obstructive pulmonary disease requiring the constant administration of oxygen. He is dyspneic at rest and can walk at the most 20 ft before needing to rest. He is scheduled to undergo an exploratory laparotomy because of a small bowel obstruction. He would be classified by the American Society of Anesthesiologists as physical status

(A) III
(B) IIIE
(C) IVE
(D) V
(E) VE

930. The first use of cocaine as a local anesthetic was performed by

(A) Karl Koller
(B) Sigmund Freud
(C) August Bier
(D) Heinrich Quincke
(E) Heinrich Braun

931. If nitrous oxide is administered at a constant concentration, the uptake into the bloodstream in milliliters per minute will

(A) be constant
(B) increase with time
(C) decrease with time
(D) depend on temperature
(E) be independent of concentration

932. The correlation of anesthetic potency with lipid solubility is known as the rule of

(A) Ferguson
(B) Michaelis and Menten
(C) Henderson and Hasselbalch
(D) Singer and Nicholson
(E) Meyer and Overton

933. One milliliter of desflurane liquid occupies what volume at 1 atm pressure and 37°C if all of the liquid is vaporized? The ideal gas constant is 0.082 $(L \cdot atm)/(K \cdot mol)$, the specific gravity of desflurane is 1.45, and its molecular weight is 168.

(A) 219 mL
(B) 238 mL
(C) 243 mL
(D) 256 mL
(E) 276 mL

934. The correct order of solubilities in blood, from greatest to least, among the volatile anesthetics is

(A) halothane > isoflurane > sevoflurane > desflurane
(B) sevoflurane > isoflurane > desflurane > halothane

(C) desflurane > isoflurane > sevoflurane > halothane
(D) desflurane > halothane > sevoflurane > isoflurane
(E) sevoflurane > halothane > desflurane > isoflurane

935. The state of general anesthesia may be reversed by

(A) the administration of a competitive antagonist
(B) increasing the atmospheric pressure
(C) increasing the ambient temperature
(D) decreasing the ambient temperature
(E) the administration of any medication which increases cerebral perfusion

936. A 55-year-old man has had a total hip arthroplasty under general anesthesia. His only significant preoperative medical problem is osteoarthritis. Which of the following is the LEAST important criterion in determining when he may be sent to his hospital room from the postanesthesia care unit?

(A) Thirty minutes have elapsed since he was given morphine 4-mg i.v. for pain.
(B) Sixty minutes have elapsed since his admission to the PACU from the operating room.
(C) He has voided 100 mL of urine in the last 30 minutes.
(D) His vital signs before the induction of general anesthesia were pulse 110 and blood pressure 140/90; these values are 75 and 110/65, respectively, at the present time.
(E) His oxygen saturation is 94% on room air.

937. A patient is to have a cholecystectomy. The anesthesiologist decides to use sevoflurane in oxygen as the sole anesthetic agent, with no other medications administered. Approximately what concentration of sevoflurane will be required to prevent movement in response to intubation?

(A) 1.75%
(B) 2.25%
(C) 3%
(D) 4%
(E) 5%

938. If 2% isoflurane in oxygen, flowing at a rate of 3 L/min, is added to a circle system, what will the concentration of isoflurane be after 6 minutes? Assume complete mixing of gas in the system, and that excess gas is scavenged. The reservoir bag has a volume of 2 L, the carbon dioxide absorber has a volume of 3 L, and the connecting hose and valves have a volume of 1 L.

(A) 1%
(B) 1.26%
(C) 1.73%
(D) 1.90%
(E) 1.96%

939. If the uptake of gaseous anesthetic in L/min is x, and the patient's cardiac output suddenly doubles, the rate of uptake will become

(A) cannot be calculated without further information
(B) $x/2$
(C) $2x$
(D) $4x$
(E) x^2

940. The likelihood of intraoperative awareness under general anesthesia is highest with the use of

(A) inadequate benzodiazepine doses
(B) high-dose opioids
(C) muscle relaxants
(D) no premedication
(E) nitrous oxide as the sole gaseous anesthetic

941. Stage 2 anesthesia can be characterized by all of the following signs EXCEPT

(A) amnesia
(B) purposeless movement
(C) hypoventilation
(D) disconjugate gaze
(E) increased airway reflexes

942. Which of the following medications is not recommended during general anesthesia for the removal of a pheochromocytoma?

(A) morphine
(B) fentanyl
(C) sevoflurane
(D) vecuronium
(E) diazepam

943. An anesthesiologist's usual practice is to administer the combination of neostigmine and atropine for reversal of neuromuscular blockade. In a patient with glaucoma, managed with the topical administration of a β-adrenergic agonist, how should this anesthesiologist's usual practice for reversing neuromuscular blockade be modified?

(A) substitution of glycopyrrolate for atropine
(B) substitution of edrophonium for neostigmine
(C) avoidance of all nondepolarizing neuromuscular blocking agents
(D) use of mivacurium followed by spontaneous recovery of neuromuscular function
(E) no modification is necessary

944. A patient is brought to the operating room for repair of an open fracture sustained from a fall from a window during a house fire. The patient was intubated at the scene by a paramedic and given 100% oxygen via Ambu bag during transport to the hospital. The most reliable method for determining whether the patient has carbon monoxide poisoning while being ventilated with 100% O_2 is

(A) routine arterial blood gas analysis
(B) pulse oximetry
(C) capnometry
(D) arterial carboxyhemoglobin level
(E) electrocardiogram (ECG) evidence of carbon monoxide-induced arrhythmias

945. Contraindications to the discharge to home of a patient who had a hernia repair under general anesthesia include all of the following EXCEPT

(A) nausea
(B) inability to drink liquids without vomiting
(C) heart rate 50% higher than the preoperative value
(D) inability to walk due to groin pain
(E) disorientation to person and place

946. You are given the honor of providing the first anesthetic in a new radiology room. The patient is a 38-year-old man with an arteriovenous malformation of the thoracic spine which is causing severe pain but no neurologic deficit. The radiologist plans to embolize the lesion and estimates that the procedure will require 10 hours. Which of the following is the LEAST important requirement for the room in which this procedure will occur?

(A) pipeline oxygen supply
(B) pipeline nitrous oxide supply
(C) adequate space to place an anesthesia machine in proximity to the patient
(D) availability of suction
(E) auxiliary lighting available to the anesthesiologist

947. A patient is brought to the operating room for repair of a fractured femur. He fell off a boat and remained in the water for a long time prior to rescue. He is hypothermic with a temperature of 33°C. Other vital signs and laboratory values are normal. It can be assumed that the MAC for isoflurane in this patient is approximately

(A) 1%
(B) 1.25%
(C) 1.5%

(D) 1.75%
(E) 2%

DIRECTIONS (Questions 948 through 968): For each of the items in this section, ONE or MORE of the numbered options is correct. Choose answer

(A) if only 1, 2, and 3 are correct
(B) if only 1 and 3 are correct
(C) if only 2 and 4 are correct
(D) if only 4 is correct
(E) if all are correct

948. Which of the following are characteristics of general anesthetics?

(1) The potencies of the optical isomers of isoflurane are identical.
(2) All general anesthetics are gases at body temperature.
(3) All general anesthetic effects may be explained by their ability to disrupt membrane lipid-protein interactions.
(4) Their anesthetic effects may be reversed by increasing the ambient pressure.

949. Which of the following is a disadvantage of the particular volatile anesthetic agent?

(1) sevoflurane: significant biotransformation
(2) sevoflurane: high blood solubility
(3) desflurane: airway irritation
(4) isoflurane: poor muscle relaxation

950. The inspired partial pressure of a volatile anesthetic is lower than the partial pressure of the anesthetic in the fresh gas because of

(1) absorption of the agent by rubber hoses
(2) dilution of fresh gas in the circle absorber
(3) adsorption of the agent by soda lime
(4) uptake of the agent by the patient

951. The factor(s) of importance in determining alveolar tension of an anesthetic is (are)

(1) ventilatory rate
(2) cardiac output
(3) inspired concentration
(4) body temperature

952. Volatile anesthetics

(1) increase electroencephalogram (EEG) voltage
(2) affect synaptic transmission equally in neurons having different neurotransmitters
(3) depress excitatory postsynaptic potentials
(4) block impulse conduction in axons at a lower concentration than required to impair synaptic transmission

953. The MAC value for isoflurane is

(1) highest in the young
(2) decreased at age 70 compared to age 20
(3) decreased at lowered body temperature
(4) decreased by administration of 70% nitrous oxide

954. Emergence from anesthesia with isoflurane

(1) is not affected by the length of the anesthetic
(2) is affected by the cardiac output
(3) is less prolonged than emergence from nitrous oxide
(4) can be demonstrated with alveolar curves that are the inverted patterns of uptake curves

955. The current theories of the mechanism of general anesthesia suggest that all volatile anesthetics

(1) act at a specific receptor
(2) act primarily in the reticular activating system
(3) depress release of neurotransmitters
(4) affect synaptic transmission

956. A patient is scheduled for outpatient knee arthroscopy for removal of loose intra-articular bodies. The patient also has a history of severe obstructive sleep apnea. Factors that would permit this patient's surgery to be performed on an ambulatory basis include

(1) a surgical time of less than 1 hour
(2) placement of femoral nerve block to reduce postoperative pain
(3) use of CPAP by the patient at home
(4) adequate management of postoperative pain with ibuprofen

957. The second gas effect

(1) has its maximum effect early in an anesthetic
(2) applies only to anesthetic gases
(3) involves a gas that must be given at high concentrations
(4) may be responsible for diffusion hypoxia

Questions 958 and 959

A 27-year-old woman is anesthetized with propofol, isoflurane, nitrous oxide, and oxygen for laparoscopy. She is placed in a steep Trendelenburg's position after insertion of the needle through the abdominal wall, and carbon dioxide is insufflated. There is sudden onset of hypotension.

958. The hypotension may be due to

(1) carbon dioxide embolism
(2) hemorrhage
(3) compression of the inferior vena cava
(4) position

959. Appropriate step(s) to take is (are) to

(1) flatten the table
(2) inform the surgeon
(3) administer epinephrine
(4) discontinue the nitrous oxide

960. At the anesthetic level associated with the alveolar concentration MAC-awake, patients

(1) respond to simple commands
(2) will not move in response to a surgical incision
(3) may manifest signs of excitement
(4) will likely be apneic

961. During surgery on the airway with a carbon dioxide laser

(1) the endotracheal tube cuff should be wrapped tightly with metallic tape
(2) a fire in an endotracheal tube made of PVC produces highly toxic gases
(3) carbon dioxide should be added to the inspired anesthetic gas to decrease the risk of an airway fire
(4) one of the best ways of avoiding an airway fire is to not intubate the patient

962. Signs of a hemolytic transfusion reaction which occurs under general anesthesia include

(1) hemoglobinuria
(2) bronchospasm
(3) hypotension
(4) thrombus formation in medium and large vessels

963. During a surgical procedure to repair a traumatized liver in a patient who was in a motor vehicle accident, the patient required 100% O_2 in order to maintain an adequate value for oxygen saturation, and each time a volatile anesthetic was given, the blood pressure dropped to an unacceptable value. Medications which might prevent the occurrence of recall for intraoperative events in the absence of nitrous oxide and a volatile agent include

(1) scopolamine
(2) fentanyl
(3) midazolam
(4) vecuronium

964. During general anesthesia with isoflurane, nitrous oxide, and cisatracurium, expected ocular effects include

(1) decreased intraocular pressure in normal individuals
(2) ablation of the oculocardiac reflex
(3) ocular akinesia
(4) angle closure glaucoma in susceptible patients

965. A patient has had a total laryngectomy in the distant past. The patient now presents for mastectomy and axillary node dissection for the management of breast cancer. Reasonable methods of managing this patient's airway during general anesthesia include

(1) inserting a low-pressure cuffed endotracheal tube via the mouth
(2) spontaneous ventilation with supplemental oxygen during total IV anesthesia
(3) inserting a laryngeal mask airway
(4) inserting a reinforced, cuffed endotracheal tube into the tracheostomy stoma

966. Nasotracheal intubation may be used safely in a patient who has

(1) fractures of the lower cervical spine and the ethmoid bone
(2) a Le Fort II fracture of the maxilla
(3) a Le Fort III fracture of the maxilla
(4) a Le Fort I fracture of the maxilla

967. A 24-year-old woman is to have diagnostic laparoscopy as an outpatient. Her medical history is significant only for symptomatic gastroesophageal reflux. She is 61 inches tall and weighs 185 pounds. Prior to the induction of general anesthesia, she should be premedicated with

(1) metoclopramide
(2) droperidol
(3) cimetidine
(4) ondansetron

968. In using general anesthesia for laparoscopic cholecystectomy,

(1) inhaled nitrous oxide will diffuse into carbon dioxide-containing spaces and increase their volume or pressure
(2) decreas in postoperative values for FVC and FEV_1 are more pronounced than after open cholecystectomy
(3) small but detectable (via Doppler or transesophageal echocardiography) carbon dioxide emboli are the rule rather than the exception
(4) minute ventilation will need to be approximately tripled to eliminate the exogenously administered carbon dioxide

Answers and Explanations

922. (C) Ralph M. Waters was appointed to the first academic position in anesthesia in America at the University of Wisconsin in 1927. Ivan Magill was a British anesthesiologist who was a pioneer of endotracheal intubation and who was responsible for the development of many anesthesia devices and techniques. Arthur Guedel of Indianapolis is remembered for his creation of the cuffed endotracheal tube. Henry K. Beecher was the first anesthetist-in-chief at the Massachusetts General Hospital and professor of anesthesia at Harvard Medical School. William T.G. Morton performed the first public demonstration of ether anesthesia at the Massachusetts General Hospital in 1846. *(1:220; 5:38)*

923. (D) The patient with chronic obstructive lung disease has a prolonged induction due to ventilation/perfusion mismatching. The cardiac output is usually not decreased. The increased PCO_2 does not directly affect the uptake of the agent. Decreased minute volume is not a factor. *(1:385–6; 4:1132–3; 5:139–40)*

924. (E) Such patients have profound peripheral vasoconstriction in an effort to conserve heat. This effect decreases the effectiveness of applying heat to the extremities and the surface of the body in warming the patient. All of the other problems listed are common in patients with severe hypothermia. *(1:1397; 4:2155–6; 5:1581–2, 1980)*

925. (B) The term MAC refers to minimum alveolar concentration. It is defined as the alveolar anesthetic concentration sufficient to prevent movement in response to surgical incision in 50% of the subjects. It is affected by age. MAC can be used to specify the potency of both volatile and gaseous anesthetics. *(1:388–90; 4:1136–8; 5:107–9)*

926. (B) Pupillary dilatation is one of the signs of light anesthesia, as are tachypnea, sweating, and somatic movement. Lack of eye movement and pupillary constriction are two determinants of adequate depth. *(4:1136; 5:1228–9)*

927. (E) Sulfur hexafluoride is 117 times less diffusible than nitrous oxide, thus the presence of nitrous oxide would cause an increase in both the size of the bubble and IOP. *(1:976; 4:2195; 5:2533)*

928. (C) Abdominal compartment syndrome often causes a decreased cardiac output and increases in both PCWP and CVP. These changes are nonspecific and may be due to other causes. The most specific test would be to demonstrate that the pressure inside the bladder was above 20 mmHg. Although increasing abdominal distension is also often present, serial measurements of abdominal girth would waste time and an increase is also not diagnostic of abdominal compartment syndrome. *(1:1289)*

929. (C) This patient would be classified as physical status IV because he has an incapacitating systemic illness. Because the patient is to undergo an emergency procedure, "E" is added to the physical status. *(1:474; 4:4; 5:2592)*

930. (A) Koller first used cocaine topically in the eye. He obtained the cocaine from his friend Freud. Quincke developed the technique of lumbar puncture, and Bier performed the first spinal anesthetic for surgery. Braun introduced the use of epinephrine to prolong spinal anesthesia. *(1:975; 5:22–3)*

931. **(C)** The uptake will decrease over time as equilibrium is reached. The uptake is dependent on concentration, being greater with a higher concentration. *(1:381–3; 4:1130–1; 5:132–4)*

932. **(E)** The Meyer-Overton rule states that anesthetic potency is proportional to lipid solubility, and this rule is valid for the majority of gaseous anesthetics. Ferguson's rule states that anesthetic potency is proportional to thermodynamic activity (ideal solubility). Singer and Nicholson proposed the lipid bilayer hypothesis of membrane structure. *(1:131–2; 4:1108; 5:115–6)*

933. **(A)** One milliliter of desflurane liquid is 1.45 g or 0.00863 mole (1.45 g/168 g/mole). Thus, by the ideal gas law, $V = nRT/P = (0.00863) \times (0.082) \times (273 + 37) = 0.219$ L = 219 mL. *(1:379; 4:1030)*

934. **(A)** The correct order of solubilities in blood, from greatest to least, among the volatile anesthetics is halothane > isoflurane > sevoflurane > desflurane. *(1:378; 5:134)*

935. **(B)** The state of general anesthesia, produced by gaseous agents or barbiturates, may be reversed by increasing the atmospheric pressure. *(4:1109–10; 5:108, 2669)*

936. **(B)** Every PACU should have definite guidelines for discharge criteria that should account for the preoperative status of the patient and the expected postoperative morbidity. An arbitrary period of observation is inappropriate. *(1:1380–1; 4:2306–7; 5:2708–9)*

937. **(C)** MAC is defined in terms of the prevention of movement in response to skin incision. The concentration necessary to prevent movement in response to a greater stimulus, such as intubation, is approximately 1.5 times the value for MAC. *(4:1124, 1138; 5:1239–40)*

938. **(D)** The time constant of the circuit is its total volume (6 L) divided by the fresh gas flow rate (3 L/min), or 2 min. The concentration at any point in time is given by the following exponential equation:

$$C = C_0(1 - e^{t/\tau})$$

where C_0 is the concentration in the fresh gas (2%), τ is the time constant (2 minutes), and t is the time in question (6 minutes). Thus, after one time constant has elapsed, the concentration in the system is 63.2% of the fresh gas concentration or 1.26% isoflurane; after two time constants, 86.5% or 1.73% isoflurane; after three time constants, 95% or 1.9% isoflurane; and after four time constants, 98.2% or 1.96% isoflurane. *(1:381; 4:1130)*

939. **(C)** Uptake by the blood of a gaseous anesthetic from the lung is proportional to cardiac output. Thus, if the cardiac output doubles, uptake will double. *(1:382–3; 4:1131–2; 5:138)*

940. **(C)** The likelihood of recall is not correlated with the use, or the lack of use, of any anesthetic agent; however, the use of muscle relaxants, which may block the observation of movement as a sign of inadequate anesthesia, is the pharmacologic risk factor of greatest importance for intraoperative awareness. *(4:1138–9; 5:42, 1229, 1237)*

941. **(C)** The second stage of anesthesia is characterized by excitement, somatic movement, increased airway reflexes, disconjugate gaze, hypertension, hyperventilation, loss of consciousness, and amnesia. *(1:407; 4:1136; 5:1228)*

942. **(A)** Circulating histamine provokes release of catecholamines from the tumor, thus medications that stimulate histamine release are best avoided. *(1:1130–1; 4:1949)*

943. **(E)** Topical administration of an anticholinergic like atropine into the eye can raise IOP in a patient with glaucoma. The rise in IOP may be dangerous in persons predisposed to narrow-angle glaucoma. IV and intramuscular administration of atropine does not result in intraocular concentrations of atropine adequate to cause mydriasis and such administration is considered safe in persons with glaucoma. *(1:985; 5:2532)*

944. (D) Unless a specific request is made, a routine arterial blood gas determination will report the pH, PCO_2 and PO_2. If oxygen saturation is reported, it is usually a calculated value and not one that is determined directly. Thus, a specific request must be made for determination of carboxyhemoglobin. *(4:2167–8; 5:2671)*

945. (A) Nausea without vomiting is very common after general anesthesia and as an isolated symptom is not a contraindication to discharge to home. Patients should be oriented and their vital signs should be near their preoperative values. Their pain should be under reasonable control and they should be able to tolerate fluids without vomiting. *(1:1234–5; 4:2260; 5:2619–21)*

946. (B) This patient requires a general anesthetic because of his pain and because of the expected duration of the procedure. General anesthesia would most likely be provided by an anesthesia machine located near the patient. If the flow of nitrous oxide is set at 2 L/min, a single full cylinder would last for about 13 hours. Conversely, the ventilator might require 10 to 20 L/min, meaning that an oxygen cylinder might last for only 30 to 60 minutes; a supply of wall oxygen is generally required whenever a ventilator is to be used. The anesthesiologist also must have the ability to suction the patient's airway and to see the patient and the anesthesia equipment, considering that much of the proposed procedure will take place with the room darkened. *(1:1327–30; 4:2288, 2294–6; 5:2640–1)*

947. (A) The effect of altered body temperature on MAC is to decrease MAC by approximately 5% for each 1°C decrease from normal body temperature. *(1:122; 4:1137; 5:108)*

948. (B) General anesthesia may be reversed by increasing the ambient pressure, regardless of the agent used to produce the anesthetic state. Some of the general anesthetics that are not gases at body temperature include halothane, isoflurane, and thiopental. While the perturbation of membrane lipid-protein interactions is an attractive hypothesis for general anesthetic action, most anesthetic agents have not been studied in terms of this action. The optical isomers of isoflurane do not have the same potency. *(4:1040–1, 1048; 5:120–1, 240; 2669)*

949. (B) Sevoflurane undergoes significant biotransformation. It is poorly soluble, permitting rapid induction and emergence. Desflurane is irritating to the airway, and commonly causes coughing and laryngospasm during light anesthesia. Isoflurane provides good muscle relaxation. *(1:387–8, 431; 5:1490, 2021, 2604)*

950. (E) All of these effects act to increase the difference between the partial pressure of the agent in the fresh gas and the partial pressure in the inspired gas. *(1:381–3; 4:1060–2; 5:132–5)*

951. (A) Alveolar ventilation, inspired concentration, and cardiac output all affect the alveolar tension of the anesthetic gas. Body temperature is not a factor. *(1:381–3; 4:1060–3; 5:132–5)*

952. (B) Volatile anesthetics block synaptic transmission at lower concentrations than are required to block axonal conduction. Furthermore, there are differential effects at neurons releasing different neurotransmitters. The characteristic EEG effect of volatile anesthetics is decreased frequency and increased voltage. *(4:1034–6; 5:111, 641–2, 1249)*

953. (E) The MAC value is highest in the very young child and decreases as one gets older. It is also decreased by lower body temperature. Nitrous oxide in 70% concentration decreases the MAC for a given agent. *(1:123, 389–90; 4:1067–70; 5:108; 1239–42)*

954. (C) Emergence from anesthesia is affected by cardiac output, since the circulation to the lung affects the gradient between the alveolar capillary and the alveolus. The alveolar curve is the inverted pattern of induction with a few minor differences. Emergence will be longer after a longer anesthetic procedure. Emergence from nitrous oxide is faster than emergence from isoflurane. *(1:386–7; 4:1060–3; 5:147–9)*

955. (D) It is thought that all volatile anesthetics affect synaptic transmission. Excitatory synapses are depressed by general anesthetics while inhibitory synapses may be depressed or potentiated. There is no specific receptor known for volatile anesthetics, and no definite location within the CNS is thought to be responsible for their actions. *(1:123–5; 4:1035–6; 5:110–2)*

956. (D) The patient with obstructive sleep apnea is at increased risk for opioid-induced apnea. The ability to manage the patient's pain without opioids may permit the patient to have safe ambulatory surgery. A perioperative nerve block does not necessarily increase safety because the patient might require opioids when the block regresses. A short surgical time does not necessarily minimize postoperative pain. CPAP may prevent obstruction-related apneic episodes, but it does not necessarily affect the magnitude of opioid-induced ventilatory depression in these patients. *(1:1039–41, 1385–6; 5:2748–9)*

957. (B) The second gas effect occurs when the administration of a high concentration of one gas increases the rise in alveolar concentration of another gas. This effect may apply to any gas. The maximum effect in anesthesia is early in the course of the anesthetic, and one of the gases must be capable of being given in high concentration. Diffusion hypoxia may occur at the end of an anesthetic when the diffusion of large volumes of nitrous oxide from the pulmonary circulation to the alveoli dilutes the alveolar concentration of oxygen. *(1:384; 1532; 4:1064–5; 5:135)*

958. (A) The patient for laparoscopic examination may be hypotensive due to carbon dioxide embolus, hemorrhage, and compression of the vena cava from increased intra-abdominal pressure. The Trendelenburg position should not cause hypotension. *(1:1058–9; 4:1974; 5:2289–90, 2296)*

959. (C) After the onset of hypotension during laparoscopic examination, the surgeon should immediately be informed and the insufflation of carbon dioxide discontinued. In the case of carbon dioxide embolism, hypotension and desaturation are the usual presenting signs. Administration of 100% oxygen may increase oxygen saturation. Placement of the patient in the left lateral position acts to trap the gas in the right ventricle and decrease the amount entering the pulmonary artery. Since carbon dioxide is very soluble, aspiration of the gas via a right atrial catheter is rarely necessary. The occurrence of hemorrhage via laceration or cannulation of a blood vessel with the insufflating needle may require laparotomy for repair. If the hypotension is due to vena cava compression, decreasing the intra-abdominal pressure should increase the blood pressure. Epinephrine is not indicated unless the hypotension persists and requires beginning advanced cardiac life support. *(1:1058–9; 4:2132; 5:2289–90)*

960. (B) MAC-awake is the alveolar concentration of an inhalational anesthetic at which 50% of the patients respond to commands. This value may be applied to patients as general anesthesia is being induced, or as they are emerging from anesthesia, and is similar to the alveolar concentrations associated with the excitement stage. This concentration is lower than that which will prevent movement in response to incision (or MAC) or result in apnea. *(1:121–2, 389–90; 4:1068; 5:102, 149, 1239)*

961. (C) There is no foolproof way of avoiding an airway fire during laser surgery on the airway. By not intubating the patient (and either using jet ventilation or intermittent apnea), one has removed one potentially flammable object from the airway. Combustion of PVC yields hydrogen chloride gas which is acidic and highly toxic and irritating. Although an endotracheal tube may be wrapped with metallic tape to decrease the likelihood that it will catch fire, there is no way to wrap the tube cuff with metallic tape. *(1:997–8; 4:2206–7; 5:2580–4)*

962. (A) Signs of a hemolytic transfusion reaction include hemoglobinuria, hypotension, and bronchospasm. Disseminated intravascular coagulation (DIC) is also common, however, this is usually manifest as a coagulopathy rather than as clotting in medium and large vessels. *(1:207–8; 4:2418–9; 5:1815–6)*

963. **(B)** Both midazolam and scopolamine produce reliable amnestic effects in most patients without adversely affecting blood pressure and cardiac output. Awareness under anesthesia is much more likely when the anesthetic technique consists of oxygen and opioid. Muscle relaxants have no action in preventing recall. *(1:555, 560; 4:2154; 5:2457)*

964. **(B)** Isoflurane decreases IOP and cisatracurium renders the eye immobile. Deep general anesthesia does not prevent the oculocardiac reflex (nor does any other maneuver or drug reliably prevent the reflex). Medications that cause mydriasis may precipitate an episode of angle closure glaucoma in susceptible patients. *(4:2184–5, 2190–1; 5:2532, 2535, 3012)*

965. **(C)** A patient who has had a total laryngectomy usually has a permanent tracheostomy stoma in the neck. Such patients have no connection to the airway via the oral route, thus oral intubation or the placement of a laryngeal mask airway are both impossible. The stoma may be intubated and a reinforced tube is a popular choice. An alternative is to supply supplemental oxygen via a tracheostomy mask while the patient breathes spontaneously during total IV anesthesia. *(4:2211–2; 5:1630, 2539)*

966. **(D)** Le Fort II and III fractures and fractures of the ethmoid bone all increase the risk of the endotracheal tube penetrating into the brain during nasotracheal intubation. This risk is not present if the patient has a Le Fort I fracture. *(1:999–1000, 1256–7; 4:2141, 2215; 5:2541–2)*

967. **(E)** Patients with symptomatic reflux should be premedicated with a histamine H_2 antagonist (such as cimetidine) and a gastrointestinal prokinetic agent (such as metoclopramide) in order to decrease the volume and increase the pH of the gastric contents. This patient also has several risk factors for postoperative nausea and vomiting, including female gender, young age, obesity, and an emetogenic operative procedure. She should receive antiemetic prophylaxis such as with droperidol and ondansetron. *(1:609–10, 1223–4; 4:2244–9; 5:2294; 2600)*

968. **(B)** Nitrous oxide will diffuse into carbon dioxide-containing spaces and increase the pressure and/or volume. Although pulmonary function may be compromised after laparoscopic cholecystectomy, it is affected to a much lesser degree than after open cholecystectomy. Carbon dioxide emboli are common (in 69% in one study) during laparoscopic cholecystectomy, however, most are fortunately of little clinical significance. Minute ventilation needs to be increased by about a third in the average patient during laparoscopic cholecystectomy in order to maintain a normal value for end-tidal carbon dioxide. *(4:1885; 5:2287, 2289, 2294, 2298)*

CHAPTER 9

Regional Anesthesia

Questions

DIRECTIONS (Questions 969 through 993): Each of the numbered items or incomplete statements in this section is followed by answers or completions of the statement. Select the ONE lettered answer or completion that is BEST in each case.

969. The dermatome level at the nipple line is

(A) C8
(B) T2
(C) T4
(D) T6
(E) T8

970. The gasserian ganglion is associated with the

(A) second cranial nerve
(B) third cranial nerve
(C) fifth cranial nerve
(D) seventh cranial nerve
(E) ninth cranial nerve

971. The superior laryngeal nerve lies

(A) superior to the hyoid bone
(B) deep to the hyoid bone
(C) immediately above the notch of the thyroid cartilage
(D) between the great cornu of the hyoid bone and the superior cornu of the thyroid cartilage
(E) deep to the cricothyroid membrane

972. If a line is drawn around the neck at the level of the lower border of the cricoid cartilage, it will mark the level of the transverse process of the

(A) second cervical vertebra
(B) third cervical vertebra
(C) fourth cervical vertebra
(D) fifth cervical vertebra
(E) sixth cervical vertebra

973. In order to provide an adequate sensory level for transurethral prostate resection, the sensory level must be

(A) T4
(B) T6
(C) T8
(D) T10
(E) T12

974. The axillary approach to the brachial plexus

(A) carries the risk of pneumothorax
(B) uses the axillary artery as a landmark
(C) is made from the lateral aspect of the arm
(D) uses the pectoralis minor insertion as a landmark
(E) is the preferable route if there is an infection in the axilla

975. In Figure 9-1, the structure labeled Y is the

(A) axillary nerve
(B) musculocutaneous nerve
(C) radial nerve
(D) ulnar nerve
(E) median nerve

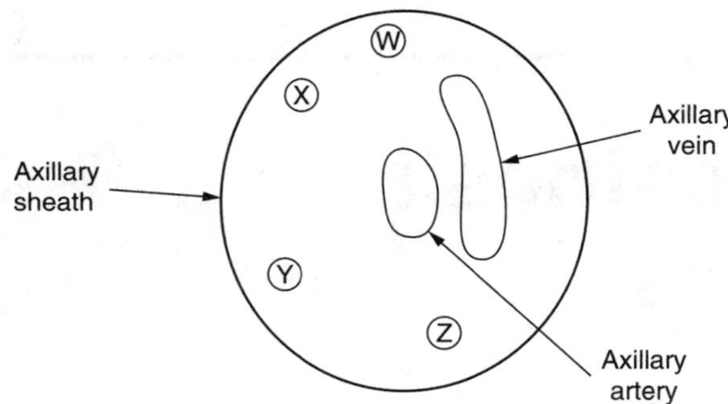

FIG. 9-1

976. An important anatomic feature in the performance of an intercostal block is

(A) the intercostal nerve serves only the thoracic area
(B) the intercostal nerve runs above the rib
(C) each intercostal nerve gives off a lateral cutaneous branch
(D) the intercostal innervation runs from the vertebral column posteriorly to the nipple line anteriorly
(E) the abdominal innervation is easily anesthetized by blocking two adjacent nerves

977. When using the caudal approach to the epidural space

(A) the needle enters through the sacral hiatus
(B) an inadvertent subarachnoid block is much less likely than when using the lumbar approach
(C) the patient must be positioned prone
(D) the technique becomes relatively more contraindicated as the patient's age decreases
(E) small volumes of agent are needed, since the volume of the canal is only 8 to 12 mL

978. The stellate ganglion

(A) is a fusion of the inferior cervical and first thoracic ganglia
(B) receives fibers whose cell bodies are located primarily at the C4-C7 levels of the spinal cord
(C) is primarily a parasympathetic ganglion
(D) lies at the level of the fifth cervical vertebra
(E) may be blocked with little danger of complications

979. When performing digital nerve blocks

(A) 5 to 10 mL of local anesthetic are used for each nerve
(B) epinephrine-containing solution should be used to decrease systemic absorption
(C) the nerves can be blocked by injecting only on the medial side of the digits where the nerves course
(D) the course of the nerve lies close to the dorsum of the finger
(E) one must remember that the digital arteries are terminal arteries

980. Postdural puncture headaches are

(A) more frequent with large-bore needles
(B) not lessened by the use of a Whitacre needle as compared to a Quincke needle
(C) aggravated by the supine position
(D) noted immediately
(E) relieved in 12 hours by an epidural blood patch

981. The pain of carcinoma of the pancreas can be blocked by

(A) celiac plexus block
(B) intercostal block
(C) lumbar sympathetic block
(D) stellate ganglion block
(E) facet block

982. Postoperative nerve deficits following regional anesthesia may be due to all of the following EXCEPT

(A) epinephrine 1:200,000 added to the local anesthetic solution
(B) positioning
(C) multiple sclerosis
(D) orthopedic casts
(E) surgical trauma

983. The dermatome level that supplies the inguinal area is

(A) T6
(B) T8
(C) T10
(D) L1
(E) L3

984. In performing a lumbar puncture by the paramedian approach, all of the following are true EXCEPT that

(A) it is necessary to flex the spine
(B) the needle penetrates the ligamentum flavum
(C) the lateral or sitting position is satisfactory
(D) the needle does not penetrate the interspinous ligament
(E) the needle enters between the laminae

985. The first modality to be lost on the onset of spinal anesthesia is

(A) touch
(B) motor power
(C) temperature
(D) vibration
(E) autonomic activity

986. The sciatic nerve is formed by all of the following nerve roots EXCEPT

(A) L3
(B) L4
(C) L5
(D) S1
(E) S2

987. The tubercle of Chassaignac is the eponym for the

(A) horn of the hyoid bone
(B) anterior portion of the cricoid
(C) transverse process of the sixth cervical vertebra
(D) tip of the mastoid
(E) medial portion of the clavicle

988. The most superior approach to the brachial plexus involves placing the needle

(A) between the lateral head of the sternocleidomastoid muscle and the anterior scalene muscle
(B) between the medial and lateral heads of the sternocleidomastoid muscle
(C) between the anterior and posterior scalene muscles
(D) between the anterior and middle scalene muscles
(E) on the first rib

989. A patient has dropped a brick on his foot, suffering open fractures of the first three toes. He is scheduled for irrigation of the wounds and K-wire fixation of the fractures. Adequate anesthesia will require blockade of each of the following nerves EXCEPT the

(A) sural nerve
(B) superficial peroneal nerve
(C) deep peroneal nerve
(D) posterior tibial nerve
(E) saphenous nerve

990. The sciatic nerve bifurcates

(A) within the gluteus maximus
(B) about halfway between the hip and the knee
(C) about 10 cm cephalad to the knee
(D) at the level of the knee
(E) about halfway between the knee and the ankle

991. The correct order of longest to shortest duration of action of local anesthetics used for epidural anesthesia is

(A) bupivacaine > lidocaine > mepivacaine > chloroprocaine
(B) mepivacaine > bupivacaine > lidocaine > chloroprocaine
(C) bupivacaine > mepivacaine > chloroprocaine > lidocaine
(D) bupivacaine > mepivacaine > lidocaine > chloroprocaine
(E) mepivacaine > bupivacaine > chloroprocaine > lidocaine

992. The obturator nerve

(A) is unimportant in arthroscopic surgery of the knee
(B) supplies motor innervation to the adductor muscles of the hip
(C) supplies sensory innervation to the skin overlying the patella
(D) is derived from the first and second sacral roots
(E) supplies sensory innervation to the lateral aspect of the knee

993. The dermatome level that supplies sensation to the heel is

(A) L4
(B) S1
(C) S3
(D) S4
(E) S5

DIRECTIONS (Questions 994 through 1024): For each of the items in this section, ONE or MORE of the numbered options is correct. Choose answer

(A) if only 1, 2, and 3 are correct
(B) if only 1 and 3 are correct
(C) if only 2 and 4 are correct
(D) if only 4 is correct
(E) if all are correct

994. The patient who absorbs fluid during transurethral resection of the prostate under spinal anesthesia is in danger of pulmonary edema as a result of

(1) increased intravascular pressure
(2) position on the operating table
(3) decreased osmotic pressure
(4) increased sodium concentration

995. A "three-in-one" block of the lumbar plexus anesthetizes the

(1) ilioinguinal nerve
(2) iliohypogastric nerve
(3) sciatic nerve
(4) obturator nerve

996. When performing a lumbar puncture using the midline approach, the needle passes through the

(1) supraspinous ligament
(2) interspinous ligament
(3) ligamentum flavum
(4) pia mater

997. The addition of 10% dextrose to a spinal anesthetic

(1) decreases the baricity
(2) increases the specific gravity of the solution relative to the cerebrospinal fluid
(3) causes a higher level when the patient is in the sitting position
(4) allows the solution to gravitate toward the most dependent portion of the subarachnoid space

998. Contraindications to spinal anesthesia include

(1) coagulopathy
(2) coronary artery disease
(3) bacteremia
(4) hypervolemia

999. The sensory distribution of the hand includes the

(1) dorsum of the thumb: radial nerve
(2) dorsum of the middle finger: median nerve

(3) dorsum of the fifth finger: ulnar nerve
(4) volar surface of the fifth finger: ulnar nerve

1000. Circumcision performed under regional block requires the infiltration of the

(1) shaft of the penis
(2) glans penis
(3) foreskin in a circular manner
(4) base of the penis

1001. Anesthesia for surgery on the medial malleolus requires block of the

(1) femoral nerve
(2) obturator nerve
(3) sciatic nerve
(4) sural nerve

1002. The musculocutaneous nerve may be blocked

(1) by an injection into the biceps muscle
(2) by an injection at the medial aspect of the antecubital fossa
(3) by intentionally entering the brachial artery and depositing local anesthetic solution deep into the artery
(4) by the supraclavicular approach to the brachial plexus

1003. The landmark(s) for a sciatic nerve block via the posterior approach is (are) the

(1) pubic tubercle
(2) greater trochanter
(3) tip of the coccyx
(4) posterior superior iliac spine

1004. An epidural block

(1) does not cause spinal headache unless the dura is punctured
(2) has a greater risk of drug overdose than does a spinal block
(3) depends on volume to attain level
(4) depends on the baricity of the agent to attain level

1005. In performing a supraclavicular block

(1) the musculocutaneous nerve is frequently not anesthetized
(2) there is a hazard of pneumothorax
(3) increased success is obtained if multiple paresthesias are obtained and multiple injections performed
(4) the first rib is an important landmark

1006. (A) structure(s) lying on the lateral aspect of the antecubital space anteriorly include(s) the

(1) radial nerve
(2) median nerve
(3) musculocutaneous nerve
(4) ulnar nerve

1007. Increased duration of anesthesia with a Bier block may be obtained by

(1) substituting bupivacaine for lidocaine
(2) adding epinephrine to the lidocaine solution
(3) intermittently deflating the tourniquet to minimize ischemic pain
(4) using a double tourniquet

1008. When administering a spinal anesthetic, the solution and technique to be used are influenced by the

(1) duration of anesthesia needed
(2) intensity of anesthesia desired
(3) surgical site
(4) position of patient during injection

1009. A patient has suffered a traumatic amputation of the thumb at the interphalangeal joint and is brought to the operating room for revision of the stump. The surgeon will not use a tourniquet and anesthesia by nerve block(s) at the wrist is chosen as the technique. Which of the following nerves must be blocked to provide adequate anesthesia?

(1) musculocutaneous
(2) radial
(3) ulnar
(4) median

1010. Prolongation of spinal anesthesia may be achieved by adding which of the following drugs to a tetracaine solution?

(1) 0.1 mg norepinephrine
(2) 25 mg phenylephrine
(3) 50 mg ephedrine
(4) 0.3 mg epinephrine

1011. The position of the patient is important in administering a spinal anesthetic. Which of the following statements is (are) TRUE?

(1) A patient in the sitting position will have a lower block if the solution is hypobaric and the patient remains erect.
(2) The normal lumbar lordosis limits the spread of hyperbaric solution in a supine patient.
(3) A patient placed supine and in the Trendelenburg position is at high risk for developing a total spinal block after injection of an isobaric solution.
(4) A patient having the lumbar puncture performed in the prone jackknife position should not have a hyperbaric solution injected.

1012. Signs and symptoms of an excessive level of spinal anesthesia include

(1) hypotension
(2) agitation
(3) nausea
(4) ventricular arrhythmias

1013. After a spinal anesthetic has been administered, the patient should be immediately observed for

(1) evidence of block level
(2) evidence of hypotension
(3) respiratory insufficiency
(4) spinal headache

1014. A 26-year-old woman had a vaginal delivery under low spinal anesthesia (saddle block). The following day, she complained of a severe headache which persisted whether she was positioned supine or prone in bed. Appropriate therapies to be instituted immediately include

(1) an epidural blood patch
(2) intravenous caffeine
(3) intramuscular morphine
(4) oral acetaminophen

1015. The epidural space

(1) extends from the base of the skull to the coccyx
(2) contains the spinal cord
(3) is traversed by spinal nerves
(4) contains a plexus of veins

1016. Which of the following statements is (are) TRUE when comparing spinal to epidural anesthesia at the same sensory level?

(1) Hypotension following epidural blockade is better treated with intravenous fluids, while that following spinal blockade should be treated with an α-adrenergic agonist.
(2) A differential block (sensory anesthesia with minimal motor blockade) is more easily achieved with an epidural anesthetic.
(3) There are fewer systemic effects of the local anesthetic drug with epidural anesthesia.
(4) Systemic effects of epinephrine are more evident with epidural anesthesia.

1017. The ulnar nerve

(1) lies dorsal to the axillary artery in the proximal arm
(2) lies in the groove of the lateral condyle at the elbow
(3) lies adjacent to the flexor carpi radialis at the wrist
(4) gives sensation to the fifth finger and part of the fourth finger

1018. Undesirable aspects of spinal anesthesia include

(1) poor control
(2) respiratory embarrassment
(3) hypotension
(4) decreased vagal activity

1019. When performing an epidural block

(1) failure to aspirate spinal fluid is assurance that the dura has not been punctured
(2) the location of the needle tip in the epidural space must be determined by the *loss of resistance* technique
(3) if aspiration is negative, a test dose is not necessary
(4) a test dose is the most important diagnostic measure to ascertain accidental dural puncture

1020. In a patient undergoing a prostatic resection under spinal anesthesia, fluid absorption in large amounts is heralded by

(1) a rise in systolic and diastolic blood pressures
(2) bradycardia
(3) mental status changes
(4) pain

1021. The relationship of the femoral nerve in the inguinal area is that it lies

(1) superficial to the inguinal ligament
(2) lateral to the femoral artery
(3) medial to the femoral vein
(4) lateral to the obturator nerve

1022. In total hip arthroplasty surgery

(1) general anesthesia with deliberate hypotension is accompanied by less blood loss than normotensive general anesthesia
(2) general anesthesia with a volatile anesthetic prevents the hemodynamic consequences of methyl methacrylate
(3) epidural anesthesia is accompanied by fewer postoperative thromboembolic phenomena than is general anesthesia
(4) there is less postoperative mortality at 1 month with spinal anesthesia than with general anesthesia

1023. Blockade of the intercostobrachial nerve may be reliably obtained by

(1) a supraclavicular block
(2) an axillary block
(3) an interscalene block
(4) local infiltration

1024. (A) structures lying medially on the anterior aspect of the wrist include(s) the

(1) median nerve
(2) ulnar nerve
(3) radial nerve
(4) ulnar artery

Answers and Explanations

969. (C) *(1:692; 4:1395)*

970. (C) The gasserian (or semilunar) ganglion is associated with the fifth (trigeminal) cranial nerve. The roots of the trigeminal nerve in the pons send sensory branches to the gasserian ganglion. *(1:718; 4:1436; 5:1704)*

971. (D) The superior laryngeal nerve lies between the great cornu of the hyoid bone and the superior cornu of the thyroid cartilage. This nerve can be blocked to provide airway anesthesia for laryngoscopy or bronchoscopy. *(1:722–3; 4:1441; 5:1708)*

972. (E) The lower border of the cricoid cartilage is at the same level as the anterior tubercle of the sixth cervical vertebra. This tubercle is an important landmark for performing a stellate ganglion block. *(1:734; 4:1464–5; 5:1710, 1742)*

973. (D) Adequate anesthesia for a transurethral prostatic resection requires a sensory level to at least T10. *(1:1021; 4:1971; 5:2190)*

974. (B) The major landmark for performing an axillary block is the axillary artery. The approach is from the medial aspect of the arm. There is no risk of pneumothorax, which is an advantage over the supraclavicular approach. The muscle inserting in the area is the pectoralis major. An infection in the axilla is a contraindication to the block. *(1:726; 4:1418–20; 5:1690–2)*

975. (C) The radial nerve is the structure identified. It lies in a posterior and inferior position relative to the landmark axillary artery. The ulnar nerve lies inferior to the artery, and the median and musculocutaneous nerves lie superior, although the musculocutaneous nerve may have exited the sheath at a point cephalad to the axilla. *(1:727; 4:1419; 5:1691–2)*

976. (C) The intercostal nerves serve the abdomen and the thorax. Running under the rib, the nerve gives off a lateral cutaneous branch near the mid-axillary line and an anterior cutaneous branch near the sternum. If abdominal surgery is to be performed, the lower six to eight intercostal nerves should be blocked. *(1:730–1; 4:1428–9; 5:1710)*

977. (A) The needle for the caudal block enters through the sacral hiatus and the patient may be positioned in the prone or lateral decubitus positions. Even though the caudal canal is of low volume, there is leakage through the foramina, requiring injection of a larger volume as compared with the lumbar approach. Inadvertent subarachnoid injection is very possible. The caudal approach is technically easier than the lumbar approach in babies, and is becoming increasingly more popular in pediatric anesthesia. *(1:1422; 4:2002; 5:1673)*

978. (A) The stellate ganglion is a fusion of the inferior cervical and first thoracic ganglia and lies at the level of the seventh cervical vertebra. The ganglion is a sympathetic ganglion and receives fibers whose cell bodies are located at the T1-T8 levels of the spinal cord. Stellate ganglion block has many potential complications, including pneumothorax, intravascular injection, and subarachnoid injection. *(1:733; 4:1464; 5:821)*

979. **(E)** When performing a digital nerve block, one must remember that the digital arteries are terminal arteries. Ischemia must be prevented; therefore, epinephrine and large volumes are avoided. The nerves lie closer to the palmar side of the hand. *(1:715; 4:1432–3)*

980. **(A)** Postdural puncture headaches are less frequent with small-bore needles and with pencil point needles (e.g., Whitacre, Sprotte, or Green needles). The typical headache is positional, being relieved by the supine position and aggravated by standing. It is relieved immediately by an epidural blood patch. *(1:692; 4:1383–6; 5:1669)*

981. **(A)** The pain of carcinoma of the pancreas is treated with a celiac plexus block. This block provides anesthesia for the abdominal, but not the pelvic, viscera. The other blocks listed are not helpful in carcinoma of the pancreas. *(1:734; 4:1470; 5:1711–2)*

982. **(A)** Postoperative nerve deficits following anesthesia are rare. There are many causes, including positioning, surgical trauma, and the application of tight casts or dressings. Epinephrine is not a known cause. Since multiple sclerosis is characterized by periods of remission and exacerbation, a worsening in symptoms may be temporally (although not causally) related to a regional anesthetic. *(1:498–9; 653–7, 701, 1104; 4:276–8,682–7; 5:585, 1151)*

983. **(D)** *(1:692; 4:1395)*

984. **(A)** It is not necessary to flex the spine when using the lateral approach. For this reason, it is a good approach for the pregnant patient and those who would have trouble flexing. The block can be done in the sitting, lateral decubitus, or prone positions. The needle does not penetrate the interspinous ligament but does penetrate the ligamentum flavum. The needle enters between the laminae. *(1:694–5; 4:1370; 5:1664)*

985. **(E)** The first function to be blocked by a spinal anesthetic is autonomic function. These nerve fibers are of small diameter (therefore having shorter critical lengths) and are the most peripherally located fibers in the nerve roots. The next sensation to be blocked is temperature, followed by pain, tactile sensation, motor power, and proprioception. *(1:704; 4:1312–3)*

986. **(A)** The third lumbar root is not involved in the composition of the sciatic nerve. The lumbosacral trunk of L4 and L5 joins the sacral fibers to form the large trunk of the sciatic nerve. *(1:736; 5:1699)*

987. **(C)** Chassaignac's tubercle is the transverse process of C6. It is an important landmark for performing a stellate ganglion block. *(1:734; 4:1464–5; 5:1704)*

988. **(D)** The most superior approach to the brachial plexus is the interscalene block in which the needle is inserted between the anterior and middle scalene muscles. *(1:724–5; 4:1412; 5:1686)*

989. **(A)** The sural nerve supplies the lateral portion of the foot and the fifth toe. All of the other nerves supply the area encompassed by the first three toes. *(1:740–1; 5:1703–4)*

990. **(C)** The sciatic nerve bifurcates in the popliteal fossa about 10 cm above the knee. *(1:740; 4:1435; 5:1702)*

991. **(D)** The correct order of longest to shortest duration of action of local anesthetics used for epidural anesthesia is bupivacaine > mepivacaine > lidocaine > chloroprocaine. *(1:458; 5:1674–5)*

992. **(B)** The obturator nerve supplies motor innervation to the adductor muscles of the hip and sensory innervation to the medial aspect of the knee. *(1:737; 5:1699)*

993. **(B)** *(1:692; 4:1395)*

994. **(B)** The syndrome of fluid absorption may lead to pulmonary edema due to increased intravascular pressure and to decreased osmotic pressure. The serum sodium will fall. The position on the table is not a factor. *(1:1019–20; 4:1972–3; 5:2191–2)*

995. **(D)** A "three-in-one" block of the lumbar plexus anesthetizes the femoral, lateral femoral cutaneous, and obturator nerves. The ilioinguinal and iliohypogastric nerves arise from above the lumbar plexus, and the sciatic nerve arises from nerve roots primarily below the lumbar plexus. *(1:739; 4:1422; 5:1744–5)*

996. **(A)** The needle does not pass through the pia mater, the covering immediately over the spinal cord. It does pass through the supraspinous ligament, the interspinous ligament (unless the paramedian approach is used), and the ligamentum flavum. *(1:689–91; 4:1365–7; 5:1655–6)*

997. **(C)** The addition of dextrose to the agent makes it heavier than the cerebrospinal fluid. The baricity is therefore increased. If the patient is sitting, the block will remain low, since the agent will gravitate toward the most dependent portion of the subarachnoid space. *(1:698–9; 4:1370–1; 5:589, 1666)*

998. **(B)** The presence of bacteremia, coagulopathy, or hypovolemia are usually considered contraindications to spinal anesthesia. The bacteremic patient may have the subarachnoid space seeded with bacteria during lumbar puncture, the patient with a coagulopathy is at risk for epidural hematoma, and the hypovolemic patient is at risk for severe hypotension in response to spinal anesthesia-induced sympathetic blockade. Spinal anesthesia is not contraindicated in the "well-prepared" patient with coronary artery disease. In fact, such a patient may be able to complain of anginal pain that might go unrecognized during a general anesthetic. *(1:709; 4:1364; 5:1654)*

999. **(E)** The sensory distribution of the hand is correct for all of the options. *(1:728–9; 4:1420; 5:1693)*

1000. **(D)** Nerve block of the penis for circumcision requires a circular block at the base of the penis. It is not necessary to block the glans, the shaft, or the foreskin. *(1:733; 4:1980–1; 5:1750)*

1001. **(B)** A femoral-sciatic block may be used for procedures below the knee that do not require the use of a tourniquet. The obturator nerve and the sural nerve need not be blocked. *(1:736–9; 4:1435; 5:2421)*

1002. **(D)** The musculocutaneous nerve may be blocked by an injection into the coracobrachialis muscle or lateral to the biceps tendon at the level of the elbow. It is reliably blocked by the interscalene and supraclavicular approaches to the brachial plexus, and less so by the axillary approach. *(1:724–8; 4:1420; 5:1689–90, 1694)*

1003. **(C)** The landmarks for a sciatic block are the greater trochanter and the posterior superior iliac spine. A line is drawn between the two, and at its midpoint, a perpendicular line is drawn. About 5 cm caudally along this line is the site of injection. *(1:738; 4:1423; 5:1700)*

1004. **(A)** An epidural block does not cause headache if the dura has not been violated. There is a greater risk of drug toxicity, since larger doses are used. The level is dependent on the volume of drug injected. Since the injection is made into a potential space, and not into a fluid-containing space, the baricity or specific gravity of the solution is irrelevant. *(1:707–9, 1416–21; 4:1399–1401, 1403–4; 5:1676, 2742–3)*

1005. **(C)** In doing a supraclavicular block, the first rib is an important landmark. If one stays on the rib, pneumothorax is less likely, but still possible. The musculocutaneous nerve is usually blocked. Because the nerve trunks are in close approximation as they cross the first rib, multiple injections are unnecessary. *(1:725–6; 4:1415–8; 5:1689–90)*

1006. **(B)** The radial nerve and the musculocutaneous nerve lie on the lateral aspect of the antecubital space anteriorly. The median nerve lies on the medial aspect of the antecubital space anteriorly. The ulnar nerve is located posterior to the medial epicondyle. *(1:728–9; 4:1431–2; 5:1693–4)*

1007. **(D)** The use of a double tourniquet permits the inflation of the distal tourniquet and deflation of the proximal tourniquet when pain caused by the latter becomes severe. Intermittently deflating the tourniquet will decrease the duration of the anesthesia. Neither the use of bupivacaine nor epinephrine will prolong the block, and the use of either is associated with potentially greater toxicity. *(1:728; 4:1482–3; 5:1695)*

1008. (E) The duration and intensity of a spinal anesthetic will vary with the solution injected: tetracaine has a longer duration than lidocaine and produces a more profound motor block than bupivacaine. The choice of a hyper-, iso-, or hypobaric solution and the position of the patient in which the solution is injected depend on the location of the surgical site. *(1:699–701; 4:1370–8; 5:1665–8)*

1009. (C) Sensory innervation to the thumb is supplied by the median and radial nerves. *(1:728–9; 4:1420; 5:1693–4)*

1010. (D) A reasonable dose of epinephrine is 0.2 to 0.3 mg, and of phenylephrine is 2 to 5 mg, to prolong the duration of a tetracaine spinal anesthetic. Norepinephrine and ephedrine are not generally used as additives to spinal anesthesia solutions. *(1:701; 4:1374; 5:1667)*

1011. (C) A patient in the sitting position will have a higher block if a hypobaric agent is used. The lumbar lordosis limits the spread of hyperbaric solutions in a supine patient. An isobaric solution should not ascend to cause a total spinal regardless of the patient's position. A patient in the prone jackknife position should have a hypobaric or an isobaric solution injected in order to prevent excessive rise in the level of the spinal anesthetic. *(1:697–9; 4:1375–8; 5:1662–7)*

1012. (A) Hypotension and nausea are common when the spinal anesthesia level is excessively high. Dyspnea is also common and may cause the patient to become agitated. Sinus tachycardia in response to hypotension may occur, as may sinus bradycardia if the sympathetic cardiac accelerator nerves are blocked. Ventricular arrhythmias are, however, uncommon. *(1:704–7; 4:1378–80, 1382–7; 5:1658–60)*

1013. (A) The patient must be observed in the postoperative period for a spinal headache, but this effect does not usually appear until postoperative days 1 or 2. Immediately after performing a spinal block, it is important that the patient be observed closely for the level of anesthesia and its cardiovascular and respiratory consequences. *(1:704–7; 4:1378–80, 1382–7; 5:1658–60)*

1014. (D) The presence of a headache in the supine position makes the diagnosis of a spinal headache unlikely. Thus, there is no indication for an epidural blood patch or intravenous caffeine. An analgesic such as acetaminophen should be given. Parenteral opioids are rarely necessary in the therapy of headache. *(1:707–8; 4:1383–5; 5:1669–70)*

1015. (E) The epidural space extends from the base of the skull to the coccyx. The space envelops the spinal cord and is traversed by the spinal nerves. There is a large plexus of veins that may be easily punctured. *(1:689–90; 4:1393–6; 5:1654–8)*

1016. (C) Hypotension following any central neural blockade is better treated with a vasopressor than with intravenous fluids. A differential block is much more easily achieved with epidural anesthesia as compared with spinal anesthesia. Since more drug (local anesthetic and vasoconstrictor) is used for epidural anesthesia, more systemic effects are to be expected. *(1:701–6; 4:1378–80, 1382–7, 1402–5; 5:1658–61, 1674–6)*

1017. (D) The ulnar nerve lies inferior to the axillary artery in the proximal arm and lies in a groove on the medial aspect of the elbow. The nerve lies adjacent to the flexor carpi ulnaris at the wrist. The ulnar nerve gives sensation to the fifth finger and part of the fourth finger. *(1:728–9; 4:1418–20, 1431–2; 5:1686, 1691–4)*

1018. (A) Undesirable effects of spinal anesthetics include poor control, respiratory embarrassment if the block is too high, and hypotension. Relative vagal activity is increased, since the sympathetic nervous system is blocked. *(1:704–7; 4:1378–80, 1382–3, 1386; 5:1658–60, 1668–70)*

1019. (D) A test dose is important in determining the placement of the needle or catheter. Failure to aspirate fluid is not an absolute sign that the dura has not been punctured. The epidural space may be located by the *loss of resistance* or the *hanging drop* techniques. *(1:695–6; 4:1397, 1401; 5:1670–3, 1677–8)*

1020. (A) There is no pain with the absorption of irrigating fluid. The blood pressure rises, and the heart rate falls, due to hypervolemia. There may also be mental status changes. *(1:1019–20; 4:1972–3; 5:2191–2)*

1021. (C) The femoral nerve lies lateral to the femoral artery and lateral to the obturator nerve. The femoral vein lies medial to the femoral nerve, and the inguinal ligament is superficial to the femoral nerve. *(1:738–9; 4:1421–4; 5:1697–8)*

1022. (B) Deliberate hypotension (induced with vasodilators and/or regional anesthesia) has been shown to decrease intraoperative blood loss in total hip arthroplasty. Regional anesthesia is associated with fewer postoperative venous thromboses and pulmonary emboli. No difference in postoperative mortality at one month has been seen between various anesthesia techniques. The hypotension which often accompanies methyl methacrylate absorption may be manifest during any anesthetic technique. *(1:1105–6, 1115–7; 4:2129–32; 5:2413, 2425–6)*

1023. (D) The intercostobrachial nerve supplies the upper arm and transmits the pain associated with a pneumatic tourniquet. The nerve will not be reliably blocked by any of the perivascular approaches to the brachial plexus, and is anesthetized by local infiltration superficial to the axillary artery. *(1:725; 4:1415; 5:1691)*

1024. (C) The ulnar nerve and ulnar artery lie medially on the anterior aspect of the wrist. The median nerve lies in the central part of the wrist, and the radial nerve lies laterally. *(1:728–9; 4:1432; 5:1693–4)*

CHAPTER 10

Anesthesia for Cardiothoracic and Vascular Surgery

Questions

DIRECTIONS (Questions 1025 through 1063): Each of the numbered items or incomplete statements in this section is followed by answers or by completions of the statement. Select the ONE lettered answer or completion that is BEST in each case.

1025. In a patient with a superior vena cava syndrome

(A) at least two large-bore catheters should be placed before induction
(B) an internal jugular central venous line is necessary
(C) intravenous lines should not be used in the upper extremities
(D) tracheal intubation may be performed with a large tube
(E) induction is facilitated by the head-down position

1026. During cardiopulmonary bypass, patients

(A) need large doses of muscle relaxants at 25°C
(B) require neuromuscular blockade during cooling and warming
(C) should have $PaCO_2$ maintained at about 30 mmHg
(D) should have their lungs hyperinflated
(E) should have continued pulmonary ventilation

1027. Nitroglycerin causes vasodilation that is markedly potentiated by

(A) metoprolol
(B) remifentanil
(C) labetalol
(D) magnesium sulfate
(E) sildenafil

1028. If a patient with a pacemaker in place requires surgery that entails the use of cautery

(A) the cutaneous electrode (skin pad) of a unipolar electrocautery unit should be as close to the pulse generator as possible
(B) electrocardiographic monitoring is necessary only if the patient is being paced
(C) electrocardiographic monitoring will give the best information
(D) little danger is present if the pacer is a demand unit
(E) constant palpation of the pulse or listening through an esophageal stethoscope is necessary

1029. The patient with a pacemaker in place may develop competing rhythms when a normal sinus rhythm is present and the unit has been converted to the asynchronous mode. If the pacing stimuli fall on the T wave of the previously conducted beats

(A) ventricular fibrillation will follow
(B) there is little danger, since the energy output is low with current pulse generators
(C) ventricular fibrillation is less likely if hypoxemia is present
(D) ventricular fibrillation is less likely with catecholamine release
(E) ventricular fibrillation is less likely with myocardial infarction

1030. Intraaortic balloon counterpulsation is a circulatory assist method that

(A) is used for patients with aortic aneurysms
(B) is used for patients with aortic insufficiency
(C) causes an intraaortic balloon to be inflated during systole
(D) increases coronary blood flow
(E) increases impedance to the opening of the left ventricle

1031. The clamping of the thoracic aorta in aneurysm repair is followed by

(A) immediate hypotension
(B) immediate hypertension
(C) cardiac standstill
(D) no change
(E) loss of blood pressure in the right arm

1032. Clamping of the distal aorta will be followed by

(A) increased cardiac output
(B) decreased arterial blood pressure
(C) decreased systemic vascular resistance
(D) increased stroke volume
(E) stable heart rate

1033. A complication of aortic surgery is paraplegia that is most commonly due to

(A) pressure on the spinal cord during surgery
(B) long periods of hypotension
(C) hypothermia associated with the surgery
(D) spinal cord ischemia
(E) loss of cerebrospinal fluid

1034. Hypovolemia may occur during abdominal aneurysm procedures as a result of all of the following EXCEPT

(A) blood loss
(B) inadequate fluid replacement
(C) use of vasodilators
(D) loss of fluid into the bowel
(E) expansion of the vascular bed during aortic cross-clamping

1035. The blood flow rate for an adult on total cardiopulmonary bypass is approximately

(A) 15 mL/kg/min
(B) 35 mL/kg/min
(C) 55 mL/kg/min
(D) 85 mL/kg/min
(E) 115 mL/kg/min

1036. The blood flow during total cardiopulmonary bypass

(A) is not adjustable
(B) is virtually nonpulsatile
(C) provides a pulsatile pressure
(D) is a pulsatile flow
(E) mimics normal flow in all respects

1037. The $PaCO_2$ of a patient on cardiopulmonary bypass

(A) is determined by the oxygen concentration of the fresh gas
(B) is generally adjusted through changes in fresh gas flow rate
(C) is generally adjusted through pulmonary ventilation
(D) should be maintained at less than 30 mmHg
(E) is determined by the type of oxygenator

1038. When an adult patient is on total cardiopulmonary bypass

(A) arterial pressure is usually maintained between 40 and 60 mmHg
(B) blood is pumped from the venae cavae and drains by gravity into the aorta for circulation
(C) the level of blood in the venous reservoir of the pump reflects the central venous pressure of the patient
(D) venous pressure elevation is of no consequence
(E) viscosity of blood decreases with progressive hypothermia

1039. Anesthesia for carotid artery surgery

(A) may cause decreased cerebral oxygen consumption
(B) involves the use of deliberate hypotension
(C) carries no danger of cerebral infarction
(D) should always be preceded by tests to show if carotid clamping can be tolerated
(E) requires general anesthesia

1040. Anesthesia for carotid endarterectomy generally involves all of the following EXCEPT

(A) hypercapnia
(B) normal or slightly increased arterial oxygen tension
(C) normal or slightly increased arterial pressure
(D) systemic heparinization
(E) normothermia

1041. The most reliable monitor for detection of intraoperative myocardial ischemia is

(A) creatine phosphokinase levels
(B) pulmonary artery catheterization
(C) transesophageal echocardiography
(D) troponin levels
(E) exhaled nitric oxide

1042. When temporary cardiac pacing is performed with a unipolar electrode

(A) the negative pole of the pulse generator is connected to the cardiac wire
(B) the negative pole of the pulse generator is connected to the ground wire
(C) the polarity of the pulse generator is irrelevant
(D) alternating current is generally employed
(E) the proper polarity depends on whether the cardiac lead is endocardial or epicardial

1043. Asymptomatic patients in normal sinus rhythm with mitral stenosis require

(A) prophylaxis against infective endocarditis
(B) diuretics
(C) anticoagulation
(D) mild tachycardia
(E) calcium channel blockade

1044. In unipolar artificial cardiac pacing, the negative pole of the pulse generator

(A) is generally connected to the cardiac tissue
(B) is generally connected to the heart via a noncardiac tissue pathway
(C) is attached to the same part of the heart as is the positive pole
(D) is the pole through which electrons flow into the pacemaker
(E) is referred to as the ground electrode.

1045. In comparing patients undergoing esophagectomy against those undergoing pulmonary resection, it is generally TRUE that esophagectomy patients

(A) have better nutritional status
(B) have less risk of aspiration
(C) have better pulmonary function
(D) are less likely to be hypoxic during single lung ventilation
(E) are less likely to need postoperative ventilation

1046. In the adult, the tracheobronchial tree

(A) divides at an uneven angle, making foreign bodies more apt to go to the left side
(B) divides into right and left bronchi, the left bronchus being narrower and longer
(C) does not move with respiration
(D) is lined with squamous epithelium
(E) is protected by circular cartilaginous rings throughout

1047. The human larynx

(A) lies at the level of the first through fourth cervical vertebrae
(B) in the adult is narrowest at the level of the cricoid cartilage
(C) is innervated solely by the recurrent laryngeal nerve
(D) is protected anteriorly by the wide expanse of the cricoid cartilage
(E) lies within the thyroid cartilage

Questions 1048 through 1051 refer to Figure 10-1.

1048. The figure shows a view of a patient with a double-lumen tube viewed from the head of the bed. To ventilate the right lung and deflate the left lung, one should

(A) clamp at 1 and uncap at 3
(B) clamp at 6 and uncap at 3
(C) clamp at 5 and uncap at 3
(D) clamp at 5 and uncap at 4
(E) clamp at 1 and 6 and uncap at 3

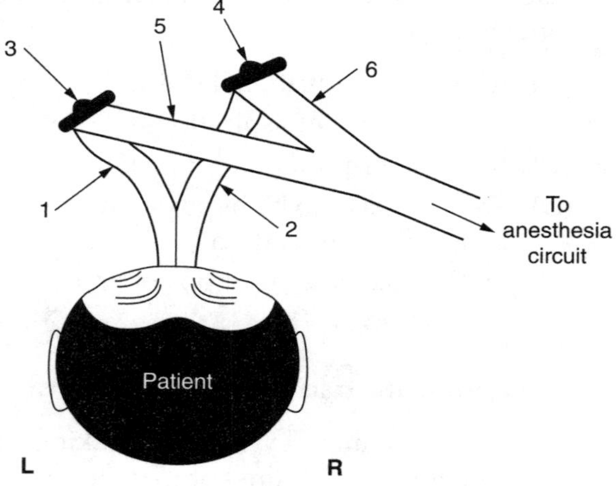

FIG. 10-1

1049. After the tube has been correctly positioned, a bronchoscope is used. By opening at 4 and looking down the lumen of 2 one should see

(A) the left upper lobe with a right-sided tube
(B) the carina with a right-sided tube
(C) the trachea with a right-sided tube
(D) the carina with a left-sided tube
(E) the left upper lobe with a left-sided tube

1050. To lavage the right lung while ventilating the left lung, one would perform all the following EXCEPT

(A) clamp at 6
(B) clamp at 2
(C) pour fluid into 4
(D) inflate the bronchial cuff
(E) inflate the tracheal cuff

1051. All of the following are true about the double-lumen tube in the figure EXCEPT

(A) lumen 1 is the bronchial lumen in a left-sided tube
(B) the pressures are equal at 1 and 2 when no clamps are applied
(C) CPAP to the right lung is applied at 4
(D) clamping at 6 isolates the left lung from the anesthesia circuit
(E) clamping at 5 and uncapping at 3 will allow the left lung to collapse

Questions 1052 and 1053 refer to Figure 10-2.

1052. The device depicted in the figure is

(A) a Univent bronchial blocker tube
(B) a Fogarty catheter
(C) a transesophageal cardiac pacemaker
(D) an independent (Arndt) endobronchial blocker
(E) an airway exchange catheter

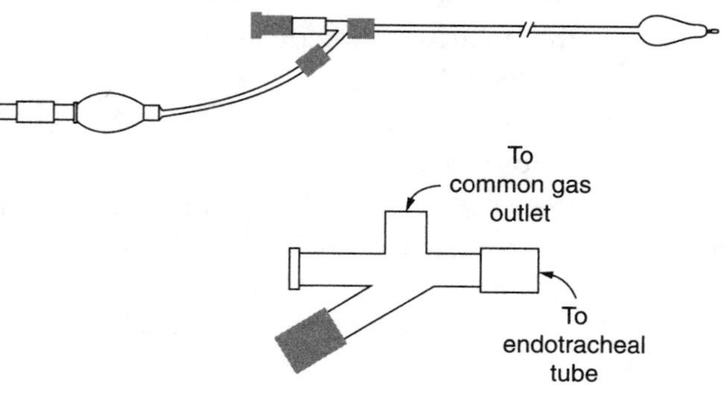

FIG. 10-2

1053. Which of the following statements about the device depicted is true?

(A) It is useful to obtain one-lung ventilation in patients who are already intubated or have a difficult airway
(B) Positive pressure ventilation has to be interrupted for proper placement
(C) There is no need for fiberoptic bronchoscopy to verify correct positioning of the device
(D) It can only be used in patients undergoing right-sided surgery
(E) With this device, the endotracheal tube has to be exchanged at the end of surgery if the patient is to remain intubated

Questions 1054 through 1059

A 65-year-old male has a right hilar mass with postobstructive pneumonia.

1054. In evaluating this patient's suitability for pneumonectomy, which of these tests would be the LEAST useful?

(A) $PaCO_2$
(B) PaO_2
(C) forced expiratory volume in 1 second (FEV_1)
(D) forced vital capacity (FVC)
(E) Ventilation/perfusion scan (V/Q scan)

1055. The patient is hypoxic on preoperative evaluation. Which therapy is LEAST likely to be effective?

(A) lateral position
(B) chest physical therapy
(C) sodium nitroprusside
(D) supplemental oxygen
(E) pneumonectomy

1056. The LEAST practical method for providing good surgical conditions for the pneumonectomy is to use a

(A) Univent tube
(B) Robertshaw right-sided double-lumen tube
(C) single-lumen tube with a Fogarty catheter
(D) laryngeal mask airway
(E) Mallinckrodt left-sided double-lumen tube

1057. For the thoracotomy, the patient is intubated with a left-sided double-lumen tube. It is possible to ventilate the left lung, but not the right lung. Possible causes include all EXCEPT

(A) the tube is advanced too far
(B) the right mainstem bronchus is blocked by tumor
(C) the right lumen is blocked by secretions
(D) the tube is not advanced far enough
(E) the tube is in the wrong side

1058. After starting one-lung ventilation with oxygen and isoflurane, a tidal volume of 500 mL, and a rate of 10 breaths per minute, the O_2 saturation of the patient starts to fall. The most effective means of improving O_2 saturation is

(A) increasing O_2 flow to the anesthesia circuit
(B) adding 10 cm H_2O PEEP to the ventilated lung
(C) clamping the right pulmonary artery
(D) starting intravenous nitroglycerin
(E) increasing the concentration of isoflurane

1059. The patient underwent uneventful right radical pneumonectomy and remains intubated at the end of the procedure. On transferring the patient from the operating room table to the bed for transfer to the intensive care unit, there is a sudden decrease in blood pressure. Pulses are faint. Appropriate maneuvers include all of the following EXCEPT

(A) placing the patient laterally on his left side
(B) applying suction to the chest tube
(C) immediately reopening the wound and exploring the surgical field
(D) providing oxygen and managing the airway
(E) supporting cardiac function with an inotropic agent

1060. A patient is undergoing a mediastinoscopy when there is a sudden loss of pulse and pressure wave being monitored at the right wrist. The mediastinoscope is withdrawn, with resumption of normal vital signs. The most likely cause of the problem is

(A) cardiac arrest
(B) superior vena cava obstruction
(C) air in the mediastinum
(D) compression of the innominate artery
(E) anesthetic overdose

1061. A 71-year-old man is admitted with a complaint of hoarseness and sore throat. On indirect laryngoscopy, a supraglottic mass is noted with edema of the cords. He is scheduled for a direct laryngoscopy under general anesthesia. The approach to this procedure should be

(A) kept simple, since it is a short procedure
(B) induction, paralysis, and laryngoscopy
(C) induction, paralysis, intubation, and laryngoscopy
(D) paralysis, intubation, induction, and laryngoscopy
(E) to establish an airway before paralysis or instrumentation

1062. A patient with esophageal obstruction is to have a general anesthetic for esophagoscopy. He has had a barium swallow on the previous day. One of the greatest dangers of the planned procedure is

(A) bleeding
(B) hypotension
(C) difficult intubation
(D) aspiration
(E) arrhythmia

1063. During apneic oxygenation

(A) the time elapsed before desaturation occurs is independent of the patient's pulmonary status
(B) all arrhythmias are due to hypoxemia
(C) the carbon dioxide tension is not important
(D) the carbon dioxide level rises about 3 mmHg/min
(E) pulse oximetry is not helpful

DIRECTIONS (Questions 1064 through 1097): For each of the items in this section, ONE or MORE of the numbered options is correct. Choose answer

(A) if only 1, 2, and 3 are correct
(B) if only 1 and 3 are correct
(C) if only 2 and 4 are correct
(D) if only 4 is correct
(E) if all are correct

1064. The bronchial arteries are drained via

(1) thebesian veins
(2) bronchial veins
(3) the coronary sinus
(4) pulmonary veins

1065. The hemodynamics of normotensive patients with aortic regurgitation may be improved on administration of

(1) atropine
(2) norepinephrine
(3) sodium nitroprusside
(4) phenylephrine

1066. Pulmonary vascular resistance is increased by

(1) sevoflurane
(2) desflurane
(3) isoflurane
(4) nitrous oxide

1067. The administration of fentanyl in large doses (0.1 mg/kg) generally results in

(1) decreased heart rate
(2) increased heart rate
(3) blunted autonomic responses to tracheal intubation
(4) more profound hypotension than is seen with morphine (1 mg/kg)

1068. In the anesthetic management of patients with coronary artery disease, aims include

(1) minimizing myocardial oxygen consumption
(2) maintaining a fast heart rate
(3) maintaining coronary blood flow
(4) maintaining light anesthesia

1069. In the presence of coronary artery disease, fast sinus tachycardia

(1) is desirable
(2) may cause hypotension
(3) decreases the energy requirements of the heart
(4) may cause pulmonary edema

1070. A patient with aortic insufficiency enters the operating room with a blood pressure of 160/40. After induction with propofol, anesthesia is maintained with isoflurane, and atracurium is administered. A fall in pressure to 80/20 and bradycardia occur. There is

(1) increased peripheral resistance
(2) decreased cardiac output
(3) decreased left atrial pressure
(4) decreased coronary filling pressure

1071. Complications attributed to ketamine when used for coronary artery surgery include

(1) hypertension
(2) hallucinations
(3) tachycardia
(4) bradycardia

1072. The receptor for nitroglycerin is

(1) shared by nitroprusside
(2) guanylate cyclase
(3) an intracellular enzyme
(4) located within endothelial cells

1073. When used for anesthesia of a patient for coronary surgery, fentanyl has the advantage over morphine of

(1) increasing the heart rate
(2) earlier emergence
(3) increased oxygen supply to the heart
(4) a shorter period of required ventilation

1074. A VVT artificial pacemaker differs from a VVI pacer in that

(1) the VVT device is relatively temporary
(2) the VVT pulses are triggered by intrinsic ventricular depolarization
(3) the VVT device is temperature sensitive
(4) R waves inhibit the VVI but activate the VVT

1075. Hemodilution performed before bypass may cause

(1) bradycardia
(2) decreased cardiac output
(3) decreased myocardial oxygen consumption
(4) improved blood flow to tissues

1076. Carbon dioxide is sometimes added to the fresh gas supply of the oxygenator during hypothermic cardiopulmonary bypass

(1) to decrease pulmonary vascular resistance
(2) to increase the affinity of hemoglobin for oxygen
(3) to dilate coronary arteries
(4) to maintain the corrected $PaCO_2$ at 40 mmHg

1077. The continuance of pulmonary ventilation during cardiopulmonary bypass

(1) leads to hypercarbia
(2) leads to improved systemic arterial blood gas values
(3) decreases time on a ventilator during recovery
(4) may interfere with the surgery

1078. Anesthetic considerations for Marfan's syndrome (arachnodactyly) include

(1) increased risk of aortic dissection
(2) aortic stenosis
(3) frequently difficult intubation
(4) mitral regurgitation

1079. During hypothermic cardiopulmonary bypass

(1) uncorrected PaO_2 is higher than corrected PaO_2
(2) uncorrected $PaCO_2$ is higher than corrected $PaCO_2$
(3) uncorrected pH is lower than corrected pH
(4) uncorrected pH is higher than corrected pH

1080. When $PaCO_2$ and pH are managed by the alpha-stat method during hypothermic cardiopulmonary bypass

(1) the corrected pH is 7.4
(2) the uncorrected pH is 7.4
(3) the corrected $PaCO_2$ is 40 mmHg
(4) the uncorrected $PaCO_2$ is 40 mmHg

1081. Cardiac assistance by means of an intra-aortic balloon pump

(1) requires a competent aortic valve
(2) augments aortic pressure during diastole
(3) decreases left ventricular afterload
(4) is helpful in case of mitral regurgitation

1082. The hemostatic drug aprotinin

(1) is a protein isolated from bovine tissue
(2) selectively inhibits various proteolytic enzymes
(3) can inhibit fibrinolysis
(4) activates plasminogen

1083. Junctional cardiac rhythm

(1) can often be terminated by temporary atrial pacing
(2) can cause cannon waves in the central venous pressure
(3) can cause cannon waves in the pulmonary artery wedge pressure
(4) is always well tolerated

1084. The right lung

(1) is the smaller of the two
(2) has three lobes
(3) has a single fissure
(4) contains a middle lobe

1085. α_1-Antitrypsin deficiency is

(1) associated with early onset of emphysema
(2) nonfamilial
(3) determined by serum assay
(4) due to a lack of an enzyme produced by the lung

1086. Which of the following is (are) an effective method(s) of analgesia after thoracotomy?

(1) intercostal blocks
(2) lumbar epidural
(3) intrapleural catheter
(4) thoracic epidural

1087. Cardiopulmonary bypass is required for a

(1) single lung transplant
(2) en-bloc double lung transplant
(3) sequential single lung transplant
(4) heart-lung transplant

1088. Bronchitis refers to a disease that

(1) is a nonneoplastic disorder of the structure or function of the bronchi resulting from infectious or noninfectious irritation
(2) may include hypertrophy of the mucus-secreting apparatus
(3) must last for more than 3 months for more than 2 successive years to be termed chronic
(4) includes abnormal enlargement of the air spaces

1089. Pneumothorax associated with surgery may

(1) be spontaneous
(2) follow tracheostomy
(3) follow attempts at subclavian catheterization
(4) be associated with decreased ventilatory pressure

1090. If a patient is known to have a mediastinal mass and there is difficulty ventilating the patient during a procedure even with an endotracheal tube in place, factors that must be considered include

(1) loss of muscle tone
(2) compression of the bronchus
(3) kinking of the tube
(4) mucus plug

1091. A 60-year-old female was admitted with a large goiter and a history of hoarseness. An incidental finding on the chest x-ray was tracheal deviation, with questionable narrowing of the tracheal lumen. The thyroid was removed with some difficulty, and at the end of the procedure, the patient was breathing spontaneously. Immediately after extubation, breathing was labored and retraction was noted. Causes of this may include

(1) recurrent laryngeal nerve injury
(2) laryngospasm
(3) tracheal collapse
(4) bronchospasm

1092. Hypoxic pulmonary vasoconstriction (HPV)

(1) occurs when regional atelectasis mechanically obstructs blood flow
(2) is primarily triggered by alveolar O_2 tension
(3) occurs mainly in the pulmonary veins
(4) can occur in a denervated lung

1093. HPV is

(1) decreased by hypocapnia
(2) increased by hypercapnia
(3) increased by hypoxia
(4) increased by alkalosis

1094. Structures that pass anteriorly to the trachea include the

(1) aortic arch
(2) thyroid isthmus
(3) innominate artery
(4) pulmonary artery

1095. Absolute indications for one-lung ventilation include

(1) infection with purulent secretions
(2) massive pulmonary hemorrhage
(3) bronchopleural fistula
(4) unilateral bronchopulmonary lavage for alveolar proteinosis

1096. The flow/volume curve in a patient with a variable intrathoracic obstruction will show

(1) scooping of the expiratory limb
(2) no change in lung volume
(3) plateau of the inspiratory limb
(4) plateau of the expiratory limb

1097. In the lateral decubitus position, the lung relationships

(1) are the same as in the semirecumbent position, i.e., the apex is in zone 1 and the bases are in zone 3
(2) ventilation is highest at the apex
(3) perfusion is greater in the upper lung
(4) compliance is unequal in the two lungs

Answers and Explanations

1025. (C) If the patient has a known superior vena cava syndrome, intravenous lines should be placed in the lower extremities. All efforts are made to facilitate drainage of the upper vessels; therefore, the patient should be head up. Upper body intravenous lines as well as jugular lines are avoided. *(1:836; 4:1805; 5:1922; 6:592)*

1026. (B) Shivering is blocked with muscle relaxants during warming and cooling but does not occur in very cold muscle. *(1:912; 2:399; 4:655; 5:1971–82)*

1027. (E) Nitroglycerin is metabolized to nitric oxide, an activator of guanylate cyclase in vascular smooth muscle. The enzyme produces intracellular cyclic-GMP. The cyclic-GMP is degraded by a specific phosphodiesterase that is inhibited by sildenafil. Profound hypotension can occur when a sublingual dose of nitroglycerin is given within several hours of an oral dose of sildenafil. Intravenous nitroglycerin must be slowly titrated in that setting. *(1:279; 3:850; 5:1947; 6:5)*

1028. (E) Constant monitoring of the heartbeat by palpation or listening is necessary. The cutaneous electrode of a unipolar cautery device should be placed away from the pacemaker generator. If the pacer is in demand mode, it may sense the cautery current as depolarization and shut off. Since the electrocardiogram usually is distorted by the cautery, palpation will give an appraisal of the heartbeat. *(1:1515; 5:3147; 6:75)*

1029. (B) With modern pacemaker units, the energy output is so low that there is little danger of fibrillation; however, ventricular fibrillation is more common in the patient with a high catecholamine titer, myocardial infarction, and hypoxemia. *(1:156; 4:1715; 5:1419; 6:75)*

1030. (D) The intra-aortic balloon pump is deflated during systole. It is rapidly inflated during diastole, thereby increasing coronary blood flow. It decreases impedance to left ventricle ejection. It is contraindicated in the patient with aortic insufficiency. *(1:916; 4:191; 5:1919; 6:9)*

1031. (B) Clamping of the thoracic aorta leads to an immediate increase in blood pressure. Assuming the heart can withstand the markedly increased afterload, there should not be standstill or hypotension. Blood pressure readings may be lost in the left arm, but the pulse in the right arm should be present, depending on the placement of the clamp relative to the take-off of the vessels to the arms. *(1:817; 4:1860; 5:2077; 6:147)*

1032. (E) Clamping of the distal aorta leads to an increased vascular resistance and blood pressure and to decreased stroke volume and cardiac output. The heart rate usually is stable. *(1:817; 4:1863; 5:2072; 6:147)*

1033. (D) A complication of aortic surgery is spinal cord ischemia due to compromise of the radicular arteries. These arteries are not constant, and the large artery to the spine, the artery of Adamkiewicz, may be compromised, leading to ischemia. Pressure on the cord can lead to ischemia. Hypothermia may be protective. Loss of cerebrospinal fluid is not a factor. Long periods of hypotension may lead to ischemia. *(1:1112; 2:425; 4:1868; 5:2072; 6:146)*

1034. (E) The expansion of the vascular bed that occurs follows the release of the clamps. It is due to a reactive vasodilatation. Other factors, such as blood loss, inadequate fluid replacement, vasoactive drugs, and extravasation of fluid, are all important. *(1:179; 2:706; 4:1868; 5:784; 6:153)*

1035. (C) The flow rate on total bypass can be varied to achieve the perfusion pressure that is desired. The flow is also adjusted to the state of peripheral resistance. *(1:912; 2:400; 5:2025)*

1036. (B) The flow during bypass is virtually nonpulsatile in both flow and pressure. While pulsatile flow on cardiopulmonary bypass would appear to be more physiologic, studies on the subject in humans were unable to demonstrate any advantages over nonpulsatile flow thus far. *(1:912; 2:400; 5:1979)*

1037. (B) Although CO_2 is sometimes added to the fresh gas during bypass, respiratory acidosis or alkalosis is generally corrected through changes in the fresh gas flow to the oxygenator. *(1:904; 2:400; 5:1980)*

1038. (A) When the patient is on bypass, the pressure is usually kept between 40 and 60 mmHg. Lower pressures can be tolerated, but there is no agreement on any specific perfusion pressure. As the patient gets more hypothermic, viscosity of blood increases, which makes hemodilution an important part of cardiopulmonary bypass with hypothermia *(1:912; 2:400; 5:1979–80)*

1039. (A) Anesthesia may cause decreased oxygen consumption. Deliberate hypotension is not a part of carotid surgery. Cerebral infarction is a possibility. General anesthesia is not mandatory, and many have advocated local or regional anesthesia. Shunting procedures are used in many cases to provide flow during clamping. *(1:943; 4:1854; 5:2100; 6:248)*

1040. (A) Hypercapnia is not necessary for carotid artery surgery. Vasodilatation has not proven to be helpful. Oxygen tension and blood pressure are kept at levels that are normal or slightly above normal. Normothermia is used, as is systemic heparinization. *(1:938; 4:1854; 5:2103; 6:248)*

1041. (C) The development of new regional wall motion abnormalities can be rapidly assessed and is highly sensitive. *(1:936; 5:2066; 6:6, 19)*

1042. (A) This configuration permits lower current outputs. *(1:1514; 4:817; 5:1416; 6:75)*

1043. (A) Though asymptomatic, mitral valve pathology carries a risk of endocarditis. Diuretics may help if symptoms develop. Tachycardia may be poorly tolerated in advanced stenosis. Anticoagulation is indicated in the event of atrial fibrillation. *(1:561; 2:421; 3:1147; 5:1080; 6:30, 575)*

1044. (A) The negative pole of the pulse generator is generally connected to the stimulating electrode in contact with the heart in unipolar pacing. The reference (ground) electrode is not in direct contact with the heart. Two electrodes contact the heart in bipolar pacing. *(1:1514; 5:1416; 6:72–3)*

1045. (C) Esophagectomy patients are generally malnourished. They are at risk for aspiration, since esophageal function is frequently compromised. Their pulmonary function is equal between their lungs, and thus they are more likely to have a significant shunt during one-lung ventilation. The surgery is more extensive than most thoracotomies, and the patients more frail, so postoperative ventilation is common. *(4:1786, 1890)*

1046. (B) The bronchial tree divides into right and left bronchi, the left bronchus being narrower and longer. Foreign bodies are more apt to go to the right side. The trachea moves during respiration and with movement of the head. It is lined with pseudostratified columnar epithelium. The rings of cartilage do not completely encircle the trachea; the posterior wall is membranous for all the rings except the cricoid. *(1:595; 4:1066; 5:1618)*

1047. (E) The larynx lies at the level of the third to sixth cervical vertebrae, and in the adult, the narrowest portion is at the level of the vocal cords. The larynx is innervated by the recurrent laryngeal nerve and the superior laryngeal nerve. It lies within the thyroid cartilage. The wide margin of the cricoid cartilage is posterior. *(1:595; 4:1066; 5:1618)*

1048. (C) By clamping at 5, gas cannot flow into the left lumen, and thus to the left lung. Uncapping at 3 allows the gas in the left lung to escape. Clamping at 1 would prevent flow to the left side, but uncapping would cause a leak in the anesthetic circuit. Assuming good tube position and isolation, the handedness of the tube does not matter. *(1:824–6; 4:1757; 5:1877–9)*

1049. (D) Through position 4, one is looking down the right lumen. From the right lumen, no left lung segmental bronchi can be seen with any type of properly positioned double-lumen tube. The right lumen will be the tracheal lumen on a left-sided tube, in which case the carina can be seen. *(1:824–6; 4:1757; 5:1879–81)*

1050. (B) All the steps are correct for ventilating via the left lumen while pouring fluid into the right lumen except that clamping at 2 will block the fluid flow. The cuffs should be up to ensure lung isolation. *(1:824–6; 4:1757; 5:1882)*

1051. (D) Clamping at 6 isolates the right lung. *(1:824–6; 4:1757; 5:1877–9)*

1052. (D) The Arndt bronchial blocker can be used in patients who are already intubated, so there is no need to change the endotracheal to a double-lumen endotracheal tube should one-lung ventilation be required. It can also be used for patients with a difficult airway, and patients with trauma who require one-lung ventilation. The Univent tube is a single-lumen endotracheal tube with an integrated, movable bronchial blocker. The Sengstaken-Blakemore tube is an esophageal device used for temporary control of intractable variceal bleeding. *(1:827–8; 5:1884, 1887–90)*

1053. (A) Positive pressure ventilation does not have to be interrupted during placement of the bronchial blocker because of the multiport adapter that will accommodate the breathing circuit, a fiberoptic bronchoscope, and the endobronchial blocker. The device has to be placed under fiberoptic bronchoscopic guidance and its use is not limited to only the left or right lung. If the patient continues on mechanical ventilation on conclusion of surgery, the endobronchial balloon is deflated and the device removed while the endotracheal tube remains in place. *(1:828; 5:1887–90)*

1054. (B) An FEV_1 of less than 800 mL and an FVC of less than 2000 mL are associated with higher morbidity and poorer functional outcome. Regional function studies, like the V/Q scan, can show what fraction of the preoperative lung function comes from the area to be resected. Patients with an elevated $PaCO_2$ are already in a state of chronic respiratory failure. Low PaO_2 can reflect either poor underlying pulmonary function, or a shunting of blood through the diseased and nonventilated lung regions. When this area is resected, hypoxemia may resolve. *(1:814–6; 4:238, 1741; 5:1852–4)*

1055. (C) Hypoxia is likely caused by high physiologic shunt in the diseased lung. Physical therapy may help clear secretions, and improve oxygenation. Supplemental oxygen will improve delivery to the good lung, and reverse some hypoxic vasoconstriction on that side. Lateral position will improve flow to the good lung, as long as it does not increase soilage from the diseased side. Pneumonectomy will remove the diseased lung and eliminate blood flow to that side. An intravenous vasodilator, like sodium nitroprusside, will exacerbate shunting by inhibiting HPV. *(1:831–3; 4:238, 1741; 5:165–6)*

1056. (D) All these methods can isolate the right (operative) lung from ventilation except the laryngeal mask. Better surgical exposure can be obtained by allowing the right lung to collapse. *(1:824–6; 4:1757; 5:1873–6)*

1057. (E) If the right side cannot be ventilated, there must be no clear channel from the right lung to the anesthesia circuit. Tumor or secretions could obstruct the lumen. The bronchial cuff could block the right side if the tube is not advanced far enough. The right tube lumen might be in the left mainstem bronchus if the tube is advanced too far. Placing the tube into the right side should not stop right-sided ventilation. *(1:826; 4:1760; 5:1877–9)*

1058. (C) As oxygen is absorbed from the nonventilated lung, it becomes atelectatic. Persistent blood flow through that lung is then not oxygenated and contributes to shunt. Methods of improving oxygenation include decreasing the shunt or increasing mixed venous oxygen concentration. The most direct method of decreasing shunt is to mechanically stop blood flow to the atelectatic lung. Increasing O_2 flow to the circuit will not increase FIO_2 so will be ineffective. Nitroglycerin, or increased isoflurane, will impair HPV. PEEP may improve V/Q matching in the ventilated lung, but may also shift blood flow to the nonventilated lung. *(1:829–30; 4:1767; 5:1894–1900)*

1059. (B) Particularly likely causes of this scenario are massive hemorrhage from the pulmonary artery stump or other great vessel, herniation of the heart through a pericardial window, or arrhythmia from cardiac irritation, hypoxia, or right heart failure. Applying suction to the chest tube could exacerbate the mediastinal shift, vascular compromise, and cardiac herniation (tension vacuthorax), as well as speed the hemorrhage. Surgical exploration is urgently needed, and the patient should also receive appropriate supportive care. *(1:847–8; 4:1824; 5:1901–4)*

1060. (D) The innominate artery passes anterior to the trachea and can be compressed by the mediastinoscope. The right subclavian artery is a branch of the innominate artery, while the left subclavian artery comes off the aorta directly and does not cross in front of the trachea. Compression of flow in the innominate will also obstruct flow to the right carotid artery, and may compromise cerebral perfusion in patients with cerebrovascular disease or an incomplete circle of Willis. Air in the mediastinum should have little hemodynamic effect. Cardiac arrest, superior vena cava obstruction, and anesthetic overdose can all cause pulselessness, but rarely as suddenly or reversibly. Having a monitor of perfusion on the other arm, such as an oximeter probe, can help differentiate innominate compression from other causes of circulatory compromise. *(1:836–4; 4:1799; 5:1911)*

1061. (E) In any patient with hoarseness and a documented supraglottic mass, it is mandatory to establish the airway before proceeding with the anesthesia. Awake fiberoptic intubation, awake tracheostomy (all with topical anesthesia), or spontaneous ventilation with an inhaled anesthetic are the only safe methods. *(4:224, 2200; 5:2542, 2546–7)*

1062. (D) Aspiration is possible, since the esophagus may contain barium from the examination as well as other oral intake. The esophagus should be suctioned before induction, although the thick barium suspension may not be completely removed. *(1:1044; 4:2204; 5:1647–8)*

1063. (D) Carbon dioxide tension rises during apneic oxygenation and may contribute to the problems seen. It will rise more quickly in individuals with a low FRC. All arrhythmias are not due to hypoxemia but may be due to increased catecholamines. These in turn may be increased due to increased carbon dioxide. Pulse oximetry has been of definite benefit in these procedures. *(1:799; 4:1777; 5:1901)*

1064. (C) The bronchial arteries are drained mainly by the bronchial veins, which drain into the superior vena cava via the azygous vein. Some of the blood from the bronchial arteries enters the pulmonary veins, and this flow also occurs during cardiopulmonary bypass. *(1:856; 4:1739; 5:2266)*

1065. (B) A faster heart rate and afterload reduction will tend to reduce the regurgitant fraction. *(1:896; 3:389; 4:1667; 5:1958; 6:39)*

1066. (D) Factors increasing pulmonary resistance include nitrous oxide, hypoxia, and hypercarbia. *(1:400; 2:398; 3:356; 4:1140; 5:166; 6:130)*

1067. (B) Administration of fentanyl will result in bradycardia, but blood pressure will be stable. The cardiac output will decrease to a mild degree. Morphine-induced changes are more pronounced. *(1:357; 2:403; 3:574; 4:103; 5:394)*

1068. (B) The anesthetic goal in patients with coronary artery disease is to minimize the oxygen requirement and to maintain coronary flow. A fast heart rate increases oxygen consumption, as does light anesthesia. *(1:884; 4:1666; 5:1496; 6:19)*

1069. (C) In the presence of coronary disease, tachycardia is undesirable, since it increases the energy requirements of the heart. This may lead to hypotension and pulmonary edema. *(1:1487; 4:1664; 5:1496; 6:19)*

1070. (C) In the case described, the fall in pressure would be due to vasodilation, and would be accompanied by decreased cardiac output, increased left atrial pressure, and decreased coronary filling pressure. *(1:896; 4:1218; 5:1958; 6:41)*

1071. (A) The major problems with ketamine are the production of hypertension and tachycardia, both of which will increase oxygen demand. Hallucinations are a potential problem. *(1:327; 3:347; 4:1218; 5:350; 6:115)*

1072. (A) Nitroglycerin and nitroprusside are both prodrugs that are converted to nitric oxide, a small free radical that activates the intracellular enzyme guanylate cyclase. The enzyme converts guanosine triphosphate (GTP) to cyclic-GMP. The receptive cells are those of the vascular smooth muscle. Endothelial cells normally synthesize nitric oxide in order to control the tone of nearby vascular muscle cells. *(5:1454)*

1073. (C) Fentanyl will lead to earlier emergence and earlier estruation. The heart rate will be lower with fentanyl. The oxygen supply to the heart will not be increased. *(1:348; 2:406; 3:574; 4:1245; 5:394; 6:14)*

1074. (C) The "T" indicates triggering while the "I" indicates inhibition. *(1:1514; 5:1431; 6:72–3)*

1075. (D) Hemodilution may lead to improved tissue flow. It does not lead to bradycardia or to decreased cardiac output or decreased myocardial oxygen consumption. *(1:210; 2:256; 4:1681; 5:1983)*

1076. (D) Carbon dioxide is sometimes added when the pH-stat strategy is used to adjust $PaCO_2$ during hypothermia. *(1:904, 4:1683; 5:1980)*

1077. (D) Continued ventilation during bypass may interfere with the surgery and may lead to depletion of surfactant. There is no evidence that it leads to better ventilation or that it improves the postoperative recovery time. *(1:912; 4:1667; 5:1971)*

1078. (B) Marfan's syndrome is associated with aortic changes, making dissection more common. Hypertension should be avoided. The high-arched palate may make intubation more difficult. Aortic stenosis is not seen, but aortic insufficiency is common. *(1:956; 5:2536; 6:534)*

1079. (A) During warming of hypothermic blood samples to 37°C for determination of uncorrected values, the hydrogen ion concentration and gas tensions increase. *(1:904; 4:1683; 4:1741; 5;1980)*

1080. (C) The other responses pertain to the pH-stat method. *(1:904; 4:1683; 4:1741; 5:1439; 5:1980)*

1081. (E) The balloon pump is counterproductive in case of aortic regurgitation. *(1:916; 4:191; 5:1991; 6:9)*

1082. (A) Clinically achieved levels of aprotinin inhibit kallikrein, plasmin, and activated protein C. *(1:232; 2:401; 3:662; 4:1681; 5:1811; 6:497)*

1083. (A) Cannon waves appear when atria contract after closure of the atrioventricular valves. Loss of the atrial component of ventricular filling may be poorly tolerated. *(1:675; 2:357; 4:1675; 5:1403; 6:84)*

1084. (C) The right lung is the larger of the two. It has three lobes (upper, middle, and lower) and two fissures (horizontal and oblique). *(1:791–2; 4:1753)*

1085. (B) α_1-Antitrypsin deficiency is familial and is determined by serum assay. The assay measures the level of a protective enzyme produced in the liver that acts to prevent autodigestion of lung tissue by the proteolytic enzymes of phagocytic cells. *(1:1361; 4:348; 5:1090–1)*

1086. (E) All of these methods can be used. Intercostal blocks and intrapleural catheters will require higher doses of local anesthetic. *(1:846–7; 4:1828; 5:1906–9)*

1087. (C) To avoid cardiopulmonary bypass the patient must be ventilated and oxygenated using a native lung while a lung is transplanted. Even some of the patients undergoing single and sequential single transplants may not tolerate one-lung anesthesia because of their severe underlying pulmonary dysfunction. *(1:1361–3; 4:1798; 5:2265–71)*

1088. (A) The first three choices are included in the definition of bronchitis; the last refers to emphysema. *(4:236; 5:1090–1; 6:177)*

1089. (A) Pneumothorax can follow tracheostomy and mediastinoscopy. It is a known complication of subclavian vein catheterization. Pneumothorax is associated with higher airway pressures, unless the chest is open (in which case there will be a leak of gases). *(1:837, 1270; 4:1826; 5:1294, 1925)*

1090. (E) Tracheal compression is a major concern in patients with an anterior mediastinal mass. Spontaneous ventilation with preservation of intrinsic muscle tone can forestall tracheal collapse. Also possible are the more common problems of mucus plugging, kinking, and malpositioning of the endotracheal tube. *(1:836; 4:224, 1801; 5:1920–2)*

1091. (E) Nerve injury is common with difficult dissections. Laryngospasm may be present due to secretions or injury to the cord. Tracheal collapse may be present due to tracheomalacia. Bronchospasm may be a reason for the dyspnea due to airway sensitivity. *(1:615; 4:224, 1801; 5:1717)*

1092. (C) HPV is a constriction of pulmonary arteries in response to alveolar hypoxia. Atelectatic lungs have identical degrees of HPV to those ventilated with nitrogen, excluding mechanical factors. HPV can occur in denervated lungs (e.g., after transplantation). *(1:831–3; 4:1765; 5:165–6)*

1093. (B) Hypocapnia decreases HPV, but hypercapnia has no effect. Hypoxia is the primary trigger of HPV. Alkalosis and acidosis both decrease HPV. *(1:831–3; 4:1768; 5:165–6)*

1094. (A) The pulmonary artery is caudal to the carina. *(1:836–7; 4:1799; 5:1909)*

1095. (E) All of the options are absolute indications for one-lung ventilation. In the setting of infection with purulent secretions, the goal is to avoid spillage and contamination of the contralateral lung. During surgery for bronchopleural fistula, the goal is to control the distribution of ventilation to the unaffected lung. Relative indications with high priority for one-lung ventilation include thoracic aortic aneurysm repair, pneumonectomy, upper lobectomy, and video-assisted thoracoscopic surgery (VATS). *(1:824; 4:1757; 5:1873–4)*

1096. (C) A variable intrathoracic obstruction limits expiratory flow, while preserving total lung volume and inspiratory flow. Expiratory flow stays constant, balancing the pressure gradient for air movement against pressure collapsing the airway. "Scooping" occurs in obstructive lung disease, with small airway closure. *(1:815; 4:224, 1814)*

1097. (D) In the lateral position, lung relationships change. The perfusion is now greater in the dependent lung. Compliance differs between the lungs because the dependent lung is at a lower FRC due to the weight of the abdominal contents and mediastinum. *(1:821–3; 4:1754; 5:1870–3)*

CHAPTER 11

Anesthesia for Neurosurgery

Questions

DIRECTIONS (Questions 1098 through 1118): Each of the numbered items or incomplete statements in this section is followed by answers or by completions of the statement. Select the ONE lettered answer or completion that is BEST in each case.

1098. Of the many factors affecting intracerebral blood flow, which of the following is a correct description?

- (A) vasomotor paralysis: vasoconstriction of vessels in or near ischemic areas
- (B) autoregulation: ability of vessels to respond in a manner consistent with maintaining homeostasis
- (C) luxury perfusion: metabolic requirements in excess of blood flow
- (D) intracerebral steal: decrease of blood flow in normal areas with increased flow to ischemic areas
- (E) inverse steal: diversion of flow to normal areas from ischemic areas

1099. When ventilating the patient with a head injury, all of the following statements are true EXCEPT

- (A) the patient should be kept supine
- (B) prolonged hyperventilation has diminished efficacy in reducing ICP
- (C) PEEP may be appropriate
- (D) hypoxia and hypercarbia should be avoided
- (E) the patient should be prevented from coughing

1100. Mannitol may lead to subdural hematoma by

- (A) causing cerebral edema
- (B) affecting clotting mechanisms
- (C) leading to cortical vein disruption
- (D) causing hypertension
- (E) leakage through the vein wall

1101. Which of these is the best agent to decrease cerebral oxygen requirement?

- (A) a muscle relaxant
- (B) a glucose solution
- (C) an anticonvulsant
- (D) a barbiturate
- (E) oxygen by mask

1102. The use of succinylcholine to facilitate endotracheal intubation in patients with increased ICP is associated with

- (A) increased ICP
- (B) no change in ICP
- (C) incomplete relaxation
- (D) conditions more satisfactory than those with the use of pancuronium
- (E) hyperkalemia

1103. The major intracranial buffer against increased ICP is the

- (A) dural sinus
- (B) CSF
- (C) white matter
- (D) gray matter
- (E) glial cells

1104. Treatment of the neurosurgical patient with mannitol may be followed by all of the following EXCEPT

(A) initial hypervolemia
(B) decreased urine volume
(C) hypovolemia
(D) decreased venous pressure
(E) a fall in arterial pressure

1105. Nitrous oxide should be avoided in patients with

(A) brain tumor
(B) subarachnoid hemorrhage
(C) closed head injury
(D) pneumocephaly
(E) subdural hematoma

1106. In neurosurgical patients, dextrose-containing solutions

(A) are the fluids of choice
(B) may cause excessive diuresis
(C) may exacerbate hyperglycemia
(D) may produce brain edema
(E) lead to water retention

1107. Signs of air embolism include all of the following EXCEPT

(A) arrhythmia
(B) hypertension
(C) heart murmur
(D) bubbles at the operative site
(E) decreased end-expired carbon dioxide

1108. As the neurosurgeon manipulates tissue in the posterior fossa, there is a sudden arrhythmia. The anesthetist should

(A) lower the head
(B) administer lidocaine
(C) inform the neurosurgeon
(D) turn off all of the anesthetics
(E) turn off the nitrous oxide

1109. To obtain maximum benefit from hyperventilation during a neurosurgical procedure, the $PaCO_2$ should be maintained at

(A) 25 to 30 mmHg
(B) 20 to 25 mmHg
(C) 15 to 20 mmHg
(D) 35 to 40 mmHg
(E) 40 to 45 mmHg

1110. Following closed head injury, systemic sequelae may include all of the following EXCEPT

(A) disseminated intravascular coagulation
(B) diabetes insipidus
(C) syndrome of inappropriate secretion of antidiuretic hormone
(D) hyperglycemia
(E) hypocarbia

1111. All of the following statements are true concerning evoked potentials EXCEPT

(A) waveform peaks are described in terms of amplitude, latency, and polarity
(B) amplitude of evoked potentials is greater than those of the EEG
(C) injury is manifested as an increase in latency and/or decrease in amplitude
(D) brain stem potentials are more resistant to anesthetic influences than cortical potentials
(E) volatile anesthetics produce dose-dependent alterations in evoked potentials

1112. An intraoperative "wake up" test performed during surgery on the spine

(A) determines if distraction of the vertebral column compromised neurologic function
(B) is not necessary if somatosensory-evoked potentials are monitored
(C) is a test of sensory function

(D) determines adequacy of fluid replacement
(E) assesses motor function in the upper extremities

1113. Electroconvulsive therapy (ECT)

(A) is relatively contraindicated in patients with known cerebral or aortic aneurysms
(B) never produces a seizure
(C) is not contraindicated in patients with intracranial mass lesions
(D) does not require hemodynamic monitoring
(E) cannot be performed with muscle relaxants

1114. Once detected, the management of the patient with venous air embolism includes all of the following EXCEPT

(A) inform the surgeon
(B) discontinue nitrous oxide
(C) induce hypertension
(D) decrease elevation of patient's head
(E) control ventilation

Questions 1115 through 1117

A patient undergoing a craniotomy in the sitting position has both a radial artery and a right atrial pressure catheter in place. The external auditory canal is 26 cm above the level of the right atrium (5 cm below the manubrium). The cranium is open and the brain exposed.

1115. With the pressure transducers located at the level of the right atrium, the mean arterial blood pressure (MAP) is 90 mmHg and the central venous pressure is 5 mmHg. What is the cerebral perfusion pressure?

(A) 95
(B) 85
(C) 70
(D) 59
(E) cannot be determined directly

1116. If the arterial catheter transducer were repositioned to the level of the external auditory canal, then

(A) the MAP would not require correction to measure perfusion pressure at the base of the brain
(B) the measured MAP would remain the same if the arm were not elevated
(C) the same effect could be accomplished by elevating the arm to the level of the external auditory canal
(D) coronary perfusion pressure (CPP) would equal measured MAP – CVP
(E) blood pressure determined with a cuff on the upper arm would be less than the measured pressure

1117. When electronically "zeroing" the transducer system, the stopcock immediately above the transducer diaphragm is opened to air and

(A) the transducer should be positioned at the point where pressure is measured
(B) the position relative to the patient is irrelevant
(C) the transducer should be positioned at the level the catheter enters the radial artery
(D) the arm must be positioned at the level of the right atrium
(E) the transducer should be rezeroed whenever the position is changed

1118. All of the following reduce cerebral blood flow EXCEPT

(A) etomidate
(B) propofol
(C) droperidol plus fentanyl
(D) diazepam
(E) ketamine

DIRECTIONS (Questions 1119 through 1144): For each of the items in this section, ONE or MORE of the numbered options is correct. Choose answer

(A) if only 1, 2, and 3 are correct
(B) if only 1 and 3 are correct
(C) if only 2 and 4 are correct
(D) if only 4 is correct
(E) if all are correct

1119. Jugular venous oxygen saturation monitoring

(1) assesses global oxygen extraction
(2) requires precise placement of the monitoring catheter in the jugular bulb
(3) detects only a single hemisphere
(4) is unaffected by changes in oxygenation of systemic blood

Questions 1120 and 1121

A 15-year-old girl had a spinal fusion with Harrington rod instrumentation. On emergence, the patient was unable to move her left lower extremity.

1120. The cause(s) of this may be

(1) overcorrection of the scoliotic curve
(2) cord compression due to hematoma
(3) surgical damage to the cord
(4) electrolyte imbalance

1121. In this patient, the proper course to follow when the loss of function is discovered is to

(1) extubate the trachea, begin blood transfusion
(2) observe for 24 hours
(3) establish baseline neurologic function and observe
(4) reexplore immediately

1122. Complications associated with the sitting position include

(1) sciatic and cranial nerve trauma
(2) pneumocephalus
(3) quadriplegia
(4) airway edema

1123. Mannitol given to a head-injured patient

(1) may lead to hyperosmolar coma
(2) removes water from the normal brain tissue
(3) may lead to cerebral edema if the blood-brain barrier is impaired
(4) is effective in doses of 0.25 g/kg

1124. Dexamethasone, given to the neurosurgical patient

(1) will reduce the edema due to brain tumor
(2) has an effect because of its osmolar property
(3) is less effective in control of edema secondary to injury
(4) is effective only in preventing Addison's disease

1125. Induction of anesthesia with only nitrous oxide, 66%, and oxygen

(1) produces increased ICP
(2) preserves the responsiveness of cerebral blood flow to carbon dioxide
(3) produces cerebrovascular dilatation
(4) induces alterations in cerebral hemodynamics that are blunted by barbiturates

1126. Attention must be given to levels of ICP on induction. Increased ICP may lead to

(1) herniation of brain tissue
(2) localized areas of brain ischemia
(3) lowering of cerebral perfusion pressure
(4) brain retraction

1127. Air embolism may be a fatal complication depending on

(1) the amount of air present
(2) the site of entry
(3) the rate of entry
(4) whether a Swan-Ganz catheter is in place

1128. Concerning magnetic resonance imaging

(1) motion artifacts are rare
(2) ferromagnetic substances are propelled toward the center of the magnetic field
(3) routine monitoring can be used
(4) large prosthetic metal implants can become hot during scanning

1129. Agents that blunt the increase in ICP during laryngoscopy include

(1) lidocaine
(2) fentanyl
(3) propofol
(4) succinylcholine

1130. Treatment of cerebral edema includes

(1) osmotic diuretics
(2) loop diuretics
(3) steroids
(4) surgical decompression

1131. Halothane, when used in the neurosurgical patient

(1) adds to cerebrovascular tone
(2) is a vasodilator
(3) decreases cerebral blood flow
(4) should be used with controlled ventilation to produce hypocapnia

1132. During hypothermia

(1) cerebral metabolism is decreased
(2) cerebral vascular resistance increases
(3) cerebral vasculature remains responsive to carbon dioxide
(4) more glucose is required by the brain for metabolism

1133. If surgery is to be performed on a patient in the sitting position

(1) the legs should be wrapped with elastic bandages
(2) the legs should be kept at a level below the heart
(3) the head should be firmly fixed
(4) the patient should be positioned as quickly as possible to avoid loss of monitors

1134. Premedication with drugs that depress ventilation should be minimized in

(1) infants
(2) patients with increased ICP
(3) comatose patients
(4) very anxious patients

1135. The use of induced hypotension in neurosurgery

(1) may be necessary for short intervals
(2) is best provided with phenoxybenzamine
(3) may increase physiologic dead space
(4) is relatively contraindicated for intracranial surgery

1136. Air embolism

(1) is common in the sitting position
(2) occurs most often with occipital craniectomy
(3) occurs through the diploic veins and venous sinuses
(4) requires a relatively negative venous pressure at the wound

1137. Treatments for air embolism include

(1) the head-up position
(2) removal of intracardiac air
(3) carotid compression
(4) elevation of venous pressure in the head

1138. Causes of arrhythmia associated with neurosurgical procedures include

(1) orbital decompression
(2) tentorial manipulation
(3) tonsillar herniation
(4) carotid artery ligation

1139. During a cerebral aneurysm procedure, sodium nitroprusside was infused. The expected results include

(1) short duration of action
(2) bradycardia
(3) acidemia
(4) nitric oxide toxicity

1140. When a precordial Doppler ultrasonic transducer is used to detect air embolus, it

(1) can detect 0.5 mL of air
(2) functions at 15 Hz
(3) should be placed over the right side of the heart
(4) is less sensitive than capnography

Questions 1141 and 1142

A comatose patient with a history of severe headaches has an ICP monitor in place. Oxygen and nitrous oxide administered to the patient lead to an increased ICP.

1141. The increase in ICP may be blocked by administration of

(1) halothane
(2) diazepam
(3) ketamine
(4) thiopental

1142. The effect of agents or drugs to modify a response of increased ICP

(1) varies with the individual
(2) is constant
(3) depends on the summation of influences on cerebrovascular tone
(4) is independent of ventilation effects

1143. Etomidate can influence CSF dynamics by

(1) decreasing secretion
(2) increasing flow
(3) increasing absorption
(4) increasing secretion

1144. Barbiturates

(1) uncouple cerebral blood flow from metabolic rate
(2) at sufficient doses produce an isoelectric EEG
(3) prevent further decreases in cerebral blood flow from hyperventilation
(4) preserve normal cerebral energy charge

Answers and Explanations

1098. (B) The definition of autoregulation is correct. Vasomotor paralysis involves vasodilatation. Luxury perfusion is perfusion in excess of requirements. Intracerebral steal involves blood flow away from ischemic areas. Inverse steal involves diversion of flow from normal areas to ischemic areas. *(1:746; 4:1620–1; 5:817, 840–1)*

1099. (A) Ventilation of the patient with a head injury requires meticulous attention. Head elevation reduces cerebral venous congestion and thereby lowers ICP. After prolonged hyperventilation the efficacy at reducing ICP is reduced as the pH of CSF returns to normal. While increases in intrathoracic pressure should be avoided, PEEP is useful when used judiciously to maintain oxygenation and avoid hypoxia and hypercarbia. The use of PEEP should be weighed against the potential to increase ICP. *(1:777–80; 4:258–60; 5:2152–5, 2470–1)*

1100. (C) Mannitol, by decreasing brain size, may produce traction and even tearing of subdural veins that result in hematoma formation. This is more of a problem in the elderly. Mannitol may also lead to cerebral edema when the blood-brain barrier is disrupted. *(1:762–3; 4:1632; 5:2134, 2172)*

1101. (D) The best agent for decreasing cerebral oxygen requirement is a barbiturate. A muscle relaxant may be useful by preventing the patient from coughing and moving. Phenytoin reduces cerebral oxygen consumption caused by seizures; in the absence of seizures it has little effect on cerebral metabolic rate. *(1:749–51; 4:1622; 5:814–6)*

1102. (A) ICP may increase after succinylcholine. The increase is attenuated with prior administration of a small dose of a nondepolarizing muscle relaxant to prevent fasciculations. Both ventilation to reduce $PaCO_2$ and administration of thiopental reduce the rise in ICP. Increased ICP does not predispose to hyperkalemia; serum potassium will not rise above the usual 0.5 to 1.0 mEq/L. *(1:424, 7640; 4:1628; 5:491; 831–2)*

1103. (B) The major buffer against increases in ICP is the CSF. The dural sinuses, which contain venous blood, are also a buffer. The other options are tissues that do not act as buffers. *(1:747–8; 4:1614–6; 5:2127–30)*

1104. (B) Urinary volume will increase with mannitol administration. At first the patient may become hypervolemic both from administration of the volume of mannitol and the increase in intravascular volume from shifting of free water from the intracellular and extravascular spaces. Hypovolemia may develop after diuresis with a fall in both venous and arterial pressure. *(1:762–3; 4:1632; 5:2134)*

1105. (D) While not an absolute contraindication to use, nitrous oxide has been reported to increase cerebral blood flow, and this may be a consideration in patients with increased ICP from head injury, intracranial hemorrhage, and tumor. Since nitrous oxide diffuses into air-containing spaces, the presence of pneumocephaly is a contraindication to use because of the potential to increase ICP. *(1:388, 392; 4:1628; 5:830–1, 2138)*

1106. (C) Generally, solutions containing dextrose are not administered during neurosurgery unless indicated for treatment of hypoglycemia. Hyperglycemia has been implicated in animal experiments to worsen neurologic outcome after ischemia. A plasma glucose levels in excess of 300 mg/dL can produce an osmotic diuresis. Dextrose-containing solutions do not lead to fluid retention. *(1:763; 4:2376–8; 5:2143)*

1107. (B) Air embolism is associated with arrhythmia, heart murmur (when there is a large amount of air), bubbles at the operative site, and decreased end-tidal CO_2. Hypotension occurs after a large volume of air is entrained. *(1:766; 4:1633–5; 5:2139–42)*

1108. (C) Traction on a number of structures in the posterior fossa may lead to arrhythmias, the most common of which is bradycardia. The surgeon should be informed immediately. If the bradycardia persists, atropine may be indicated. Air embolism is more commonly associated with atrial and ventricular irritability and lidocaine may be necessary to treat frequent premature ventricular systoles or ventricular tachycardia associated with air embolism. Although it may be necessary to change the anesthetic or lower the head if air embolism is suspected, the first maneuver is to inform the surgeon of the problem and suspected diagnosis. *(1:765–7; 5:2157–8)*

1109. (A) The cerebral vasoconstrictive effect is diminished by reducing $PaCO_2$ below 25 mmHg. In addition, at a $PaCO_2$ below 20 mmHg, the effect is self-defeating with potential development of cerebral ischemia. *(1:763; 4:1618; 5:2131–3)*

1110. (E) There are many systemic sequelae of closed head injury. Coagulopathy and disseminated intravascular coagulation may result from cerebral trauma, possibly from the release of brain thromboplastin into the systemic circulation. Posterior pituitary dysfunction is common and manifests as disturbances in antidiuretic hormone secretion. Hyperglycemia with nonketotic hyperosmolar coma can also occur. These patients often suffer from respiratory compromise and are hypoxemic and hypercarbic. *(1:780–1; 4:258–60; 5:2152–7; 6:254–6)*

1111. (B) Because the amplitude of evoked potential responses is small (0.1 to 20 mV) as compared to the standard EEG (>50 mV), signal averaging is required to eliminate background noise. Multiple factors influence evoked responses including anesthetics and temperature. Resistance to these effects varies among the various potentials measured. *(1:754–6,; 4:888–97; 5:1527–37)*

1112. (A) Awakening the patient intraoperatively to assess motor function in the lower extremities after distraction of the vertebral column is common in spine surgery. Compromise of blood supply may occur from surgical manipulation or straightening of the cord. The use of somatosensory-evoked potentials assesses only sensory function and does not rule out a motor deficit. *(1:1112; 4:2121; 5: 2419)*

1113. (A) In order to be effective in treating depression, ECT must induce a seizure. Because of the rapid and unpredictable changes in blood pressure accompanying the seizure, patients with aneurysms may not be candidates for this therapeutic modality. In addition, the increases in cerebral metabolic activity and concomitant rise in blood flow can increase ICP and, therefore, may cause cerebral herniation in patients with decreased intracranial compliance from mass-occupying lesions. Therefore, these conditions represent relative contraindications to ECT. *(1:1334–5; 4:264–6; 5:2654–6)*

1114. (C) The treatment of venous air embolism is directed at preventing further entraining of air at the operative site, minimizing the size of the embolism, and supporting the circulation until the nitrogen is absorbed and excreted through the lungs. Increasing blood pressure will neither prevent entraining more air nor speed-up resolution. Negative intrathoracic pressure generated during spontaneous ventilation may increase the entrainment of air. *(1:766–7; 4:1633–7; 5:2138–42)*

1115. (C) The cerebral perfusion pressure is mean arterial blood pressure less the intracranial pressure (MAP – ICP). In this case the ICP is

0 since the cranium is open. Since the MAP is measured at the level of the right atrium, the pressure would be less at the circle of Willis (located at the level of the external auditory canal) because the head is elevated. Thus, the MAP at the base of the brain would be 70 mmHg; 90 mmHg – (26 cm ÷ 1.3 mmHg/cm) in order to correct for the height. *(1:765–6; 5:1279–81; 6:253)*

1116. (A) With the transducers positioned at the level of the circle of Willis, the measured pressure is the arterial pressure at the base of the brain. Position of the arm containing the arterial catheter with respect to the transducer will not alter the value of the measured pressure. Arterial pressure of any structure below the level of the transducer will be higher than the measured value. *(1:765–6; 5:1279–81; 6:253)*

1117. (B) Electronic zeroing merely sets a pressure of 0 mmHg to the electrical output of the transducer when the zeroing maneuver is performed. Opening the stopcock to ambient pressure (air pressure) exerts a pressure on the transducer diaphragm equal to the height of the column of fluid between the opened stopcock and the diaphragm. By convention, the stopcock opened is the one directly above the diaphragm and the column height is 1 to 2 cm. The position of the transducer relative to the patient will not affect the electronic zero and is irrelevant to the procedure (opening the stopcock disconnects the patient from the system). Once the electronic zero is established, there is no need to rezero the instrument unless to correct for electronic drift which can occur with time. *(1:673–4; 5:1279–80)*

1118. (E) Ketamine produces large increases in cerebral blood flow in humans. Both etomidate and propofol reduce CBF and cerebral metabolic rate. The reduction in CBF produced by diazepam is small and probably of no clinical importance. The combination of droperidol and fentanyl reduces CBF and ICP in humans, whereas the two administered separately can increase CBF. *(1:336–7, 749–50; 4:1623; 5:2130)*

1119. (A) Jugular bulb venous oximetry detects changes in brain oxygen extraction from blood. If the monitoring catheter is not positioned within the jugular bulb, contamination with noncerebral venous blood dilutes its effectiveness as a monitor. The catheter is positioned in a single jugular bulb and reflects venous drainage from a single hemisphere. Changes in arterial blood oxygen capacity alter oxygen delivery and are reflected in the percent saturation of the draining venous blood. *(1:759; 5:2155–6)*

1120. (A) The deficit described may be the result of the first three options. An electrolyte imbalance is not a consideration. *(1:1110–3; 4:2121; 5:2418–9)*

1121. (D) If the patient awakens with a deficit that cannot be explained, the surgical site should be explored for potential cord compression from a hematoma. The patient should be left intubated, pending a decision. Since there is no advantage in waiting, surgical reexploration and correction should occur as early as possible. *(1:1110–3; 4:2121; 5:2418–9)*

1122. (E) All are potential problems with a patient in the sitting position. The sciatic nerve is at risk for compression from inadequate padding or traction injury from improper positioning. Traction of cranial nerves, especially the abducens (VI), can result from caudal displacement of the brain. Air trapped over the superior surface of the brain produces pneumocephalus. Quadriplegia has been reported due to cord compression from extreme flexion of the neck. Similarly, airway edema can occur from obstruction of venous drainage. *(1:661–5, 765, 1112; 4:1633–7; 5:1164–6, 2136–7; 6:243)*

1123. (E) Mannitol may lead to hyperosmolar coma if given in large doses. The drug has its effect by removing water from normal tissues. If the blood-brain barrier is impaired, mannitol may pass into the cells and cause cerebral edema. The drug is effective in doses as low as 0.25 g/kg. *(1:7778–80; 4:1632; 5:2134)*

1124. (B) Dexamethasone is effective in the treatment of cerebral edema in patients with brain tumors. It has not been shown to be effective in the patient with brain injury from acute closed head trauma. The mechanism is probably from a decrease in inflammatory processes caused by the tumor. *(1:763; 4:247; 5:2133–4; 6:238)*

1125. (E) Induction of anesthesia with nitrous oxide as the sole agent has been shown to increase cerebral blood flow and ICP. Although nitrous oxide produces cerebrovascular dilatation, the cerebral blood flow response to carbon dioxide is preserved. These effects on cerebral hemodynamics are altered by the addition of other anesthetic agents; barbiturates and benzodiazepines blunt the increase in ICP. *(1:392, 765; 4:1628; 5:830–1, 2130–1; 6:240)*

1126. (A) If ICP increases on induction, areas of focal ischemia may occur from herniation and/or decreased perfusion pressure. Brain retraction may cause ischemia during surgery. *(1:741; 4:1631; 5:2127–8; 6:235)*

1127. (B) Both the amount and rate of entry of air are important factors in determining the consequences of venous air embolism. The site of entry is not important. Although a pulmonary artery catheter may be useful for monitoring, the port is too small to be helpful in the withdrawal of air. *(1:766–7; 4:1635–6; 5:2138–42; 6:243–5)*

1128. (C) All ferromagnetic material can create artifacts in the scan and can be drawn into the magnetic field. This is especially a problem with metal implants such as aneurysm clips that can move and large prostheses that can become heated. The patient must remain still during scanning, which can take several minutes. *(1:1332–3; 4:2292–4; 5:2161–2)*

1129. (A) Adequate induction of general anesthesia will blunt the increase in ICP associated with laryngoscopy. Lidocaine and fentanyl are effective in blunting the rise in ICP. Propofol decreases ICP effectively. Succinylcholine is associated with an increase in ICP when administered without additional induction agents. *(1:764–5; 5:2130–1; 6:241)*

1130. (E) Treatment of cerebral edema includes all of the options: osmotic diuretics, loop diuretics, steroids, and surgical decompression. *(1:762–3; 4:1631–3; 5:2127–34; 6:236–8)*

1131. (C) Halothane, when used in the neurosurgic patient, should be combined with controlled ventilation to decrease ICP. It is a potent cerebral vasodilator. *(1:747; 4:1624; 5:825–30; 6:240–242)*

1132. (A) During hypothermia, the brain requires less glucose and oxygen. With reduced metabolic demand, cerebral blood flow decreases and cerebral vascular resistance increases. Responsiveness to changes in $PaCO_2$ is unchanged. *(1:750; 4:2370–1; 5:816)*

1133. (B) Positioning the patient in the sitting position requires planning. In order to minimize venous pooling in the lower extremities, the legs should be wrapped and positioned level with the heart. The head must be firmly fixed in a head rest. Positioning should be done slowly to avoid hypotension. *(1:765–7; 4:1639; 5:1164–6, 2136–9)*

1134. (A) Sedating premedication is used, if at all, in small doses in the very young, patients with increased ICP, and comatose patients. Sedation of the anxious patient, if possible, is preferable. *(1:761; 4:1630; 5:2146; 6:240)*

1135. (B) Induced hypotension may be necessary for short periods of time during a neurosurgic procedure. Thus, a short-acting drug, such as sodium nitroprusside, is ideal. The use of hypotension may lead to increases in dead space. Hypotension is not contraindicated in the neurosurgic patient. *(1:773–7; 4:1644; 5:2149)*

1136. (E) All of the options are true concerning air embolism. In order for air to reach the heart, there must be a relatively negative pressure in the veins at the site of surgery. This does not require a great elevation of the head. *(1:766–73; 4:1633–7; 5:2138–42; 6:243–5)*

1137. (C) Treatment of air embolism requires stopping the further influx of air and removal of the entrained air. Positioning the patient head-down elevates venous pressure at the wound site and prevents entrainment. Venous pressure can also be elevated by compression of the jugular veins. Rapid entry of a large volume of air can accumulate as bubbles in the heart and produce a mechanical obstruction to cardiac output. Entrained air can be removed from the right atrium by aspiration through a central venous catheter. Most of the air, though, passes into the pulmonary vasculature and is excreted by exhalation. *(1:766–7; 4:1635–8; 5:2138–42; 6:243–5)*

1138. (E) Any one of these manipulations may be associated with arrhythmias. The oculocardiac reflex may be initiated by manipulation of any of the structures of the afferent pathway, including the globe, orbital contents, ophthalmic division of cranial nerve V, and trigeminal ganglion and nerve. Tonsillar herniation, resulting from cerebral edema or hematoma in the posterior fossa, can produce traction or compression of the brain stem and generate arrhythmias. Tentorial manipulation can produce bradycardia and asystole. Manipulation of the carotid sinus activates afferents of the baroreflex and can induce profound alterations in heart rate. *(1:766; 4:2184; 5:2157; 6:243)*

1139. (B) Sodium nitroprusside has a short duration of action. It may cause an acidemia and is usually associated with tachycardia. Cyanide toxicity may also occur. *(1:7731; 4:164; 5:818)*

1140. (B) Precordial Doppler is one of the most sensitive monitors of venous air embolism clinically available and can detect air in quantities as small as 0.5 mL. The ultrasonic probe, which functions at 2.0 MHz, is placed over the right side of the heart. *(1:766; 4:1636; 5:2139; 6:244)*

1141. (C) Diazepam and thiopental may block increases in ICP. Halothane and ketamine are potent cerebral vasodilators and increase ICP. *(1:749–50; 4:1622–7; 5:820–30; 6:240)*

1142. (B) Drug effects on ICP are not constant but vary among individuals and with ventilation. The total effect on cerebrovascular tone will determine the effect on blood volume and ICP. *(1:749–50; 4:1622–8; 5:820–30)*

1143. (B) Inhalational and intravenous anesthetic agents have a small impact on the rate of formation and absorption of CSF. Etomidate alone amongst these drugs has been described to both decrease secretion and increase absorption of CSF. Although this combination may reduce total CSF volume, the overall effect is small and would be significant, if at all, only in cases of patients with increased ICP undergoing a long anesthetic treatment. *(4:709–10; 5:832)*

1144. (C) At increasing doses barbiturates produce an isoelectric EEG associated with a 50 to 60% decrease in cerebral metabolic rate. Cerebral vasoconstriction is probably the result of metabolic depression, since blood flow remains coupled to metabolism. Even at maximum EEG suppression, hyperventilation during barbiturate anesthesia will further reduce CBF. Normal cerebral metabolism and energy charge are preserved during barbiturate anesthesia. *(1:332–3, 750; 4:1622–3; 5:821)*

CHAPTER 12

Obstetric Anesthesia

Questions

DIRECTIONS (Questions 1145 through 1184): Each of the numbered items or incomplete statements in this section is followed by answers or by completions of the statement. Select the ONE lettered answer or completion that is BEST in each case.

1145. Magnesium sulfate

(A) does not cross the placenta
(B) has therapeutic levels between 6 and 8 mEq/L
(C) may produce respiratory depression in the neonate
(D) is a CNS stimulant
(E) decreases sensitivity to NMB agents

1146. Which of the following statements regarding drug effects in the peripartum is true?

(A) Benzodiazepines can prolong labor.
(B) Barbiturates cause no neonatal depression.
(C) Ketamine adversely affects uterine blood flow.
(D) An advantage of butorphanol is absence of respiratory depression.
(E) Morphine produces more respiratory depression of the newborn than does meperidine in equianalgesic doses.

1147. True statements regarding nonobstetric surgery during pregnancy include all of the following except

(A) nitrous oxide is best avoided in early pregnancy
(B) regional anesthesia, when appropriate, is a reasonable alternative
(C) aspiration prophylaxis should be accomplished with either a nonparticulate antacid and/or an H_2-blocker combined with a gastric motility agent
(D) benzodiazepines are contraindicated due to the increased incidence of cleft palate
(E) left uterine displacement (LUD) for the prevention of aortocaval compression is not necessary in the first trimester

1148. An infant born with a heart rate of 70, a weak cry, good muscle flexion, poor reflexes, and a blue-pale complexion would be given an Apgar score of

(A) 2
(B) 4
(C) 5
(D) 6
(E) 7

1149. The following statements regarding pregnancy and mitral stenosis are true except

(A) mitral stenosis is the most common acquired cardiac lesion presenting during pregnancy
(B) epidural analgesia is indicated to attenuate the increased cardiac output and tachycardia associated with the pain of labor
(C) ephedrine is preferred as a vasopressor
(D) invasive hemodynamic monitoring is indicated in cases of symptomatic mitral stenosis
(E) maternal expulsive efforts should be avoided during the second stage of labor

1150. Preeclampsia is associated with

(A) hypovolemia
(B) hypernatremia
(C) low hematocrit reading
(D) hyperkalemia
(E) hypotension

1151. A parturient is diagnosed with retained placenta after delivery and requires manual exploration of the uterus. All of the following are acceptable alternatives except

(A) intravenous analgesia
(B) epidural analgesia
(C) saddle block
(D) intravenous nitroglycerine, 400 mcg
(E) induction of general anesthesia with 1 mg/kg ketamine

Questions 1152 and 1153 refer to Figure 12-1.

1152. The tracing in the graph shows a pattern referred to as

(A) late deceleration
(B) variable deceleration
(C) early deceleration
(D) maximal deceleration
(E) late acceleration

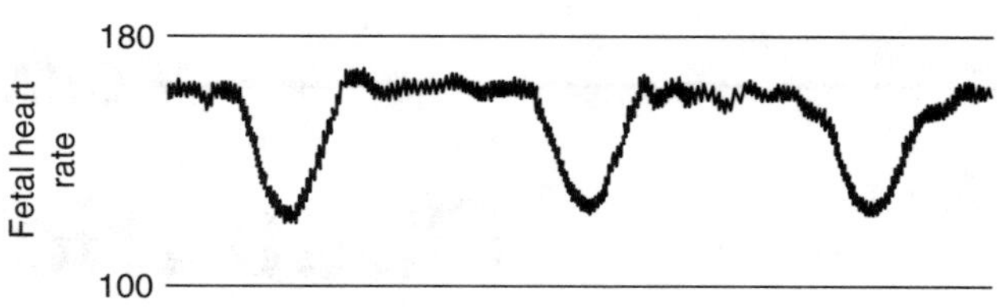

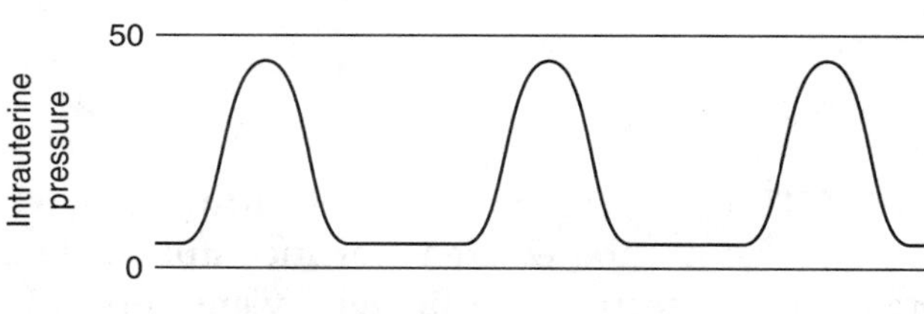

FIG. 12-1

1153. The type of heart rate tracing shown in the graph is usually associated with

(A) cord compression
(B) placental insufficiency
(C) head compressions
(D) acute fetal asphyxia
(E) tetanic contraction

1154. True statements regarding breech presentation include all of the following except

(A) a higher incidence of congenital abnormalities
(B) a lower frequency of prolapsed umbilical cord
(C) fetal head entrapment may necessitate the need for rapid induction of general anesthesia
(D) causes of breech presentation include preterm delivery, multiple gestation, and uterine anomalies
(E) accounts for approximately 3 to 4% of all pregnancies

1155. Regional anesthesia techniques that can be used for forceps deliveries include all of the following except

(A) bilateral pudendal block
(B) paracervical block
(C) subarachnoid block
(D) caudal block
(E) epidural block

1156. True statements regarding antepartum hemorrhage include all of the following except

(A) vaginal delivery may be attempted in select cases of placental abruption
(B) blood loss may be underestimated in the case of placental abruption
(C) placenta previa characteristically presents as painful vaginal bleeding
(D) a case of placenta previa can usually be diagnosed by ultrasound
(E) may present acutely during VBAC (vaginal birth after cesarean section) secondary to uterine rupture

1157. Measures to prevent aspiration pneumonia include all of the following except

(A) administration of an antacid before delivery
(B) rapid intubation
(C) rapid induction
(D) determination of the last time the patient has eaten
(E) cricoid pressure

1158. Neural pathways responsible for the transmission of pain during the first and second stages of labor include

(A) T10 to L1 and S2 to S4
(B) T8 to L2 and S1 to S3
(C) T6 to T12 and S1 to S4
(D) T10 to L5
(E) T12 to L3 and S2 to S5

1159. The usual blood loss with an uncomplicated vaginal delivery of twins or with cesarean section is approximately

(A) 400 mL
(B) 600 mL
(C) 800 mL
(D) 1000 mL
(E) 1200 mL

1160. Prophylactic measures taken to prevent maternal hypotension following spinal anesthesia include all of the following except

(A) administration of 500 to 1000 mL of fluid before induction
(B) lateral displacement of the uterus
(C) head-down tilt immediately after injection
(D) placing the patient on her side after block is established
(E) infusion of vasopressor

1161. In the Friedman curve shown in Figure 12-2, area C is the

(A) latent phase
(B) acceleration phase
(C) phase of maximum slope
(D) deceleration phase
(E) second stage of labor

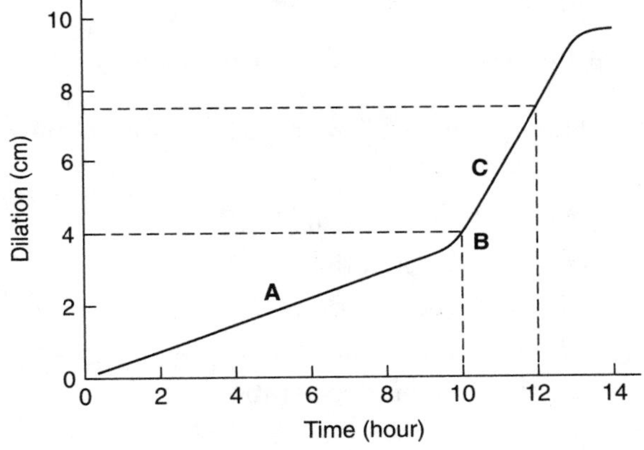

FIG. 12-2

1162. All of the following statements are true regarding fetal blood capillary pH testing except

(A) it is a method that cannot be used to assess fetal well-being during labor of a breech presentation
(B) a fetal scalp pH of 7.25 or higher is considered normal
(C) a fetal scalp pH of less than 7.20 is indicative of significant asphyxia and the need for immediate delivery
(D) a fetal scalp pH of 7.20 to 7.24 is intermediate and requires close monitoring and repeat sampling
(E) it requires adequate dilation of the cervix

1163. During parturition

(A) vital capacity (VC) decreases
(B) minute ventilation is unchanged
(C) functional residual capacity (FRC) decreases
(D) forced expiratory volume in 1 second (FEV_1) decreases
(E) tidal volume (TV) decreases

1164. The local anesthetic that attains the lowest fetal concentration relative to maternal concentration is

(A) lidocaine
(B) procaine
(C) 2-chloroprocaine
(D) mepivacaine
(E) bupivacaine

Questions 1165 and 1166 refer to Figure 12-3.

1165. The tracing in the graph shows a pattern referred to as

(A) early deceleration
(B) late deceleration
(C) mid deceleration
(D) variable deceleration
(E) sinusoidal deceleration

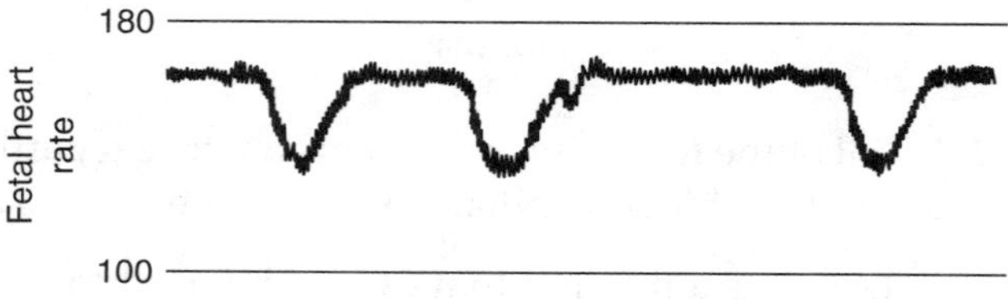

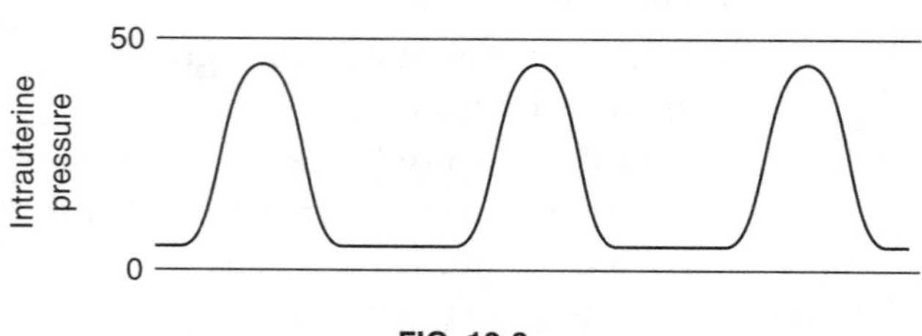

FIG. 12-3

1166. This type of fetal heart rate pattern is usually associated with

(A) uteroplacental insufficiency
(B) head compression
(C) cord compression
(D) severe fetal asphyxia
(E) prematurity

1167. Cholinesterase levels during pregnancy are

(A) highest at term
(B) unchanged from normal levels
(C) increased
(D) decreased resulting in a clinically significant prolongation of amide type local anesthetics
(E) decreased to a level not resulting in a clinically significant prolongation of the action of succinylcholine in the doses generally given

1168. During delivery, a lower dose of local anesthetic is required for regional anesthesia because

(A) pregnant women have greater pain tolerance
(B) the pain fibers are more superficial in the spinal canal
(C) the epidural and subarachnoid spaces are larger
(D) the epidural and subarachnoid spaces are decreased in size
(E) maternal hyperventilation decreases pain

1169. During parturition

(A) cardiac output is decreased
(B) cardiac output is increased
(C) stroke volume is decreased
(D) central venous pressure is decreased
(E) cardiac output remains constant

1170. Figure 12-4 shows the change in cardiac output with pregnancy. The discrepancy in the two lines is due to the effect

(A) on respiration
(B) on uterine blood flow
(C) on venous return
(D) of pressure on the aorta
(E) on the CNS

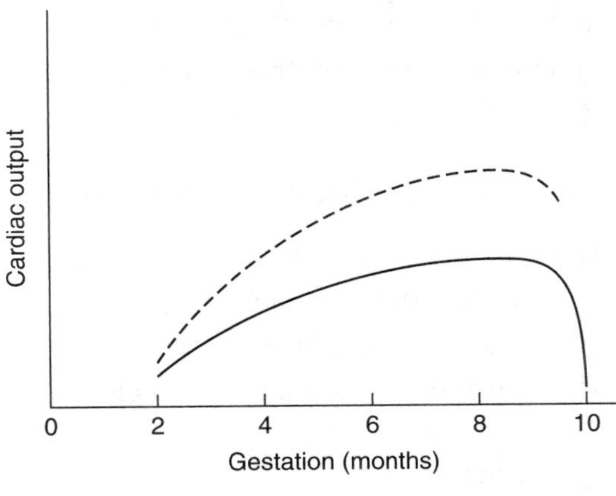

FIG. 12-4

1171. The pregnant woman at term

(A) shows no anatomic changes in the respiratory system
(B) is primarily a nose breather
(C) is primarily a mouth breather
(D) is less susceptible to hypoxia
(E) has decreased oxygen consumption

1172. In a pregnant woman at term, you would expect an increase in all of the following values except

(A) FRC
(B) VC
(C) peak flow rate
(D) lung compliance
(E) inspiratory capacity

1173. All of the following are indicative of fetal well-being except

(A) good long-term variability
(B) biophysical profile score of 8/10
(C) reactive nonstress test
(D) a positive oxytocin contraction stress test
(E) fetal scalp pH of 7.27

1174. In preeclamptic or eclamptic patients

(A) regional anesthesia is contraindicated
(B) hyporeflexia is common
(C) fluid restriction is necessary because of presence of edema
(D) resistance to vasopressors is common
(E) significant coagulopathies may occur

1175. Parturients who have been on terbutaline to halt preterm labor may exhibit all of the following except

(A) cardiac arrhythmias
(B) pulmonary edema
(C) tachycardia
(D) hypotension
(E) hypoglycemia

1176. The one finding present in eclamptic patients and not in preeclamptics is

(A) hyperreflexia
(B) decreased uteroplacental perfusion
(C) presence of seizure activity
(D) treatment with magnesium sulfate
(E) general vasoconstriction

1177. Nerve injury during labor and delivery can result from all of the following except

(A) compression of lumbosacral trunk by the head of fetus
(B) peroneal nerve injury by lithotomy stirrup
(C) epidural hematoma secondary to block
(D) femoral nerve compression by the lithotomy stirrup
(E) chemical contamination of the subarachnoid space

1178. Which statement is correct regarding epidural blood patches?

(A) The success rate is approximately 50%.
(B) The presence of backache after the procedure implies improper technique.
(C) The incidence of infection following the procedure is approximately 2%.
(D) Epidural saline injection has the same success rate as an epidural blood patch.
(E) If the site of dural rent is uncertain, the lowermost interspace should be used.

1179. Which of the following statements is true regarding neuraxial opioids?

(A) Spinal opioids augment sympathetic blockade produced by local anesthetics.
(B) Naloxone does not reverse nausea and vomiting caused by intrathecal opioids.
(C) Epidural opioids are as effective as a bupivacaine-opioid mixture in relieving pain.
(D) Intrathecal morphine produces approximately 24 hours of analgesia.
(E) Intrathecal fentanyl produces approximately 16 hours of analgesia.

1180. True statements regarding total spinal block include all of the following except

(A) it is a rare complication that can result from both intrathecal or epidural administration of local anesthetic
(B) it cannot result from subdural administration of local anesthetic
(C) supportive measures to provide oxygenation and prevent aspiration, including endotracheal intubation may be necessary
(D) preventative measures may include waiting for a partial spinal to wear off and then administering a second spinal in the case of elective cesarean section
(E) conversion to a general anesthetic may be necessary

1181. A 35-year-old heroin-addicted parturient in labor is requesting pain relief. Which of the following options is LEAST desirable?

(A) meperidine
(B) continuous epidural analgesia
(C) nitrous oxide
(D) butorphanol
(E) lumbar sympathetic block

1182. Uterine atony may be treated with all the following except

(A) uterine massage
(B) intramuscular methergonovine
(C) intrauterine prostaglandin $F_{2\alpha}$
(D) intravenous methergonovine
(E) intravenous oxytocin

1183. Regional anesthesia is contraindicated in

(A) vaginal birth after cesarean
(B) abruptio placentae
(C) diagnosis of placenta previa
(D) thrombocytopenia and an elevated bleeding time
(E) eclampsia

1184. The following techniques can all be employed to relieve pain during first stage of labor except

(A) paracervical block
(B) epidural anesthesia
(C) spinal anesthesia
(D) lumbar sympathetic block
(E) pudendal block

DIRECTIONS (Questions 1185 through 1216): For each of the items in this section, ONE or MORE of the numbered options is correct. Choose answer

(A) if only 1, 2, and 3 are correct
(B) if only 1 and 3 are correct
(C) if only 2 and 4 are correct
(D) if only 4 is correct
(E) if all are correct

1185. A normal newborn has which of the following values?

(1) RR 35 breaths per minute
(2) PO_2 60 mmHg
(3) HR 120 bpm
(4) systolic BP 50 mmHg

1186. Anesthesia for in vitro fertilization can be provided with

(1) pudendal block
(2) intravenous sedation
(3) local infiltration
(4) spinal anesthesia

1187. Increased neonatal depression has been observed after cesarean section in which of the following circumstances?

(1) general anesthesia compared to regional anesthesia
(2) 8 minutes between induction and delivery
(3) use of a volatile agent
(4) 4 minutes between uterine incision and delivery

1188. Predisposing conditions for uterine rupture include

(1) multiple gestation
(2) previous uterine surgery
(3) multiparity
(4) prolonged intrauterine manipulation

1189. Which of the following cardiovascular changes occur in pregnancy?

(1) No change in arterial BP during pregnancy.
(2) Greatest rise in cardiac output is postdelivery.
(3) Red blood cell mass decreases resulting in anemia of pregnancy.
(4) Central blood volume increases.

1190. Which of the following CNS changes occur in the parturient?

(1) Nerve fibers have increased susceptibility to local anesthetics.
(2) MAC for inhalational agents is decreased.
(3) Activation of endorphin system.
(4) No change in spread of local anesthetics in subarachnoid space during first trimester.

1191. Factors which may contribute to the hypotension seen with aortocaval compression include

(1) thiopental
(2) epidural block
(3) halothane
(4) hypervolemia

1192. After a spinal anesthetic, a parturient becomes hypotensive. The drug(s) of choice for treatment is (are)

(1) norepinephrine
(2) phenylephrine
(3) dopamine
(4) ephedrine

1193. Gastric changes associated with pregnancy include

(1) decreased acid secretion
(2) decreased gastric emptying time
(3) downward displacement of the pylorus
(4) incompetence of the gastroesophageal sphincter

1194. A woman has been in labor for 2 hours when her membranes rupture and the umbilical cord is prolapsed. Her treatment should include

(1) breech extraction or forceps extraction if the cervix is fully dilated
(2) knee-chest position
(3) constant monitoring of fetal heart tones
(4) immediate cesarean section if delivery is not imminent

1195. The administration of meperidine to a parturient may lead to

(1) decreased Apgar scores
(2) decreased neonatal minute ventilation
(3) respiratory acidosis
(4) highest exposure in the fetus 2 to 3 hours after administration

1196. Maternal changes associated with preeclampsia include

(1) increased cardiac output
(2) decreased renal blood flow
(3) decreased cerebral blood flow
(4) increased hepatic blood flow

1197. Anesthesia for emergency cesarean section can be provided by

(1) general anesthesia
(2) local infiltration
(3) subarachnoid block
(4) epidural block (if epidural catheter not already in place)

1198. In multiple deliveries, anesthetic considerations must include the fact(s) that

(1) the infants are usually postmature
(2) complications, such as prolapsed cord, are more frequent
(3) the first infant has a higher incidence of complications
(4) the mother has a higher incidence of complications

1199. Serious risks of general anesthesia in obstetric patients include

(1) aspiration pneumonia
(2) depression of the newborn
(3) impaired uterine contractility
(4) hemorrhage

1200. A patient scheduled for an elective cesarean section prefers spinal anesthesia. Which options are acceptable?

(1) 8 to 10 mg of tetracaine
(2) 20 to 30 mg of lidocaine
(3) 8 to 12 mg of bupivacaine
(4) 30 to 40 mg chloroprocaine

1201. True statements regarding amniotic fluid embolism include

(1) mortality remains around 40 to 50%
(2) classical signs and symptoms include abrupt onset of dyspnea, hypoxemia, cyanosis, seizures, loss of consciousness, and hypotension
(3) usually occurs prior to labor
(4) disseminated intravascular coagulation (DIC) and cardiac arrest are common

1202. True statements regarding methods employed to decrease the risk of toxicity from intravascular injection of local anesthetic during administration of an epidural include

(1) aspiration of the catheter prior to injection
(2) the use of 3 mL of 1.5% lidocaine with 1:200,000 epinephrine (test dose) to detect an increase in maternal heart rate or subjective symptoms
(3) fractionation of the dose of local anesthetic
(4) rapid administration of 1 to 2 mL of air with simultaneous Doppler monitoring of the precordium

1203. When compared to spinal anesthesia, epidural anesthesia

(1) permits less control
(2) causes more hypotension
(3) is more apt to cause spinal headache
(4) carries less risk of arachnoiditis

1204. Pudendal block for obstetrics

(1) requires more drug than a saddle block
(2) can provide good relief from cervical dilatation
(3) does not cause vasomotor blockade
(4) provides a sensory level to T11

1205. Disadvantages of paracervical block include

(1) fetal acidosis
(2) increases in uterine contractions
(3) fetal bradycardia
(4) inadvertent subarachnoid block

1206. A 25-year-old gravida 2, para 1 woman is scheduled for repeat cesarean section. Anesthetic options for epidural anesthesia include

(1) 1.5 to 2% lidocaine
(2) 0.5% bupivacaine
(3) 3% chloroprocaine
(4) 1% tetracaine

1207. Important considerations for breech presentation include

(1) uterine relaxation
(2) increased need for complete breech extraction with regional anesthesia
(3) possible lengthening of second stage of labor with epidural
(4) increased neonatal depression with regional anesthesia

1208. In the anesthetic management of a patient with preeclampsia, which of the following statements is (are) true?

(1) Magnesium sulfate may potentiate the duration and intensity of both depolarizing and nondepolarizing muscle relaxants.
(2) General anesthesia and tracheal intubation may be more difficult due to edema of the airway and friability secondary to impaired coagulation.
(3) Ketamine and ergot alkaloids should be avoided.
(4) Fluid retention and oliguria result in a hypervolemic state.

1209. Fetal effects of drugs administered to the mother may be modified by

(1) blood flow from the umbilical vein passing through the ductus venosus
(2) drug uptake by the fetal liver
(3) dilution of blood in the right atrium
(4) shunting across the foramen ovale

1210. Inhalation agents pass through the placental membrane rapidly as a result of

(1) rapid rate of diffusion
(2) low fat solubility
(3) low molecular weight
(4) low concentration gradient

1211. Placental circulation may be decreased by

(1) uterine contractions
(2) maternal hypotension
(3) aortoiliac compression
(4) hemorrhage

1212. A normal pregnancy is associated with increased

(1) minute volume
(2) tidal volume
(3) respiratory rate
(4) ventilation only at term

1213. The newborn infant

(1) has a respiratory acidosis
(2) has a high PCO_2
(3) is hypoxic
(4) has a metabolic acidosis

1214. The newborn infant who remains apneic

(1) has an exaggerated acidosis
(2) develops a PCO_2 rise of 10 mmHg/min
(3) has a fall in pH of 0.1 unit/min
(4) can tolerate this condition without permanent damage

1215. Magnesium sulfate

(1) decreases excitability of muscle membranes
(2) decreases sensitivity of the motor endplate to acetylcholine
(3) potentiates both depolarizing and nondepolarizing neuromuscular junction blocking agents
(4) acts as a mild vasodilator

1216. True statements concerning continuous epidural infusion include

(1) there is an increased incidence of motor block compared to intermittent boluses
(2) bupivacaine 0.125% is frequently used
(3) there is no advantage of adding opioids to the local anesthetic solution
(4) a potential complication is migration of the catheter

Answers and Explanations

1145. (C) Magnesium sulfate is given to prevent seizures. A loading dose of 4 g is given followed by an infusion of 1 to 2 g/h to achieve a blood level of 4 to 8 mEq/L. It reduces the muscle membrane excitation, decreases sensitivity to acetylcholine, and potentiates all NMB agents. It crosses the placenta and can cause respiratory depression in the newborn. *(1:1154; 5:2332)*

1146. (C) Opioids may inhibit labor if given prior to the active phase, benzodiazepines have no effects on labor, though they are often avoided because of their amnestic properties as well as potential for neonatal depression. Barbiturates may have protracted effects on the newborn. Agonist-antagonist opioids have a ceiling effect on respiratory depression after a certain dose. *(1:1146; 5:2317)*

1147. (D) Benzodiazepines are no longer believed to cause facial cleft defects and therefore may be used safely in pregnancy. Nitrous oxide is controversial regarding its teratogenic potential in humans, especially with single, limited exposure. In animal studies it has been shown to interfere with DNA synthesis. More recent data refute the teratogenic potential of nitrous oxide in humans; however, its use is optional as part of a general anesthetic. Although the teratogenic potential of nitrous oxide may be less in question, it is probably still best avoided, especially in early pregnancy. Incompetence of the LES (lower esophageal sphincter) and anatomic changes associated with pregnancy both increase the risk of aspiration pneumonia. Although it is unclear at what stage of pregnancy this becomes significant, aspiration prophylaxis is prudent in the pregnant patient. Regional anesthesia is a reasonable alternative to general anesthesia and potentially limits drug exposure during surgery. The uterus does not become an intra-abdominal organ until 12 to 13 weeks gestation and therefore LUD is not necessary in the first trimester. *(1:1163–7; 4:2019–20; 5:2337–8)*

1148. (C) The baby would get 1 point for heart rate, 1 for a weak cry, 2 for muscle tone, 1 for reflexes, and no points for color. *(1:1162; 5:2348)*

1149. (C) Phenylephrine is the preferred vasopressor as maternal tachycardia is poorly tolerated. Invasive monitoring is often necessary in symptomatic patients to guide fluid management. Maternal expulsive efforts are best avoided with the use of an epidural and elective instrumental delivery to prevent the deleterious effects of the Valsalva maneuver on venous return. *(1:1156; 4:2013; 6:665)*

1150. (A) In spite of the fact that patients may exhibit edema and weight gain, they are hypovolemic due to the vasoconstriction that is part of the disease. They have hypertension, hyponatremia, increased hematocrit (due to the vasoconstriction and hypovolemia), and hypokalemia. *(1:1153; 4:2011)*

1151. (D) Small doses of nitroglycerine (50 to 100 mcg) have been shown to relax the uterus to allow for manual extraction of the placenta. Caution should be used with the use of regional anesthesia in the presence of significant postpartum bleeding to avoid hypotension. Ketamine is a good induction drug in the presence of maternal bleeding, but has been shown to produce a dose-related increase in uterine tone. *(4:2010–11; 6:677)*

1152. (C) The pattern represented is that of early deceleration, in which the heart rate decreases with the onset of contraction, reaches the low point with the peak of contraction, and then returns to baseline as the uterus relaxes. In contrast, late decelerations start after the contraction is underway, and the low point occurs after the contraction is over. Variable decelerations are variable in shape and onset. *(1:1159; 4:1998)*

1153. (C) The early deceleration pattern is seen with fetal head compression, which leads to increased vagal tone. Cord compression leads to variable decelerations, whereas placental insufficiency leads to late decelerations. *(1:1159; 4:1997)*

1154. (B) There is a higher frequency of prolapsed cord, especially with either a double footling or complete breech presentation because the fetal head no longer occupies the lower uterine segment. A higher incidence of congenital abnormalities has been found. The fetal head is the largest presenting part and it is last to present, therefore, fetal head entrapment can occur and is a life-threatening complication of vaginal breech delivery, which may require rapid induction of general anesthesia. *(4:2017; 6:675)*

1155. (B) A paracervical block is good for the first stage of labor, since it helps with the pain associated with cervical dilatation. A pudendal block will provide anesthesia for the second stage and is appropriate for low forceps delivery and episiotomies. A subarachnoid block, caudal block, or epidural block also is appropriate. *(1:1149; 5:2319)*

1156. (C) Placenta previa classically presents as painless, bright red bleeding during pregnancy. Small placental abruptions may deliver vaginally as long as there is no evidence of fetal distress. The amount of blood loss may be underestimated in placental abruption because it is often concealed behind the placenta. Unlike placental abruption, placenta previa can usually be detected by ultrasound. Uterine rupture is a rare but potentially catastrophic event requiring immediate cesarean section, possible hysterectomy and full resuscitative measures of both mother and fetus. *(4:2009–11; 5:2333–6)*

1157. (D) The obstetric patient is treated as a patient with a full stomach, so the history of no recent food intake is not important. If a general anesthetic is needed, the patient should have antacids and rapid sequence induction with cricoid pressure. *(1:1150; 5:2326)*

1158. (A) Neural pathways responsible for the transmission of pain during the first stage of labor are visceral in nature and involve afferent pathways from T10 to L1. The pathways involved for the second stage of labor are somatic and produced by the distension of the perineum and stretching of the fascia, skin, and subcutaneous tissues and involves afferent pathways from S2 to S4 via the pudendal nerves. *(1:1146; 5:2315; 6:656–7)*

1159. (D) The usual blood loss in an uncomplicated vaginal delivery of twins or in a cesarean section is approximately 1000 mL. *(1:1150)*

1160. (C) Methods to prevent maternal hypotension include fluid administration, lateral displacement of the uterus, placing the patient on her side, and infusion of a vasopressor. A head-down tilt will increase the level of the block and make the hypotension worse. *(1:1148; 5:2326)*

1161. (C) The area labeled C is the phase of maximum slope. The first, relatively flat area is the latent phase. This is followed by an acceleration phase, the phase of maximum slope, and then the deceleration phase. The second stage of labor follows. Knowing the normal progression of labor helps one to decide when to provide anesthesia. *(5:2314)*

1162. (A) Fetal blood capillary testing may be used to assess fetal well-being during labor. It requires adequate cervical dilation to allow for sampling of the presenting part. It may be used in both vertex and breech presentations. If the testing is intermediate, continued close monitoring and repeat sampling within 30 minutes is required if delivery has not been accomplished. *(1:1160; 4:1998; 5:2313)*

1163. (C) Oxygen demand increases with pregnancy. Minute ventilation doubles as a result of increased respiratory rate and TV. There is no or minimal change in FEV_1 and VC, and there is a decrease in FRC. *(1:1141; 5:2310)*

1164. (C) 2-Chloroprocaine is metabolized very rapidly; therefore, it does not attain a high concentration in the fetus. The other agents are broken down more slowly and are able to cross the placenta. *(1:1143, 1149; 5:2323)*

1165. (D) There are three major types of fetal decelerations. Early decelerations demonstrate a slow drop in heart rate beginning with the uterine contraction (UC) with the nadir coinciding with the peak of the UC. It returns to baseline by the end of the UC. It is a result of vagal stimulation secondary to head compression and is not indicative of fetal asphyxia. Late decelerations begin after the UC and return to baseline after the end of the UC. They are often repetitive and associated with decreased fetal heart rate variability. They are associated with uteroplacental insufficiency. Variable decelerations are variable in configuration and bear no consistent temporal relationship to the onset of the UC. They are thought secondary to cord compression and unless severe and repetitive are not thought to be indicative of fetal compromise. *(1:1159; 4:1997–8; 5:2313; 6:680–1)*

1166. (C) See the explanation for Question 1165.

1167. (E) Plasma cholinesterase levels are typically decreased by 25% during pregnancy. The decreased levels do not result in clinically significant effects on ester-type local anesthetics or succinylcholine in the doses generally used. *(4:1990)*

1168. (D) Less local anesthetic is needed in the later stages of pregnancy, since the epidural space is decreased in size. This is due to the venous congestion caused by the weight of the uterus. In addition, there are studies showing that the nerves are more sensitive to the local anesthetics. This is thought to be a hormonal effect. *(1:1142; 5:2311)*

1169. (B) During parturition, the cardiac output is increased. This may be offset by the effects of the uterine weight on the vena cava, leading to decreased return and decrease in cardiac output. Central venous pressure increases. *(1:1141; 5:2308)*

1170. (C) There is an increase in cardiac output as pregnancy progresses until about the 8th month, at which time the increase is attenuated. When studying the cardiac output in the pregnant patient, one must know whether the study was made with the patient in the supine or lateral position. In the supine position, cardiac output will be decreased because of the weight of the uterus on the great veins of the abdomen. *(1:1141; 5:2309)*

1171. (C) The pregnant patient at term is a mouth breather, since nasal congestion and swelling make it more difficult to breathe through the nose. Placement of nasal catheters or airways may lead to bleeding. The FRC is decreased at term, making the patient more prone to hypoxia. There is an increase in oxygen consumption. *(1:1141; 5:2310)*

1172. (A) FRC decreases at term. VC is unchanged or slightly increased, and there is an increase in peak flow rate, lung compliance, and inspiratory capacity. *(1:1141; 5:2310)*

1173. (D) An oxytocin contraction test (OCT) involves the stimulation of UCs with either oxytocin or breast stimulation. The presence of repetitive late decelerations represents a positive OCT, indicative of uteroplacental insufficiency and fetal compromise. Long-term variability implies an intact sympathetic/parasympathetic system. A biophysical profile (BPP) is an ultrasound, which incorporates fetal movement, tone, breathing motion, and amniotic fluid combined with a nonstress test to assess fetal well-being. A BPP of eight or more is reassuring. A nonstress test involves the observation of two fetal heart rate accelerations within a 20 to 30-minute period to assess fetal well-being. A reactive nonstress test is reassuring. Fetal scalp pH testing involves the assessment of fetal blood pH during labor. A fetal pH greater than or equal to 7.25 is reassuring. *(4:1996–8; 5:2313–4)*

1174. (E) Preeclampsia can involve all organ systems with associated decreased platelets, abnormal clotting studies, and abnormal liver function tests. Widespread vasoconstriction results in decreased intravascular volume and extravascular water and sodium retention. Patients are usually hyperreflexic and hypertensive but intravascularly depleted with increased sensitivity to vasopressors. If no coagulopathy exists and the patient is properly monitored and hydrated, continuous epidural analgesia may improve placental perfusion. *(1:1152–5; 5:2330–3)*

1175. (E) Tocolytic agents, such as ritodrine and terbutaline, relax uterine smooth muscle by direct stimulation of β-adrenergic receptors. These patients are at increased risk for fluid overload, tachycardia, dysrhythmia, hyperglycemia, decreased BP, and pulmonary edema. The etiology of pulmonary edema is controversial and has been reported to be both cardiac and noncardiac in origin. *(1:1157)*

1176. (C) All the findings are present in preeclampsia and eclampsia, except the presence of seizure activity defines eclampsia. *(1:1152)*

1177. (D) The most common cause of neurologic injury during labor and delivery is secondary compression of the lumbosacral trunks by the fetal head. Incorrect positioning can cause peroneal nerve injury secondary to the lithotomy stirrups if the legs are positioned to the inside. Femoral nerve compression can occur secondary to excessive flexion of the hip in the lithotomy position. Epidural hematomas and chemical contamination of subarachnoid space are fortunately rare complications. *(1:1151; 5:2329)*

1178. (E) The success rate of epidural blood patching is over 90% and the infection rate is less than 0.01% if sterile techniques are used. The use of saline is much less successful but incidence of backache is not unusual after blood patching. Because the spread of solution is usually greater in a cephalad direction, the lowermost interspace should be chosen if the site of dural puncture is uncertain. *(1:708; 5:2328)*

1179. (D) The addition of opioids to local anesthetic solutions used in major conduction block enhances the quality of peripartum analgesia without causing a sympathetic block; however, the combination is more effective than either class of agents used alone in relieving pain. Morphine is longer acting than fentanyl and provides up to 24 hours of analgesia compared to 8 to 12 hours for fentanyl. Much smaller doses of naloxone than needed to reverse analgesia are effective in reversing nausea and vomiting. *(4:2005)*

1180. (B) Total spinal block is a very serious and rare complication of intrathecal, epidural, or subdural administration of local anesthetic. Several mechanisms have been proposed after failed epidural including expansion of the epidural space resulting in compression of the spinal canal and further cephalad spread of intrathecal anesthetics, rapid transfer of anesthetic from the epidural space through the dural hole, and sufficient coverage of the neural roots which decreases the dose requirements of subsequent spinal anesthesia. Supportive measures including vasopressors and fluids should be instituted. Endotracheal intubation and general anesthesia may be required. *(1:1151; 4:2003–4; 5:2329)*

1181. (D) Agonist-antagonists may trigger withdrawal symptoms when administered to opioid-tolerant patients. *(5:2318)*

1182. (D) Methergonovine is not approved for intravenous injection; all other agents are useful to treat uterine atony. Rapid infusion of oxytocin can cause hypotension, methergonovine can cause hypertension, and prostaglandin $F_{2\alpha}$ can cause bronchoconstriction and hypertension. *(4:2011; 5:2336)*

1183. (D) Even though there had been some concern about masking symptoms and signs of uterine rupture in patients at risk during a vaginal birth after cesarean, this has not been the case if fetal heart rate and uterine activity are continuously monitored. DIC has been associated with abruptio placentae, but if clotting studies and platelet counts are normal, regional anesthesia may be performed. If bleeding is minimal or absent in a patient with placenta previa, regional anesthesia is an option. In a patient with eclampsia and normal clotting studies and platelet count, epidural anesthesia is an option as long as blood pressure is well maintained. *(1:1147; 4:2000, 2010)*

1184. (E) Pain during the first stage of labor is caused by uterine contractions and cervical dilatation. All the techniques listed can be beneficial except pudendal block, which only supplies the lower vaginal canal and perineum through sacral roots S2 to S4. *(1:1148; 5:2322)*

1185. (A) The normal newborn breathes approximately 30 to 40 breaths per minute, has a heart rate approximately 120 bpm, a systolic BP of 60 to 70 mmHg, and a PO_2 of 50 to 70 mmHg. *(1:1162; 4:2017)*

1186. (C) Anesthesia for IVF can be provided with various regional techniques including spinal and epidural, as well as general anesthesia or intravenous sedation. Pain is experienced when the peritoneum and the ovarian capsule are punctured. There is concern over effects of various medications on fertilization rates, and further investigation is necessary. *(4:2020)*

1187. (D) There is no difference in neonatal condition between general anesthesia and regional anesthesia. No neonatal depression is demonstrated if delivery is within 10 minutes of induction with 50% N_2O and less than 3 minutes between uterine incision and delivery. *(1:1150; 4:2007)*

1188. (C) Predisposing conditions for uterine rupture include previous uterine surgery, especially vertical incisions and prolonged intrauterine manipulation. There has been concern about regional anesthesia masking signs of rupture, especially in patients attempting vaginal birth after cesarean. The experience seems to support the safety of epidural anesthesia if continuous electronic monitoring of uterine activity and the fetal heart rate is performed. *(4:2010)*

1189. (C) Cardiac output increases with each contraction secondary to autotransfusion with the greatest rise in cardiac output immediately post-delivery. Arterial BP falls during pregnancy but transiently rises with uterine contractions both secondary to pain and autotransfusion with increase in central blood volume. RBC mass increases but plasma volume rises to a greater extent resulting in a relative anemia of pregnancy. *(4:1988; 5:2308)*

1190. (A) Anesthetic requirements are decreased in pregnancy, with 25% less local anesthetic needed for regional anesthesia. Anatomic effects, such as distended epidural veins, increase spread of local anesthetics. Hormonal changes alter nerve response. Increased progesterone levels may be responsible for the 25 to 40% reduction in MAC to general anesthetics. *(4:1991; 5:2311)*

1191. (A) Medications or techniques that decrease cardiac output or vascular resistance, such as thiopental, epidural block, or halothane, will contribute to the hypotension. Hypervolemia will counteract the problem. *(1:1151; 5:2326)*

1192. (C) Ephedrine increases blood pressure without decreasing uterine blood flow. Drugs with primarily α-adrenergic activity may increase uterine vascular resistance and decrease uterine blood flow. Recent studies in humans have shown phenylephrine to be as safe as ephedrine. *(1:1151; 5:2326)*

1193. (D) Gastric changes seen with pregnancy include incompetence of the gastroesophageal sphincter, increased acid secretion, increased gastric emptying time, and upward displacement of the pylorus. These make the pregnant patient more susceptible to aspiration. *(1:1142; 5:2310)*

1194. (E) Prolapse of the umbilical cord is a true obstetric emergency. Steps should include all of the options. This can occur without warning, and one must always be prepared for such an emergency. *(4:2017; 5:2336)*

1195. (E) The administration of meperidine may lead to decreased Apgar scores due to fetal depression and decreased fetal minute ventilation. This may lead to respiratory acidosis. The highest exposure in the fetus is 2 to 3 hours after administration to the mother. This makes timing of the drug to the mother important. *(1:1146; 5:2317)*

1196. (A) Maternal changes seen with preeclampsia include increased cardiac output, decreased renal blood flow, and decreased cerebral blood flow. The last two are reasons for some of the complications. Hepatic flow is decreased in the preeclamptic patient. *(1:1153; 5:2329)*

1197. (A) Anesthesia for an emergency cesarean section requires fast onset. Therefore, an epidural block is not appropriate. Since time is of the essence in these situations, a room with all equipment is kept ready at all times. *(1:1148–50; 5:2323–5)*

1198. (C) The multiple delivery birth has the potential for many problems. The infants are frequently premature. Complications, such as prolapsed cord, are more frequent. The first infant has a lower incidence of complications. The mother also has a higher incidence of problems with multiple births. *(4:2017; 5:2336; 6:676)*

1199. (E) All of these factors may be problems in the obstetric patient for general anesthesia. Steps must be taken to avert the problems when possible. In some patients, general anesthesia is the best approach. One must use the lowest dose of anesthesia possible and take steps to avoid aspiration. The possibility of hemorrhage must be borne in mind; all volatile anesthetics impair uterine contractility and may therefore potentiate postpartum bleeding. *(1:2325; 4:2007)*

1200. (B) A sensory level of T4 is necessary for a cesarean section, but pregnant patients have a reduced requirement for local anesthetic. Typically, 65 to 75 mg of 5% lidocaine in 7.5% dextrose gives a duration of 45 to 75 minutes, 8 to 10 mg of 1% tetracaine in an equal volume of 10% dextrose provides 120 to 180 minutes, and 8 to 12 mg of 0.75% bupivacaine in 8.25% dextrose gives 120 to 180 minutes. Chloroprocaine is not used for spinal anesthesia. *(1:1149; 5:2324)*

1201. (C) AFE is a rare and catastrophic complication of pregnancy. It is secondary to a disrupted barrier between amniotic fluid and maternal circulation and usually occurs during tumultuous labor. The signs and symptoms are dramatic and abrupt, classically presenting as dyspnea, hypoxemia, seizures, loss of consciousness, and hypotension. Fetal distress is present and coagulopathy is common. The majority of patients experience cardiac arrest. Treatment is supportive and may include intubation, pressors, fluids, and blood products. Despite aggressive treatment, mortality remains around 80%. *(6:678–80)*

1202. (E) Large doses of local anesthetics may be given during regional anesthesia for obstetrics, which can result in adverse reactions. The use of a test dose containing 15 mcg of epinephrine is still advocated by some. Objections include false positives due to uterine contractions and the theoretical concern that epinephrine may reduce uteroplacental perfusion. Fractionation helps to reduce potential toxicity. Epidural catheters should always be aspirated prior to injection. Some authors advocate the use of air as a means to detect intravascular injection. *(1:1149; 4:2002; 5:2320)*

1203. (D) The epidural technique is more controllable and causes less hypotension than the spinal technique. There is much less risk of postdural puncture headache. The incidence of arachnoiditis is decreased with the epidural approach. *(1:1149; 4:2006)*

1204. (B) The pudendal block is useful for relieving pain for low forceps application and for episiotomy. It requires more drug than that used for a saddle block. There is no vasomotor blockade. The sensory loss is in the area of the perineum only. *(1:1148; 4:2004; 5:2322)*

1205. (B) A paracervical block can lead to fetal bradycardia and subsequent fetal acidosis. The block does not affect uterine contractions. There is no risk of subarachnoid injection. *(1:1148; 4:2004; 5:2322)*

1206. (A) Usually 15 to 25 mL of 1.5 to 2% lidocaine, 0.5% bupivacaine, or 3% 2-chloroprocaine will give a sensory level of T4. Tetracaine is typically used only for spinal anesthesia. *(1:1147; 4:2000–1; 5:2324)*

1207. (B) There had been some concern about regional anesthesia preventing pregnant patients from pushing effectively, but the incidence of complete breech extraction is not increased. The second stage of labor may be slightly prolonged, but regional anesthesia provides pain relief and maximal perineal relaxation for the aftercoming head delivery. *(4:2017; 6:675)*

1208. (E) Magnesium potentiates the effect of muscle relaxants by diminishing the excitability of the motor end plate to acetylcholine and depressing the excitability of the skeletal muscle membrane. Generalized edema occurs in preeclampsia secondary to maternal plasma protein depletion from endothelial damage and proteinuria. Thrombocytopenia can result in coagulopathy. This can contribute to edema and friability of the airway structures and increase the risk of difficult airway. Ketamine and ergot alkaloids can both exacerbate hypertension and are contraindicated in preeclamptic patients. Despite the presence of increased fluid and sodium retention and hypertension, patients with preeclampsia are intravascularly volume depleted due to a shift of protein and fluid from the intravascular to the extravascular compartments. *(1:1152–5; 4:2011–3; 6:658–64)*

1209. (E) These factors all protect the fetus from receiving the full effect of drugs administered to the mother. The blood is diverted or diluted at several loci, decreasing the drug concentration in the brain of the fetus. *(1:1144; 5:2313)*

1210. (B) Inhalation agents pass through the placental membrane rapidly as a result of a rapid diffusion rate, high fat solubility, low molecular weight, and a high concentration gradient. *(1:1142–3; 5:2312–3)*

1211. (E) Placental circulation is decreased by uterine contractions. This is part of normal labor. If tetanic contractions occur, the fetal circulation may be seriously impaired. Hemorrhage may lead to maternal hypotension, which may decrease uterine blood flow. Aortoiliac compression may decrease uterine blood flow directly, and if vena cava compression occurs with it, the hypotension will aggravate the problem. *(4:1991; 5:2312)*

1212. (A) Respiratory effects of pregnancy are noted early and are progressive during the pregnancy. Minute volume increases, as does TV and respiratory rate. Hyperventilation begins very early, reaches its maximum at about the end of the first trimester, and remains elevated at term. *(1:1141; 5:2310)*

1213. (E) The newborn infant is prone to respiratory problems, since he or she has acidosis on both metabolic and respiratory bases and is hypoxic. *(1:1161; 5:2346)*

1214. (A) The newborn has an acidosis as noted previously. Any apnea at birth will aggravate this problem. Immediate resuscitation is mandatory, since the PCO_2 will rise rapidly and the pH will fall. *(1:1161; 5:2346)*

1215. (E) Magnesium sulfate is administered to prevent seizures. The loading dose is 4 g over 10 minutes followed by continuous infusion of 1 to 2 g/h to achieve a blood level of 4 to 6 mEq/L. In addition to all the properties listed, magnesium sulfate decreases acetylcholine release. *(1:1154; 5:2332)*

1216. (C) Continuous epidural infusions are very popular because of the consistent pain relief with minimal side effects. Low concentrations of bupivacaine (0.0625 to 0.125%) with 2 mcg/mL of fentanyl is probably the most common solution in use. The addition of opioids allows for a lower concentration of local anesthetic and provides better analgesia than local anesthetic alone. *(4:2000–2; 5:2320–2)*

CHAPTER 13

Pediatric Anesthesia

Questions

DIRECTIONS (Questions 1217 through 1249): Each of the numbered items or incomplete statements in this section is followed by answers or by completions of the statement. Select the ONE lettered answer or completion that is BEST in each case.

1217. When assessing the fetus at birth

(A) 90% of all infants have 1-minute Apgar scores of 8 or more
(B) the 5-minute Apgar score correlates well with acidosis and survival
(C) it is abnormal for a neonate to have cyanosed extremities beyond 60 seconds of age
(D) the normal heart rate of a term neonate is between 100 and 120 beats per minute
(E) the Apgar scoring system was devised by Dr. Virginia Apgar in 1973

1218. Concerning perinatal respiratory physiology, all of the following are true EXCEPT

(A) at term, the fetal lung contains about 30 mL/kg of an ultrafiltrate of plasma
(B) the fetal lung does not usually contain amniotic fluid
(C) approximately two-thirds of the lung fluid is expelled during the birth process
(D) babies born by cesarean section have about the same amount of lung fluid as babies born vaginally
(E) small preterm infants born vaginally frequently have excess lung water

1219. Concerning cardiovascular physiology after birth, all of the following are true EXCEPT

(A) the pulmonary vascular resistance decreases
(B) pressures on the left side of the heart increase
(C) completion of closure of the ductus arteriosus requires adequate arterial muscle tissue
(D) mechanical closure of the ductus arteriosus occurs after 2 to 3 days
(E) events during anesthesia may cause a return to fetal circulation

1220. The first sign of total spinal anesthesia in a neonate is most likely to be

(A) a decrease in oxygen saturation
(B) agitation or irritability
(C) hypotension
(D) an increase in heart rate
(E) loss of consciousness

1221. A 2-year-old child (weight 13 kg) is scheduled for circumcision. The most suitable dose of local anesthetic for a dorsal penile block is

(A) bupivacaine 0.25% 8 mL
(B) lidocaine 1% 8 mL
(C) lidocaine 1.5% with epinephrine 1:200,000 8 mL
(D) bupivacaine 0.25% 15 mL
(E) bupivacaine 0.125% 15 mL

1222. The systolic blood pressure of the normal full-term neonate is usually

(A) 50 to 60 mmHg
(B) 60 to 70 mmHg
(C) 70 to 80 mmHg
(D) 90 to 100 mmHg
(E) 100 to 110 mmHg

1223. The most appropriate circulatory/hematologic variables for a 6-month infant born at full term are

(A) HR 120, BP 90/60, stroke volume 7.5 mL/beat
(B) HR 140, BP 90/60, stroke volume 15 mL/beat
(C) HR 140, BP 110/75, stroke volume 15 mL/beat
(D) HR 120, O_2 consumption 10 mL/kg, Hb 11.5 g/dL
(E) O_2 consumption 5 mL/kg, Hb 16.5 g/dL, stroke volume 7.5 mL/beat

1224. A 3-year-old healthy child is in preop holding for elective repair of an umbilical hernia. The mother informs you that the child had 4 oz of apple juice 2 hours ago. You respond

(A) cancel surgery
(B) delay surgery 2 hours
(C) delay surgery 4 hours
(D) delay surgery 6 hours
(E) proceed with surgery now

1225. Changes in cardiovascular and hemodynamic parameters at birth include all of the following EXCEPT

(A) closure of the foramen ovale
(B) ductus arteriosus closure
(C) decreased afterload on the left ventricle
(D) increased pulmonary return
(E) large increase in volume load on left ventricle

1226. An infant is born with Apgar scores of 2 and 5 at 1 minute and 5 minutes of life, respectively. Initial resuscitation should include all of the following EXCEPT

(A) oxygen
(B) radiant heat
(C) intubation
(D) glucose
(E) bicarbonate

1227. A 5-month-old infant is anesthetized for correction of an eye condition. Immediately after intubation bilateral breath sounds and chest excursion are noted and there is 100% oxygen saturation with an FIO_2 of 0.5. After positioning for surgery, the oxygen saturation is noted to have dropped to 94%, no other changes having been made. The most likely cause for this fall in oxygen saturation is

(A) a kinked endotracheal tube
(B) bronchospasm
(C) migration of the endotracheal tube into the right mainstem bronchus
(D) inspissated secretions plugging the tube
(E) a failure in the anesthesia machine

1228. The normal serum glucose in a newborn is

(A) 10 to 20 mg/dL
(B) 20 to 30 mg/dL
(C) 40 to 60 mg/dL
(D) 60 to 70 mg/dL
(E) 80 to 90 mg/dL

1229. An infant under anesthesia loses body heat by all of the following routes EXCEPT

(A) the metabolism of brown fat
(B) breathing dry gases
(C) conduction to cold surroundings
(D) cold skin preparation solutions
(E) exposure of abdominal contents

1230. Urinary output during anesthesia in a child should be

(A) 1 mL/kg/h
(B) 2 mL/kg/h
(C) 3 mL/kg/h

(D) 4 mL/kg/h
(E) 5 mL/kg/h

1231. The first sign of malignant hyperthermia in an anesthetized infant is frequently

(A) rapid rise in body temperature
(B) tachycardia
(C) hot skin
(D) arrhythmia
(E) hot circle absorber

1232. The major factor associated with the closure of patent ductus arteriosus (PDA) in the newborn is

(A) increased $PaCO_2$
(B) decreased $PaCO_2$
(C) increased PaO_2
(D) decreased PaO_2
(E) increased pulmonary artery pressure

1233. The normal respiratory dead space in an infant is

(A) 0.5 mL/kg of body weight
(B) 1 mL/kg of body weight
(C) 2 mL/kg of body weight
(D) 3 mL/kg of body weight
(E) 4 mL/kg of body weight

1234. If a heated nebulizer humidifier is used during anesthesia, a small child

(A) will not require an alteration in fluid management
(B) will require an increased amount of intravenous fluid due to extra insensible water loss resulting from an increased body temperature
(C) will require less sodium to be infused
(D) can receive 100% humidified inspired gases
(E) will require more fluid to replace increased urinary losses

1235. The most important factor in maintaining normothermia in the operating room is

(A) body temperature at the beginning of the procedure
(B) room temperature
(C) use of a warming blanket
(D) use of warm fluids
(E) temperature of prep solutions

1236. In the newborn infant, the spinal cord extends to the

(A) first lumbar vertebra
(B) second lumbar vertebra
(C) third lumbar vertebra
(D) fourth lumbar vertebra
(E) fifth lumbar vertebra

1237. The typical 2-year-old child should be intubated with an endotracheal tube having an internal diameter of

(A) 3 mm
(B) 3.5 mm
(C) 4.5 mm
(D) 5.5 mm
(E) 6.5 mm

1238. The number of alveoli in a full-term infant as compared to an adult is

(A) the same
(B) 8%
(C) 25%
(D) 50%
(E) 75%

1239. If a child who is admitted with an incarcerated inguinal hernia has a mild upper respiratory tract infection

(A) the surgery should be cancelled
(B) the surgery should be allowed to proceed, but the child should not be intubated
(C) the child should be started on antibiotics, and the surgery should proceed
(D) the surgery should proceed with careful monitoring
(E) the patient should be operated on only under spinal anesthesia

1240. Classic signs of congenital diaphragmatic hernia include all of the following EXCEPT

(A) decreased movement of the hemithorax
(B) rounded abdomen
(C) mediastinal shift
(D) bowel sounds in chest
(E) bowel pattern in chest by x-ray

1241. Preparation for surgery of the patient with sickle cell disease should include all of the following EXCEPT

(A) transfuse to a hemoglobin level of 15 g/dL
(B) treat infection
(C) maintain good hydration
(D) provide good pulmonary care
(E) avoid stasis of blood flow

1242. A 6-week-old baby born at 32 weeks gestation presents in holding for elective repair of an inguinal hernia. The parents believe that they will be taking their child home today after surgery. You inform them

(A) they may take their child home today
(B) the child may have to stay for several hours
(C) the surgery will be postponed until the child reaches 60 weeks postconceptual age
(D) the child will be admitted for 23 hours apnea monitoring
(E) the child will need apnea monitoring at home tonight

1243. A 7-year-old 35-kg girl is scheduled for excision of a large intra-abdominal mass. Her starting hematocrit is 36% and the minimally acceptable hematocrit is 24%. How much blood could the patient lose before transfusion is necessary?

(A) 250 mL
(B) 450 mL
(C) 650 mL
(D) 950 mL
(E) 1100 mL

1244. The infant with hypothyroidism presents many difficulties for the anesthesiologist, including

(A) hyperthermia
(B) hypoventilation
(C) resistance to opioids
(D) small mouth and tongue
(E) hyperkinetic myocardium

1245. A 16-year-old patient with Down syndrome is admitted for dental extractions. In the preoperative preparation

(A) atropine should be avoided
(B) opioids should be avoided
(C) neck mobility should be documented
(D) heavy sedation is required
(E) anticonvulsants should be withheld

1246. A 5-year-old boy is admitted with an open eye secondary to severe globe laceration. He had eaten 1 hour before his accident. General anesthesia is required for the repair. The intubation should be accomplished

(A) by an awake intubation
(B) after injection of 100 mg of succinylcholine
(C) after rocuronium followed by succinylcholine
(D) after vecuronium administration
(E) after deep halothane by inhalation induction

1247. The best method for prevention of the oculocardiac reflex is

(A) preoperative atropine
(B) a retrobulbar block
(C) IV atropine during the procedure
(D) administration of vecuronium
(E) administration of neostigmine

1248. The efferent limb of the oculocardiac reflex is the

(A) ciliary nerve
(B) trigeminal nerve
(C) vagus nerve

(D) facial nerve
(E) ophthalmic nerve

1249. Maintenance fluid requirements for a 24-kg child are

(A) 24 mL/h
(B) 30 mL/h
(C) 54 mL/h
(D) 64 mL/h
(E) 72 mL/h

DIRECTIONS (Questions 1250 through 1288): For each of the items in this section, ONE or MORE of the numbered options is correct. Choose answer

(A) if only 1, 2, and 3 are correct
(B) if only 1 and 3 are correct
(C) if only 2 and 4 are correct
(D) if only 4 is correct
(E) if all are correct

1250. A 1-day-old child presents with coughing and choking at his first feed. Following investigation, a diagnosis of tracheoesophageal fistula (TEF) is made. Which of the following statements is (are) TRUE?

(1) Esophageal atresia is associated with TEF in 10% of cases.
(2) Air leak through the fistula is minimized with paralysis.
(3) Postoperative intubation is necessary to protect the airway from aspiration.
(4) Sump suction is maintained in the esophageal pouch to lessen the risk of aspiration.

1251. Which of the following statements is (are) TRUE of fetal hemoglobin?

(1) It is comprised of two alpha chains and a delta chain.
(2) It has an oxygen dissociation curve shifted to the right relative to that of adult hemoglobin.
(3) The transfer of oxygen from mother to fetus is facilitated by the high affinity of fetal hemoglobin for 2,3-DPG.
(4) The high affinity of fetal hemoglobin for oxygen impairs oxygen delivery to the tissues.

1252. Congenital lobar emphysema

(1) most commonly affects the left upper lobe
(2) usually presents with cardiovascular collapse due to mediastinal shift
(3) coexists with congenital heart disease in about 15% of cases
(4) should be treated with assisted ventilation as soon as possible in order to improve gas exchange

1253. When the amniotic fluid is heavily meconium stained, which of the following statements is (are) TRUE?

(1) Fifteen percent of neonates develop respiratory difficulties in the first few days of life.
(2) Ten percent of neonates have radiographic evidence of pneumothorax or pneumomediastinum.
(3) Meconium is best removed by suction via an endotracheal tube.
(4) Absence of meconium in the mouth and pharynx precludes the presence of meconium in the trachea.

1254. Concerning renal function in the neonate,

(1) maturation of renal function is more rapid in premature infants
(2) the glomerular filtration rate of the neonate is less than 20% of the adult value
(3) glomerular filtration rate does not affect the neonate's ability to handle free water
(4) glomerular filtration rate reaches the adult value by about 1 year

1255. Concerning Apgar scores,

(1) the maximum Apgar score is 9
(2) Apgar scores are conventionally done at 1 and 10 minutes after birth
(3) an infant who has cyanosed extremities, a heart rate of 105, poor respiratory effort, flexes limbs, and grimaces to insertion of a nasal catheter, has an Apgar score of 5
(4) Apgar scores of neonates born to women who smoke are generally lower than those of neonates born to mothers who do not

1256. For mechanical ventilation of the neonate

(1) a 3.0-mm endotracheal tube is a suitable first choice for a neonate of 2.0 kg
(2) pressures of 30 cm H_2O are usually required for adequate IPPV
(3) PEEP should be maintained at 1 to 3 mmHg
(4) positive pressure ventilation should be conducted at a rate of about 20 breaths per minute

1257. The cardiac output of the neonate

(1) may increase significantly by increases in stroke volume
(2) is very sensitive to changes in afterload
(3) is relatively insensitive to volume loading
(4) is reflected by a leftward displacement of the cardiac function curve as compared to the adult

1258. Which of the following statements is (are) TRUE of spinal anesthesia in the neonate?

(1) It is suitable as the sole technique of anesthesia for procedures lasting 2 hours or more.
(2) The apex of the conus medullaris is usually at L2 to L3.
(3) Epinephrine should never be added to local anesthetics.
(4) Tetracaine 0.4 mg/kg is a suitable dose for subarachnoid block.

1259. A 3-year-old child is brought to the emergency room at 1 a.m. She was put to bed in apparently good health, but awoke 4 hours later crying and having difficulty breathing. Physical examination reveals that the child is flushed, drooling, sitting upright and has severe inspiratory stridor. Which of the following statements is (are) TRUE?

(1) The most likely diagnosis is acute laryngotracheobronchitis.
(2) A possible diagnosis is inhaled foreign body.
(3) Rectal temperature should be checked.
(4) The child should be taken straight to the operating room for intubation/emergency tracheostomy.

1260. Neonates have MAC values for volatile agents than adults because of

(1) an immature nervous system
(2) an immature blood-brain barrier
(3) elevated progesterone levels
(4) elevated blood levels of β-endorphins

1261. Omphalocele

(1) results from a defect in the development of the anterior abdominal wall
(2) is a congenital defect originating in the first trimester of pregnancy
(3) is usually associated with infection and loss of extracellular fluid
(4) is associated with a high incidence of congenital abnormalities

1262. Compared with adults, the rate of oxyhemoglobin desaturation in neonates and infants when apneic is due to

(1) high heart rate
(2) high metabolic rate
(3) small endotracheal tube
(4) small functional residual capacity (FRC)

1263. A 2-year-old child presents for repair of a ventricular septal defect. Preoperative evaluation reveals congestive heart failure, failure to thrive, and a pulmonary: systemic flow ratio greater than 2:1. Which of the following statements is (are) TRUE?

(1) Halothane in 100% oxygen is the most appropriate induction technique.
(2) An FIO_2 of 1.0 is used during induction of anesthesia to prevent rapid decreases in pulmonary vascular resistance.
(3) If preoperative right ventricular pressures are similar to systemic pressures, reversal of shunt flow is unlikely during induction of anesthesia.
(4) If shunt reversal occurs, with ensuing cyanosis, treatment consists of 100% oxygen and α-adrenergic agonists.

1264. In the average full-term neonate

(1) total body water (TBW) constitutes 75% of body weight
(2) ECF represents 40% of body weight
(3) adipose tissue represents 15 to 20% of body weight
(4) intracellular fluid (ICF) constitutes a greater proportion of TBW than ECF

1265. The infant's heart undergoes structural, enzymatic, and electrophysiologic changes from birth through early childhood. Which of the following statements is (are) correct concerning the infant's cardiovascular system?

(1) At birth, the left ventricle is only slightly heavier than the right ventricle.
(2) At birth, approximately 30% of the myocardium is composed of contractile elements.
(3) The electrocardiogram shows a left-to-right shift of the QRS forces in the precordial leads during infancy.
(4) Purkinje fiber action potentials are significantly shorter.

1266. A 5-year-old boy presents for elective outpatient repair of an inguinal hernia. His parents inform you that he has a runny nose with yellow nasal discharge and a mild occasional wet cough. They deny any fevers. You inform the parents that you will

(1) proceed with surgery today
(2) start on antibiotics
(3) admit to the floor postoperatively for overnight monitoring
(4) postpone surgery for 2 weeks

1267. Pulmonary vascular resistance in newborns decreases because of

(1) closure of the ductus arteriosus
(2) lung expansion
(3) decreased pH
(4) improved oxygenation

1268. Drugs are handled differently by neonates because of differences in

(1) absorption
(2) TBW
(3) albumin levels
(4) metabolism

1269. Less muscle relaxant is needed by a neonate because

(1) musculature is poorly developed
(2) muscle mass is less
(3) the myoneural junction is not well developed
(4) TBW is greater

1270. The Mapleson D system

(1) is useful only for children over 20 kg
(2) can be used for spontaneous or controlled ventilation
(3) provides moist gas flows
(4) has a pop-off valve at the end of the expiratory tube

1271. The administration of sodium bicarbonate to a newborn infant with an Apgar score of 2 at 2 minutes

(1) will increase the PaO_2
(2) is contraindicated
(3) may cause hepatic necrosis if given through a venous catheter whose tip is in the liver
(4) should be given intramuscularly

1272. The work of breathing in a newborn infant

(1) involves overcoming elastic forces
(2) involves overcoming resistive forces
(3) is least when the infant breathes at 35 times per minute
(4) is less if the child is breathing very rapidly

1273. The systolic blood pressure in infants

(1) is less than that of adults
(2) is of less importance than in an adult
(3) is between 60 and 70 mmHg at birth
(4) is equal to adult pressure by age 2

1274. The metabolic activity of a child

(1) is lower than that of an adult
(2) is highest in the first 2 years of life
(3) is lowered by a febrile illness
(4) rises with onset of puberty

1275. The airway of the newborn, as compared to the adult, has

(1) a more cephalad-placed larynx
(2) an epiglottis that has the same shape in each
(3) vocal cords slanted upward and backward
(4) the most narrow area at the rima glottidis

1276. A 4-year-old child develops postintubation laryngeal edema after a tonsillectomy. The treatment for this may include

(1) inhalation of mist
(2) steroids
(3) racemic epinephrine
(4) sedation

1277. The child with epiglottitis typically

(1) lies on the right side
(2) has a sudden onset of symptoms
(3) has a hacking cough
(4) is febrile

1278. The typical child with laryngotracheobronchitis has

(1) a gradual onset of symptoms
(2) a barking cough
(3) a low-grade fever
(4) a subglottic obstruction

1279. Postintubation laryngeal edema in a child

(1) is most common in the newborn period
(2) can be decreased by the use of an appropriate lubricant
(3) is noted particularly on exhalation
(4) should be treated with oxygen and mist

1280. The premature newborn infant who requires surgery

(1) needs no anesthesia, since pain fibers are not developed
(2) will not react to pain
(3) should have an anesthetic course of oxygen and relaxant
(4) should be evaluated for anesthesia using the same criteria as any patient

1281. Adolescents may be involved in substance abuse. The patient who presents with a history of solvent abuse may exhibit which of the following physiologic derangements?

(1) hepatic dysfunction
(2) peripheral neuropathy
(3) renal dysfunction
(4) neutropenia

1282. If anesthesia is needed for the patient described in Question 1281, acceptable anesthetic techniques include

(1) halothane by inhalation
(2) sevoflurane by inhalation
(3) spinal anesthesia with tetracaine
(4) isoflurane by inhalation

1283. The Pierre Robin syndrome presents anesthetic difficulty because of the prevalence of

(1) heart problems
(2) lung problems

(3) renal problems
(4) intubation problems

1284. The oculocardiac reflex is associated with

(1) blood pressure changes
(2) bradycardia
(3) traction
(4) cardiac arrhythmias

1285. General anesthesia is indicated for complex radiologic procedures when the patient

(1) is very young
(2) is mentally handicapped
(3) has involuntary movements
(4) is febrile

1286. A child who undergoes outpatient anesthesia should

(1) not be intubated
(2) be watched for at least 1 hour postoperatively
(3) be kept NPO for at least 6 hours after surgery
(4) be admitted if croup develops

1287. Retinopathy of prematurity

(1) is related to incomplete vascularization of the retina at birth
(2) is directly related to the FIO_2
(3) is related to the oxygen tension of the retinal artery blood
(4) occurs only after exposure to hyperoxemia for at least 24 hours

1288. Which of the following maneuvers is (are) recommended in the prevention of subglottic edema in children?

(1) gentle intubation
(2) use of an anesthetic cream on the tube
(3) use of a sterile endotracheal tube
(4) use of steroid cream on the tube

Answers and Explanations

1217. (A) The 1-minute Apgar score correlates well with acidosis and survival, and the 5-minute score may be predictive of neurologic outcome. Infants very frequently have a bluish tinge at the moment of birth, but at 60 seconds of age, most are entirely pink except for their hands and feet, which remain blue. The normal heart rate of a term neonate is 120 to 160 beats per minute. Dr. Virginia Apgar first introduced her scoring system in 1953. *(1:23, 1150, 1160, 1162; 2:12–16; 4:1994–95; 5:2352)*

1218. (D) About 50 to 150 mL/kg/day of fluid is produced by the fetal lung at term (an ultrafiltrate of plasma). This fluid is expelled into the mouth where it is either swallowed or released into the amniotic fluid. Normally the lung fluid contains no amniotic fluid, but if the depth of fetal breathing increases (such as during stress) some may be drawn in. During the normal birth process the muscles of the vagina and pelvic floor squeeze the fetal chest and expel about two-thirds of the lung fluid, the remainder being removed by capillaries, lymphatics, and breathing. If this vaginal squeeze does not take place, infants may have excess lung water and respiratory difficulties. This may occur after cesarean section (especially if there has been no previous labor) and in small preterm infants. *(2:11–14; 4:430; 5:2345)*

1219. (D) At birth, the pulmonary vascular resistance decreases dramatically in response to lung expansion, increased pH, and a rise in alveolar oxygen tension. This reduces the pulmonary artery pressure and increases pulmonary blood flow. The increased amount of blood returning to the left atrium raises the pressures in the left atrium closing the foramen ovale. The ductus arteriosus closes primarily in response to increased oxygen tension. This is functional closure, but complete mechanical closure requires adequate arterial muscle and occurs after 10 days or more. *(1:1171–73; 2:353–55; 4:2017; 5:2347)*

1220. (A) Total spinal anesthesia, produced either with a primary spinal technique or secondary to an attempted epidural anesthetic, presents as respiratory insufficiency rather than as hypotension. The reason for this is the lack of sympathetic tone in neonates. The first indication of trouble is falling oxygen saturation rather than a falling blood pressure. *(1:1181; 2:646–47; 5:1731)*

1221. (A) Bupivacaine 0.25% is the most suitable choice of agent because it has the longest duration of action and may therefore give some postoperative pain relief. It may be used in doses up to 1 mL/kg, but a volume of 8 mL is quite sufficient. Epinephrine-containing solutions should never be used for this type of block because of the risk of ischemia to the penis. *(2:639–41; 5:1733)*

1222. (B) *(2:15–16; 5:2834)*

1223. (A) At the age of 6 months, a normal infant might be expected to have a heart rate of 120 ± 20, blood pressure 90/60 ± 30/10, cardiac stroke volume of 7.5 ± 2 mL/beat, O_2 consumption of 5 ± 0.9 mL/kg, and hemoglobin concentration of 11.5 g/dL. *(2:15–16; 5:2834)*

1224. (E) Proceed with surgery. Most institutions allow a 2-hour fast for clear liquids, 4-hour fast for breast milk, 6-hour fast for formula or light meal, and 8-hour fast for solids. *(2:41–42; 5:2381)*

1225. (C) At birth, pulmonary vascular resistance declines rapidly in response to lung expansion and exposure of pulmonary resistance vessels to alveolar oxygen. At the same time, systemic vascular resistance increases. Pulmonary blood flow and venous return to the left atrium increase, and closure of the foramen ovale occurs when mean left atrial pressure exceeds mean right atrial pressure. Functional closure of the ductus arteriosus occurs in response to a rise in arterial oxygen saturation in the first 24 hours after birth. Anatomic closure of both the ductus arteriosus and the foramen ovale occurs much later. *(1:1171–73; 2:354–56; 4:1994, 2017; 5:2369–70)*

1226. (D) In the newborn infant, the glucose level is normally lower than in an older child, the value normally being 30 mg/dL. Resuscitation will require intubation, oxygen, heat, and bicarbonate. *(2:19, 223; 4:2017–19; 5:2361)*

1227. (C) When a small but persistent change in oxygen saturation is noted, the position of the endotracheal tube must be reassessed. The other causes noted above are also possibilities in this situation but are less likely. *(2:93; 4:2065, 2084; 5:2384–85)*

1228. (C) The glucose concentration is lower in the newborn than the adult and lower in the premature than in the full-term infant. Full-term infants are considered as being hypoglycemic if the serum glucose is less than 30 mg/dL and premature infants as hypoglycemic if serum glucose is less than 20 mg/dL. *(2:223; 4:434)*

1229. (A) The metabolism of brown fat is a heat-producing mechanism. All of the other options are common methods of losing heat in an operating room. The temperature must be monitored and methods initiated to prevent heat loss. A warm blanket placed under the infant minimizes heat loss by conduction. *(2:612–15; 4:2101; 5:1592)*

1230. (A) Urinary output for a child should be 1 mL/kg/h. This should be monitored by a urinary catheter if appropriate for the procedure. *(2:528; 4:2102–03; 5:2836)*

1231. (B) Tachycardia is an early sign of malignant hyperthermia. If one waits for the development of fever, hot skin, and a hot absorber, the syndrome has already progressed significantly. *(2:620; 4:2426–50; 5:1169–86)*

1232. (C) The major factor that causes closure of the PDA is the PaO_2. Hypoxia in the early newborn period may delay closure. The level of $PaCO_2$ is not a factor. *(2:10–11, 296–97, 355–56; 4:1709–10; 5:2007)*

1233. (C) The normal dead space in an infant is 2 mL/kg. This has always been one of the problems inherent in the equipment used for pediatrics. If the dead space of the equipment exceeds the infant's dead space, providing effective ventilation is more difficult. *(1:1177; 2:12)*

1234. (D) Heated humidifiers provide fluid to the child, and this must be considered in the fluid plan for the procedure. Normal plans for replacement include the insensible losses from the respiratory tract. This loss will be greatly reduced with humidification. The fluid infusion should be reduced by the amount of insensible loss. Heated nebulizers can produce up to 100% humidity at body temperature. *(2:223, 629; 5:1585)*

1235. (B) Most operating rooms are kept at a low temperature for the comfort of the staff. The infant loses heat by conduction to the cold table, by convection from the cold air, by evaporation, and by radiation. Once the child is covered, some of the heat loss decreases, but if a body cavity is opened, more surface area is available for heat loss, particularly evaporative. *(2:116–17, 613–15; 4:2064; 5:1592)*

1236. (C) The spinal cord extends to the L3 level in the newborn. When performing spinal anesthesia, this must be kept in mind. In the adult, the spinal cord ends at the level of the upper edge of L2. *(2:642; 5:1720)*

1237. (C) The typical 2-year-old child should be intubated with an endotracheal tube of 4.5 to 5-mm internal diameter. *(2:719–20; 4:2084; 5:2384)*

1238. (B) At birth, the infant has about 24 million primitive alveoli for air exchange. An adult has about 300 million (this number is usually reached by the age of 8 years). If a disease intervenes, complete development may not occur. *(2:10; 5:2369)*

1239. (D) The child who has an incarcerated hernia is going to need surgery to reduce the hernia lest strangulation occur. An upper respiratory infection is not an ideal situation, but since the procedure must be done, it should be done with careful attention to detail, including monitoring. *(2:43–44; 4:2106–07; 5:2381–82)*

1240. (B) The child with a diaphragmatic hernia classically has a scaphoid abdomen. There is decreased movement of the hemithorax and a mediastinal shift. Bowel sounds may be heard in the chest, and bowel may be seen in the thorax on x-ray. *(1:1183–85; 2:304–05; 5:2396)*

1241. (A) Studies have demonstrated that patients transfused to hemoglobin values of 10 mg/dL have lower incidence of sickle cell-related complications. Many centers are now transfusing (direct, partial exchange, or exchange) to lower the percentage of sickle cells present. The preparation should strive for good hydration, good pulmonary care, and during the procedure, avoidance of hypothermia and stasis. *(2:44–45; 4:466–67, 5:1112)*

1242. (D) Former premature infants are at risk for postanesthesia apnea if they are younger than 55 to 60 weeks PCA. The incidence of apnea is inversely proportional to both gestational age and PCA. In addition, the risk of postanesthesia apnea increases if the infant has a history of apnea or if anemia is present. The incidence of apnea is quite low in ages greater than 45- or 50-week PCA, however, most institutions use 55 or 60 weeks PCA as a cutoff for day surgery in this population. *(2:45–47, 55–56; 5:2397–98, 2593)*

1243. (D) To calculate the allowable blood loss (ABL), one must determine the patients' blood volume using 70-mL/kg body weight. The most common formula is *(2:236; 5:2390)*

$$\text{ABL} = \text{blood volume} \times \frac{\text{Hct}_I - \text{Hct}_T}{\text{Hct}_{\text{mean}}}$$

where Hct_I = initial hematocrit, Hct_T = target hematocrit, Hct_{mean} = mean of Hct_I and Hct_T. Thus

$$\text{ABL} = (35 \text{ kg} \times 70 \text{ mL/kg}) \times 70 \text{ mL/kg} \times \frac{12}{30}$$

$$= 2450 \text{ mL} \times \frac{12}{30} = 980 \text{ mL}$$

1244. (B) Hypoventilation, hypothermia, and difficult intubation are potential problems for the anesthesiologist in treating the baby with myxedema. The children are sensitive to opioids. The mouth is of normal size, and the tongue may appear large. The heart is not hyperkinetic as in hyperthyroidism. *(2:117; 4:428, 461; 5:1048)*

1245. (C) About 13 to 18% of patients with Down syndrome have an atlantoaxial subluxation. Sensitivity to atropine was thought to be present in these children, but that was later disproved. The children can receive sedatives and opioids if needed. Anticonvulsants should be given as usual. *(2:120, 279, 362; 4:427–28, 461; 5:1099)*

1246. (C) This is a difficult management problem. Succinylcholine or an awake intubation will both cause an increase in intraocular pressure, possibly causing expulsion of the contents of the globe. The use of succinylcholine preceded by a defasciculating dose of rocuronium will provide good intubating conditions in the least time and has not been associated with cases of expulsion of the globe contents. The practitioner should be aware of the side effects and implications associated with the use of succinylcholine in children, especially young boys in whom hyperkalemic cardiac arrests have been reported. The use of this drug in pediatric patients has largely been restricted to emergencies. An inhalational induction may be required in the extremely uncooperative child; however, the risk of aspiration may be increased compared with a rapid-sequence induction. *(2:1299–202, 489–90; 4:2192; 5:481–82, 489–91, 2378–79, 2534)*

1247. (C) Atropine is recommended by most anesthesiologists, but there are those who believe that the administration of atropine may lead to an increased incidence of arrhythmias and that prevention is not warranted. IV atropine is more effective than intramuscular or oral atropine in preventing this reflex. It is important to inform the surgeon as soon as bradycardia is seen. The reflex does fatigue early and usually is not persistent. *(2:480; 4:2184, 5:739)*

1248. (C) The oculocardiac reflex is a trigeminovagal reflex, with the efferent nerve being the vagus. *(2:480; 4:2184; 5:739)*

1249. (D) Maintenance fluids are calculated as follows: for the first 10 kg of body weight, the hourly fluid is 4 mL/kg. The second 20 kg of body weight has an hourly fluid need of 2 mL/kg. For each kg of body weight above 20 kg, the maintenance fluid need is 1 mL/kg. Thus for the 24-kg child in this question, the maintenance fluid requirement is 64 mL/h (40 mL + 20 mL + 4 mL). *(2:222–23; 5:2388)*

1250. (D) In 90% of cases, esophageal atresia is associated with TEF, but esophageal atresia may be an isolated occurrence. Careful positioning of the endotracheal tube with the tip just above the carina but below the fistula may help to minimize gastric insufflation, as will a gastrostomy. Positive pressure ventilation with or without paralysis is likely to increase the air leak through the fistula but is often necessary. Early extubation after surgery is desirable because it prevents prolonged pressure of the endotracheal tube on the suture line. Constant sump suction from the upper esophageal pouch decreases the accumulation of saliva and reduces the potential for aspiration. *(1:1187–88; 2:302–04; 4:2069–71; 5:2396)*

1251. (D) Fetal hemoglobin has two alpha chains and two gamma chains. Fetal hemoglobin has a lower affinity for 2,3-DPG than adult hemoglobin and an oxygen dissociation curve shifted relatively to the left. This means that fetal hemoglobin has a higher affinity for oxygen than adult hemoglobin. Oxygen transfer from mother to fetus is therefore facilitated, but oxygen release at the tissues is impaired. *(2:20; 4:432–33; 5:2843)*

1252. (B) Congenital lobar emphysema most commonly affects the left upper lobe but may involve the whole lung. It usually presents with progressive respiratory failure but mediastinal shift and cardiovascular collapse may occur. Anesthetic care is aimed at minimizing expansion of the emphysema; spontaneous respiration should be maintained if possible until the thorax is open, low peak inspiratory pressures should be used if assisted ventilation is required, and nitrous oxide should be avoided. *(2:305)*

1253. (A) If the airway is not suctioned adequately before or shortly after the onset of breathing, meconium in the airways will move distally into the small airways and alveoli, causing respiratory difficulties and intrathoracic gas leaks. Fortunately, these gas leaks are usually small, but large pneumothoraces can occur. Meconium is best removed by endotracheal intubation and suction using a specially designed device. It is also useful to suction the stomach to remove meconium, which might be regurgitated and aspirated later. *(1:1173; 2:19; 4:2019; 5:2354–55)*

1254. (C) Renal function is markedly diminished in the neonate because of low perfusion pressure and immature glomerular filtration and tubular function. The ability to handle antibiotics (largely excreted by glomerular filtration) is therefore reduced. Glomerular filtration rate is about 20 mL/min/1.73 m^2 at birth (adult value ≈ 120) but develops rapidly thereafter reaching near complete maturation at 5 months of age and complete maturation at 1 to 2 years; however, maturation may be delayed in premature infants. *(1:1174; 2:16–17; 4:426; 5:2370)*

1255. (D) The Apgar scoring system involves five variables, which are evaluated individually and scored from 0 to 2 (heart rate, respiratory effort, color, reflex irritability, and muscle tone). Thus, the maximum score is 10. Scores are conventionally done at 1 and 5 minutes after birth. A heart rate of 105 scores 2, but all the other variables score 1 each, giving a total score of 6 in this patient. *(1:1162; 4:2018; 5:2352–53)*

1256. (B) Endotracheal intubation should be performed with a tube of an appropriate size for the baby. Positive pressure ventilation should normally be commenced at a rate of 30 to 60 breaths per minute and a PEEP of 1 to 3 mmHg maintained. Most asphyxiated neonates do not have lung disease and rarely require inflation pressures greater than 25 cm H_2O. *(1:613; 2:719–24; 4:2018; 5:2384)*

1257. (C) The myocardial structure of the heart, particularly the volume of cellular mass devoted to contractility, is significantly less in the neonate than in the adult. This difference and others produce a leftward displacement of the cardiac function curve and less compliant ventricles. This accounts for the tendency toward biventricular failure, sensitivity to volume loading, poor tolerance to increased afterload, and rate-dependent cardiac output. *(1:1176–77; 2:356–57; 5:2368–69)*

1258. (D) The duration of any spinal anesthetic drug is shorter in the neonate than the adult, and for this reason epinephrine is usually added; however, even with tetracaine 0.4 to 0.5 mg/kg with epinephrine, the block usually lasts only about 90 minutes, therefore spinals are unsuitable as the sole technique for longer procedures. The spinal cord often ends as low as L3 to L4 in neonates so blocks should be performed at L4 to L5 or L5 to S1 in patients of this age. *(1:1181; 2:644–47; 5:1719–21, 1728)*

1259. (C) The most likely diagnosis is acute epiglottitis, which characteristically has a swift onset without a preceding URI. The differential diagnosis includes acute laryngotracheobronchitis (slower onset over days, not hours) and inhaled foreign body. This is a true pediatric emergency, and securing the airway is the first priority. The child is quite toxic, unable to swallow her own saliva, and has severe stridor. The child should not undergo any procedure which might provoke crying and complete obstruction of the airway (such as inserting a thermometer into the rectum or insertion of an IV line) until facilities are ready for emergency intubation in an OR setting, and tracheostomy should this fail. *(2:319–20; 4:2103–05)*

1260. (E) The neonate has an immature central nervous system with attenuated responses to nociceptive cutaneous stimuli. These responses mature in the first few months of an infant's life, along with an increase in MAC. Progesterone has been shown to reduce the MAC of the pregnant mother. The newborn infant has elevated progesterone levels, and animal studies have suggested that these are related to reduced MAC values. High levels of β-endorphins are present in the first few days of postnatal life and may cross the immature blood-brain barrier of the neonate, thus increasing the pain threshold and reducing MAC values. *(1:1198–99; 2:134–41; 4:2088–89; 5:2372–73)*

1261. (C) Omphalocele and gastroschisis should not be confused. Between the 5th and 10th weeks of fetal life, abdominal contents are extruded into the extraembryonic coelom. Failure of part of these contents to return to the abdomen at about week 10 results in an omphalocele, which is covered by a membrane—the amnion. This protects the abdominal contents from infection and loss of ECF. In contrast, gastroschisis develops later in fetal life. It results from interruption of the omphalomesenteric artery at the base of the umbilical cord. The gut herniates out through this tissue defect and the degree of herniation may be slight, or almost all of the abdominal contents may be outside the peritoneal cavity. The intestines and viscera are not covered by any membrane and so are highly susceptible to infection and fluid loss. There is a high incidence of other congenital abnormalities with omphalocele but not with gastroschisis. *(1:1185–87; 2:309–10; 4:2067–68; 5:2395–96)*

1262. (C) The metabolic rate in neonates and infants is nearly double that of adults. In combination with a small dynamic FRC, they desaturate extremely quickly when apneic compared with adults. *(2:12–13; 5:2369)*

1263. (D) The preoperative evaluation suggests that this child's VSD is large and that there may be problems encountered on induction of anesthesia. These patients without significant systemic hypotension can tolerate only low concentrations of potent inhalational anesthetic agents. Therefore, a halothane induction is probably less safe than an induction with IV agents such as ketamine and fentanyl which maintain systemic arterial pressure. It may also be important to use an FIO_2 of less than 1.0 in order to avoid a rapid decrease in pulmonary vascular resistance and subsequently increased left-to-right shunting. If right ventricular pressures are close to systemic pressures, relatively mild systemic hypotension (such as during induction of anesthesia) may cause the shunt to reverse (now becoming right to left) with systemic desaturation and further myocardial dysfunction. If shunt reversal does occur, 100% oxygen and use of an α-adrenergic agonist such as phenylephrine to increase systemic vascular resistance usually results in resumption of left-to-right shunting. *(2:404–05; 4:1699–1730; 5:2010–11)*

1264. (A) TBW accounts for ≈ 75% of weight at birth, and falls to ≈ 65% of weight by the age of 1 year. TBW is divided into two compartments: ICF and ECF. In the neonate, ECF constitutes ≈ 53% of TBW and ≈ 40% of weight; ICF constitutes ≈ 47% of TBW and ≈ 35% of weight. Adipose tissue affects water content. It accounts for roughly 16% of a newborn's weight and ≈ 23% of weight in a 1-year-old. *(2:218–22; 4:2059)*

1265. (C) At birth, the left and right ventricles are roughly equal in size and wall thickness, but the right ventricle is slightly heavier. The relative growth of the left ventricle and regression of the right ventricle causes a right-to-left shift of the ECG in infancy. The proportion of muscle mass composed of contractile elements is 30% in the newborn compared with 60% in the adult. The resting potential of Purkinje fibers is less negative in the neonate, and repolarization is faster. The Purkinje action potentials are significantly shorter in neonates than adults. *(1:1177; 2:356–57; 5:2368–69)*

1266. (D) The child with a URI is a dilemma. It has consistently been observed that the likelihood of laryngospasm, bronchospasm, and desaturation is increased when a patient has a mild URI, especially if an endotracheal tube is used. In the case of an acute URI the risk of the above may be greater. Most institutions will postpone elective surgery if the patient has signs and symptoms of an acute URI, especially if nasal discharge is purulent and a productive cough is present. The duration of postponing is another dilemma, however, 2 to 3 weeks is the usual time to wait for rescheduling surgery. *(2:447; 5:2381–82)*

1267. (C) Pulmonary vascular resistance decreases with lung expansion and improved oxygenation. A decrease in pH will increase pulmonary vascular resistance. The closure of the ductus does not affect vascular resistance. *(2:354–55; 4:2017; 5:2369–70)*

1268. (E) Infants differ from adults in drug absorption. The infant stomach has a higher pH, leading to increased absorption of some drugs. TBW is greater in the infant, making distribution different. The albumin levels are lower, leading to more unbound drug. The immaturity of the liver leads to decreased drug metabolism of some drugs. *(2:122–32; 4:2058–59; 5:2371–72)*

1269. (A) Lower doses of muscle relaxant are needed for the neonate because of poorly developed musculature, smaller muscle mass, and an immature myoneural junction. The greater TBW causes a dilutional effect, which leads to a requirement for larger doses. *(2:122–23, 208; 4:2061–62; 5:2378–79)*

1270. (C) The Mapleson D system is most useful in the small child under 10 kg. The flows are dry. The circuit can be used for spontaneous or controlled ventilation. The pop-off valve is placed near the end of the expiratory tube. *(1:1200; 2:721–22; 4:2079–82; 5:2393)*

1271. (B) Administration of sodium bicarbonate to a newborn may increase the PaO_2 through a decrease in pulmonary vascular resistance if the pH is increased above 7.10 to 7.20. Sodium bicarbonate is very hypertonic; care must be taken to avoid liver damage from injection into the liver. The drug must be given intravenously. *(2:229; 4:2018; 5:2357–58)*

1272. (A) The work of breathing involves overcoming both elastic and resistive forces. As the child breathes very fast, the work is increased, since turbulent flow is more prevalent. The optimal rate is 35 breaths per minute. *(1:1176–77; 2:85–87)*

1273. (B) The systolic blood pressure in an infant is lower than that in an adult, but it is of equal importance. At birth, the systolic pressure is between 60 and 70 mmHg and rises slowly. It remains relatively low until age 6, when it starts to increase and approach adult levels. *(2:16; 5:2834)*

1274. (C) Metabolic activity rises with puberty but is highest in the first 2 years of life. The metabolic rate is higher in a child than in an adult. Febrile illnesses cause a rise in metabolic rate. *(2:87; 4:433–34; 5:2385, 2394, 2834)*

1275. (B) The infantile airway has a more cephalad-placed larynx and vocal cords that are slanted. The epiglottis is long and narrow. The narrowest portion of the infant airway is at the level of the cricoid cartilage. *(2:81–82; 4:2084–85; 5:1646–47, 2353–54, 2385)*

1276. (A) Posttraumatic or postintubation croup is treated with mist inhalations, steroids, and racemic epinephrine. Steroids are not immediately effective, and some question their use, but they are usually given. Sedation of the child is based on the decision of whether or not the child's anxiety may be contributing to the problem. Sedation should not be a routine order. *(2:93; 4:2103–05; 5:2384)*

1277. (C) The child with epiglottitis typically is sitting up to allow better handling of the secretions. Since the child has pain on swallowing, drooling is common. The onset is sudden, and the child is febrile. *(2:319–20; 4:2105; 5:2352, 2384)*

1278. (E) The child with laryngotracheobronchitis (croup) has a gradual onset of symptoms, has a barking cough, a low-grade (if any) fever, and a subglottic obstruction. *(2:320–21; 4:2105; 5:2352, 2384)*

1279. (D) Postintubation croup is uncommon in the newborn unless the endotracheal tube is in place for a long period of time. This may result in subglottic stenosis. Lubricants are of no help in reducing croup. Oxygen and mist are appropriate therapy. Croup is manifest on inspiration. *(1:1202–03; 2:93, 705; 4:2105; 5:2384)*

1280. (D) The newborn infant requires anesthesia and should receive anesthesia. Studies have demonstrated that the newborn reacts to pain. If the child has an unstable cardiovascular system and cannot tolerate anesthesia, resuscitation should proceed, and the anesthesia should be resumed after stability is restored. This does not differ from the treatment of adults. *(1:1178; 2:403; 4:2058)*

1281. (A) Patients who have a history of solvent abuse may have a peripheral neuropathy (common after prolonged exposure to hexane or methyl ethyl ketone). Halogenated hydrocarbons may cause acute or chronic hepatic and renal disease, depending on the amount inhaled and the duration of the abuse. Neutropenia is unlikely. *(1:1195; 4:277–78; 5:751)*

1282. (D) A patient with hepatic dysfunction secondary to toxin exposure should probably not receive halothane, since it decreases hepatic blood flow more than other volatile anesthetics. Similarly, sevoflurane should probably be avoided in the patient with decreased renal function because it is metabolized to fluoride ion, which is nephrotoxic. A regional anesthetic is relatively contraindicated in a patient with a peripheral neuropathy. Isoflurane by inhalation is probably the best choice, since its administration results in the preservation of hepatic blood flow and in the formation of only minute quantities of fluoride. *(4:515, 355, 1145, 1919; 5:241–51)*

1283. (D) The Pierre Robin syndrome presents problems due to the recessed chin, making intubation very difficult. Tracheostomy may be necessary to avoid airway obstruction. There usually are no associated renal, cardiac, or lung problems. *(2:119; 4:429, 5:2386–87)*

1284. (E) The oculocardiac reflex is associated with blood pressure changes if the heart rate drops to low levels. Traction on the extraocular muscles is not the only cause. Pressure on the eye and retrobulbar blocks have caused the reflex. Cardiac arrhythmias other than bradycardia also are seen. In some reports, other arrhythmias are seen more frequently after atropine has been given to prevent the reflex. *(1:973–74; 2:480–81; 4:2184; 5:739, 2530)*

1285. (A) General anesthesia is usually required for children less than 8 years of age and for children with a mental handicap, to prevent movement during radiologic procedures such as CT or MRI scans and angiography. A fever is not a reason for general anesthesia. *(2:44)*

1286. (C) An outpatient child can be intubated if the procedure calls for intubation. The child should be given fluids as soon as a desire is expressed. If croup develops, the child should be observed for a period of time after treatment to make sure there are not going to be progressive symptoms. Admission may be necessary. *(2:63–64)*

1287. (B) There is controversy concerning the role of oxygen in retinopathy of prematurity. The retina is at risk until the vascularization is complete at about 44 weeks. It is not related to FIO_2 but is related to oxygen tension. It is also related to other factors besides oxygen tension. Retinopathy has been reported in full-term infants and in premature infants never given oxygen therapy. *(2:45,142–43; 4:2193; 5:2534–35)*

1288. (B) Subglottic edema may be prevented by the use of an appropriate-sized endotracheal tube and by a gentle intubation. The use of sterile, prepackaged, and disposable vinyl plastic tubes is recommended. The head should be kept still to prevent movement of the tube. The use of creams on the tubes has not been shown to be effective. *(1:1120, 1122, 1202–03; 2:93–94; 5:11650, 2723)*

CHAPTER 14

Respiratory Therapy and Critical Care

Questions

DIRECTIONS (Questions 1289 through 1321): Each of the numbered items or incomplete statements in this section is followed by answers or by completions of the statement. Select the ONE lettered answer or completion that is BEST in each case.

1289. A patient breathing oxygen at 5 L/min via a mask without a reservoir will have an FIO_2 of approximately

(A) 30%
(B) 40%
(C) 50%
(D) 60%
(E) 70%

1290. A patient with heparin-induced thrombocytopenia type II who presents with acute pulmonary embolus should receive the following anticoagulant medication:

(A) enoxaparin
(B) warfarin
(C) aspirin
(D) argatroban
(E) no anticoagulant is necessary

1291. Aerosol therapy is useful in providing all of the following EXCEPT

(A) improved bronchial hygiene
(B) humidification
(C) a means to deliver medication
(D) systemic medications
(E) expectoration promotion

1292. The following values are obtained from a patient receiving treatment in an intensive care unit:

Heart rate	110 bpm
Blood pressure	95/60 mmHg
Cardiac output	3.0 L/min
Arterial oxygen saturation	90%
Mixed venous oxygen saturation	70%
Arterial pH	7.36
PaO_2	60 mmHg
$PaCO_2$	40 mmHg
Hemoglobin	10 g/dL

Which of the following will result in the greatest oxygen delivery?

(A) use of an inotrope to double cardiac output
(B) transfusion of packed red blood cells to increase the hemoglobin concentration to 15 g/dL
(C) increase in FIO_2 to oxygenation saturation equal to 100% and PaO_2 equal to 150 mmHg
(D) increase in heart rate to 130 bpm
(E) hyperventilation to pH equal to 7.40 and $PaCO_2$ equal to 35 mmHg

1293. A partial rebreathing mask contains a small reservoir bag that

(A) requires high oxygen flow
(B) leads to increased carbon dioxide
(C) leads to decreased carbon dioxide
(D) provides a high FIO_2 at low oxygen flows
(E) has no effect on blood gases

1294. For the following clinical scenarios, noninvasive positive pressure ventilation should be considered as a possible intervention EXCEPT

(A) a 75-year-old woman with COPD presenting to the emergency room with tachypnea and wheezing
(B) an 80-year-old obtunded man presenting to the emergency room with intracerebral hemorrhage and hypoventilation
(C) a 40-year-old morbidly obese man with obstructive sleep apnea recovering from ankle surgery under general anesthesia
(D) a 20-year-old immunosuppressed woman who develops fever and pulmonary infiltrates following bone marrow transplantation
(E) a 65-year-old man with a nonischemic cardiomyopathy presenting to the emergency room with rales on chest examination after skipping his daily furosemide dose for 2 days

Question 1295 refers to Figure 14-1.

1295. In the figure, "A" indicates a metered-dose inhaler (MDI), and "B" indicates the location for placement of the patient's mouth. The purpose of the object labeled with the "X" is to

(A) measure the dose of drug delivered
(B) heat the incoming medication
(C) improve aerosolized delivery
(D) obstruct gas flow
(E) sterilize the inhaled gas

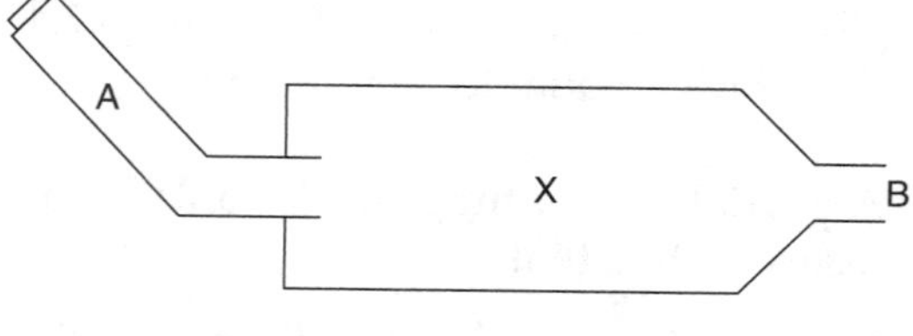

FIG. 14-1

1296. A technique most useful in treatment of bronchopleural fistula is

(A) spontaneous breathing with a mask
(B) assisted ventilation with positive pressure
(C) controlled ventilation with a mechanical ventilator
(D) positive end-expiratory pressure
(E) high-frequency jet ventilation

1297. All of the following statements regarding suctioning of the airway in the patient with an artificial airway are true EXCEPT it

(A) should be performed at regular and frequent intervals
(B) may lead to bradycardia
(C) may lead to hypoxemia
(D) may cause trauma to the mucosa
(E) may cause atelectasis

1298. Absorption atelectasis

(A) is clinically unimportant
(B) occurs primarily with low-flow oxygen therapy
(C) occurs when high partial pressures of nitrogen are present
(D) occurs more with early airway closure
(E) is never seen when denitrogenation has been used

1299. The pressure within a tracheal tube cuff that will interrupt mucosal blood flow is approximately

(A) 5 mmHg
(B) 10 mmHg
(C) 15 mmHg
(D) 20 mmHg
(E) 30 mmHg

1300. A patient given supplemental oxygen at 4 L/min via a nasal cannula will have an FIO_2 of approximately

(A) 24%
(B) 26%
(C) 30%
(D) 32%
(E) 36%

1301. The patient with a full stomach is no longer at risk for aspiration

(A) once fully relaxed with a nondepolarizing muscle relaxant
(B) after the stomach has been decompressed with a nasogastric tube
(C) after proper placement of a cuffed endotracheal tube in the trachea
(D) if he or she is gagging on the endotracheal tube prior to extubation
(E) none of the above

1302. Pulmonary aspiration is most reliably diagnosed by

(A) tachypnea
(B) rales on chest auscultation
(C) infiltrate on chest radiograph
(D) arterial hypoxemia
(E) visualization of gastric contents in the oropharynx

1303. Standard total parenteral nutrition solutions typically contain all of the following EXCEPT

(A) dextrose
(B) phosphate
(C) fat emulsion
(D) glutamine
(E) zinc

1304. In pressure support ventilation, inspiration ceases when

(A) the preset time is reached
(B) the preset tidal volume is delivered
(C) the preset pressure level is achieved
(D) the flow diminishes to 25% of peak value
(E) vital capacity is reached

1305. Application of appropriate PEEP in the patient with ARDS achieves all of the following EXCEPT

(A) decreased total lung water
(B) improved lung compliance
(C) decreased intrapulmonary shunting
(D) diminished alveolar overdistention
(E) improved arterial oxygenation

1306. An elderly patient with unresectable cholangiocarcinoma and obstructive jaundice is admitted to the critical care unit following percutaneous biliary stenting. The temperature is 39.5°C. The hemodynamic values are heart rate 120 bpm, blood pressure 80/40 mmHg, central venous pressure 1 mmHg, pulmonary artery pressure 20/5 mmHg, and pulmonary artery occlusion pressure 4 mmHg. The cardiac index is 2.0 L/min/m^2. The intervention that should be performed FIRST is

(A) calculation of systemic vascular resistance
(B) volume resuscitation
(C) administration of dopamine
(D) administration of phenylephrine
(E) administration of broad-spectrum antibiotics

1307. Possible associated factors with prolonged weakness following administration of nondepolarizing muscle relaxants in the critically ill patient include all of the following EXCEPT

(A) immobilization
(B) corticosteroid treatment
(C) inadequate analgesia
(D) vecuronium
(E) neuropathy

1308. A 60-year-old patient with pulmonary hypertension develops massive hemoptysis immediately following PA catheter balloon inflation. The appropriate intervention is

(A) immediate transfusion of type O negative packed red blood cells
(B) reversal of heparin anticoagulation with protamine
(C) immediate transfer to the operating room for surgical repair
(D) administration of lorazepam intravenously to calm the patient
(E) endobronchial intubation

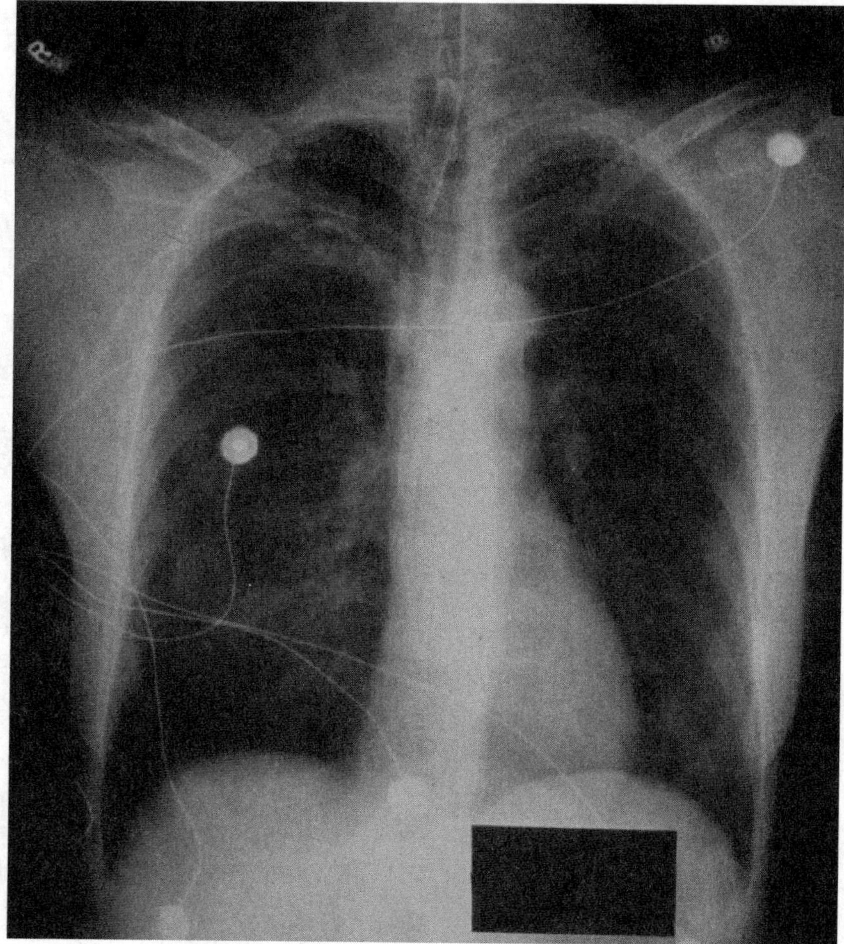

FIG. 14-2

1309. Treatment of the patient pictured in the Figure 14-2 should be

(A) flexible fiberoptic bronchoscopy
(B) thoracostomy tube insertion
(C) endotracheal tube withdrawal to 2 cm above the carina
(D) increased PEEP
(E) extubation

Questions 1310 and 1311

A 30-year-old construction worker is admitted to the ICU with multiple traumatic injuries suffered when he was crushed by a crane. Two days following admission, he becomes anuric with a serum creatine phosphokinase level equal to 30,000 U/L.

1310. Expected findings include all of the following EXCEPT

(A) hyperkalemia
(B) metabolic acidosis
(C) hypophosphatemia
(D) muscle necrosis
(E) positive urine myoglobin

1311. Treatment of this patient includes all of the following EXCEPT

(A) urine alkalinization
(B) osmotic diuresis
(C) calcium administration
(D) fluid restriction
(E) dialysis

1312. A 21-year-old male is admitted to the ICU following a severe closed head injury. Intracranial pressure (ICP) as measured by ventriculostomy is 28 mmHg. Treatment should consist of all of the following EXCEPT

(A) hyperventilation
(B) administration of mannitol
(C) hypothermia
(D) drainage of cerebral spinal fluid
(E) head down position

1313. The most common laboratory abnormality in the patient receiving TPN is

(A) hypoglycemia
(B) hyperlipidemia
(C) hyperphosphatemia
(D) elevated hepatic transaminases
(E) hypocalcemia

Questions 1314 and 1315

A 25-year-old woman is admitted to the ICU following a motor vehicle collision in which she sustained multiple rib fractures. Two hours following admission, she is intubated and placed on controlled mechanical ventilation because of respiratory failure.

1314. Expected cardiopulmonary changes following endotracheal intubation and positive pressure ventilation with positive end-expiratory pressure include all of the following EXCEPT

(A) decreased work of breathing
(B) increased dead space to tidal volume ratio
(C) decreased pulmonary vascular resistance
(D) increased risk of barotrauma
(E) reexpansion of atelectatic alveoli

1315. The patient becomes hypotensive immediately after institution of mechanical ventilation. The appropriate first intervention is

(A) stat chest radiograph
(B) intravenous fluid bolus
(C) pericardiocentesis
(D) hyperventilation with 100% oxygen
(E) thoracotomy and aortic crossclamp

1316. The neurologic symptom in a patient with a blood carboxyhemoglobin level of 50% is

(A) headache
(B) nausea
(C) confusion
(D) coma
(E) death

1317. Risk factors associated with perioperative renal failure include all of the following EXCEPT

(A) hypervolemia
(B) diabetes mellitus
(C) amikacin therapy
(D) sepsis
(E) hypertension

1318. The best method for preventing central venous line infection in the critically ill patient is to

(A) prepare the skin with a povidone-iodine solution
(B) always select the subclavian vein as the cannulation site
(C) schedule routine line changes over a guidewire every 3 days
(D) initiate systemic antibiotics on placement of the central line
(E) use sterile barrier precautions (mask, cap, sterile gown, sterile gloves, large sterile drape) when placing the line

1319. During cardiopulmonary resuscitation (CPR), the critical coronary perfusion pressure associated with return of spontaneous circulation is

(A) 10 mmHg
(B) 20 mmHg
(C) 30 mmHg
(D) 40 mmHg
(E) 50 mmHg

Questions 1320 and 1321

An 18-year-old man is declared brain dead in the ICU following severe head injury suffered from a fall. Organ donation is planned. He has polyuria and is hypotensive. Laboratory values reveal that serum sodium is 159 mEq/L, serum potassium is 3.2 mEq/L, BUN 32 mg/dL, serum creatinine is 1.9 mg/dL, serum osmolarity is 320 mOsm/L, and urine osmolarity is 250 mOsm/L.

1320. The patient's clinical condition is best explained by

(A) diabetes mellitus
(B) massive hemorrhage
(C) nephrogenic diabetes insipidus
(D) syndrome of inappropriate antidiuretic hormone (SIADH)
(E) central diabetes insipidus

1321. Treatment of the patient should consist of

(A) insulin
(B) transfusion
(C) diuresis
(D) free water restriction
(E) vasopressin

DIRECTIONS (Questions 1322 through 1347): For each of the items in this section, ONE or MORE of the numbered options is correct. Choose answer

(A) if only 1, 2, and 3 are correct
(B) if only 1 and 3 are correct
(C) if only 2 and 4 are correct
(D) if only 4 is correct
(E) if all are correct

1322. Racemic epinephrine administered by aerosol

(1) is a mixture of epinephrine and norepinephrine
(2) is a decongestant
(3) causes mucosal edema
(4) is a mild bronchodilator

1323. The pulmonary changes characteristic of the postoperative period include

(1) ventilation/perfusion mismatch
(2) decreased functional residual capacity
(3) hypoventilation
(4) increased compliance

1324. Incentive spirometry increases

(1) FRC
(2) inspiratory transpulmonary pressure
(3) oxygen saturation
(4) expiratory transpulmonary pressure

1325. The mark "Z-79" on an endotracheal tube

(1) refers to the size of the tube
(2) refers to a committee of the American National Standards Institute
(3) refers to the manufacturer of the tube
(4) identifies the tube as nontoxic

1326. A patient being maintained by mechanical ventilation develops pneumonia caused by a gram-negative rod. Contributing factors to this infection may include

(1) a contaminated nebulizer
(2) high FIO_2
(3) decreased host defense mechanisms
(4) ranitidine therapy

1327. Intermittent mandatory ventilation (IMV)

(1) allows spontaneous ventilation
(2) gives a prescribed number of ventilations per minute
(3) ventilates the patient independent of his or her own breathing
(4) is the same as assisted ventilation

1328. PEEP

(1) causes decreased FRC
(2) may cause decreased blood pressure
(3) causes decreased compliance
(4) may cause barotrauma

1329. Supportive therapies for a patient diagnosed with early ARDS should include

(1) IV corticosteroids
(2) low tidal volume (6 mL/kg) ventilation
(3) prophylactic antibiotics
(4) PEEP set to achieve alveolar recruitment without overdistention

1330. Therapies which may diminish mortality rate in severe sepsis include

(1) activated protein C
(2) strict glycemic control
(3) early fluid resuscitation
(4) hydrocortisone

1331. Tension pneumothorax

(1) may result from positive pressure ventilation
(2) may be followed with serial chest radiographs if the size of the pneumothorax is small
(3) causes hypotension
(4) occurs when the gas in the pleural space is at a pressure lower than atmospheric pressure

1332. Following successful CPR, a patient suffers from recurrent ventricular fibrillation. He should be treated with

(1) vasopressin
(2) biphasic defibrillation
(3) amiodarone
(4) high-dose (0.1 mg/kg) epinephrine

1333. Effective and recommended techniques for diminishing postoperative respiratory complications in thoracotomy patients include

(1) incentive spirometry
(2) intermittent positive pressure breathing
(3) preoperative smoking cessation
(4) routine postoperative bronchoscopic suctioning

1334. A healthy 18-year-old male is receiving inhalation anesthesia via a mask for an elective knee arthroscopy. Midway through the procedure, he regurgitates liquid gastric contents into the mask. Oxygen saturation falls to 85%; chest auscultation reveals diffuse rhonchi. Appropriate therapeutic interventions include

(1) endotracheal intubation and suctioning
(2) administration of clindamycin
(3) positive pressure ventilation
(4) administration of methylprednisolone

1335. Contributing to hyperglycemia in sepsis are increased levels of

(1) epinephrine
(2) cortisol
(3) glucagon
(4) thyroid hormone (T_3)

1336. In the critically ill trauma patient, a respiratory quotient (R/Q) greater than 1.0 indicates

(1) overfeeding
(2) an anabolic state
(3) lipogenesis
(4) error in measurement

1337. Potential beneficial effects of inhaled nitric oxide in treatment of acute lung injury include

(1) selective pulmonary vasodilation
(2) diminished PA pressure
(3) diminished intrapulmonary shunt
(4) patient amnesia

1338. True statements about ventilator-associated pneumonia include

(1) the associated mortality is less than 5%
(2) contamination of inhaled, humidified gases is the predominant cause
(3) is preventable by the use of antibiotics
(4) is preventable by head of bed elevation

1339. The airway pressure tracing in Figure 14-3 represents

(1) synchronized intermittent mandatory ventilation (SIMV) with spontaneous breaths
(2) intermittent mandatory ventilation (IMV) without spontaneous breaths
(3) assist/control ventilation (ACV) with spontaneous breaths
(4) controlled mechanical ventilation (CMV)

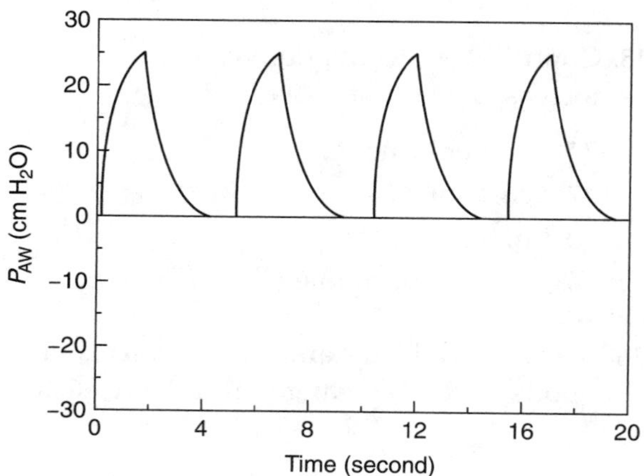

FIG. 14-3

1340. Appropriate treatment of pulseless electrical activity (PEA) includes

(1) needle decompression of tension pneumothorax
(2) administration of epinephrine
(3) volume resuscitation
(4) defibrillation

1341. Expected complications of a traumatic T1 spinal cord transection include

(1) hypotension
(2) hypothermia
(3) bradycardia
(4) apnea

Questions 1342 and 1343

The patient depicted in Figure 14-4 has the following values:

Heart rate	120 bpm
Blood pressure	95/60 mmHg
Central venous pressure	6 mmHg
PA pressure	45/20 mmHg
PAOP	8 mmHg
Cardiac output	7.5 L/min
PaO_2	60 mmHg
FIO_2	80%

1342. Pathophysiologic findings of ARDS include

(1) pulmonary leukocyte aggregation
(2) increased pulmonary capillary membrane permeability
(3) pulmonary vascular microemboli
(4) homogenous pattern of lung injury

1343. Effects of "auto-PEEP" include

(1) increased venous return
(2) increased work of breathing
(3) easier triggering of the ventilator
(4) alveolar overdistention

1344. Factors contributing to overestimation of left ventricular end-diastolic volume from PA catheter include

(1) mitral stenosis
(2) high PEEP
(3) large **v** waves on PAOP tracing
(4) measuring PAOP at end inspiration in the patient on CMV

1345. A 60-year-old man with chronic stable angina undergoes uneventful total hip replacement under spinal anesthesia. A postoperative electrocardiogram in the recovery room reveals new ST segment depression in the lateral leads. He is awake, alert, and denies pain. His temperature is 36°C, heart rate 97 bpm, blood pressure 160/80 mmHg, respiratory rate is 14, and oxygen saturation is 94% on room air. Appropriate interventions include

(1) administration of nitroglycerin
(2) administration of oxygen
(3) administration of esmolol
(4) nothing since he is asymptomatic

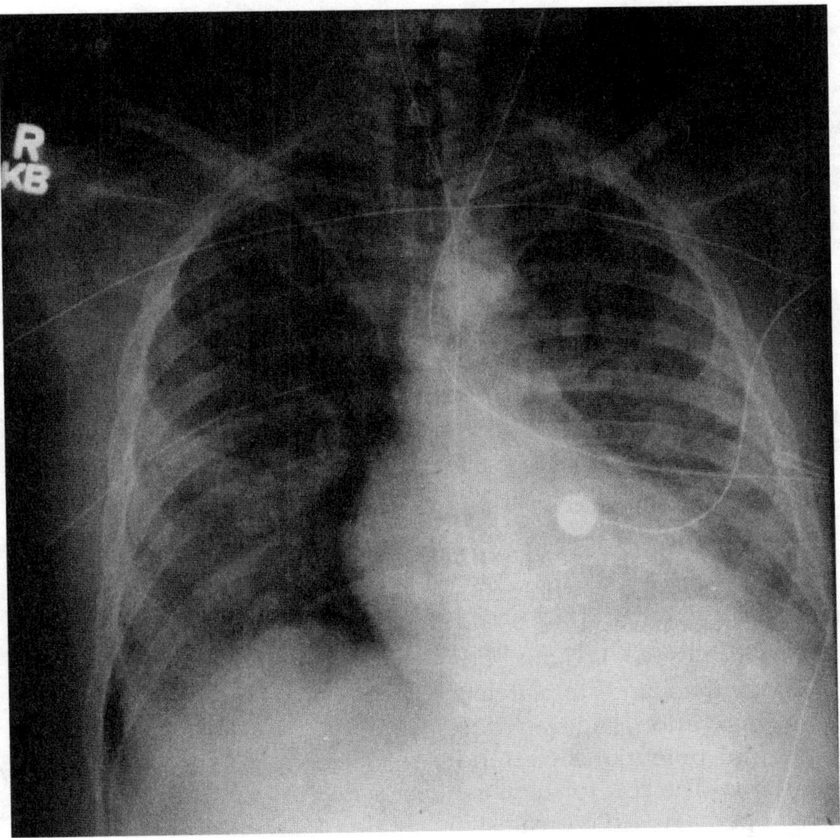

FIG. 14-4

1346. Possible etiologies for the patient's condition include

(1) pancreatitis
(2) trauma
(3) aspiration
(4) congestive heart failure

1347. Appropriate respiratory management of the patient depicted includes

(1) large (>15 mL/kg) tidal volume and low respiratory rate
(2) bronchial toilet
(3) withholding PEEP
(4) inverse ratio pressure control ventilation

Answers and Explanations

1289. (B) A patient breathing oxygen at 5 L/min via a mask without a reservoir will have an FIO_2 of approximately 40%. *(5:2813)*

1290. (D) Heparin-induced thrombocytopenia (HIT) type II is an immune-mediated phenomenon where heparin bound to the platelet factor 4 (PF 4) receptor on platelets induces an IgG antibody response. IgG antibody then binds to platelets, inducing platelet activation, aggregation, thrombocytopenia, and potential for arterial and venous thrombosis. The risk for heparin-induced thrombocytopenia type II is greatest in patients exposed to unfractionated heparin and is markedly reduced by use of low molecular weight heparin (e.g., enoxaparin dalteparin). Once a patient has developed antibodies to the heparin-PF 4 complex, however, low molecular weight heparins are contraindicated due to cross-reactivity with the antibody. Anticoagulation with a direct thrombin inhibitor (e.g., argatroban or hirudin) should be initiated due to the high risk for thrombosis in these patients. Warfarin is contraindicated in the acute setting because the drug produces an initial hypercoagulable state due to suppression of protein C preceding the antiprothrombin effect. Prophylactic platelet transfusions are contraindicated in HIT type II because, despite the thrombocytopenia, HIT type II places the patient at very high risk for thrombosis with platelet transfusion providing more substrate for platelet aggregation and subsequent thrombosis. *(1:228–9, 930–1; 5:1116, 1974)*

1291. (D) An advantage to aerosol therapy is direct drug delivery to the lungs, minimizing systemic effects of medications. *(5:2816–8)*

1292. (A) Oxygen delivery (DO_2) is equal to cardiac output (CO) multiplied by the arterial oxygen content of blood (CaO_2). Thus

$$DO_2 = CO\ (L/min) \times CaO_2\ (mL/dL)$$

$$CaO_2 = \text{hemoglobin concentration (Hb, g/dL)} \times 1.39\ mL\ O_2/g\ Hb \times \text{arterial oxygen saturation } (SaO_2, \%) + 0.003\ mL\ O_2/mmHg/dL \times PaO_2\ (mmHg).$$

$$CaO_2 = Hb \times 1.39\ mL\ O_2/g\ Hb \times SaO_2 + 0.003\ mL\ O_2/mmHg/dL \times PaO_2$$

Doubling the cardiac output results in the greatest oxygen delivery, 761 mL/min. *(1:201; 5:699–700)*

1293. (D) The partial rebreathing mask provides a high FIO_2 and requires low oxygen flows. Its use should result in an increased oxygen tension in the blood. A partial rebreathing mask has no effect on carbon dioxide tension. *(5:2813)*

1294. (B) Criteria for consideration of NIPPV include an awake, cooperative patient. Obtunded patients who may have an impaired gag and cough reflex may be at increased risk of aspiration with this ventilatory modality. Other relative contraindications to the use of NIPPV include copious respiratory secretions, facial trauma or anatomic abnormality (e.g., burns) impairing tight mask fit, recent gastroesophageal surgery, and hemodynamic instability. *(5:2826–7)*

1295. (C) The object depicted with an X is a spacer, an accessory which may be added to a MDI to improve aerosolized drug delivery to the lungs.

This may facilitate proper patient technique when using an MDI, minimizing oropharyngeal drug deposition. MDIs are pressurized canisters that do not require an external gas source to deliver aerosolized medications. They are most commonly used for administration of inhaled beta-agonists, but anticholinergics, corticosteroids, and various combinations may also be delivered in this fashion. *(5:2815–7)*

1296. (E) The patient with a bronchopleural fistula requires a method of ventilation that does not increase the flow across the fistula. Spontaneous ventilation with a mask is seldom sufficient with a large fistula, and methods employing positive pressure may not be able to accomplish ventilation because of loss of inspired air through the fistula. In these patients, the use of high-frequency jet ventilation may be effective. *(1:840; 5:1925)*

1297. (A) Suctioning of the artificial airway is not a benign procedure. Stimulation of the trachea or carina increases vagal activity leading to bradycardia and hypotension. Suctioning of gas out of the lung may cause hypoxia and atelectasis. Trauma to the airway can also occur. For these reasons, suctioning should be done only when indicated. There should be some evidence of secretions that can be cleared, and aseptic technique should be employed. *(5:2819–20)*

1298. (D) Absorption atelectasis occurs with high oxygen concentrations and is more likely when the nitrogen concentration is low. Early airway closure increases the risk of absorption atelectasis. This is clinically important, since shunting can occur around the atelectatic units. *(4:1763; 5:711)*

1299. (E) The pressure within the cuff that will interrupt mucosal flow is approximately 30 mmHg. *(5:1630)*

1300. (E) Oxygen delivered at 4 L/min via a nasal cannula provides an FIO_2 of approximately 36% in a patient with a normal ventilatory pattern. *(5:2813)*

1301. (E) Nondepolarizing muscle relaxants may prevent active vomiting but not passive regurgitation of gastric contents into the esophagus and pharynx. Inadequate laryngeal closure may then lead to aspiration, even in the setting of a properly inflated endotracheal tube cuff. Because a nasogastric tube (even if placed to suction) does not reliably empty the stomach, it provides no guarantee against aspiration. Gagging on the endotracheal tube during emergence of anesthesia may occur during stage 2 of anesthesia, prior to the awake state. Extubation during stage 2 may result in aspiration. *(1:1392; 5:1635, 1647)*

1302. (D) Arterial hypoxemia as measured by pulse oximetry or PaO_2 is the most reliable way to diagnose aspiration in the proper clinical setting. Tachypnea and tachycardia are common signs. Many patients who have aspirated will not develop rales or any radiographic infiltrate. The presence of gastric contents in the oropharynx does not necessarily indicate that aspiration has occurred since the larynx may be competent. *(1:1392)*

1303. (D) Glutamine, a nonessential amino acid, is an important fuel source for small intestinal cells and probably plays a role in maintaining intestinal villous integrity; however, its instability in aqueous solution prevents its routine availability in standard parenteral nutrition formulations. *(5:2898–9)*

1304. (D) In pressure support ventilation, the patient initiates a breath and the ventilator adjusts inspiratory flow to rapidly achieve the preset pressure level throughout inspiration. Inspiration ends when flow falls to a preset percent of initial maximum, typically 25%. *(5:2824)*

1305. (A) PEEP actually increases total lung water but acts to redistribute interstitial lung water away from areas of gas exchange, thus improving oxygen diffusion from alveolus to capillary. PEEP improves arterial oxygenation and lung compliance and decreases shunt and alveolar overdistention. *(5:2820)*

1306. (B) Initial treatment of septic shock consists of volume resuscitation to restore intravascular volume depleted via leaky capillary endothelium. Controversy exists as to whether colloid or crystalloid should be used for volume resuscitation, although colloid has no demonstrable outcome benefit. Once appropriate intravascular volume has been achieved, ongoing hypotension is best addressed by use of a vasopressor, often in combination with inotropic support. *(1:1477; 4:2159; 5:2795)*

1307. (C) Immobilization, concomitant steroid therapy, use of vecuronium (which has a pharmacologically-active metabolite), and development of a motor neuropathy may all lead to prolonged weakness in the critically ill patient treated with muscle relaxants. While inadequate analgesia and anxiolysis are frequent complications to the use of muscle relaxants in the critical care unit, they have not been implicated in prolonged neuromuscular weakness. *(5:530–2)*

1308. (E) The first intervention in the setting of massive hemoptysis following PA rupture should be endobronchial intubation to protect the unaffected lung and prevent asphyxiation. Other maneuvers recommended for this frequently fatal complication include placing the patient in the lateral position with the unaffected lung up, reversal of anticoagulation, decreasing PA pressure, administration of PEEP, and consideration of emergent surgical repair. *(1:818–9; 5:1307)*

1309. (B) This patient has a large right pneumothorax with mild mediastinal shift to the left. Proper therapy is thoracostomy tube (chest tube) placement. If the patient were hypotensive, indicative of tension pneumothorax, treatment should consist of immediate needle decompression followed by thoracostomy tube. *(1:840; 4:1826; 5:1294, 1902)*

1310. (C) Rhabdomyolysis is produced by massive muscle tissue injury resulting in release of cell contents into the circulation. Myoglobin, the heme moiety which carries oxygen in muscle, is released from necrotic muscle and delivered to the glomerulus where it is excreted. Acidic urine pH (<5.6) promotes dissociation of myoglobin into ferrihematin, a nephrotoxin which precipitates in proximal renal tubular cells and may lead to oliguric or anuric renal failure. Hyperkalemia, hyperphosphatemia, hypocalcemia, and anion-gap metabolic acidosis are characteristic laboratory findings of rhabdomyolysis. Urine myoglobin is often positive, even if the urine is not red, because of the low threshold for excretion of myoglobin at the glomerulus. *(1:1016; 5:804–5)*

1311. (D) Treatment of rhabdomyolysis is aimed at maintaining adequate intravascular volume, glomerular blood flow, and renal tubular flow to promote excretion of myoglobin and prevent tubular ferrihematin precipitation. Thus, IV infusion of crystalloids and maintenance of urine flow greater than 100 mL/h with use of osmotic diuretics like mannitol are recommended. Alkalinization of the urine with sodium bicarbonate or acetazolamide to maintain urine pH >5.6 is also recommended provided that this intervention does not result in significant impairment in systemic acid-base homeostasis. Calcium administration is indicated for treatment of hyperkalemia. Some patients, nevertheless, may require dialysis for treatment of refractory hyperkalemia. Fluid restriction is not appropriate. *(1:1016; 5:804–5)*

1312. (E) The head should be elevated to 30° in the patient with increased ICP to enhance venous drainage and caudad movement of cerebrospinal fluid (CSF). Care should be taken to assure that the patient is euvolemic when elevating the head to avoid hypotension, which will further compromise cerebral perfusion pressure. Other therapies useful in lowering ICP in the setting of head trauma include temporary hyperventilation to allow the initiation of other more definitive therapies (e.g., surgical decompression), osmotic therapy (mannitol or hypertonic saline), drainage of CSF, surgical intervention, and barbiturate coma. *(1:1263–6; 5:2471–3)*

1313. (D) The most common abnormality in patients receiving TPN is elevated serum hepatic transaminases levels. Serum glutamic oxaloacetic transaminase (SGOT) and serum glutamic pyruvate transaminase (SGPT) are frequently increased to two times normal levels; in addition alkaline phosphatase may be mildly elevated. More profound rises in alkaline phosphatase, accompanied by elevated bilirubin levels, may indicate a cholestatic jaundice, which may be the result of lipid emulsion administration. This condition may be managed by either diminishing the amount of lipid administered to approximately 1 g/kg/day or placing the patient on enteral nutrition. Hypoglycemia is usually a result of interruption of TPN. Hypophosphatemia and hypercalcemia are frequently seen in patients on TPN because phosphate requirements (40 to 100 mEq/day) are usually underestimated while calcium requirements (3 to 8 mEq/day) are overestimated. *(5:2911–3)*

1314. (C) PPV causes an increase in PVR and an increase in right ventricular afterload. PPV also causes diminished venous return. Both effects reduce cardiac output with resultant hypotension, especially in hypovolemic patients. Dead space to tidal volume ratio (V_D/V_T) increases in the patient on mechanical ventilation. Thus, more inefficient (dead space) ventilation requires a higher minute ventilation in the mechanically ventilated patient to maintain the same $PaCO_2$. A patient on CMV performs less work of breathing than the spontaneously ventilating patient, particularly one in respiratory distress with ventilatory pump failure. Barotrauma is a risk of PPV, particularly if high peak airway pressures or hyperinflation are present. PEEP improves oxygenation via alveolar recruitment, reexpansion of atelectatic alveoli. *(1:803; 5:2820–1)*

1315. (B) IV fluid bolus is the appropriate first intervention since the most likely etiology of the patient's hypotension is a low cardiac output due to diminished venous return from positive intrathoracic pressure. In this setting, fluid administration can often correct the cardiac output to normal. Other effective measures include elevation of the legs to 30° to increase venous return, substituting partial ventilatory support for CMV, and temporary hemodynamic support with an inotrope and/or vasopressor. Other less likely potential aetiologies of hypotension, which should be investigated, include tension pneumothorax, cardiac tamponade, cardiac contusion, and hypovolemia due to bleeding. Hyperventilation may prove deleterious by causing hyperinflation and worsened hypotension. *(1:1469; 5:2820–1)*

1316. (D) Carboxyhemoglobin produces coma at a blood level of approximately 50%. Levels of 10% produce headache, 20% nausea, 30% visual disturbances, 40% confusion, and 60% death. *(1:1276)*

1317. (A) Hypovolemia, not hypervolemia, is a risk factor for development of perioperative renal failure in addition to aminoglycoside exposure, sepsis, diabetes mellitus, and hypertension. *(1:1010, 1014; 5:1492)*

1318. (E) Use of maximal sterile barrier precautions has been demonstrated to diminish the rate of central line infection in critically ill patients. Skin preparation with chlorhexidine solution has been shown to be superior in site decontamination as compared to povidone-iodine solution and is therefore recommended. While some studies have shown that subclavian cannulation is associated with a lower infection rate than internal jugular or femoral vein, others have demonstrated no statistically significant superiority. In addition, subclavian venous cannulation carries a higher risk of pneumothorax compared to other sites and is not practical in every critically ill patient. Systemic antibiotics are not recommended for routine "prophylactic" coverage of central venous catheters, but antibiotic impregnated central venous catheters have been demonstrated to diminish the risk of central line infection and are recommended for critically ill patients who are likely to require central venous access for at least 3 days. Routine scheduled central venous line changes over a guidewire have not been shown to provide an outcome benefit and are not recommended. *(5:2797–8)*

1319. (B) Experimental evidence suggests that the critical coronary perfusion pressure associated with successful restoration of spontaneous circulation is 15 to 25 mmHg. *(1:1491)*

1320. (E) Central diabetes insipidus, which frequently occurs following severe head injury or brain death, best explains the presence of polyuria, hypernatremia, serum hyperosmolarity, and urine hyposmolarity. *(1:1137–8, 1348; 5:1767–8, 2960)*

1321. (E) Appropriate therapy for diabetes insipidus includes replacement of free water deficit and administration of aqueous vasopressin or desmopressin acetate (DDAVP). *(1:1137–8, 1348; 5:1767–8, 2960)*

1322. (C) Racemic epinephrine is a decongestant and mild bronchodilator. It is a 50:50 mixture of the levo- and dextroisomers of epinephrine. Because it causes mucosal vasoconstriction, it is useful in treating airway edema. *(1: 1202–3; 5:2816)*

1323. (A) The postoperative period is usually marked by a decreased FRC and decreased ventilation. Mismatch of ventilation and perfusion is also usually present. Compliance is decreased. *(1:809–10; 5:2712–5)*

1324. (A) Incentive spirometry increases inspiratory, but not expiratory, transpulmonary pressure. The most effective method for increasing expiratory transpulmonary pressure is continuous positive airway pressure by mask. The use of incentive spirometry also leads to an increase in FRC and oxygen saturation. *(1:847; 5:818)*

1325. (C) The mark "Z-79" on an endotracheal tube refers to a committee of the American National Standards Institute. The mark identifies the tube as nontoxic and in compliance with the standards set down. The initials "I.T." stand for implant tested and also show that the tube is nontoxic. *(5:1629)*

1326. (E) Gram-negative pneumonia is common in mechanically ventilated patients. The bacteria may colonize any portion of the ventilator circuit, especially where there is excess moisture. An increased FIO_2 decreases mucociliary function. In addition, an artificial airway and tracheal suctioning decrease host defenses. The use of histamine H_2 receptor antagonists has become common in ventilated patients to prevent ulceration induced by stress. The increased gastric pH makes bacterial colonization of the stomach more likely. Since aspiration is also common in the ventilated patient, the stomach then becomes one additional source of bacteria. *(1:1378–9; 5:2425–6)*

1327. (A) IMV allows the patient to maintain his or her own breathing and receive a prescribed number of breaths per minute from the ventilator. It differs from assisted ventilation in that the breaths from the ventilator do not augment the patient's breaths. *(1:1469; 5:2823)*

1328. (C) The use of PEEP may cause decreased blood pressure, especially in the hypovolemic patient. Since it causes increased intrathoracic pressure, barotrauma may occur. *(1:1469; 5:2820–1)*

1329. (C) The NIH sponsored ARDSnet trial reported in 2000 definitively established the concept of low tidal volume ventilation for improving mortality in ARDS. The study compared traditional tidal volumes of 12 mL/kg to a "lung-protective" strategy of 6 mL/kg and demonstrated a 9% improvement in mortality. Corticosteroids have been used with variable results in the later "fibroproliferative" phase of ARDS but have not been proven helpful in early ARDS. Whereas treatment of established infection is indicated in patients with ARDS, there is no role for prophylactic antibiotics. The use of PEEP is helpful in ARDS to recruit atelectatic portions of the lung, thereby diminishing shear forces associated with repeated opening and closing of lung units. By opening atelectatic alveoli, PEEP also serves to reduce shunt. *(1:1469; 5:2790–3, 2820)*

1330. (E) Recent studies have demonstrated improved outcome from severe sepsis with (1) the anti-inflammatory and anticoagulant medication activated protein C, (2) early fluid and vasopressor use directed to physiologic goals, (3) use of a combination of corticosteroid and mineralocorticoid treatment in patients determined to be adrenally insufficient, and 4) tight glucose control with use of insulin infusion to maintain blood glucose 80 to 110 mg/dL. *(5: 2794–2800)*

1331. (B) Tension pneumothorax occurs when there is gas in the pleural space at a pressure higher than atmospheric. It may occur as a result of barotrauma from PPV, and commonly causes hypotension. A tension pneumothorax should be decompressed; a needle inserted into the pleural space may be used emergently; however, tube thoracostomy is the definitive treatment. *(1:840; 4:1826; 5:1294, 1902)*

1332. (A) First-line therapy for treatment of VF should consist of biphasic defibrillation at a dose of 150 J for up to three shocks. Either vasopressin 40 U times one dose or epinephrine 1 mg every 5 minutes should be administered intravenously if the first three attempts at defibrillation fail. Both vasopressin and epinephrine serve to increase coronary perfusion, increasing the likelihood of subsequent successful defibrillation. High-dose (0.1 to 0.2 mg/kg) epinephrine has not been shown to improve outcome from cardiac arrest and thus is no longer recommended. Amiodarone is a recommended antiarrhythmic agent for use in cases of recurrent VF and has been shown to be more effective in terminating recurrent VF than lidocaine. The dose of amiodarone for VF is 300 mg IV push. *(5:2935–8)*

1333. (B) Incentive spirometry, when properly performed, expands the lungs and opens airways, preventing atelectasis and improving secretion clearance. IPPB gives, as the term implies, intermittent short duration positive inspiratory pressure without an endotracheal tube. Because the results of IPPB are no greater than those of incentive spirometry and because the technique has been shown to increase gastric distention with a risk for aspiration, it cannot be routinely recommended. Smoking cessation occurring more than 4 weeks prior to surgery has substantial beneficial effects, including improved mucociliary function, decreased sputum production, and improvement in pulmonary function tests. Routine postoperative bronchoscopy for pulmonary toilet has not been shown to diminish pulmonary complications, but may be helpful for a subset of patients who develop lobar collapse. *(5:1858–9, 2818–20)*

1334. (B) Appropriate treatment of aspiration of liquid gastric contents is endotracheal intubation, suctioning, and PPV to open atelectatic alveoli. Prophylactic antibiotics are not indicated in cases of simple aspiration since the initial pulmonary insult is a chemical burn caused by the gastric acid. Antibiotics administered at this stage may be harmful by selecting for resistant microbial organisms if secondary bacterial infection occurs. Corticosteroids have not been shown to improve outcome from aspiration pneumonitis and may induce deleterious side effects. Bronchoalveolar lavage is indicated only in the setting of aspiration of particulate matter. *(1:1392)*

1335. (A) Elevations in circulating levels of epinephrine, cortisol, and glucagon during sepsis contribute to increased serum glucose. Circulating thyroid hormone (T_3) is decreased in sepsis. *(5:2897)*

1336. (B) The respiratory quotient is the ratio of carbon dioxide production to oxygen consumption (V_{CO_2}/V_{O_2}). A normal *R/Q* is approximately 0.8 with a range from 0.7 to 1.0, depending on diet. Metabolism of carbohydrates yields an *R/Q* of 1.0 while metabolism of the more efficient fats yields an *R/Q* closer to 0.7. An *R/Q* greater than 1.0 in a critically ill patient indicates excessive caloric intake with lipogenesis and may impair weaning from the ventilator. *(5:2906–7)*

1337. (A) Inhaled nitric oxide causes vascular smooth muscle relaxation by activating the enzyme guanylate cyclase. Because nitric oxide is rapidly bound to hemoglobin and inactivated in the systemic circulation, inhaled nitric oxide is a selective pulmonary vasodilator. Improved *V/Q* matching, improved oxygenation, and diminished intrapulmonary shunt are other potential beneficial effects of inhaled nitric oxide in acute lung injury. Inhaled nitric oxide does not produce amnesia. *(1:1469; 5:165, 685–6, 1454–5)*

1338. (D) The simplest, most cost-effective way to prevent ventilator-associated pneumonia (VAP) is continuous head of bed elevation at 30°.

The mortality attributable to VAP is estimated at 5 to 20%. While inhalation of contaminated, humidified gases may cause pneumonia in the intubated patient, the predominant mechanism for development of VAP involves microaspiration of oropharyngeal and gastric secretions. Rather than preventing VAP, antecedent antibiotics appear to be a risk factor for its development. *(5:2796–7)*

1339. (C) The diagram demonstrates either IMV without spontaneous breaths or CMV. A spontaneous breath in SIMV or ACV would be preceded by a slight negative deflection in the pressure tracing, indicative of patient inspiratory effort. *(1:1470; 5:2823)*

1340. (A) PEA refers to any electrical activity of the heart, excluding ventricular tachycardia and VF. There are many potential causes of PEA, including hypovolemia, hypoxia, cardiac tamponade, tension pneumothorax, hypothermia, pulmonary embolism, myocardial infarction, drug overdose, hyperkalemia, and acidosis. Treatment of PEA is supportive while the underlying cause is identified and definitively treated. Defibrillation is not indicated. *(1:1494; 5:2945)*

1341. (A) High thoracic spinal cord transection produces hypotension because sympathetic efferents which mediate peripheral vasoconstriction are interrupted, causing vasodilatation. Cardiac accelerator sympathetic innervation (T1-T4) is similarly interrupted, resulting in unopposed vagal innervation to the heart and bradycardia. Finally, thermoregulation is impaired by autonomic denervation, frequently resulting in hypothermia because of uncontrolled peripheral vasodilatation and no compensatory capability to shiver. Spinal cord transection below C5 would not be expected to produce apnea. *(1:1268; 5:2475–6)*

1342. (A) The pathophysiology of ARDS includes pulmonary leukocyte aggregation, increased pulmonary capillary membrane permeability, and pulmonary vascular microemboli. Despite the homogenous appearance of ARDS on chest radiograph, however, pathologic and CT findings indicate a heterogenous distribution of disease in gravity dependent portions of the lung. *(1:1468–9; 5:2790–1, 2821–5)*

1343. (C) Auto-PEEP or intrinsic PEEP refers to positive intra-alveolar pressure at end expiration. It differs from extrinsic PEEP or applied PEEP in that it is unintentional. Factors which may contribute to auto-PEEP include increased airway resistance, bronchospasm, airway secretions, tachypnea, prolonged lung emptying, and shortened expiratory time. Typically, patients with severe COPD, who require long expiratory times for complete expiration, and who are on mechanical ventilatory support, are at risk for development of significant auto-PEEP. Any patient on mechanical ventilatory support, however, may develop auto-PEEP depending on the pulmonary physiology and strategy of ventilatory support. Auto-PEEP may increase a spontaneously ventilating patient's work of breathing by providing positive end-expiratory alveolar pressure, which must be overcome in order to initiate inspiratory flow. Clues to the presence of auto-PEEP in the mechanically ventilated patient include patient-ventilator asynchrony and patient inspiratory efforts, which do not trigger the ventilator. As does extrinsic PEEP, auto-PEEP diminishes venous return and may cause hypotension. *(5:1466)*

1344. (E) All of the factors mentioned may result in overestimation of LVEDV from PAOP. PAOP provides a good correlation to LVEDV if certain criteria are met. First, the catheter must be in appropriate position in Zone III of the lung, as described by West. Second, use of PAOP to estimate LVEDV assumes a continuous, uninterrupted communication between the tip of the catheter in the pulmonary capillary and the left ventricle. Mitral stenosis imposes an obstruction to flow at the level of the mitral valve with elevated left atrial pressure and elevated PAOP not necessarily correlating with elevated LVEDV. High PEEP may convert lung Zone III to Zone II, thereby reflecting alveolar pressure in the PAOP and leading to overestimation of LVEDV. **V** waves on a PAOP trace may indicate mitral regurgitation, which may result in overestimation of the PAOP and LVEDV. Measuring

PAOP at end inspiration in a patient on PPV may cause overestimation of LVEDV because alveolar pressure can be transmitted across the thorax, converting Zone III to Zone II and erroneously high reading of the PAOP. *(1:676–7; 5:1322–3)*

1345. (A) Silent ischemia is a common perioperative phenomenon. Even patients with a preoperative history of typical angina may develop asymptomatic ischemia on ECG in the perioperative period. Appropriate interventions include treatment with antianginal medications, including β-blockers and nitroglycerin, to diminish myocardial oxygen demand. Additionally, oxygen should be administered to enhance myocardial supply. The patient should not be discharged from the recovery room, nor should the ECG changes be ignored. *(1:935–7; 5:2065–8)*

1346. (A) Pancreatitis, trauma, and aspiration may all be etiologic factors in the development of ARDS. Clinical features of ARDS are marked hypoxemia, diffuse bilateral infiltrates on chest radiograph, and low PAOP. The presence of congestive heart failure as an explanation for the large A-a gradient excludes the diagnosis of ARDS. In this patient, the low PAOP and high cardiac output are not consistent with congestive heart failure. *(1:1468–9; 5:2790–1)*

1347. (C) The syndrome of ARDS is marked by poor lung compliance. Large tidal volumes are not recommended because of the risk of alveolar overdistention and ventilator-induced lung injury. PEEP, however, is indicated in the treatment of ARDS because it improves pulmonary compliance, diminishes alveolar overdistention when administered in appropriate amount, and improves arterial oxygenation, thereby allowing lower FIO_2. Inverse ratio pressure control ventilation, a mode of pressure-limited ventilation where inspiratory time is greater than expiratory time, is useful in the treatment of ARDS to increase mean airway pressure, expand atelectatic alveoli, and improve arterial oxygenation without deleterious cardiovascular effect. Bronchial toilet is an important component of therapy for ARDS since these patients are at high risk for nosocomial pneumonia. *(1:1468–9; 5:2790–1, 2821–5)*

CHAPTER 15

Acute and Chronic Pain

Questions

DIRECTIONS (Questions 1348 through 1366): Each of the numbered items or incomplete statements in this section is followed by answers or by completions of the statement. Select the ONE lettered answer or completion that is BEST in each case.

1348. An example of a central pain state is

(A) postoperative incision pain
(B) gallbladder pain
(C) phantom limb pain
(D) bone fracture pain
(E) headache

1349. Occipital neuralgia involves

(A) the greater occipital nerve
(B) the cervical plexus
(C) a pain distribution confined to the occipital area
(D) the scapular nerve
(E) trophic lesions of the skull

1350. Postherpetic neuralgia

(A) is common in children and adolescents
(B) is best treated with opioids
(C) never responds to local application of counterirritants
(D) usually responds to tricyclic antidepressants
(E) is a difficult syndrome to treat, and success is limited

1351. Signs of complex regional pain syndrome may include all of the following EXCEPT

(A) hyperesthesia
(B) coldness or hyperthermia
(C) pain that is of short duration
(D) brittle nails
(E) osteoporosis

1352. The effectiveness of a neurolytic agent is dependent on all of the following EXCEPT

(A) location of the injection
(B) concentration
(C) histology of the nerve
(D) volume
(E) needle size

1353. Inhibitory substances that are believed to modulate the transmission of nociceptive signals in the dorsal horn of the spinal cord include the following EXCEPT

(A) substance P
(B) β-endorphins
(C) dopamine
(D) epinephrine
(E) adenosine

1354. If a patient undergoing thoracotomy receives intercostal blocks with bupivacaine, his or her postoperative period will

(A) be little different from controls
(B) show marked improvement in respiratory function over controls
(C) show little difference in vital capacity but marked pain relief
(D) be marked by hyperventilation
(E) be marked by increased incidence of atelectasis

1355. Which of the following is a nociceptor?

(A) Meissner's corpuscles
(B) pacinian corpuscles
(C) Merkel's disks
(D) free nerve endings on A-delta and C-fibers
(E) Golgi-Mazzoni endings

1356. The one best block for pain secondary to pancreatic cancer is a

(A) stellate ganglion block
(B) Bier block
(C) block of the hypogastric plexus
(D) celiac plexus block
(E) intrathecal neurolysis

1357. Advantages of patient-controlled analgesia (PCA) include all of the following EXCEPT

(A) high patient satisfaction
(B) elimination of painful injections
(C) no need to adjust dosing parameters with increasing age
(D) more consistent levels of analgesia
(E) the ability of the patient to titrate pain relief to painful procedures such as chest physical therapy

1358. The most specific and immediate treatment for nausea and vomiting thought secondary to epidural morphine is

(A) droperidol 0.625 mg IV
(B) a transdermal scopolamine patch
(C) naloxone 0.4 mg by mouth every 2 hours as needed
(D) a transdermal clonidine patch
(E) naloxone 5 mcg/kg/h IV

1359. Which of the following is TRUE about postdural puncture headache?

(A) Using large gauge cutting needles is the best preventive measure when performing a spinal anesthetic.
(B) It results from decreased CSF pressure secondary to loss of CSF at the puncture site.
(C) Gender has no influence on incidence.
(D) Early ambulation will increase the incidence.
(E) An epidural blood patch is always indicated.

1360. Indications for lumbar sympathetic blockade include all of the following EXCEPT

(A) acute herpes zoster
(B) phantom limb pain
(C) complex regional pain syndrome
(D) lumbar facet syndrome
(E) vascular insufficiency

1361. All of the following drugs are useful in the treatment of postoperative pain in a cancer patient who was on slow release morphine preoperatively EXCEPT

(A) methadone
(B) oxycodone
(C) butorphanol
(D) hydromorphone
(E) morphine

1362. A 40-year-old woman is seen in the pain clinic. She complains of a burning pain in her right hand. This pain is exacerbated by the slightest touch, and she says that the hand sometimes turns blue. The one best nerve block to elucidate a diagnosis is a(n)

(A) axillary block
(B) Bier block
(C) lumbar sympathetic block
(D) stellate ganglion block
(E) interscalene block

1363. The lumbar sympathetic chain

(A) is visible by fluoroscopy
(B) is anterolateral to the L2, L3, and L4 vertebrae
(C) is anterior to the abdominal aorta
(D) is the needle position for a celiac plexus block
(E) is a unilateral structure

1364. You are asked to discontinue an epidural catheter on a 68-year-old patient who is 2 days post a sigmoid resection. On review of his medications you see that he received a dose of low-molecular-weight heparin (LMWH) 2 hours ago. You should

(A) wait 12 hours after the last dose of LMWH
(B) remove the catheter after confirming that the activated partial thromboplastin time is normal
(C) pull the catheter immediately
(D) discontinue the catheter after waiting 24 hours
(E) give a unit of fresh frozen plasma, then pull the catheter

1365. All of the following statements are true about myofascial pain EXCEPT:

(A) It is dermatomal in distribution.
(B) It should be treated early in the course of the disease.
(C) Injection of local anesthetic may provide relief.
(D) The pain can occur in the back, neck, and shoulders.
(E) It can sometimes be relieved by simply needling the affected area.

1366. All of the following are major anatomic structures in the transmission and relay of nociceptive information EXCEPT

(A) spinothalamic tract
(B) locus coeruleus
(C) thalamus
(D) reticular formation
(E) sensory cortex

DIRECTIONS (Questions 1367 through 1388): For each of the items in this section, ONE or MORE of the numbered options is correct. Choose answer

(A) if only 1, 2, and 3 are correct
(B) if only 1 and 3 are correct
(C) if only 2 and 4 are correct
(D) if only 4 is correct
(E) if all are correct

1367. Nociceptive information is transmitted in

(1) C-fibers
(2) A-beta fibers
(3) A-delta fibers
(4) A-gamma fibers

1368. Complex regional pain syndrome

(1) may result from a gunshot wound
(2) can respond to sympathectomy
(3) is often accompanied by dystrophic changes of bone
(4) can be exacerbated by anxiety

1369. In the use of intrapleural catheters for postoperative pain

(1) pneumothorax can occur
(2) the usual dose is 20 to 30 mL 0.5% bupivacaine
(3) the mechanism of action is unilateral intercostal nerve block
(4) the use of parenteral opioids is contraindicated

1370. Guidelines concerning the management of patients with cancer pain include

(1) anticipating and treating side effects
(2) dosing analgesics only on an as needed basis
(3) considering the use of adjuvant drugs
(4) avoiding opioid drugs due to the potential for respiratory depression

1371. You are asked to see a patient who 2 days ago had lower-extremity peripheral vascular surgery under uneventful epidural anesthesia. The patient is complaining of severe back pain. The pain is possibly due to

(1) anterior spinal artery thrombosis
(2) epidural hematoma
(3) adhesive arachnoiditis
(4) epidural abscess

1372. Compared to IV dosing of opioids, intrathecal opioids

(1) may cause urinary retention
(2) are more potent at equal doses
(3) can cause delayed respiratory depression
(4) seldom cause pruritus

1373. Nonpharmacologic methods of pain control include

(1) transcutaneous electrical nerve stimulation (TENS)
(2) acupuncture
(3) biofeedback
(4) behavior modification

1374. Active agents useful in the performance of a neurolytic block include

(1) 40% potassium hydroxide
(2) 50% alcohol
(3) 10% glycerin
(4) 10% phenol

1375. Advantages of TENS for postoperative pain include

(1) no patient instruction is needed
(2) absence of opioid-induced side effects
(3) it can be used by patients with pacemakers
(4) it provides an element of patient participation

1376. An elderly gentleman in pain clinic complains of right-sided thoracic pain. He reports having a very painful rash about 6 months ago. The lesions are gone, but he insists that the pain persists. Beneficial method(s) of therapy has (have) been demonstrated to include

(1) gabapentin
(2) TENS
(3) tricyclic antidepressants
(4) opioids

1377. Epidural steroid injections may be effective in which of the following conditions?

(1) herniated nucleus pulposus without neurologic deficit
(2) ankylosing spondylitis
(3) herniated nucleus pulposus with nerve root compression
(4) functional low back pain

1378. Pharmacologic methods for treating patients with chronic pain may include

(1) nonsteroidal anti-inflammatory agents
(2) anticonvulsants
(3) tricyclic antidepressants
(4) opioids

1379. Spinal cord stimulation

(1) always requires an external generator
(2) involves an electrical stimulator placed in the epidural space
(3) is useful for postoperative pain management
(4) may be helpful in patients with intractable back pain

1380. Complications of spinal cord stimulation include

(1) lead migration
(2) infection
(3) lead breakage
(4) bleeding

1381. A patient with obstructive lung disease develops a complex regional pain syndrome involving the right arm after an injury. Treatment of the syndrome may involve

(1) stellate ganglion block
(2) physical therapy
(3) surgical sympathectomy
(4) TENS

1382. The placebo effect occurs

(1) only in patients who are emotionally unstable
(2) with normal individuals
(3) only in those who are faking pain
(4) with pharmacologically active substances

1383. Indications for a lumbar sympathetic block include

(1) acute herpes zoster
(2) complex regional pain syndrome
(3) peripheral vascular insufficiency
(4) causalgia

1384. Nerve blocks may be used to aid in the diagnosis of chronic pain. Such blocks

(1) may temporarily relieve pain
(2) use 6% phenol as the agent
(3) are not contraindicated, especially if the patient refuses
(4) seek a permanent interruption of the pathways

1385. Of the two principal neurolytic agents, alcohol and phenol,

(1) alcohol has the greater tendency to produce neuritis
(2) alcohol is used in 6% concentration
(3) alcohol has the more rapid onset of action
(4) phenol is used in a 50% concentration

1386. Pain theories currently being proposed postulate a

(1) straight stimulus-to-nervous system path
(2) system involving large fibers only
(3) system with inhibition exerted by small fibers
(4) system involving excitatory and inhibitory input with resulting sensation

1387. Which of the following are possible advantages of acupuncture over other methods of pain therapy?

(1) safety, since the physiology of the patient is not disturbed
(2) a tonic or regulatory effect on the body
(3) no residual effects
(4) an anti-inflammatory effect

1388. In the neurolytic treatment of pain in a patient with pain from cancer, which of the following is TRUE?

(1) Alcohol injections around a peripheral nerve may produce an uncomfortable neuritis.
(2) Loss of bowel or bladder function may occur.
(3) Such treatment should be used only in those with terminal disease.
(4) Absence of sensation may be perceived by the patient to be worse than the pain.

Answers and Explanations

1348. (C) Phantom limb pain is an example of central pain. Postherpetic neuralgia can be considered to be a type of neuropathic pain. The other options are examples of somatic pain. Central pain can be very difficult to treat. *(4:2346)*

1349. (A) Occipital neuralgia involves the greater occipital nerve and leads to a chronic headache that may extend to the shoulder or forward to the area around the eye. It may be due to compression of the occipital nerve within the skull. The scapular nerve is not involved nor is the cervical plexus. The occipital nerve block may be useful in diagnosis and treatment. *(1:721)*

1350. (E) Postherpetic neuralgia most frequently occurs in elderly patients due to reactivation of the varicella zoster virus. It is usually refractory to opioids. Some patients respond to tricyclic antidepressants or to the topical administration of capsaicin, which causes depletion of substance P. *(1:1448)*

1351. (C) The signs of complex regional pain syndrome are hyperesthesia, extremities that may be cold or warm, pain that is long lasting, and trophic changes. These may include brittle nails and osteoporosis. *(1:1446; 5:2775)*

1352. (E) The size of the needle used to deposit the neurolytic agent is not important. What is important are the location of the injection (proximity to the nerve), the concentration of the neurolytic agent, the volume of the agent, and the histology of the nerve. Smaller nerves are easier to block than larger nerves. *(1:1453; 5:2776)*

1353. (A) All of choices listed are thought to be inhibitory modulators in the dorsal horn of the spinal cord with the exception of substance P. Substance P is found in the synaptic vesicles of unmyelinated C-fibers. It has been shown to aggravate pain. *(1:1404)*

1354. (B) There is some controversy concerning the usefulness of intraoperative intercostal blocks. Some reports did not demonstrate any difference in postoperative ventilation. Most current authors have found that the postoperative course is easier and shorter if blocks are done at the time of the thoracotomy. Most use bupivacaine. There is still a danger of a total spinal block and of administration of a toxic dose of the local anesthetic drug. *(1:730; 5:2743)*

1355. (D) Cutaneous nociceptors are defined by the fiber type and the type of stimuli they respond to. Merkel's disks and Meissner's corpuscles are touch receptors. Pacinian corpuscles and Golgi-Mazzoni endings sense pressure. *(1:1435; 4:2346)*

1356. (D) The celiac plexus block is useful for pain associated with upper abdominal malignancy. The most common complication is hypotension. *(1:735; 5:1711)*

1357. (C) The demand dose, lockout interval, and 1-hour limit should be individually adjusted according to the patient's physiologic status and requirement for analgesia. *(4:2330; 5:2734)*

1358. (E) A continuous IV infusion of naloxone has been shown to be effective in relieving nausea and vomiting secondary to epidural morphine. Droperidol or transdermal scopolamine may be effective, but they are not as specific or rapid acting. *(5:2740)*

1359. (B) Postdural puncture headache is thought to result from decreased CSF pressure due to CSF lost from the dural puncture site. Early ambulation is not a risk factor; however, the size of the needle used and female gender are risk factors. An epidural blood patch, although a highly effective means of treating a postdural puncture headache, is not always necessary. *(1:707; 5:1383)*

1360. (D) Lumbar sympathetic blockade has been shown to be effective in all of the pain syndromes listed, except lumbar facet syndrome. Heat, nonsteroidal anti-inflammatory agents, and facet injections may be useful. *(1:1445–8; 5:1486)*

1361. (C) Butorphanol is a mixed agonist-antagonist and as such can precipitate a withdrawal or abstinence syndrome in a patient who is tolerant to opioids. The other drugs listed are pure μ-opioid agonists. *(1:368; 5:419)*

1362. (D) The most likely diagnosis in this case is complex regional pain syndrome. The interruption of sympathetic stimulation via a stellate ganglion block will aid in providing a diagnosis. *(1:1445; 5:2775)*

1363. (B) The lumbar sympathetic chain is a bilateral structure anterolateral to the lumbar vertebra. It is posterior to the abdominal aorta and not visible by fluoroscopy. The target area for a celiac plexus block is the anterolateral aspect of the T12–L1 junction. *(1:735; 5:2775)*

1364. (A) The general consensus is that removal of an epidural catheter occurs 12 hours after the last dose of LMWH. Unlike unfractionated heparin, measuring the activated clotting time or the activated partial thromboplastin time cannot monitor the effects of LMWH. *(1:1422)*

1365. (A) Symptoms of myofascial pain include pain in a nondermatomal distribution. The pain can be elicited by palpating the muscles near the affected area. A mainstay of treatment is the injection of local anesthetic into the trigger point. *(1:1445; 5:2772)*

1366. (B) Nociceptive information travels via the spinothalamic tract. The cell body of the primary afferent is located in the dorsal horn. From there the impulse travels via the spinothalamic tract. Branches may synapse in areas of the brain stem, including the periaqueductal gray and the nucleus raphe magnus. The thalamus and sensory cortex are also important in the pain pathway. *(1:1404; 5:2706)*

1367. (B) Nociceptive input is transmitted from free nerve endings to A-delta and C-fibers. *(1:1406; 5:2706)*

1368. (E) Causalgia is the term that is sometimes used to indicate sympathetically maintained pain in association with a major nerve injury. Complex regional pain syndrome is now the current accepted term. *(1:1444)*

1369. (A) The use of an intrapleural catheter can be effective for unilateral postoperative pain such as cholecystectomy. A pneumothorax can occur during catheter placement. Since only local anesthetic is injected, the use of parenteral opioids is not contraindicated. *(1:1327; 5:1829)*

1370. (B) Most published guidelines for the treatment of cancer pain include the concept of dosing medications, including opioids, to adequately treat pain on an around-the-clock basis. Respiratory depression is unusual in opioid-tolerant patients. The use of adjuvant drugs including antidepressants and anticonvulsants may be useful. Side effects such as nausea and constipation should be anticipated and treated. *(1:1451; 5:2770)*

1371. (C) A presenting symptom of a space-occupying lesion in the spinal canal is backache. This could possibly be due to an epidural hematoma or abscess. Anterior spinal artery thrombosis presents as painless paraplegia. Adhesive arachnoiditis results in a progressive loss of nerve function. *(1:709; 5:2702)*

1372. (A) The amount of opioid administered intrathecally to provide analgesia is typically much smaller than that which is given intravenously or intramuscularly. The delayed onset of respiratory depression is due to opioid transport from the CSF to brain stem respiratory centers. The frequency of pruritus can be as high as 90%. Urinary retention can also occur. *(1:1477; 5:2739–41)*

1373. (E) All of the nonpharmacologic modalities outlined above have been shown to offer relief to certain select patients with chronic pain. *(1:1455)*

1374. (C) Alcohol (50 to 100%) and phenol (5 to 20%) are both neurolytic agents. Glycerin is often added to phenol to make its specific gravity greater than that of cerebrospinal fluid when used for a hyperbaric spinal technique. *(1:1453; 5:2775)*

1375. (C) TENS provides a "tingling" sensation that is not usually perceived as pain. Patients should be instructed in the use of the stimulator. It has the advantages of allowing the patient to participate in his or her therapy and the lack of opioid-induced side effects. The presence of a cardiac pacemaker is considered a relative contraindication. *(1:1424; 4:2777)*

1376. (B) This patient has postherpetic neuralgia. Tricyclic antidepressants are often the most effective in treating this difficult condition. The anticonvulsant gabapentin has also been shown to be useful. The use of TENS or opioids is helpful in the occasional patient, but usually the results are disappointing. *(1:1448)*

1377. (B) An indication for epidural steroid injections is nerve root irritation and accompanying inflammation. The use of epidural steroids in low back pain is an area of controversy; however, most experts agree that a trial of epidural steroids may be indicated in radicular pain of less than 12 months duration. Epidural steroid injections have been found to be ineffective in the relief of ankylosing spondylitis and functional low back pain. *(1:1443; 1:1448)*

1378. (E) All of the aforementioned classes of drugs have been shown to be effective in the management of chronic pain states. The use of opioids in chronic nonmalignant pain is controversial, but it has been shown to be effective in select patients. *(1:1448; 5:2769)*

1379. (C) Spinal cord stimulation involves the placement of stimulating electrodes in the epidural space. Typically the generators are implanted. With proper patient selection, it has been shown to be useful in several chronic pain conditions, including ischemic pain and "failed back syndrome" with a radicular component. *(1:1457; 5:2774)*

1380. (E) All items listed are potential complications of spinal cord stimulators. *(1:1457; 5:2774)*

1381. (E) All of the options listed are important for the patient with a complex regional pain syndrome. In the patient with obstructive lung disease, one may want to avoid the stellate ganglion approach, since there is a possibility of pneumothorax which could be life threatening. *(1:1446)*

1382. (C) The placebo effect is present in all therapeutic situations. It is not seen only in those who are unstable or who are faking symptoms, but is seen in normal patients. *(1:52; 5:2744)*

1383. (E) All of the syndromes listed can be treated or diagnosed with a lumbar sympathetic block. *(5:2775)*

1384. (B) In many cases of chronic pain, one can do a diagnostic block to establish the type of pain present. This usually involves the use of a local anesthetic. Diagnostic blocks are not done with a neurolytic agent, such as phenol, which causes permanent effects. Patient refusal is an absolute contraindication. *(1:145; 5:2775)*

1385. (B) Alcohol has a faster onset of action but a greater tendency to produce neuritis. It is used in concentrations of 50 to 100%. Phenol is used in concentrations of 5 to 20%. *(1:1453)*

1386. (D) Current pain theories involve the small and large nerves and a path through the dorsal root. Both inhibitory and excitatory nerves are involved. *(1:1403; 4:2346; 5:2730)*

1387. (B) Acupuncture, the insertion of needles into specific points in the body, has the advantage of being a relatively benign mode of therapy for pain. It has been shown to have effects on the brain and endocrine system. *(4:2366; 5:612)*

1388. (E) In treating chronic pain, one must remember that the patient's perception of the pain may not be as bad as the loss of bowel and urinary control and the absence of sensation. This must be thoroughly discussed with the patient prior to proceeding. Most pain specialists believe that neurolytic blocks should be reserved for the patient with a short life expectancy. *(1:1453–4)*

CHAPTER 16

Complications of Anesthesia and Quality Assurance

Questions

DIRECTIONS (Questions 1389 through 1417): Each of the numbered items or incomplete statements in this section is followed by answers or by completions of the statement. Select the ONE lettered answer or completion that is BEST in each case.

1389. A factor NOT involved in the stretch mechanism responsible for brachial plexus injury is the

(A) pectoralis minor tendon attachment to the coracoid process
(B) clavicle
(C) first rib
(D) subclavian artery
(E) head of the humerus

1390. After induction of anesthesia, an armored tube is inserted into the patient's trachea, and the cuff is inflated. However, it is not possible to ventilate the patient with a pneumatic ventilator. Which of the following is NOT a possible cause of this?

(A) the anesthesia machine is not connected to the wall electrical outlet
(B) herniation of the cuff
(C) intraluminal herniation
(D) impingement of the tip against the trachea
(E) kinking

1391. The most common eye injury sustained under anesthesia is

(A) corneal perforation
(B) conjunctivitis
(C) uveitis
(D) corneal abrasion
(E) retinal artery thrombosis

1392. A patient who develops a corneal abrasion during anesthesia may be treated or diagnosed with all of the following EXCEPT

(A) an eye patch
(B) a topical anesthetic to relieve pain
(C) an antibiotic ointment
(D) a cycloplegic
(E) fluorescein

1393. In the intensive care unit, a patient with a pulmonary artery catheter suddenly develops hemoptysis. The most likely explanation is

(A) pneumonia
(B) postoperative hemorrhage
(C) disconnected intravenous (IV) tubing
(D) pulmonary artery rupture
(E) sepsis

1394. Complications of cricothyroid membrane puncture include all of the following EXCEPT

(A) subcutaneous emphysema
(B) bleeding
(C) pneumothorax
(D) hoarseness
(E) bronchial tear

1395. Hypoxia secondary to failure of the oxygen supply may be due to all of the following EXCEPT

(A) a cracked flowmeter
(B) transposition of pipes during construction
(C) failure of the reducing valve
(D) an accidental change of flowmeter setting
(E) a shift of the oxyhemoglobin dissociation curve

1396. A high-volume, low-pressure cuff

(A) may not prevent aspiration
(B) has more tendency to leak fluids in patients on controlled ventilation
(C) will protect from aspiration if filled to specifications
(D) is better if vaginations are present along the wall
(E) will seal better in patients who are breathing spontaneously

1397. A patient with upper gastrointestinal tract bleeding is intubated with a high-volume, low-pressure cuff endotracheal tube. There is blood present in the tracheal aspirate each time she is suctioned. The most likely cause of this is

(A) tracheoesophageal fistula
(B) tracheal bleeding
(C) a bleeding diathesis
(D) aspiration of regurgitated blood
(E) burst cuff on endotracheal tube

1398. The differential diagnosis of cyanosis during or immediately following anesthesia includes all of the following EXCEPT

(A) asphyxia from diffusion anoxia
(B) central respiratory stimulation
(C) upper respiratory tract obstruction
(D) atelectasis
(E) respiratory paralysis from high spinal

1399. The most common cause of anesthetic disasters is

(A) aspiration pneumonia
(B) hepatitis
(C) circulatory instability
(D) malignant hyperthermia
(E) hypoxemia

1400. You are working in a newly renovated operating room. After induction of anesthesia with thiopental, you deliver a mixture of nitrous oxide and oxygen, each at 2 L/min. The nitrous oxide is turned off before intubation, and the patient becomes cyanotic. Flushing with oxygen does not help. The problem that must be ruled out is

(A) the patient may be obstructed
(B) the patient may have had a myocardial infarction
(C) the patient may be febrile
(D) the oxygen and nitrous oxide lines may have been switched
(E) there is no flow from the machine

1401. The one check that is mandatory to detect the problem in Question 1400 is

(A) check for tightness of the circuit
(B) check tanks for fullness
(C) have a functional oxygen analyzer
(D) check endotracheal tube for patency
(E) check flush valve on machine

1402. All of the following are complications of tracheostomy EXCEPT

(A) infection of the tracheobronchial tree
(B) tracheal erosion
(C) arytenoid cartilage damage
(D) stomal stricture
(E) plugged tube

1403. The highest incidence of intraoperative awareness occurs during which type of surgery?

(A) cardiac surgery
(B) obstetrical emergencies
(C) neurosurgery
(D) pediatric surgery
(E) trauma surgery

1404. The treatment of suspected transfusion reaction includes all of the following EXCEPT

(A) stopping the transfusion
(B) infusion of 500 mL of 20% mannitol
(C) treating hypotension
(D) giving fluids at a rate sufficient only to keep IV lines open
(E) alkalinizing the urine

1405. In morbidly obese patients, obstructive sleep apnea

(A) symptoms correlate with daytime oxygen saturation
(B) symptoms improve with weight reduction
(C) can be reliably predicted by body mass index (BMI)
(D) is strongly associated with hypertension
(E) correlates with neck circumference

1406. All of the following are true concerning the National Practitioner Data Bank (NPDB) EXCEPT

(A) a practitioner may make a query about his or her file at any time
(B) it is a central repository for medical licensing information
(C) limitations to a physicians' hospital privileges are reported
(D) medical malpractice payments are reported
(E) voluntary resignations of staff privileges are reported

1407. The statistical term for the spread of individual values in data with a normal distribution is

(A) mean
(B) median
(C) variance
(D) standard deviation
(E) standard error of the mean

1408. Patients who undergo outpatient surgery with isoflurane should not drive or operate machinery for at least

(A) 30 minutes
(B) 1 hour
(C) 4 hours
(D) 8 hours
(E) 16 hours

1409. The motto on the seal of the American Society of Anesthesiologists (ASA) is

(A) Safety
(B) Vigilance
(C) Excellence in Medicine
(D) Ever Forward
(E) Concern

1410. In obtaining informed consent, an anesthesiologist should

(A) have the surgeon explain the anesthesia procedures to the patient
(B) tell the patient only the common problems
(C) not paint too gloomy a picture lest the patient be frightened
(D) disclose the potential for death or serious harm
(E) answer only specific questions

1411. Once a fire is discovered during laser airway surgery, the FIRST thing that the anesthesiologist must do is

(A) extinguish the burning material
(B) ventilate with 100% oxygen
(C) perform a bronchoscopy
(D) stop ventilation
(E) call for help

1412. A 65-kg 75-year-old woman is undergoing a cataract extraction under monitored anesthesia care (conscious sedation). You administer 20 mg of methohexital intravenously prior to the performance of a retrobulbar block by the surgeon. As the block is performed, the heart rate, as measured by the electrocardiogram, is 40. The proper course of action to take next is

(A) endotracheal intubation
(B) administer atropine 0.4 mg IV
(C) cancel the surgery pending pacemaker insertion
(D) request that the surgeon stop manipulation
(E) start chest compressions

1413. You are asked to anesthetize a woman for an open cholecystectomy. She is 5 ft tall and approximately 285 lb. Potential problems than you might encounter related to her known obstructive sleep apnea include all of the following EXCEPT

(A) difficult mask ventilation
(B) associated polycythemia
(C) difficult laryngoscopy
(D) associated systemic hypertension
(E) sensitivity to the depressant effects of opioids

1414. The most common anesthesia-related reason for admission to the hospital following an outpatient surgical procedure is

(A) pain
(B) inability to void
(C) lack of escort
(D) nausea and vomiting
(E) somnolence

1415. Oxygen toxicity

(A) develops after breathing 100% oxygen for 12 hours
(B) is not dose related
(C) develops after 36 hours of exposure to 25% oxygen
(D) is due specifically to the oxygen molecule
(E) is so important that 100% oxygen should never be used

1416. The nervous structures most likely involved in improper positioning are nerves of the

(A) cervical plexus
(B) brachial plexus
(C) lumbar plexus
(D) sciatic system
(E) autonomic system

1417. Drugs useful in the treatment or prevention of postoperative nausea and vomiting include the following EXCEPT

(A) prochlorperazine
(B) ondansetron
(C) cimetidine
(D) dexamethasone
(E) droperidol

DIRECTIONS (Questions 1418 through 1450): For each of the items in this section, ONE or MORE of the numbered options is correct. Choose answer

(A) if only 1, 2, and 3 are correct
(B) if only 1 and 3 are correct
(C) if only 2 and 4 are correct
(D) if only 4 is correct
(E) if all are correct

1418. Complications of retrobulbar blockade include which of the following?

(1) simulation of the oculocardiac reflex arc
(2) central retinal artery occlusion
(3) retrobulbar hemorrhage
(4) brain stem anesthesia

1419. Which of the following statements concerning anesthesia for laser airway surgery are TRUE?

(1) The cuff of a foil-wrapped endotracheal tube is invulnerable to the laser.
(2) The endotracheal tube cuff should be filled with saline.

(3) A fire cannot occur with a properly shielded tube.
(4) Endotracheal tube fires are the most common complication of CO_2 laser laryngeal surgery.

1420. Risk factors involved in perioperative ischemic optic neuropathy include

(1) blood loss
(2) decreased systolic blood pressure
(3) retrobulbar hemorrhage
(4) decreased intraocular pressure

1421. Factors contributing to the incidence of postdural puncture headache include

(1) gender
(2) needle bevel
(3) needle size
(4) early ambulation

1422. Quality assurance (QA) in anesthesia

(1) aims to improve the quality of anesthesia care
(2) blames individual anesthesiologists for poor outcomes
(3) should assess patient satisfaction
(4) is designed to discipline anesthesiologists

1423. If a venous air embolism is suspected, which of the following maneuvers should be performed?

(1) flooding the surgical field with saline and packing it with surgical sponges
(2) aspiration of air via RA catheter
(3) compression of neck veins
(4) discontinuation of nitrous oxide

1424. Immediately after the uneventful induction of general anesthesia and placement of a laryngeal mask airway (LMA), you note gastric contents in the airway tubing. The best course of action includes

(1) suctioning the patient's airway
(2) lateral head positioning
(3) endotracheal intubation
(4) immediate positive pressure ventilation

1425. Bacteremia may be hazardous for those with heart lesions or prosthetic valves. If an operative procedure is required

(1) the patient should always be anesthetized under mask
(2) oral intubation is more likely to cause bacteremia than is nasal intubation
(3) intubation is not a source of bacteremia
(4) nasotracheal intubation should be avoided if possible

1426. In the prone position, points that must be protected from pressure and compression include

(1) eyes
(2) male genitalia
(3) breasts
(4) toes

1427. A patient has right-sided internal jugular catheterization done while under anesthesia. Complications that may occur include

(1) hematoma
(2) arrhythmias
(3) air embolism
(4) chylothorax

1428. The needle used for nerve block is more likely to break if

(1) it is inserted into the skin to the hub
(2) its direction is changed without withdrawal to the subcutaneous tissue
(3) it has been bent previously
(4) it has a security bead

1429. A cachectic 27-year-old woman is undergoing anesthesia for carcinoma of the cervix. Blood loss required the use of blood transfusion under pressure via two large-bore cannulae. During this time, systolic arterial pressure decreased from 90 to 75 mmHg. The reason(s) for this drop in pressure may be

(1) anesthetic overdose
(2) acidosis
(3) cardiac arrhythmia
(4) hypocalcemia

1430. The patient with a full stomach

(1) should have regional anesthesia to protect against aspiration
(2) can have light general anesthesia by mask with no danger of aspiration
(3) should be intubated in the head-down position
(4) should remain intubated until the reflexes have returned

1431. Anaphylaxis occurring during anesthesia

(1) always begins with laryngeal edema
(2) always includes laryngeal, circulatory, and respiratory symptoms
(3) is always of short duration
(4) must always be treated vigorously

1432. Complications of nasotracheal intubation include

(1) sinusitis
(2) nosebleed
(3) nasal necrosis
(4) broken teeth

1433. Hypoxia secondary to failure of the oxygen supply may be due to

(1) depletion of the oxygen cylinder
(2) insufficient opening of the cylinder to permit free flow of gas
(3) failure of gas pressure
(4) failure to open the valve of the system

1434. Factors that predispose to aspiration include

(1) old age
(2) debilitation
(3) alcoholic stupor
(4) impaired swallowing function

1435. Hypocalcemia occurring during blood transfusion

(1) follows rapid transfusion
(2) is of no clinical significance
(3) is aggravated by hypothermia
(4) need not be treated

1436. At the end of a procedure for dental extraction, the patient is "crowing" on ventilation. This

(1) may signify complete laryngospasm
(2) may be relieved by gentle assistance to ventilation
(3) will require administration of succinylcholine
(4) may signify partial glottic closure

1437. The following are considered to be "higher risk" operations for the development of ischemic optic neuropathy

(1) transurethral resection of prostate (TURP)
(2) open heart surgery
(3) orthopedic surgery
(4) spinal surgery

1438. The prerequisites for an explosion in the operating room include

(1) an explosive gas, not necessarily an anesthetic
(2) a supply of oxygen
(3) a relatively confined space
(4) a source of ignition

1439. Statistics useful in interpreting continuous numerical data include

(1) the mean
(2) variance
(3) standard deviation
(4) Student's *t*-test

1440. The doctrine of *res ipsa loquitur* is applicable where the

(1) accident is of a kind that usually does not happen without negligence
(2) apparent cause of the accident is within the control of the defendant
(3) plaintiff could not have contributed to the accident
(4) defendant wants to settle out of court

1441. If it is discovered at the end of a procedure that a tooth is missing, the anesthesiologist should

(1) awaken the patient as usual
(2) check the oropharynx and nasopharynx
(3) ask the family about the tooth when the patient is in the recovery room
(4) get x-rays of the head, chest, and abdomen

1442. You are requested to perform anesthesia for laser excision of condylomata. If you are in the vicinity of the laser plume, the following protective equipment should be worn

(1) high-efficiency filter mask
(2) latex gloves
(3) goggles
(4) respirator

1443. The consequences of intraoperative awareness and subsequent recall may include which of the following?

(1) nightmares
(2) sleep disturbances
(3) flashbacks
(4) daytime anxiety

1444. Hazards to personnel known to be present in operating rooms include an increase in

(1) the incidence of spontaneous abortions
(2) the number of congenital abnormalities in the children of exposed males and females
(3) liver disease
(4) cancer

1445. The morbidity and mortality of patients undergoing corrective surgery for hip fractures

(1) are significantly lower with spinal anesthesia
(2) depend on the agent used
(3) are significantly higher with general anesthesia
(4) are essentially the same for spinal and general anesthesia

1446. Reactions to transfusions are

(1) immediate
(2) fatal in most cases
(3) always of a hemolytic nature
(4) to be anticipated and must be monitored carefully

1447. The incidence of hepatitis after blood transfusion

(1) is usually due to hepatitis C
(2) is one case per 100 units transfused
(3) is lower if blood donation is limited to volunteers
(4) is higher after transfusion of red cells

1448. A patient who has a bullous type of skin disorder requires an anesthetic technique that avoids

(1) the use of tape
(2) intubation
(3) friction
(4) deep anesthesia

1449. Complication(s) of a celiac plexus block include(s)

(1) postural hypotension
(2) local anesthetic toxicity
(3) impairment of hip flexion
(4) total spinal blockade

1450. Symptoms referable to positioning may involve the

(1) circulatory system
(2) respiratory system
(3) nervous system
(4) neuromuscular system

Answers and Explanations

1389. (D) The subclavian artery is not involved in brachial plexus injuries. The other options are important. One should check to determine if there is any tension in the position. The arm should be as free of tension as possible, and this should be checked frequently during the procedure. *(1:646; 5:1154)*

1390. (A) There are many causes of ventilatory problems during a procedure. It is important to have an approach in mind when this type of situation occurs. An advantage of a pneumatic ventilator is that it does not require electricity to operate. Thus, not having the machine connected to the wall electrical outlet will not cause this specific problem, although it may be associated with serious consequences. Although an armored endotracheal tube is unlikely to kink, a kink may still be present in tubing leading to or from the carbon dioxide absorber which may lead to an inability to ventilate. *(1:583; 5:298)*

1391. (D) The most common eye problem under anesthesia is corneal abrasion. It is important to have the eyes protected at all times. Corneal abrasion may occur in any position, although it is more common if the head is draped. Retinal artery thrombosis is a very serious problem, usually caused by pressure on the eyes. *(1:1396; 5:1157)*

1392. (B) Corneal abrasions are very painful and should be adequately treated. The use of topical anesthesia is not recommended because of its tendency to retard regeneration of the corneal epithelium. In addition, if a topical anesthetic is placed in the eye and the patient rubs the eye as he or she emerges from anesthesia, further damage may be incurred. Cycloplegic agents are used to reduce ciliary spasm and photophobia. Fluorescein staining is useful in detecting corneal abrasions. *(1:1396; 5:1157)*

1393. (D) Pulmonary artery rupture is a serious complication of pulmonary artery catheterization. The nonbleeding lung must be protected by the insertion of a double lumen endotracheal tube. The patient must be volume resuscitated and if it exists, the coagulopathy should be corrected. Surgical exploration and lung resection may be required. *(1:818–9)*

1394. (E) Bronchial tear should not occur since the bronchi are more distal. The other options are possible. *(1:630; 5:2547)*

1395. (E) Anesthesia machines must be checked before use, since the oxygen path may be interrupted at many stages. Many of the leaks are hard to detect. These defects do not affect the oxyhemoglobin dissociation curve. *(4:216; 5:274)*

1396. (A) A high-volume, low-pressure cuff has been associated with cases of aspiration. The presence of vaginations decreases pressure but allows a track for fluids. The cuff is still helpful in protecting the trachea from injury but it is not full protection from regurgitation. Other methods should be used with it (e.g., nasogastric suctioning and antacids). *(1:1391–2; 5:1630)*

1397. (D) As discussed in the answer to Question 1396, regurgitation may still occur. Although this is the most likely problem, the other causes should be ruled out. *(1:1391–2; 5:1630)*

1398. (B) Central respiratory stimulation will not lead to cyanosis. Diffusion anoxia may lead to cyanosis, as may obstruction of the airway. Atelectasis and respiratory paralysis may also cause compromised oxygenation. *(5:2712)*

1399. (E) The most common cause of anesthetic disasters is hypoxemia. Hypoxemia is the endpoint of other problems (e.g., esophageal intubation, low blood pressure, and machine problems). For this reason, it is important to monitor carefully and do machine checks carefully. *(1:670)*

1400. (D) Any time there is construction activity in an operating room suite, one must check the gases being delivered. The most common problems are with the patient (e.g., obstruction or cardiovascular disease), but one must make sure the oxygen is getting to the patient. *(5:276)*

1401. (C) A functional oxygen analyzer is mandatory to check the gases being delivered. All of the other checks are important, but an analyzer is a must. *(1:589; 5:307)*

1402. (C) Arytenoid damage does not occur with tracheostomy. Infection may occur. Tracheal erosion may occur as a long-term problem. Stricture of the stoma may occur and require revision. The tube may be plugged with secretions. *(5:2546)*

1403. (E) The incidence of intraoperative awareness in trauma is estimated to be between 11 and 43%. In contrast, the incidence of awareness in nonobstetric or noncardiac cases is estimated to be approximately 0.2%. *(5:1239)*

1404. (D) If transfusion reaction occurs, fluids should be given in large amounts to maintain urine flow. Mannitol is given for the same reason. The rest of the options are true. *(1:207; 5:1818)*

1405. (B) Morbid obesity (BMI > 30 kg/m^2) is a common, but not absolute, risk factor for obstructive sleep apnea. It is defined as the presence of 5 or more apneic periods per hour or 15 episodes per 7-hour study. Due to the activation of the sympathetic nervous system by repeated arousal these patients are often hypertensive. Obstructive sleep apnea cannot be reliably predicted by BMI or daytime oxygen saturation. There is no direct correlation with neck circumference. The symptoms usually improve after weight loss. *(5:1031)*

1406. (E) The NPDB is a nationwide repository of information concerning physicians' licensing, medical malpractice payments, and hospital privileges. The NPDB does not include voluntary resignations of hospital privileges. All the other choices are true. *(1:93)*

1407. (D) The statistical term for the spread of individual values of data with a normal distribution is the standard deviation. The mean is the average of the data, and the standard error of the mean is a precision factor. Variance is the difference of each value from the mean. *(1:56; 4:760)*

1408. (E) Most anesthesiologists tell patients not to drive or operate machinery until the next day, but there are some data suggesting that 2 days may be more appropriate. *(1:1234; 5:2621)*

1409. (B) This motto is seen on the cover of every issue of *Anesthesiology*. It is just as important today when we have many monitors as it was when the ASA was founded. *(1:68; 5:2621)*

1410. (D) In the preanesthetic discussion, one must be honest with the patient. The common problems must be discussed, and the possibility of death or serious problems should be explained. It is not necessary to be exhaustive in the discussion of problems, but potential problems related to the particular procedure or the particular patient should be discussed. *(1:89; 5:3179)*

1411. (D) Once a fire is discovered in the airway during laser surgery, the first thing the anesthesiologist must do is stop ventilation. One should also disconnect the circuit to stop the supply of oxygen-enriched gas. Once the fire is extinguished, the patient should be ventilated with 100% oxygen. Direct laryngoscopy should be performed initially, followed by bronchoscopy, to examine the airway. *(5:2584)*

1412. (D) The oculocardiac reflex may be precipitated by the performance of a retrobulbar block. The afferent branch is via the trigeminal nerve. The efferent nerve is the vagus nerve. If severe or persistent bradycardia exists, the administration of IV atropine is recommended. *(1:973; 5:739)*

1413. (B) There is no associated correlation with polycythemia and obstructive sleep apnea. All the other choices are correct. *(5:1031)*

1414. (D) Studies have shown that nausea and vomiting are leading anesthesia-related reasons for admission after ambulatory surgery. *(1:1233; 5:2620)*

1415. (A) Oxygen toxicity develops after 12 hours of exposure to 100% oxygen. One can breathe 25% oxygen for an indefinite period of time. In some circumstances 100% oxygen is required. It should be used as long as needed, but steps should be taken to decrease the concentration as soon as possible. *(1:1390–1; 2676–7)*

1416. (B) The brachial plexus is the most frequent source of nerve injuries, although this may be a reflection of the numbers of procedures in which this structure and its nerves are at risk. It is important to remember that all nerves should be protected. *(1:607; 5:1154)*

1417. (C) Cimetidine alone was shown to be no better than a placebo in the prevention of postoperative nausea and vomiting. The other medications all have antiemetic effects. *(1:1396; 5:2710)*

1418. (E) All of the complications listed above are possible. Due to the potential for airway and cardiac difficulty, persons skilled in airway management and cardiac resuscitation should be immediately available. *(1:977; 5:1709)*

1419. (C) Endotracheal tube fires are the most common complication of CO_2 laser laryngeal surgery. Fires can still occur, even though a properly shielded endotracheal tube is used. The endotracheal tube cuff is left unwrapped and is therefore vulnerable to the laser. It is recommended that the endotracheal tube cuff be filled with saline to act as a "heat sink" if the cuff is perforated. *(1:997; 5:2543)*

1420. (A) The risk factors for perioperative ischemic optic neuropathy are ill-defined. They can include blood loss, anatomic variation of the blood supply to the optic nerve, emboli, vasopressors, the presence of systemic disease, and retrobulbar hemorrhage. Increased intraocular or orbital venous pressure has been associated with ischemic optic neuropathy. *(1:984–5; 5:3005–11)*

1421. (A) Females have a greater incidence of postdural puncture headaches than males. The use of a large needle will increase the likelihood of a postdural puncture headache, as will the insertion of the bevel of the needle perpendicular to the longitudinally directed dural fibers. Early ambulation will not increase the incidence of postural puncture headache. *(1:707; 5:2328)*

1422. (B) Quality assurance in anesthesia is a process ensuring the continual improvement of anesthetic practice and outcome. It is not intended to blame or discipline individual anesthesiologists. *(1:93; 4:21)*

1423. (E) All of the actions listed should be done; in addition, circulatory support should be provided if needed. *(1:766–7; 5:2139–42)*

1424. (A) The best course of action when aspiration occurs at induction would include (after removing the LMA) turning the patient's head to the side, suctioning the oropharynx, and since this patient is unable to protect his or her airway, endotracheal intubation. If possible, before positive pressure ventilation is begun, the trachea should be suctioned. *(1:1392)*

1425. (D) Nasotracheal intubation is a source of bacteremia. The procedure should be done without the nasotracheal tube unless it is needed. The patient should receive prophylactic antibiotics as recommended by the American Heart Association. *(4:1431; 5:1078)*

1426. (E) The prone position provides many opportunities for positioning problems. The pressure points must be carefully checked and padded as necessary. *(1:1111; 5:2412)*

1427. (A) Chylothorax may occur with left (but not with right) internal jugular cannulation. Air embolism may occur if air is allowed to enter the catheter. Hematoma is a possibility. Arrhythmias may occur if the catheter enters the heart and stimulates the sensitive areas of the right ventricle. *(1:675; 4:1176; 5:1293)*

1428. (A) Needles with security beads cannot be inserted to the hub and, therefore, are less likely to break. Needles should never be inserted to the hub. If a change in needle direction is attempted without withdrawing it to the skin, the needle may be weakened. Bending weakens the needle. *(1:715)*

1429. (E) All of these options are possible causes of hypotension, however, decreased levels of ionized calcium occur at high rates of transfusion. *(1:205; 5:1815)*

1430. (D) Regional anesthesia does not guarantee that aspiration will not occur. The head-down position is not appropriate, since passive regurgitation is more likely to occur in that position. Once intubated, the patient should remain intubated until awake and able to protect the airway. *(1:1044; 5:1635)*

1431. (D) Anaphylaxis is treated immediately. It may begin with bronchospasm or an intense rash. The first symptom may be hypotension and no wheezing may be heard. The duration may be prolonged. *(1:1300; 5:1092)*

1432. (E) Nasotracheal intubation can be complicated by sinusitis, nosebleed, and nasal necrosis. It is possible for dental damage to occur, especially if endotracheal tube placement is aided by direct oral laryngoscopy. *(5:1633)*

1433. (E) Hypoxia due to delivery failure of the oxygen supply has many causes. The only way to protect our patients is to check and monitor carefully. *(1:670; 5:275)*

1434. (E) Aspiration may follow sedation, debility, impaired swallowing, or any factor that affects the ability to protect the airway. *(1:1248; 4:1441)*

1435. (B) Hypothermia will aggravate the problem of hypocalcemia by decreasing the rate of metabolic clearance of exogenous citrate. This is another good reason to warm blood when it is given rapidly. *(1:201; 5:1814)*

1436. (C) The fact that the patient is crowing on ventilation means that the patient has partial glottic closure. With full glottic closure, the patient will have retraction but will not be moving air that would make any sounds. Gentle positive pressure will usually relieve the phonation. *(1:612; 5:2539)*

1437. (C) The risk factors for perioperative ischemic neuropathy are still ill-defined. Statistically, the incidence occurs more frequently with open heart and spinal surgery. *(5:3010)*

1438. (E) All of these elements are prerequisites for an explosion in the operating room. *(5:3139)*

1439. (E) All of the options are useful in interpreting continuous numerical data. *(1:56; 4:758; 5:882)*

1440. (A) The doctrine of *res ipsa loquitur* (the thing speaks for itself) is invoked if the accident usually does not happen without negligence, the plaintiff was not a factor, and the defendant usually has control of whatever caused the accident. Settling out of court is not a factor. *(1:90)*

1441. (C) If it is discovered that a tooth is missing, immediate steps should be taken to ascertain the location. Examine the oropharynx and nasopharynx. This can be done visually and by x-ray. If the tooth is in the trachea, a bronchoscopy should be performed immediately. *(1:1396; 4:2452; 5:1647)*

1442. (A) Many studies have shown that live virus can be recovered from the plume produced by carbon dioxide and argon lasers. It is recommended that in addition to the careful evacuation of the laser smoke, personnel that are near the plume wear gloves, goggles, and a high-efficiency filter mask. *(1:77–8)*

1443. (E) The ability to recall intraoperative events can lead to disturbing consequences for the patient. All of the listed responses have been reported as postoperative events in this context. In some patients it can lead to a posttraumatic stress disorder including repetitive nightmares and a preoccupation with death. *(5:1239)*

1444. (A) Studies have shown that there is an association between fetal anomalies and employment in an operating room. These anomalies occur in the offspring of both male and female employees. In addition, liver disease was reported more often in male anesthesiologists but the prevalence of malignancy was no different than in controls. *(5:3152)*

1445. (D) There is controversy regarding the relative merits of each type of anesthesia. Studies are available for each viewpoint. If one looks at the mortality data at 1 month postoperatively, there is little difference. *(5:2417)*

1446. (D) Reactions to transfusions are to be anticipated. Most are not fatal. Many do not occur immediately. The reaction may be hemolytic, but it may also be an allergic reaction. *(5:1815)*

1447. (B) Hepatitis may follow transfusion of blood products; however, with the advent of better screening methods for the blood supply, the incidence of transfusion-related hepatitis has declined. The incidence of hepatitis C is about three times more likely than hepatitis B. The incidence is lower if the donor pool is volunteers. Transfusion of red cells does not carry a higher risk. Current data cite the risk of hepatitis C as approximately one case per 200,000 units transfused. *(5:1818)*

1448. (B) Patients with skin bullae present difficult anesthesia management problems. Tracheal intubation should be used if needed. The tube should be secured with wide cord or umbilical tape to avoid the use of adhesive tape. To do the procedure under mask requires pressure on the face, and that may cause skin deterioration. Any friction may lead to loss of skin. *(1:513)*

1449. (E) All of the listed complications may occur after a celiac plexus block. This block can be useful in controlling the pain associated with pancreatic cancer. Due to the potentially devastating consequences, needle placement should be confirmed with radiographic means prior to performing a neurolytic block. *(1:735; 5:1711)*

1450. (E) Symptoms produced by positioning may involve the circulatory system, e.g., hypotension; the nervous system, e.g., nerve damage; the respiratory system, e.g., restriction of ventilation; and the neuromuscular system, e.g., pain from muscle damage due to the position. *(1:639)*

PART III

Practice Test

CHAPTER 17

Practice Test

Questions

PART 1

DIRECTIONS (Questions 1 through 86): Each of the numbered items or incomplete statements in this section is followed by answers or by completions of the statement. Select the ONE lettered answer or completion that is BEST in each case.

1. The following may indicate need for artificial cardiac pacing EXCEPT
   (A) atrial fibrillation with rapid ventricular rate
   (B) atrial fibrillation with slow ventricular rate
   (C) complete heart block
   (D) junctional rhythm
   (E) sinus bradycardia

2. The following changes in lab values may be present during pregnancy EXCEPT
   (A) decreased hematocrit
   (B) decreased PCO_2
   (C) increased pH
   (D) decreased creatinine
   (E) increased factors VII, VIII, X, and fibrinogen

3. The amount of gas absorbed by a liquid is proportional to the pressure of the gas above the liquid. This is a statement of
   (A) Dalton's law
   (B) Boyle's law
   (C) Charles' law
   (D) Henry's law
   (E) Ferguson's rule

4. Heparin inhibits blood coagulation
   (A) by binding calcium ions
   (B) through interactions with protamine
   (C) by activating antithrombin III
   (D) by activating plasmin
   (E) by activating von Willebrand factor

5. At functional residual capacity (FRC)
   (A) the pressure difference between the alveoli and atmosphere is zero
   (B) chest wall elastic forces are greater than the lungs' elastic recoil
   (C) the pressure difference between the alveoli and the intrapleural space is zero
   (D) the total pulmonary vascular resistance is very high
   (E) no further inspiration is possible

6. A 6-year-old boy presents to holding for elective repair of an inguinal hernia. He is a mild asthmatic and has not taken any daily asthma medications for 5 weeks. He is currently wheezing throughout on physical examination. You inform the mother
   (A) proceed with surgery without treatment
   (B) postpone until tomorrow after two doses of montelukast
   (C) refer to pediatrician for evaluation and treatment
   (D) proceed with surgery after albuterol
   (E) perform surgery under sedation and analgesia with local infiltration

7. The nerve most likely to be injured in the lithotomy position is the

(A) obturator nerve
(B) femoral nerve
(C) saphenous nerve
(D) peroneal nerve
(E) tibial nerve

8. Problems that may be encountered in scleroderma include all of the following EXCEPT

(A) limited mouth opening
(B) arterial dilatation
(C) pulmonary fibrosis
(D) contractures
(E) pericardial effusion

9. In the artificially ventilated neurosurgical patient, PEEP

(A) should be used routinely
(B) should be used on selected patients, and the patient's head should always be down
(C) has no effect on ICP
(D) should be withheld in all cases
(E) should be titrated against requirements for oxygenation and neurologic status

10. Halothane is

(A) insoluble in rubber
(B) soluble in rubber, and this may be a factor in induction with high flow rates
(C) soluble in rubber but eliminated very rapidly
(D) soluble in rubber, and the retained halothane may affect a second patient exposed to the same apparatus
(E) soluble in rubber, and the rubber reacts with halothane to change its composition

11. Continuous spinal anesthesia with a small caliber subarachnoid catheter is no longer a recommended anesthetic technique because of

(A) the high incidence of postdural puncture headache
(B) the risk of arachnoiditis
(C) some occurrences of cauda equina syndrome
(D) difficulty in controlling the spread of the local anesthetic
(E) some occurrences of the catheter tearing during removal

12. One gram of oxygen occupies what volume at 1 atm pressure and 20°C? The ideal gas constant is 0.082 (L·atm)/(K·mol).

(A) 751 mL
(B) 7.51 L
(C) 75.1 L
(D) 751 L
(E) 7510 L

13. Preeclampsia is characterized by all of the following EXCEPT

(A) intravascular volume depletion
(B) proteinuria
(C) occurrence anytime during pregnancy
(D) no change in placental perfusion
(E) decreased production of renin

14. Normally the systolic blood pressure is highest in the

(A) ascending aorta
(B) descending aorta
(C) femoral artery
(D) dorsalis pedis artery
(E) pulmonary artery

15. The stimulus for the temperature-regulatory center is

(A) the norepinephrine level of the blood
(B) a hormone released from muscle cells
(C) an integrated input from temperature-sensitive cells throughout the body
(D) the cerebral cortical input
(E) the afferents from muscle spindles

16. The hardness of soda lime

(A) is increased by the addition of silica
(B) is of little importance

(C) is undesirable because as the granule disintegrates, its surface area increases
(D) causes channeling
(E) imparts a uniform-sized granule

17. A TRUE statement regarding drug action in a parturient is

(A) placental transfer is minimal with muscle relaxants
(B) opioids do not cross the placenta
(C) inhalational anesthetics increase uterine muscle tone
(D) nitrous oxide is contraindicated for cesarean section secondary to interference with vitamin B_{12} synthesis
(E) thiopental does not cross the placenta

18. Afterload is reduced while diastolic perfusion pressure is increased by

(A) dopamine
(B) epinephrine
(C) nitroglycerin
(D) nitroprusside
(E) intra-aortic balloon counterpulsation

19. Hurler's syndrome (gargoylism) is a disturbance of mucopolysaccharide metabolism. Anesthesia is complicated by all of the following EXCEPT

(A) dwarfism
(B) macroglossia
(C) hypertelorism
(D) hepatosplenomegaly
(E) short neck

20. The central chemoreceptors responsible for respiratory control are

(A) located in the midbrain
(B) responsive primarily to change in carbon dioxide
(C) responsive primarily to change in hydrogen ion concentration
(D) responsive primarily to hypoxemia
(E) activated by chronic hypercapnia

21. A 72-year-old female with a past medical history of osteoporosis underwent open reduction and internal fixation of her left hip. She was previously diagnosed with heparin-induced thrombocytopenia type II and has a platelet count of 56,000. Which of the following agents can be administered for postoperative prophylaxis of deep vein thrombosis?

(A) low-molecular weight heparin
(B) warfarin
(C) tirofiban
(D) unfractionated heparin
(E) lepirudin

22. Concerning a 3-month-old 5-kg full-term infant

(A) hematocrit has reached its lowest post-delivery value
(B) MAC is lowest at this age
(C) blood volume is 300 mL
(D) maintenance fluid is 40 mL/h
(E) separation anxiety is usually present

23. During the performance of a stellate ganglion block, the patient becomes apneic. This is most likely due to

(A) vertebral artery injection of local anesthetic
(B) injection of local anesthetic into the periosteum
(C) phrenic nerve paralysis
(D) subarachnoid injection of local anesthetic
(E) pneumothorax

24. After an accidental needlestick with a needle contaminated with blood from a patient with the acquired immunodeficiency syndrome (AIDS), the likelihood of developing an infection with the human immunodeficiency virus is approximately

(A) 0.01%
(B) 0.05%
(C) 0.2%
(D) 2%
(E) 10%

25. Ischemia from a right coronary artery lesion would most likely be evident on electrocardiographic lead

(A) I
(B) II
(C) AVR
(D) AVL
(E) V5

26. A 2-year-old child is brought to the emergency room with a 6-day history of increasing cough and wheezing. Symptoms have not responded to bronchodilator therapy and broad spectrum antibiotics. On examination, the child is agitated, tachypneic, and has a tachycardia of 150 bpm. There are decreased breath sounds in the right lower zone of the lung fields. This clinical presentation is most consistent with

(A) right lower lobe pneumonia
(B) exacerbation of underlying asthma
(C) inhaled foreign body
(D) acute laryngotracheobronchitis
(E) ruptured congenital bulla

27. The dermatome level of the xiphoid is

(A) T6
(B) T8
(C) T10
(D) T12
(E) L2

28. After upper abdominal surgery, FRC is

(A) unaffected
(B) normal within 24 hours
(C) decreased for days
(D) transiently increased by 33%
(E) permanently impaired

29. Sevoflurane

(A) can cause hepatotoxicity
(B) is flammable at a concentration of 6%
(C) gives poor muscle relaxation
(D) undergoes hepatic metabolism
(E) causes myocardial irritability

30. A full tank of oxygen with an internal pressure of 1900 psi at 20°C is stored in an outdoor, poorly ventilated shed. On a hot summer day, the interior temperature of the shed reaches 155°F. The internal pressure of the oxygen gas will increase to

(A) 2008 psi
(B) 2085 psi
(C) 2154 psi
(D) 2176 psi
(E) 2213 psi

31. Which of the following statements concerning management of difficult intubation in parturients is accurate?

(A) In an elective cesarean section, proceed under mask ventilation if patient has had nothing by mouth.
(B) Never attempt blind nasal intubation because of the risk of bleeding from engorged airways.
(C) If able to ventilate with mask and the fetus is in serious distress, no additional steps should be taken to obtain a secure airway and delivery should proceed immediately.
(D) Patients should have a repeat airway assessment prior to anesthesia for cesarean section.
(E) Never attempt cricothyrotomy in a preeclamptic patient because of potential coagulopathies.

32. Side effects of intrathecal morphine include the following EXCEPT

(A) pruritus
(B) hypotension
(C) nausea and vomiting
(D) urinary retention
(E) respiratory depression

33. Which portion of the brachial plexus is blocked by the supraclavicular approach?

(A) roots
(B) cords

(C) divisions
(D) trunks
(E) peripheral nerves

34. A-gamma fibers

(A) are not myelinated
(B) conduct both afferent and efferent information
(C) are the afferent fibers from pacinian corpuscles
(D) are the efferent nerve fibers to muscle spindles
(E) cannot be blocked by local anesthetics

35. The main value of epinephrine in cardiopulmonary resuscitation (CPR) is

(A) vasoconstriction of peripheral vessels other than those supplying the brain and heart
(B) increased myocardial oxygen demand
(C) dilation of renal blood vessels
(D) increased intensity of ventricular fibrillation
(E) bronchial dilation

36. Compliance refers to

(A) a change in volume per unit change in pressure
(B) a change in pressure per unit change in volume
(C) the reciprocal of conductance
(D) a change in volume per unit change in pressure per unit of time
(E) a measurement of tracheal movement

37. Desflurane

(A) is highly lipid soluble
(B) has a MAC of 5.5%
(C) is highly water soluble
(D) is not irritating to the airway
(E) metabolism releases significant quantities of fluoride ion

38. When properly positioned, a left-sided Robertshaw double-lumen tube will have its lumens ending

(A) in the left bronchus and the right bronchus
(B) in the left bronchus and in the trachea
(C) in the right bronchus and in the trachea
(D) in the left upper lobe bronchus and in the left lower lobe bronchus
(E) both in the trachea

39. An 85-year-old patient has a normal value of 1.0 mg/dL for serum creatinine. Compared with a 20-year-old patient of the same weight and with the same serum creatinine value, the 85-year-old patient has approximately what fractional value of creatinine clearance?

(A) 0.1
(B) 0.2
(C) 0.5
(D) 0.7
(E) 0.9

40. One milliliter of desflurane liquid occupies what volume at 1 atm pressure and 37°C if all of the liquid is vaporized? (The ideal gas constant is 0.082 (L·atm)/(K·mol), the specific gravity of desflurane is 1.47, and its molecular weight is 168.)

(A) 182 mL
(B) 190 mL
(C) 198 mL
(D) 210 mL
(E) 222 mL

41. When calculating the percentage of body surface area involved in an extensive burn in a 2-year-old child, the head constitutes

(A) 8% of total surface area
(B) 10% of total surface area
(C) 13% of total surface area
(D) 15% of total surface area
(E) 19% of total surface area

42. In a patient with severe pain, administration of thiopental 80 mg will

(A) produce sleep
(B) will have an antianalgesic action
(C) will have no effect
(D) will sedate the patient but not induce sleep
(E) be more effective if given alone

43. A patient in the critical care unit is mechanically ventilated in a pressure control mode. The minute ventilation is determined by all of the following EXCEPT the

(A) respiratory rate
(B) lung compliance
(C) airway resistance
(D) set tidal volume
(E) chest wall compliance

44. Injection of thiopental will lead to a dose-dependent

(A) decrease in arterial blood pressure
(B) increase in stroke volume
(C) increase in cardiac output
(D) bradycardia
(E) decrease in coronary blood flow

45. Which of the following nerve blocks would be the most effective for the relief of pain associated with carcinoma of the pancreas?

(A) stellate ganglion
(B) celiac plexus
(C) lumbar sympathetic
(D) thoracic paravertebral
(E) hyperbaric spinal

46. Noncardiac mechanisms of pulseless electrical activity (electromechanical dissociation) include the following EXCEPT

(A) hypovolemia
(B) pericardial tamponade
(C) tension pneumothorax
(D) pulmonary embolism
(E) ventricular fibrillation

47. Administration of ketamine causes

(A) decreased heart rate
(B) increased heart rate
(C) decreased cardiac output
(D) no change in cardiac output
(E) no change in heart rate

48. Concerning the baroreceptor reflex

(A) the sensor is located in the aortic arch
(B) afferent information is conducted via the glossopharyngeal nerve
(C) the central integrating center is located in the midbrain
(D) all volatile anesthetics depress the reflex equally
(E) the afferent stimulus is cardiac output

49. The most common arrhythmia seen on injection of a second dose of succinylcholine is

(A) sinus arrest
(B) sinus tachycardia
(C) nodal rhythm
(D) atrial fibrillation
(E) bigeminy

50. If a current of 6 A flows for 3 seconds, the amount of energy expended

(A) cannot be calculated without further information
(B) is 18 joules
(C) is 2 joules
(D) is 18 coulombs
(E) is 2 coulombs

51. The cardiovascular changes seen after administration of atracurium in the anesthetized patient may be due to all of the following EXCEPT

(A) the speed of injection
(B) ganglionic blockade
(C) histamine release
(D) intravascular volume status
(E) the dose injected

52. Which of the following statements is TRUE regarding congenital diaphragmatic hernia (CDH)?

(A) CDH occurs in 1 in 10,000 live births.
(B) Seventy percent of all herniations occur through the foramen of Bochdalek.
(C) The neonate presents with respiratory distress and a distended abdomen.
(D) Ninety percent of infants with CDH have an accompanying cardiac lesion.
(E) Approximately 5% of infants with CDH present with the symptoms of bowel obstruction.

53. The LEAST volume of local anesthetic is usually required when performing a(n)

(A) axillary block
(B) supraclavicular block
(C) interscalene block
(D) femoral-sciatic block
(E) intercostal block, bilaterally, at six ribs

54. A 3-month-old black baby is scheduled for elective repair of an inguinal hernia. He has an older brother with sickle cell anemia, but he has not had any diagnostic tests for sickle cell anemia. His hematocrit is 30%. This baby

(A) almost certainly has sickle cell anemia
(B) should receive a preoperative transfusion
(C) should undergo a screening test for HbS prior to anesthesia
(D) may undergo anesthesia safely without further testing
(E) has a 50% chance of having sickle cell anemia

55. A laryngeal mask airway

(A) may be used to treat postoperative airway obstruction
(B) protects against aspiration of gastric contents
(C) is usually inserted blindly
(D) requires the patient to breathe spontaneously
(E) is an effective treatment for laryngospasm

56. Each of the following drugs has opioid agonistic actions EXCEPT

(A) butorphanol
(B) dezocine
(C) pentazocine
(D) naloxone
(E) buprenorphine

57. The person who first administered ether in a public demonstration was

(A) Horace Wells
(B) Crawford W. Long
(C) William T.G. Morton
(D) John C. Warren
(E) Edward G. Abbott

58. Indications for stellate ganglion block include the following EXCEPT

(A) phantom limb pain
(B) anesthesia for digit amputation
(C) postherpetic neuralgia
(D) complex regional pain syndrome
(E) Raynaud's disease

59. A properly performed and successful interscalene block may provide inadequate anesthesia for

(A) tendon repair of the fifth finger
(B) creation of a radiocephalic arteriovenous fistula for dialysis access
(C) repair of recurrent shoulder dislocation
(D) axillary dissection
(E) carotid endarterectomy

Questions 60 through 64

Many patients are unaware of the reasons for which they are taking their medications. When provided with a list of medications by a patient, an anesthesiologist must be able to recognize the medications and know the likely pathologic states for which the medications are indicated. For Questions 60 through 64, a medication is followed by five diseases or pathologic states. Choose the ONE disease for which the medication may be indicated.

60. Primidone

(A) depression
(B) psychosis
(C) grand mal epilepsy
(D) breast cancer
(E) malaria

61. Sodium etidronate

(A) petit mal seizures
(B) hypercholesterolemia
(C) Paget's disease of the bone
(D) urinary tract infection
(E) rheumatoid arthritis

62. Gemfibrozil

(A) hypercholesterolemia
(B) claudication
(C) type II diabetes mellitus
(D) atrial fibrillation
(E) congestive heart failure

63. Zidovudine

(A) cytomegalovirus retinitis
(B) genital herpes
(C) AIDS
(D) shingles
(E) cutaneous papillomavirus infection

64. Fluoxetine

(A) influenza
(B) manic depressive disorder
(C) Graves' disease
(D) Parkinson's disease
(E) Hodgkin's disease

65. Net right-to-left shunt is a feature of

(A) atrial septal defect
(B) ventricular septal defect
(C) patent ductus arteriosus (PDA)
(D) tetralogy of Fallot
(E) endocardial cushion defect

66. Under normal physiologic conditions cerebral blood flow (CBF) is

(A) 1 mL/100 g/min
(B) 10 mL/100 g/min
(C) 25 mL/100 g/min
(D) 50 mL/100 g/min
(E) 100 mL/100 g/min

67. Which of the following is NOT considered a recommended, evidence-based practice in critical care medicine?

(A) Maintaining the hemoglobin concentration at 10 g/dL or greater in all critically ill patients
(B) Optimal sterile technique (maximal sterile-barrier precautions) when inserting central venous catheters
(C) Institution of sedation protocols that include daily sedation interruption and patient targeted goals for sedation and analgesia administration
(D) Limiting tidal volumes to 6 mL/kg of predicted body weight in patients with a diagnosis of ARDS
(E) Mechanical ventilator weaning protocols driven by non-physician health care providers

68. General anesthesia as compared to regional anesthesia in parturients is associated with all of following EXCEPT

(A) less hypotension
(B) less cardiovascular instability
(C) less uterine relaxation
(D) more rapid induction
(E) better control of airway

69. Measurements that can be obtained by spirometry include all of the following EXCEPT

(A) tidal volume
(B) closing volume
(C) expiratory reserve volume
(D) inspiratory reserve volume
(E) vital capacity

70. Increased amplitude of **v** waves in a central venous pressure recording indicates

(A) junctional rhythm
(B) atrial fibrillation
(C) tricuspid regurgitation
(D) hypovolemia
(E) heart block

71. A patient undergoes anesthesia for removal of a bronchial carcinoma. His anesthesia consists of thiopental, halothane, and N_2O/O_2, after intubation facilitated with 80 mg of succinylcholine. He was given vecuronium, 10 mg, immediately after intubation. At the end of the procedure, he is apneic. This may be due to all of the following EXCEPT

(A) vecuronium
(B) succinylcholine
(C) bronchial carcinoma
(D) hyperventilation
(E) thiopental

72. The principal extracellular anion is

(A) Na^+
(B) K^+
(C) Cl^-
(D) HCO_3^-
(E) Mg^{2+}

73. The blood component with the least risk of transmitting hepatitis C is

(A) cryoprecipitate
(B) fresh frozen plasma
(C) packed red blood cells
(D) frozen washed red blood cells
(E) 5% albumin

74. The pin index safety system prevents

(A) the connection of the wrong gas to the cylinder manifold
(B) the delivery of the wrong gas from the wall
(C) the delivery of a hypoxic mixture from the flowmeter
(D) starting a case with empty tanks
(E) improper pipe fittings from the source of gas to the wall

75. Propranolol is useful in the treatment of

(A) sinus bradycardia
(B) chronic bronchitis
(C) type II diabetes mellitus
(D) atrioventricular (AV) block
(E) atrial flutter

76. A double-lumen tube is used for anesthesia in patients with severe

(A) asthma
(B) hemoptysis
(C) emphysema
(D) pneumonia
(E) tracheal stenosis

77. Regarding necrotizing enterocolitis,

(A) it is an anomaly found predominantly in premature infants
(B) umbilical artery catheterization should be performed in order to monitor hematologic and metabolic abnormalities
(C) the mortality is about 50%
(D) cardiovascular collapse usually occurs early in the course of the illness
(E) metabolic abnormalities include hypoglycemia resulting from intestinal malabsorption

78. The feature of electrocautery, which minimizes the risk of ventricular fibrillation is

(A) low voltage
(B) low amperage
(C) high frequency
(D) short duration
(E) low power

79. In order for an epidural to relieve the second stage of labor, which of the following segments must be blocked?

(A) T11–T12
(B) S2–S4
(C) L3–L4
(D) T10–L1
(E) L2–L4

80. Procainamide

(A) is eliminated chiefly through the liver
(B) slows conduction of cardiac impulses
(C) may not be administered orally
(D) is easily hydrolyzed by plasma cholinesterase
(E) increases ventricular automaticity

81. A diuretic that exerts its effect on the ascending loop of Henle is

(A) hydrochlorothiazide
(B) ethacrynic acid
(C) chlorthalidone
(D) triamterene
(E) benzthiazide

82. The interaction of rocuronium and acetylcholine at the myoneural junction is one of

(A) synergism
(B) competition for binding sites
(C) chemical combination
(D) alteration of metabolism
(E) altered protein binding

83. The best statistical test to compare the mean values of two populations given $N = 30$ is

(A) analysis of variance
(B) Student's t-test
(C) chi-squared statistic
(D) sum of squares
(E) linear regression

84. The physical interaction of a mixture of thiopental and meperidine is one of

(A) synergism
(B) interaction at site of absorption
(C) altered protein binding
(D) precipitation
(E) altered metabolism

85. The binding of oxygen to hemoglobin is enhanced by

(A) carbon dioxide
(B) hydrogen ions
(C) 2,3-diphosphoglycerate
(D) hypothermia
(E) nitrous oxide

86. Rocuronium

(A) is a ganglionic blocker
(B) at 4 × ED95 causes significant histamine release
(C) may be administered intramuscularly
(D) is eliminated by the kidney
(E) is not suitable for continuous infusion

DIRECTIONS (Questions 87 through 175): For each of the items in this section, ONE or MORE of the numbered options is correct. Choose answer

(A) if only 1, 2, and 3 are correct
(B) if only 1 and 3 are correct
(C) if only 2 and 4 are correct
(D) if only 4 is correct
(E) if all are correct

87. Following the uneventful vaginal delivery of a 7 lb 6 oz baby, you are asked to remove the epidural catheter that was providing analgesia for labor. Inspection of the catheter reveals that part of it remains in the patient. Appropriate actions include

(1) consulting a neurosurgeon
(2) obtaining a magnetic resonance imaging scan
(3) antibiotic prophylaxis
(4) informing the patient

88. Endogenous catecholamines include

(1) epinephrine
(2) norepinephrine
(3) dopamine
(4) dobutamine

89. Isoflurane, in contrast to halothane

(1) abolishes autoregulation
(2) can produce an isoelectric EEG at clinically relevant doses
(3) decreases CBF
(4) lowers ischemic threshold for EEG changes

90. Increased dead space ventilation may occur during anesthesia due to

(1) deliberate hypotension
(2) short rapid ventilations
(3) pulmonary embolus
(4) increased cardiac output

91. Issues important to induction of general anesthesia during pregnancy include

(1) increased risk for aspiration
(2) decreased oxygen reserve secondary to decreased FRC
(3) capillary engorgement of the airway requiring a smaller size endotracheal tube
(4) uterine displacement to prevent hypotension

92. The whole blood activated clotting time

(1) involves the intrinsic pathway
(2) involves the extrinsic pathway
(3) is a measure of heparin activity
(4) is a measure of warfarin activity

93. The flow rate of fluid flowing smoothly through a tube is

(1) proportional to the pressure gradient
(2) proportional to the fourth power of the radius of the tube
(3) inversely proportional to the viscosity of the fluid
(4) proportional to the absolute temperature of the fluid

94. Milrinone

(1) causes vasodilation
(2) inhibits Na,K-ATPase
(3) has little effect on heart rate
(4) stimulates phosphodiesterase

95. Concerning tracheoesophageal fistula (TEF) and esophageal atresia (EA)

(1) the incidence of TEF or EA is about 1 in 3000 live births
(2) the incidence in premature babies is higher than in term infants
(3) the associated congenital anomalies occur in 30 to 50% of patients
(4) patients with TEF are frequently nursed prone

96. Lamotrigine

(1) is an atypical antipsychotic drug
(2) has a prolonged half-life when pentobarbital, carbamazepine, or phenytoin is administered concomitantly
(3) is not effective in absence epilepsy
(4) is effective against a broader spectrum of seizures than phenytoin and carbamazepine

97. Toxic effects of local anesthetics

(1) are more prone to affect the cardiovascular system than the nervous system
(2) occur at about the same dose for all agents
(3) may be manifest as a shortened P-R interval
(4) may involve depression of the SA node

98. The glomerular capillaries

(1) originate from efferent arterioles
(2) are an extension of Bowman's capsule
(3) are many layers thick
(4) are the only capillaries in the body that drain into arterioles

99. When epinephrine is employed as a vasoconstrictor, regimen(s) that is (are) relatively safe when halothane is employed include(s)

(1) concentration no greater than 1:100,000
(2) adult dose not greater than 10 mL of 1:100,000 within 10 minutes
(3) adult dose not greater than 20 mL of 1:200,000 within 10 minutes
(4) adult total dose not to exceed 30 mL of 1:100,000 within 1 hour

100. A patient with peripheral vascular disease has a gangrenous fifth toe. Which of the following nerves must be blocked at the ankle to provide adequate anesthesia for an amputation of this toe?

(1) sural nerve
(2) superficial peroneal nerve
(3) posterior tibial nerve
(4) deep peroneal nerve

101. Hepatic blood flow is decreased by

(1) halothane
(2) etomidate
(3) sevoflurane
(4) nitrous oxide

102. Normal mechanisms by which the brain accommodates a slowly expanding mass lesion include

(1) shift of cerebrospinal fluid (CSF) to the spinal subarachnoid space
(2) removal of free water from brain tissue
(3) decrease in venous blood volume
(4) decrease in arterial blood volume

103. The Macintosh laryngoscope blade

(1) has a straight spatula
(2) has no flange
(3) has a pointed tip
(4) is manufactured in various sizes

104. The metabolic breakdown of halothane results in

(1) trifluoroacetic acid
(2) chloride ion
(3) fluoride ion
(4) bromide ion

105. The blood volume of an average 70-kg adult could be processed into

(1) 10 units of fresh frozen plasma
(2) 10 units of packed red blood cells
(3) 10 units of cryoprecipitate
(4) 10 units of platelets

106. A patient who has been on oral steroids for the year preceding surgery

(1) may not have any perioperative problems
(2) has little additional perioperative risk if exogenous steroids are given
(3) is at risk of experiencing a stress reaction if no perioperative steroids are given
(4) should be given perioperative steroids

107. When comparing the infant's airway to the adult's

(1) the glottis is higher
(2) the narrowest point of the airway is lower
(3) the chest is more compliant
(4) airway resistance is lower

108. Vasopressin

(1) decreases urine flow
(2) increases the permeability of the collecting duct to water
(3) is secreted by the posterior pituitary
(4) acts throughout the nephron system

109. When meconium aspiration occurs

(1) the most commonly affected neonates weigh less than 2000 g at birth
(2) the long-term outcome is often poor
(3) chest physical therapy is of negligible benefit
(4) a major cause of death is persistent fetal circulation

110. Vital capacity includes

(1) tidal volume
(2) inspiratory reserve volume
(3) expiratory reserve volume
(4) FRC

111. The reaction of carbon dioxide and soda lime produces

(1) $CaCO_3$
(2) Na_2CO_3
(3) H_2O
(4) HCl

112. Protamine sulfate

(1) is a normal component of blood
(2) exhibits anticoagulant activity in the absence of heparin
(3) fails to inhibit heparin in vitro
(4) binds to heparin through ionic interactions

113. Mannitol

(1) is filtered at the glomerulus
(2) is an osmotic diuretic
(3) causes an immediate increase in intravascular volume
(4) is of value in congestive failure

114. For the open repair of a laceration of the extensor tendon of the fifth finger between the proximal and distal interphalangeal joints, which of the following nerves must be blocked at the wrist?

(1) median nerve
(2) musculocutaneous nerve
(3) radial nerve
(4) ulnar nerve

115. If no intravenous access is available during CPR, which of the following drugs may be administered via the endotracheal tube?

(1) atropine
(2) lidocaine
(3) epinephrine
(4) calcium chloride

Questions 116 and 117 refer to Figure 17-1.

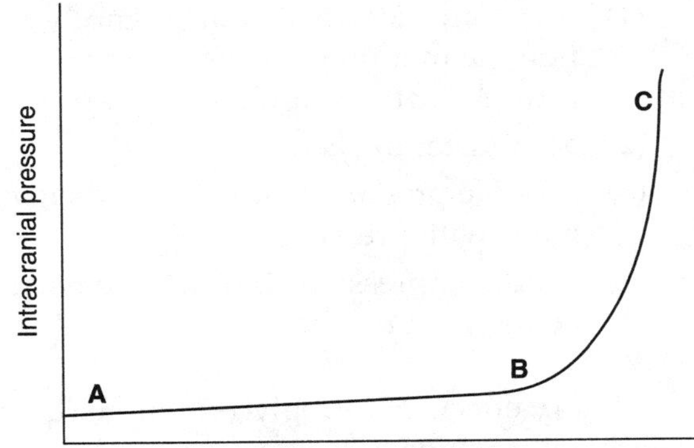

FIG. 17-1

116. At position C on the curve, a patient

(1) has exhausted compensatory mechanisms
(2) moves toward position A with coughing
(3) may benefit from IV mannitol
(4) may be hypertensive and tachycardic

117. A patient at position B on the curve would be expected to

(1) have clinical manifestations of increased ICP
(2) move toward position C with hyperventilation
(3) have unilateral mydriasis
(4) move toward position C with administration of halothane

118. Drugs that may suppress the symptoms of opioid withdrawal include

(1) butorphanol
(2) chlorpromazine
(3) nalbuphine
(4) clonidine

119. Initial resuscitative measures of the moderately depressed infant include

(1) warm and dry the infant promptly, suction the mouth, pharynx, and nose, and provide tactile stimulation
(2) administer oxygen
(3) positive pressure ventilation by bag and mask with pressures of 15 to 20 cm H_2O
(4) chest compressions if heart rate remains below 100 bpm

120. Salutary effects of nitroglycerin in myocardial ischemia include

(1) inhibition of coronary spasm
(2) increased preload
(3) decreased preload
(4) increased afterload

121. Aerosol therapy is used to

(1) humidify inspired gases
(2) deliver medications
(3) mobilize secretions
(4) induce bronchospasm

122. Disadvantage(s) of the use of halothane for deliberate hypotension is (are)

(1) seizure activity at deep levels of anesthesia
(2) increased myocardial blood flow due to coronary vasodilation
(3) increased myocardial work
(4) loss of autoregulation of CBF

123. During transurethral resection of the prostate, uptake of the 1.5% glycine irrigant through venous sinuses may lead to

(1) increased circulating fluid volume
(2) hemolysis of red cells
(3) hemodilution of electrolytes
(4) increased serum sodium concentration

124. Tissue perfusion will affect thiopental concentrations in the blood. The patient in shock is an example of this, since

(1) CBF is poorly maintained in shock
(2) vasoconstriction occurs in skin and skeletal muscles in shock
(3) the fractional dose of thiopental going to the brain is lower in shock
(4) cardiac output is reduced and CBF is relatively higher in the shock patient

125. The treatment of postdural puncture headache includes

(1) orders to keep NPO
(2) reassurance
(3) subarachnoid injection of the patient's blood
(4) codeine or aspirin

126. The patient with myxedema may have

(1) decreased cardiac contractility
(2) psychosis
(3) oliguria
(4) tachycardia

127. Hypoxemia may occur under anesthesia because of

(1) blood loss
(2) depressed myocardial function

(3) shunting
(4) airway obstruction

128. Ocular effects of ketamine anesthesia include

(1) open lids with an adequate level of anesthesia
(2) diplopia occurring in the postoperative period
(3) nystagmus
(4) decreased intraocular pressure

129. Emergence sequelae with ketamine include delirium and dreaming, which

(1) is always of an unpleasant nature
(2) occurs equally in men and women
(3) is not remembered because of the amnestic effects of ketamine
(4) can be minimized by giving benzodiazepines

130. Diffusion hypoxia

(1) is due to a large volume of nitrous oxide in the lungs
(2) is due to dilution of alveolar carbon dioxide
(3) is due to oxygen displacement
(4) occurs only in the first 5 to 10 minutes of recovery

131. The blocking property of a local anesthetic

(1) is greatly affected by pH
(2) involves the penetration of the nerve membrane by the uncharged form of the agent
(3) is decreased if injected into inflamed tissue
(4) is enhanced by the addition of alkaline buffer

132. Two drugs, A and B, may interact by which of the following mechanisms?

(1) drug A may interfere with the gastrointestinal absorption of drug B
(2) drug A may interfere with the protein binding of drug B
(3) drug A may interfere with the renal excretion of drug B
(4) drug A may alter electrolyte patterns, which, in turn, may affect drug B

133. Complications of monoamine oxidase (MAO) inhibitors include

(1) orthostatic hypotension
(2) meperidine-induced hypertension and convulsions
(3) hypertension precipitated by tyramine-rich foods
(4) potentiation of ephedrine

134. Aortocaval compression syndrome includes which of the following?

(1) decreased venous return secondary to inferior vena caval compression
(2) decreased uterine blood flow
(3) aortic compression
(4) occurrence during first trimester

135. Which of the following is (are) benzodiazepine antagonists?

(1) fluoxetine
(2) fluphenazine
(3) flunisolide
(4) flumazenil

136. Properties of the calcium entry blocker verapamil include

(1) depressed contractility (negative inotropy)
(2) depressed conduction velocity (negative dromotropy)
(3) depressed heart rate (negative chronotropy)
(4) vasoconstriction

137. Prevention of postoperative atelectasis can be achieved by

(1) avoidance of alveolar collapse
(2) adequate coughing
(3) frequent change in body position
(4) use of sustained maximum inflation techniques

138. Pneumothorax may be due to

(1) alveolar rupture
(2) chest wall trauma
(3) connection between the distal airway and the pleural space
(4) a break in the parietal pleura

139. Which of the following medications is active primarily at β_1-adrenergic receptors?

(1) nadolol
(2) timolol
(3) labetalol
(4) atenolol

140. A regulator used on a compressed gas cylinder

(1) reduces the pressure of the gas from the cylinder
(2) is an instrument of measurement
(3) permits the expansion of the compressed gas to a lower pressure
(4) determines the pressure of the gas in pounds per square inch (psi)

141. Hypertrophic obstructive cardiomyopathy is hemodynamically worsened by

(1) hypovolemia
(2) junctional rhythm
(3) β-adrenergic agonists
(4) α-adrenergic agonists

142. Which of the following antihypertensive medications may be administered intravenously?

(1) prazosin
(2) methyldopa
(3) nifedipine
(4) enalapril

143. Common cyanotic congenital heart lesions include

(1) tetralogy of Fallot
(2) total anomalous pulmonary venous drainage
(3) transposition of the great vessels
(4) ventricular septal defect

144. Pathways in the lung for collateral ventilation include

(1) the intraalveolar pores of Kohn
(2) connections between respiratory bronchioles and terminal bronchioles
(3) interlobar connections
(4) interbronchial connections

145. Administration of an inhibitor of angiotensin-converting enzyme (ACE) to a patient with congestive heart failure usually causes

(1) a decrease in afterload
(2) a decrease in sodium excretion
(3) an increase in cardiac output
(4) an increase in preload

146. A patient with a thyroid tumor develops hypertension during the procedure. This may be due to

(1) light anesthesia
(2) pheochromocytoma
(3) hypercarbia
(4) hypocalcemia

147. Omeprazole

(1) decreases gastric acid secretion
(2) facilitates gastric emptying
(3) inhibits the metabolism of some medications by cytochrome P450
(4) decreases the volume of gastric juice

148. Features of cardiac tamponade include

(1) dyspnea and orthopnea
(2) tachycardia
(3) hypotension
(4) loud heart sounds

149. PEEP may improve oxygenation by

(1) prevention of airway closure
(2) recruitment of collapsed alveoli
(3) redistribution of extravascular lung fluid
(4) decreasing cardiac output

150. Protocols for suctioning the patient with an artificial airway should include

(1) preoxygenation
(2) use of a catheter as close to the size of the airway as possible
(3) application of suction for less than 10 seconds
(4) ventilating the patient with Ambu bag and room air after suctioning is completed

151. Management of the airway during induction of general anesthesia in a patient in a halo brace for a nondisplaced fracture of C6 incurred in a high-speed motor vehicle accident include(s)

(1) assessment of other injuries of the face
(2) awake fiberoptic intubation
(3) adequate anesthesia of the trachea
(4) removal of the cervical brace for intubation

152. A 42-year-old school teacher undergoes transabdominal hysterectomy for endometrial carcinoma. Her past medical history is significant for vitamin B_{12} deficiency diagnosed 3 years ago and otherwise unremarkable. According to the patient, she is in excellent health because of her strictly vegan diet. The case is conducted under general anesthesia with thiopental, fentanyl, and vecuronium for induction, followed by isoflurane, nitrous oxide, and oxygen for maintenance of anesthesia. After 4 hours of operating time, the patient is extubated uneventfully in the operating room and transferred to the PACU. On the 3rd postoperative day, you are called to evaluate the patient on the floor because she has developed signs of lower extremity paresthesias and has an unsteady gait. Which of the anesthetics used for induction and/or maintenance of anesthesia could have a role in this patient's symptoms?

(1) thiopental
(2) isoflurane
(3) vecuronium
(4) nitrous oxide

153. When assessing the acutely hypoxemic patient, causes that may be important are

(1) hypoventilation
(2) hypoperfusion
(3) ventilation/perfusion (V/Q) mismatch
(4) abnormal diffusion

154. Hypoglycemia occurring during an anesthetic

(1) can be prevented by the infusion of glucose
(2) is readily detected
(3) can be diagnosed accurately only by blood glucose determination
(4) cannot be detected if a glucose solution is running

155. Brain tumors affect ICP by

(1) increasing intracranial tissue content
(2) increasing interstitial fluid volume
(3) altering cerebral blood volume
(4) obstructing CSF flow

156. Stage 3 of general anesthesia, i.e., surgical anesthesia, is characterized by

(1) slow, irregular respiration
(2) hypotension
(3) active eyelid reflex
(4) conjugate gaze

157. Oxygen toxicity to the lung

(1) depends on the length of exposure
(2) is more apt to occur in critically ill patients
(3) is more apt to occur if the FIO_2 is over 0.5
(4) depends on oxygen fraction, not partial pressure

158. Causes of failure to pace a heart by an implanted pacemaker intraoperatively may include

(1) acute myocardial infarction
(2) decrease in pacing threshold of the heart
(3) electromagnetic interference
(4) heart rate faster than pacemaker rate

159. A 3-year-old child presents with acutely raised ICP from a blocked ventriculoperitoneal shunt. Surgical alleviation of the problem is planned. True statements regarding anesthesia in this patient include

(1) premedication with intramuscular opioids is beneficial, minimizing further increases in ICP due to agitation and induction of anesthesia
(2) induction of anesthesia with IV ketamine 0.5 to 1 mg/kg is a suitable technique
(3) rapid sequence induction should be avoided
(4) concentrations of isoflurane of up to 1% result in minimal increases in CBF

160. Pulmonary vascular resistance is increased by

(1) hypoxia
(2) hypercarbia
(3) acidosis
(4) atelectasis

161. Shunt pathways include

(1) those through atelectatic areas
(2) those through the bronchial circulation
(3) a patent foramen ovale
(4) the thebesian veins

162. Laboratory tests which are truly indicative of liver function include

(1) prothrombin time
(2) alanine amino transferase (ALT or SGPT)
(3) serum albumin
(4) γ-glutamyl transpeptidase (GGTP)

163. The precordial thump

(1) may convert sudden ventricular tachycardia to normal sinus rhythm
(2) may produce a ventricular contraction in complete heart block with ventricular asystole
(3) may induce ventricular fibrillation
(4) is likely to terminate ventricular fibrillation

164. Total pulmonary compliance

(1) is measured by dividing pressure by volume
(2) involves the lung only
(3) is independent of previous breaths
(4) is increased by surfactant

165. Hypercarbia occurring under anesthesia may be due to

(1) increased dead space ventilation
(2) exhaustion of soda lime
(3) pulmonary embolism
(4) hypoventilation

166. Consequence(s) of the prone position in anesthesia can include

(1) corneal abrasion
(2) impaired cerebral perfusion
(3) ulnar nerve injury
(4) venous distention

167. Cerebral blood volume is increased by

(1) elevated right heart pressure
(2) increased intrathoracic pressure
(3) head-down position
(4) direct pressure applied to jugular veins

168. Pressure-support ventilation

(1) is a pressure-limited, time-cycled mode of ventilation
(2) terminates the inspiratory phase when a preselected airway pressure is reached
(3) results in increased work of breathing compared to assist-mode ventilation
(4) delivers a tidal volume which is dependent on the patient's pulmonary compliance

169. Assuming a constant rate of carbon dioxide formation, the relationship between alveolar ventilation and carbon dioxide and hydrogen ion concentration is

(1) carbon dioxide varies inversely with ventilation
(2) doubling ventilation will lead to respiratory alkalosis
(3) reducing ventilation to one-fourth of normal will lead to a profound respiratory acidosis
(4) alveolar ventilation only responds to endogenous CO_2

170. Dopamine activates

(1) dopaminergic receptors
(2) α-adrenergic receptors
(3) β-adrenergic receptors
(4) nicotinic receptors

171. Postoperative hypothermia can

(1) increase oxygen consumption
(2) delay awakening
(3) increase peripheral vascular resistance
(4) initiate nonshivering thermogenesis in adults

172. Drugs that can be used to elicit seizures include

(1) methohexital
(2) enflurane
(3) etomidate
(4) propofol

173. Reabsorption of water occurs in the

(1) proximal tubule
(2) ascending loop of Henle
(3) distal tubule
(4) collecting duct

174. The normal pulmonary arterial circulation responds to increased cardiac output by

(1) increased pulmonary artery pressure
(2) zone 3 increases in size
(3) decreased pulmonary vascular resistance
(4) decreased physiologic dead space

175. A 36-week pregnant multipara is admitted in labor with history of vaginal bleeding. Preparations for delivery include

(1) two IV lines
(2) two units of blood available in the operating room
(3) examination only with a "double setup"
(4) induction of general anesthesia with thiopental favored over ketamine

PART 2

DIRECTIONS (Questions 176 through 261): Each of the numbered items or incomplete statements in this section is followed by answers or by completions of the statement. Select the ONE lettered answer or completion that is BEST in each case.

176. A patient underwent a thyroidectomy at 08:00 hours, and at 22:00 hours he complained to the nurse of difficulty in breathing. She took his blood pressure, which was moderately elevated above previous determinations, but she also noticed that his wrist flexed when the blood pressure cuff remained inflated. The cause of the stridor is probably

(A) vocal cord paralysis
(B) partial vocal cord paralysis
(C) laryngeal edema
(D) cervical hematoma
(E) hypocalcemia

177. Etomidate is a(n)

(A) isopropylphenol
(B) butyrophenone
(C) eugenol
(D) steroid
(E) imidazole

178. Carbon dioxide transport involves all of the following EXCEPT

(A) water
(B) bicarbonate ion
(C) carbonic anhydrase
(D) hemoglobin
(E) carboxyhemoglobin

179. Pierre Robin syndrome is associated with

(A) congenital heart anomalies
(B) choanal atresia
(C) maxillary hypoplasia
(D) true macroglossia
(E) posterior location of the larynx relative to the normal

180. Vasodilators which are likely metabolized to nitric oxide include the following EXCEPT

(A) sodium nitroprusside
(B) nitroglycerin
(C) nifedipine
(D) amyl nitrite
(E) isosorbide dinitrate

181. Transdermal scopolamine

(A) will not produce behavioral side effects
(B) is not effective for emesis prophylaxis in children
(C) is a muscarinic agonist
(D) does not cross the blood-brain barrier
(E) can be effective in postoperative nausea in adult inpatients

182. A patient with chronic renal failure who is receiving dialysis three times weekly suffers a fracture of the humerus in a fall. He is brought to the operating room for an open reduction of the fracture on the evening of a day on which he missed his dialysis due to the injury. Expected abnormalities in this patient include all of the following EXCEPT

(A) metabolic acidosis
(B) hyperkalemia
(C) uremia
(D) thrombocytopenia
(E) hypervolemia

183. The highest plasma concentration of a local anesthetic occurs after which of the following regional anesthetic techniques?

(A) interscalene
(B) lumbar epidural
(C) intercostal
(D) axillary
(E) caudal

184. During a craniotomy, after the dura mater is opened, the ICP

(A) increases
(B) equals zero
(C) changes directly proportional to blood flow
(D) decreases only if the head is elevated
(E) is unchanged

185. A drug that is a pure opioid antagonist is

(A) butorphanol
(B) levallorphan
(C) naloxone
(D) neostigmine
(E) edrophonium

186. The unit of capacitance is the

(A) ohm
(B) ampere
(C) volt
(D) farad
(E) henry

187. In the kidney position, proper positioning technique includes

(A) arms at side
(B) kidney rest under the lower rib cage
(C) lower leg flexed
(D) upper leg flexed
(E) head kept flat on the table

188. Absolute contraindications to major conduction anesthesia in parturients include all of the following EXCEPT

(A) preexisting neurologic disease of spinal cord
(B) patient refusal
(C) infection at the site of needle insertion
(D) hypovolemic shock
(E) coagulopathy

189. Umbilical vessel catheterization may result in all of the following EXCEPT

(A) rupture of the bladder
(B) arteriovenous fistula
(C) infection
(D) thrombophlebitis
(E) hepatic necrosis

190. Milrinone

(A) decreases cyclic-AMP levels
(B) is metabolized by MAO
(C) inhibits phosphodiesterase
(D) is a β-adrenergic agonist
(E) increases peripheral vascular resistance

191. Bupivacaine

(A) is an ester local anesthetic with short duration
(B) should be used in a concentration of 0.75% for epidural anesthesia during labor
(C) is associated with cardiac toxicity due to its effect on sodium channels
(D) toxicity is not a problem unless it is injected intravenously
(E) cardiotoxicity is noted by the onset of tachycardia

192. Acidosis may accompany the rapid transfusion of blood to control hemorrhage, thus

(A) the blood should be warmed to 120°F
(B) the blood should be warmed to 38°C
(C) each unit has an acid load of approximately 40 mEq
(D) sodium bicarbonate should be administered at a rate of 25 mEq/unit of blood transfused
(E) the acidosis should be treated with lactate

193. A 75-year-old man is having a transurethral resection of prostate (TURP) performed under spinal anesthesia. About 90 minutes into the procedure, the patient thrashes about and complains of nausea and an inability to see. The most likely cause is

(A) hypothermia
(B) glycine toxicity
(C) bladder perforation
(D) hemorrhage
(E) bacteremia

194. Of the cranial contents contributing to ICP, which has the smallest volume?

(A) brain tissue
(B) tissue water
(C) venous blood
(D) arterial blood
(E) CSF

195. The correct ranking of local anesthetic toxicity, from most toxic to least toxic, is

(A) tetracaine > bupivacaine > lidocaine > chloroprocaine
(B) bupivacaine > tetracaine > chloroprocaine > lidocaine
(C) bupivacaine > tetracaine > lidocaine > chloroprocaine
(D) bupivacaine > lidocaine > chloroprocaine > tetracaine
(E) chloroprocaine > bupivacaine > lidocaine > tetracaine

196. Boyle's law

(A) states that pressure is directly proportional to volume
(B) states that pressure is directly proportional to temperature
(C) states that pressure is inversely proportional to volume
(D) states that pressure is inversely proportional to temperature
(E) states that volume is directly proportional to temperature

197. Cromolyn sodium

(A) has a direct effect on bronchioles
(B) stimulates cyclic-AMP synthesis
(C) releases catecholamines
(D) prevents histamine release from mast cells
(E) decreases bronchiole reactivity

198. In the sympathetic nervous system

(A) preganglionic fibers synapse only in paravertebral sympathetic ganglia
(B) preganglionic cell bodies are located throughout the spinal cord
(C) the preganglionic neurotransmitter is acetylcholine
(D) target organ receptors are only adrenergic
(E) effects are mediated entirely through cyclic AMP

199. The total dosage of lidocaine for a child should not exceed

(A) 1 mg/kg
(B) 3.5 mg/kg
(C) 5 mg/kg
(D) 7 mg/kg
(E) 10 mg/kg

200. All of the following are true of the preterm infant EXCEPT

(A) increased depression from maternally-administered opioids
(B) incomplete bundle branch block
(C) less protein available for drug binding
(D) more sensitive to CNS toxicity of lidocaine
(E) immature enzyme systems for drug metabolism

201. Agents which inhibit AV conduction include the following EXCEPT

(A) adenosine
(B) verapamil
(C) isoproterenol
(D) digoxin
(E) esmolol

202. A Univent or bronchial blocker can be used in all of the following procedures EXCEPT

(A) bronchopulmonary lavage
(B) transthoracic esophagogastrectomy
(C) video thoracoscopy
(D) left lower lobe segmentectomy
(E) right pneumonectomy

203. Hypoxemia in the recovery room may be due to all of the following EXCEPT

(A) increased FRC
(B) shivering
(C) decreased cardiac output
(D) intrapulmonary shunt
(E) decreased vital capacity

204. Actions following the discovery of an airway fire include the following EXCEPT

(A) clamp the endotracheal tube
(B) extinguish fire with water or saline
(C) continue to ventilate with 100% O_2
(D) admit patient to ICU for observation
(E) ventilate by mask

205. A 50-year-old man with a long history of tobacco abuse and who is chronically dyspneic at rest is in the recovery room after an emergency appendectomy. He is receiving supplemental oxygen by mask and appears to be hypoventilating. An arterial blood gas is obtained, which shows

$pH = 7.19$, $PO_2 = 85$ mmHg, $PCO_2 = 90$ mmHg. This state can best be described as

(A) pure respiratory acidosis
(B) combined respiratory acidosis and metabolic acidosis
(C) respiratory acidosis with compensating metabolic alkalosis
(D) metabolic acidosis with compensating respiratory alkalosis
(E) pure metabolic acidosis

206. Major side effects of amiodarone include all of the following EXCEPT

(A) hepatitis
(B) exacerbation of arrhythmias
(C) thyroid dysfunction
(D) pneumonitis
(E) renal failure

207. The blood volume of an infant with a normal hemoglobin is estimated to be

(A) 30 mL/kg
(B) 40 mL/kg
(C) 50 mL/kg
(D) 60 mL/kg
(E) 80 mL/kg

208. At 17°C

(A) EEG activity is unchanged
(B) cellular integrity is lost
(C) CBF increases
(D) $CMRO_2$ is less than 10% of normothermic value
(E) the brain switches to anaerobic metabolism

209. Hetastarch

(A) is a crystalloid solution
(B) is a synthetic colloid
(C) is free of allergenic potential
(D) does not interfere with coagulation
(E) is metabolized in the liver to glucose monomers

210. Atrial natriuretic peptide

(A) is produced by the kidney
(B) increases intravascular volume
(C) increases blood pressure
(D) is a polypeptide
(E) causes sodium retention

211. All of the following decrease gastric acid secretion EXCEPT

(A) cimetidine
(B) glycopyrrolate
(C) metoclopramide
(D) ranitidine
(E) scopolamine

212. The English physician who popularized chloroform was

(A) Ivan Magill
(B) John Snow
(C) Robert Macintosh
(D) Charles Suckling
(E) Charles Jackson

213. Which of the following induction agents increases ICP?

(A) thiopental
(B) methohexital
(C) ketamine
(D) propofol
(E) midazolam

214. Sensitive methods for detecting venous air embolism include all of the following EXCEPT

(A) precordial Doppler
(B) mass spectrometry
(C) capnograph
(D) ECG
(E) transesophageal echocardiography

215. All of the following statements concerning local anesthetic toxicity in parturients are true EXCEPT

(A) most cases of cardiac arrest occurred with 0.75% bupivacaine
(B) all local anesthetics can cause seizures
(C) chloroprocaine has a high maternal:fetal ratio
(D) bupivacaine is more cardiotoxic than mepivacaine
(E) current preparations of chloroprocaine contain bisulfite which may be neurotoxic

216. Etomidate

(A) has anticonvulsant activity
(B) releases histamine
(C) has analgesic properties
(D) has an antiemetic effect
(E) increases intraocular pressure

217. As the temperature of the brain decreases

(A) MAC increases
(B) autoregulation of blood flow is lost
(C) $CMRO_2$ decreases 6 to 7% per °C
(D) cerebral Q_{10} decreases
(E) brain oxygen extraction increases

218. Which of the following statements about regional anesthesia in parturients is TRUE?

(A) Spinal anesthesia is not useful for vaginal delivery secondary to motor block.
(B) Lumbar sympathetic block does not result in interruption of pain impulses during labor.
(C) Hypotension is the most common complication of major conduction anesthesia.
(D) Caudal anesthesia is contraindicated secondary to the high incidence of fetal injury during placement of the block.
(E) Pudendal blocks will not provide adequate anesthesia for an episiotomy.

219. The agent with the highest ratio of β-adrenergic agonist to α-adrenergic agonist activity is

(A) isoproterenol
(B) dobutamine
(C) epinephrine
(D) norepinephrine
(E) phenylephrine

220. A hypotensive, comatose patient is brought to the operating room after being struck by a bus. He is emaciated and has a long history of alcoholism. It is important to administer thiamine to this patient because thiamine deficiency

(A) may be the cause of the observed coma
(B) may be the cause of the observed hypotension
(C) may be precipitated by the administration of glucose-containing solutions
(D) may cause rhabdomyolysis if succinylcholine is given
(E) may potentiate the cardiovascular depressant effects of volatile anesthetics

221. A properly performed and successful axillary block may provide inadequate anesthesia for

(A) open reduction of a fracture of the humerus
(B) creation of a radiocephalic arteriovenous fistula for dialysis access
(C) repair of a gamekeeper's thumb
(D) carpal tunnel release
(E) laceration of the flexor carpi ulnaris tendon

222. Glucuronide metabolites of medications will be excreted in the urine more rapidly than the parent compounds by which of the following processes?

(A) tubular secretion
(B) glomerular filtration
(C) facilitated diffusion
(D) passive nonionic diffusion
(E) tubular reabsorption

223. In the normal upright lung

(A) the blood flow is greatest at the apex
(B) the ventilation is greatest at the apex
(C) the ventilation/perfusion (V/Q) ratio is higher at the apex

(D) ventilation is uniform
(E) the PO_2 is lower at the apex

224. The correct ranking of the potencies of the volatile anesthetics, from most potent to least potent, is

(A) sevoflurane > desflurane > isoflurane > halothane
(B) desflurane > sevoflurane > isoflurane > halothane
(C) isoflurane > halothane > desflurane > sevoflurane
(D) halothane > isoflurane > sevoflurane > desflurane
(E) halothane > sevoflurane > desflurane > isoflurane

225. Esmolol is inactivated by

(A) monoamine oxidase
(B) catechol *O*-methyltransferase
(C) erythrocyte esterase
(D) plasma pseudocholinesterase
(E) acetylcholinesterase

226. Antineoplastic chemotherapy with which of the following agents may cause a peripheral neuropathy?

(A) methotrexate
(B) doxorubicin
(C) busulfan
(D) bleomycin
(E) vincristine

227. A child is admitted for general anesthesia for closure of a severe scalp laceration. He had eaten 2 hours before his accident. He should

(A) always have a rapid sequence induction with thiopental and succinylcholine
(B) have a nasogastric tube passed to remove gastric contents before induction
(C) not be operated on for 6 hours
(D) have vomiting induced
(E) be allowed to awaken with the endotracheal tube in place at the end of the procedure

228. The administration of meperidine to a patient taking an MAO inhibitor may cause

(A) CNS excitation and seizures
(B) cardiogenic shock
(C) decreased cardiac conduction or third-degree AV block
(D) ventricular tachycardia
(E) hepatic necrosis

229. The following are true statements about ranitidine EXCEPT

(A) it can decrease gastric pH
(B) it has a half-life of about 9 hours
(C) a typical dose for aspiration prophylaxis in adults is 120 mg PO
(D) it is a histamine H_2 receptor blocker
(E) it usually has fewer CNS side effects than cimetidine

230. As one moves from the apex of the lung to the dependent portions

(A) the alveoli become larger
(B) the caliber of the air passages becomes larger
(C) pleural pressure decreases
(D) compliance becomes greater
(E) ventilation of the alveoli becomes less due to decreased compliance

231. A patient given supplemental oxygen at 2 L/min via a nasal cannula will have an FIO_2 of approximately

(A) 22%
(B) 25%
(C) 28%
(D) 32%
(E) 36%

232. When a lethal amount of ethanol is ingested, the toxic effect of ethanol usually leading to death is

(A) hypotension due to vasodilation
(B) hypotension due to decreased cardiac output
(C) seizure
(D) apnea
(E) ventricular arrhythmia

233. A patient has insulin-dependent diabetes mellitus. She has peripheral vascular disease manifested as claudication which appears after she has walked a few blocks. She is scheduled to undergo vaginal hysterectomy because of postmenopausal bleeding. She would be classified by the American Society of Anesthesiologists (ASA) as physical status

(A) I
(B) II
(C) III
(D) IIIE
(E) IV

234. All of the following agents are used in the treatment of epilepsy EXCEPT

(A) benztropine
(B) valproic acid
(C) ethosuximide
(D) phenytoin
(E) carbamazepine

235. A patient is undergoing implantation of a hip prosthesis under general anesthesia. She has been taking prednisone for rheumatoid arthritis for 6 months. During the procedure, there is a sudden drop in blood pressure. The first step to be taken is to

(A) administer hydrocortisone 100 mg intravenously
(B) establish the cause of the hypotension
(C) cancel the procedure
(D) begin an infusion of phenylephrine
(E) discontinue all anesthetic agents

236. The best drying effect before fiberoptic endoscopy is performed is achieved by the administration of

(A) neostigmine
(B) pyridostigmine
(C) edrophonium
(D) atropine
(E) glycopyrrolate

237. The position of the vocal cords after bilateral cutting of the recurrent laryngeal nerve is

(A) due to unopposed action of the superior laryngeal nerve
(B) in complete abduction
(C) the same whether the nerve is completely or incompletely severed
(D) due to paralysis of the cricothyroid muscle
(E) both aligned to the left, since the left is the dominant branch of the nerve

238. A 16-year-old girl with suspected Hodgkin's disease is admitted for a short diagnostic procedure. She has massively enlarged hilar nodes, conjunctival edema, facial swelling, and difficult ventilation. The LEAST appropriate anesthetic plan would be

(A) local anesthesia
(B) awake fiberoptic intubation
(C) inhalation induction and rigid bronchoscopy
(D) propofol/succinylcholine induction and endotracheal intubation
(E) inhalation induction and laryngeal mask airway

239. At sea level, if a patient without any cardiopulmonary pathology is administered 50% O_2, after an appropriate time to reach equilibrium, the PaO_2 will be approximately

(A) 288 mmHg
(B) 300 mmHg
(C) 357 mmHg
(D) 370 mmHg
(E) 380 mmHg

240. The leading cause of death in the preeclamptic patient is

(A) cardiac arrest
(B) renal failure
(C) hepatic rupture
(D) respiratory arrest after inability to intubate
(E) cerebral hemorrhage

241. Arrhythmias that may respond to electrical countershock include the following EXCEPT

(A) ventricular tachycardia
(B) ventricular fibrillation
(C) atrial fibrillation
(D) supraventricular tachycardia
(E) complete heart block

242. A patient is scheduled for an elective cardioversion. A single IV injection of a short-acting agent is planned. Which of the following agents is most likely to cause vomiting after the procedure when used in this patient?

(A) ketamine
(B) thiopental
(C) etomidate
(D) methohexital
(E) propofol

243. Given the following set of data, 2, 2, 2, 6, 4, 2, the mode is

(A) 6
(B) 2
(C) 18
(D) 3
(E) cannot be determined by the information given

244. Palpation of an artery is not important in the performance of which of the following blocks?

(A) femoral
(B) axillary
(C) stellate ganglion
(D) supraclavicular
(E) sciatic

245. Which of the following is the LEAST common cause of postpartum hemorrhage?

(A) lacerations
(B) retained placenta
(C) coagulopathy
(D) uterine atony
(E) uterine inversion

Questions 246 and 247 refer to Figure 17-2.

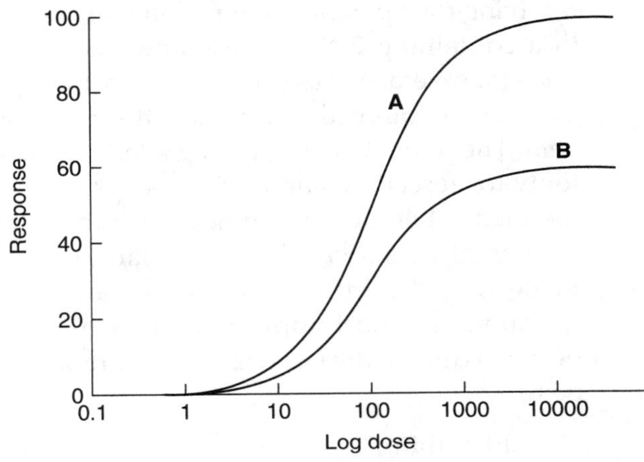

FIG. 17-2

246. If A and B are two different medications producing the same effect, then the figure shows that

(A) A has higher efficacy than B
(B) A has lower efficacy than B
(C) A and B must act via different receptors to produce their effects
(D) A is more potent than B
(E) A has a higher ED_{50} than B

247. If A depicts the dose-response relationship of drug X acting alone, and B depicts the dose-response relationship of drug X in the presence of drug Y, then it can be said that drug Y

(A) is a competitive antagonist of drug X
(B) is a noncompetitive antagonist of drug X
(C) must act via a different receptor than drug X
(D) increases the efficacy of drug X
(E) decreases the potency of drug X

248. Temporary atrial pacing can be achieved by means of the following EXCEPT

(A) transvenous endocardial leads
(B) epicardial leads
(C) external noninvasive electrodes
(D) esophageal electrodes
(E) a unipolar configuration

249. A patient with esophageal carcinoma has been receiving total parenteral nutrition with a solution containing 20% glucose and 4% amino acids for several weeks prior to surgery through a central venous catheter in the left subclavian vein. The patient is undergoing a left thoracotomy for resection, and during the procedure the medical student holding the rib retractor accidentally removes the subclavian catheter. Regarding the infusion of total parenteral nutrition, the most appropriate maneuver to perform during the surgical procedure would be to

(A) do nothing
(B) position the patient supine, insert a right subclavian catheter, and restart the infusion of parenteral nutrition
(C) restart the parenteral nutrition solution via a peripheral IV catheter
(D) administer 5 units of regular insulin intravenously and measure the serum glucose concentration in 15 minutes
(E) begin an infusion of 10% glucose via a peripheral vein

250. Acute pulmonary vasoconstriction will first elevate

(A) central venous pressure
(B) pulmonary artery pressure
(C) pulmonary capillary wedge pressure
(D) left atrial pressure
(E) aortic pressure

251. Cellular mediators synthesized from arachidonic acid include all of the following EXCEPT

(A) prostaglandins
(B) prostacyclins
(C) leukotrienes
(D) interleukins
(E) thromboxanes

252. Which of the following diseases is NOT transmitted to health care workers via needles or surgical instruments contaminated with blood or tissue from an infected patient?

(A) Creutzfeldt-Jakob disease
(B) hepatitis A
(C) hepatitis C
(D) cytomegalovirus
(E) syphilis

253. Patients who are expected to have a prolonged duration of neuromuscular blockade after succinylcholine include all of the following patients EXCEPT the patient with

(A) atypical cholinesterase
(B) a massive burn
(C) glaucoma managed with topical isoflurophate
(D) decreased bladder tone who takes neostigmine
(E) myasthenia gravis treated with pyridostigmine

254. All of the following statements about the diaphragm are true EXCEPT that it

(A) is innervated by the phrenic nerve
(B) gets its principal innervation from the third, fourth, and fifth cervical nerves
(C) has equal populations of fast- and slow-twitch muscle fibers in the adult
(D) has an excursion that changes with changes of posture
(E) has a peripheral tendon by which it attaches to the rib cage

255. When dextran 40 is to be used after endarterectomy to prevent thrombosis, a dose of dextran 1 is administered a few minutes before beginning the infusion of dextran 40

(A) as a test dose to check for allergic reactions
(B) to block mast cell receptors for dextran

(C) as a loading dose because it has a more rapid onset
(D) because it binds to coagulation factor XII
(E) as an inhibitor of dextran metabolism

256. The following statements are true concerning substance abuse among anesthesiologists EXCEPT

(A) alcohol and fentanyl are the drugs most likely to be abused
(B) a proper drug accountability system will prevent misuse
(C) individual confrontations with the drug user are not recommended
(D) the relapse rate of anesthesiology residents allowed to reenter a program is 66%
(E) it is the policy of the ASA to treat substance abuse disorders as a disease

257. All of the following are used as bronchodilators EXCEPT

(A) ipratropium
(B) cromolyn sodium
(C) theophylline
(D) terbutaline
(E) albuterol

258. All of the following are true of the calculation of Reynold's number in tube flow EXCEPT

(A) a value above 2000 has significance
(B) it involves density
(C) it involves length
(D) it involves viscosity
(E) it involves velocity

259. The dermatome level corresponding to the umbilicus is

(A) T2
(B) T4
(C) T6
(D) T8
(E) T10

260. Mixed venous oxygen saturation tends to be increased by the following EXCEPT

(A) arteriovenous fistulae
(B) cyanide poisoning
(C) right-to-left intracardiac shunts
(D) sepsis
(E) cirrhosis

261. In the case of surgery involving electrocautery for a patient with an automatic implantable cardioverter/defibrillator (ICD), it is prudent to

(A) deactivate the ICD
(B) employ bipolar electrocautery
(C) place the skin electrode of the electrocautery device far from the ICD generator
(D) reprogram the ICD to an asynchronous mode
(E) infuse lidocaine intraoperatively

DIRECTIONS (Questions 262 through 350): For each of the items in this section, ONE or MORE of the numbered options is correct. Choose answer

(A) if only 1, 2, and 3 are correct
(B) if only 1 and 3 are correct
(C) if only 2 and 4 are correct
(D) if only 4 is correct
(E) if all are correct

262. In the Tec 6 vaporizer, desflurane vapor

(1) is present at a temperature higher than room temperature
(2) is present at a pressure higher than atmospheric pressure
(3) will not leak out from the vaporizer if the vaporizer is filled while in use
(4) is mixed with carrier gas utilizing the variable bypass principle

263. Agent(s) which have selectivity at the β_2-adrenergic receptor include

(1) metaproterenol
(2) isoetharine
(3) albuterol
(4) metoprolol

264. As the depth of anesthesia under halothane increases, one would expect

(1) tidal volume to increase
(2) $PaCO_2$ to increase
(3) respiratory rate to decrease
(4) oxygen consumption to decrease

265. Droperidol exerts antagonistic actions at which of the following receptors?

(1) dopaminergic
(2) muscarinic cholinergic
(3) α-adrenergic
(4) H_1-histaminergic

266. The patient with acromegaly is likely to have

(1) a difficult airway
(2) glucose intolerance
(3) hypertension
(4) hyperthyroidism

267. Effective means for delivery of opioids for the relief of postoperative pain include

(1) oral administration
(2) sublingual administration
(3) transdermal application
(4) IV infusion

268. The irrigating fluids for prostatic resection should be

(1) nonhemolytic
(2) close to water in composition
(3) isosmolar
(4) electrolyte solutions

269. Inhibitors of MAO include

(1) phenelzine
(2) isocarboxazid
(3) tranylcypromine
(4) selegiline

270. Patients who have suffered a subarachnoid hemorrhage may

(1) have ST segment elevation consistent with myocardial ischemia
(2) develop hydrocephalus
(3) have elevated cardiac enzymes
(4) require hypertension for cerebral vasospasm

271. Heparin inhibits the activity of which of the following factors in the clotting cascade?

(1) thrombin
(2) XII
(3) X
(4) VII

272. Assuming a normal oxyhemoglobin dissociation curve position and a normal hematocrit, a PaO_2 of

(1) 90 mmHg corresponds to a saturation of approximately 96.5%
(2) 90 mmHg corresponds to a dissolved oxygen of approximately 0.27 mL O_2/100 mL blood
(3) 90 mmHg corresponds to a combined oxygen of approximately 19.8 mL/100 mL blood
(4) 30 mmHg corresponds to an oxygen saturation of 30%

273. Side effects of neurolytic blockade may include

(1) sphincter impairment
(2) postneurolytic neuralgias
(3) muscle weakness
(4) seizures

274. Following a 50% burn in a 3-year-old child,

(1) within 4 days an amount of albumin equal to about the total body plasma content is lost through the wound
(2) immediately after injury, cardiac output is increased

(3) the use of pulse oximetry is mandatory in monitoring the patient if carbon monoxide poisoning is suspected
(4) evaporative fluid losses are approximately 4 L for each square meter of burned surface per day

275. Agents which readily cross the blood-brain barrier and block the muscarinic cholinergic receptor include

(1) glycopyrrolate
(2) diphenhydramine
(3) cimetidine
(4) scopolamine

276. Important considerations in the parturient with heart disease include

(1) minimizing increases in cardiac output that occur with labor and delivery in patients with mitral stenosis
(2) epidural anesthesia is preferable for cesarean section in the patient with aortic stenosis
(3) epidural anesthesia is preferable for labor and delivery in patients with aortic insufficiency
(4) decreased blood pressure should be managed with large volumes of IV fluids

277. The vapor pressure of a liquid

(1) is a function of its temperature
(2) decreases with an increase in temperature
(3) is not affected by the total barometric pressure above the liquid
(4) decreases with an increase in density of the liquid

278. In the technique of one-lung anesthesia, the method used to ventilate the dependent lung is important. It is true that

(1) hyperventilation of the dependent lung may inhibit hypoxic pulmonary vasoconstriction
(2) high airway pressure to the dependent lung may be harmful
(3) high inspired oxygen concentration may lead to absorption atelectasis
(4) high inspired oxygen concentration may lead to vasoconstriction

279. Drugs considered unsafe for the patient known to be susceptible to malignant hyperthermia include

(1) succinylcholine
(2) nitrous oxide
(3) isoflurane
(4) ketamine

280. Overdrive pacing may be used to terminate

(1) paroxysmal atrial tachycardia
(2) atrial fibrillation
(3) atrial flutter
(4) ventricular fibrillation

281. If a healthy adult breathes 100% oxygen for several days, he is likely to develop

(1) absorption atelectasis
(2) blindness
(3) pulmonary edema
(4) cerebral edema

282. Patients may be protected from microshock currents delivered by wires or catheters that enter the heart by ensuring that the equipment to which the wires or catheters are attached is connected to

(1) an isolated power supply
(2) a ground fault circuit interrupter
(3) a line isolation monitor
(4) earth ground

283. Which of the following statements is (are) true regarding hyperbaric oxygen therapy when used in cases of severe carbon monoxide poisoning?

(1) An adequate concentration of dissolved oxygen can be achieved in the blood regardless of the carboxyhemoglobin concentration.
(2) Oxygen toxicity is related to the PO_2 in the inspired gas.
(3) Bullous emphysema is a contraindication to hyperbaric oxygen therapy.
(4) Although hyperbaric oxygen therapy increases oxygen delivery to tissues, it has no effect on the rate of excretion of carbon monoxide.

284. A 38-week gestational-age female with an Apgar score of 4 requires which of the following?

(1) clearing of the airway with a bulb syringe
(2) intubation immediately
(3) long positive inflation of lungs at 60 breaths per minute
(4) epinephrine if the heart rate is less than 100

285. Which of the following agents is (are) high-ceiling diuretics?

(1) ethacrynic acid
(2) furosemide
(3) bumetanide
(4) metolazone

286. Fetal hemoglobin

(1) has a dissociation curve shifted to the left relative to that for adult hemoglobin
(2) constitutes about 75% of total hemoglobin in the neonate
(3) results in decreased oxygen delivery at any given oxygen tension in the neonate
(4) has a lower affinity for oxygen than adult hemoglobin at any given partial pressure

287. Seizures

(1) produce cerebral acidosis
(2) can be stopped by isoflurane anesthesia
(3) increase both CBF and metabolism
(4) are suppressed by most general anesthetics

288. A patient with mitral valve prolapse who has a history of paroxysmal supraventricular tachycardia and who is scheduled for cholecystectomy should receive which of the following prophylactically in the perioperative period?

(1) anticoagulants
(2) a temporary transvenous cardiac pacemaker
(3) antiarrhythmics
(4) antibiotics

289. Mucociliary flow will be altered during halothane anesthesia in response to

(1) concentration of halothane
(2) airway humidity
(3) body temperature
(4) FIO_2

290. Which of the following are mixed opioid agonist-antagonist agents?

(1) dezocine
(2) buprenorphine
(3) nalbuphine
(4) butorphanol

291. Elevation of the ST segment occurs

(1) in Prinzmetal's angina (coronary spasm)
(2) in acute pericarditis
(3) in myocardial infarction
(4) during normal exercise

292. A 20-year-old female student with anorexia nervosa presents for emergency orthopedic surgery. Likely abnormalities associated with this condition include

(1) ascites due to hypoalbuminemia
(2) decreased cardiac output in response to stress
(3) hyperkalemia
(4) metabolic acidosis

293. Which of the following statements is (are) TRUE regarding anion gap?

(1) An anion gap of 12 mEq/L is considered normal.
(2) Diabetic ketoacidosis is associated with a low anion gap.
(3) Metabolic acidosis accompanying diarrhea has a normal anion gap.
(4) Anion gap may be estimated by subtracting the serum chloride concentration from the serum sodium concentration.

294. Which of the following statements is (are) TRUE regarding postoperative nausea and vomiting?

(1) Drugs which cross the blood-brain barrier better are more likely to cause postoperative nausea and vomiting.
(2) Patients who have never had anesthesia or surgery before but who have a history of motion sickness should be considered at high risk for postoperative nausea and vomiting.
(3) All available antiemetic agents act at the emetic center in the medulla.
(4) There is an inverse relationship between the patient's age and the risk of postoperative nausea and vomiting.

295. Alprostadil (prostaglandin E_1) tends to

(1) maintain patency of the ductus arteriosus
(2) maintain patency of the foramen ovale
(3) decrease pulmonary vascular resistance
(4) increase systemic vascular resistance

296. Signs of a properly performed stellate ganglion block may include

(1) ptosis
(2) miosis
(3) hoarse voice
(4) conjunctival injection

297. Therapy for patients with closed head trauma may include

(1) CNS depressants to reduce ICP
(2) use of pressors to maintain cerebral perfusion pressure at 80 mmHg or above
(3) endotracheal intubation and mechanical ventilation for hypoxemia
(4) routine hyperventilation

298. If a gas at a given temperature is in a closed container, the pressure exerted by the vapor will change under which of the following conditions?

(1) the container is heated
(2) the container is compressed
(3) the container is cooled
(4) the container is agitated

299. Flowmeters used in anesthesia machines

(1) are variable orifice devices
(2) may be interchanged among gases of the same density
(3) have a fixed pressure drop across the float regardless of flow rate
(4) utilize glass tubes having a precisely constant caliber from end to end

300. Adenosine

(1) occurs as a human metabolite
(2) is inactivated by catechol O-methyltransferase
(3) decreases AV conduction
(4) is a β-adrenergic agonist

301. Administering humidified gases and mist therapy to a patient may result in

(1) lung infection
(2) loss of heat
(3) water load
(4) increased airway resistance

302. Norepinephrine

(1) is derived from tyrosine
(2) is inactivated by MAO
(3) is actively recovered from the synaptic cleft
(4) binds to nicotinic receptors

303. Antihypertensives which interact with the renin-angiotensin system include

(1) losartan
(2) prazosin
(3) enalapril
(4) minoxidil

304. Closing volume

(1) is measured by nitrogen washout
(2) is effort dependent
(3) is measured with trace gases
(4) decreases with age

305. Which of the following medications is (are) useful in the therapy of schizophrenia?

(1) fluphenazine
(2) chlorpromazine
(3) haloperidol
(4) doxepin

306. In the patient with sickle cell anemia

(1) both hypo- and hyperthermia may lead to sickling
(2) hypovolemia is better tolerated than hypervolemia
(3) pneumatic tourniquets should be avoided
(4) FIO_2 should be adjusted to maintain normal, but not elevated, values for PaO_2

307. Metoclopramide has antagonistic effects at which of the following receptors?

(1) dopaminergic D_2
(2) β-adrenergic
(3) 5-HT_3 (serotonin$_3$)
(4) muscarinic cholinergic

308. The difference between the alveolar pressure and the ambient pressure is the sum of

(1) transpulmonary pressure
(2) esophageal balloon pressure
(3) chest wall transmural pressure
(4) pulmonary artery occlusion pressure

309. The mask shown in Figure 17-3

(1) makes use of the Venturi effect
(2) is used to regulate the concentration of oxygen
(3) uses oxygen to entrain room air
(4) is dependent on the patient's minute ventilation to determine oxygen concentration

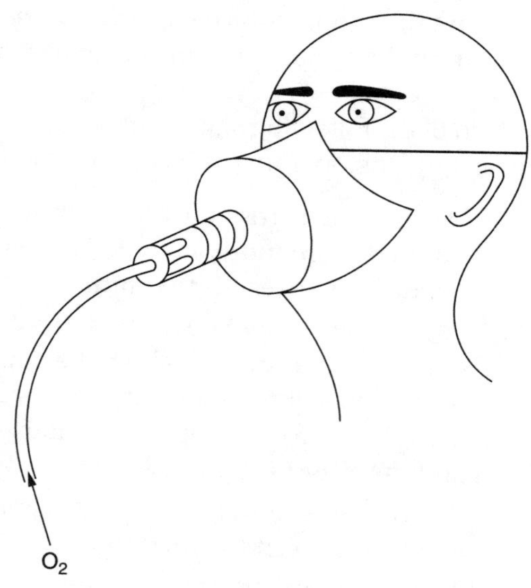

FIG. 17-3

310. Captopril

(1) activates ACE
(2) increases levels of angiotensin II
(3) decreases venous capacitance
(4) decreases arteriolar resistance

311. CBF-metabolism coupling

(1) is the major mechanism for modulation of regional delivery of nutrients to brain tissue
(2) reflects parallel changes in CBF as metabolic rate changes
(3) represents a constant ratio
(4) is abolished by general anesthetics

312. The optimal PEEP level depends on the

(1) degree of hypoxemia
(2) type of lung disease

(3) state of hydration
(4) state of left ventricular function

313. β_1-Adrenergic agonists

(1) dilate coronary vessels
(2) increase cardiac rate
(3) increase cardiac conduction velocity
(4) increase contractility

314. Erroneous values for oxygen saturation may be displayed by a pulse oximeter if

(1) the probe is exposed to room light
(2) the patient becomes tachycardic
(3) methylene blue is administered to the patient
(4) helium is administered along with oxygen and nitrous oxide

315. Compared to epidural morphine, epidural fentanyl

(1) is not enhanced by the addition of local anesthetics
(2) is less likely to cause delayed onset of respiratory depression
(3) is less lipophilic than morphine
(4) has a more rapid onset

316. MAC-BAR

(1) cannot be determined by observing patients' responses during anesthesia
(2) is less than MAC-awake for all anesthetics
(3) is about 50% higher than MAC
(4) refers to an alveolar concentration sufficient to blunt all responses during anesthesia

317. Cervical spine instability should be considered in patients with

(1) ankylosing spondylitis
(2) Down's syndrome
(3) Marfan's syndrome
(4) rheumatoid arthritis

318. An infant aged 3 months (but only 46 weeks postconceptual age) requires an inguinal hernia repair. The child was born by a spontaneous vaginal delivery, had an uncomplicated perinatal course, and was discharged home 2 weeks later. Which of the following statements is (are) TRUE?

(1) In order to minimize disruption of feeding and sleep, the infant should be operated on as an ambulatory surgical patient.
(2) Spinal anesthesia is a suitable technique.
(3) Induction of anesthesia with halothane is inadvisable because of the relatively slow uptake of inhalational agents in infants.
(4) Halothane may cause depression of the chemoreceptor pathways, but is a suitable choice of inhalational agent.

319. The calculated shunt fraction (Q_s/Q_t)

(1) may change with changes in cardiac output
(2) increases with atelectasis
(3) will be decreased with 100% oxygen
(4) increases in patients with pulmonary embolus

320. The circle of Willis is formed by branches of the

(1) internal carotid artery
(2) external carotid artery
(3) subclavian artery
(4) occipital artery

321. Labetalol blocks

(1) α-adrenergic receptors
(2) dopaminergic receptors
(3) β-adrenergic receptors
(4) muscarinic receptors

322. General anesthesia for cesarean section should consist of which of the following?

(1) Predelivery O_2 @ 1 L/min, N_2O @ 2 L/min, and 0.5% isoflurane
(2) Adequate IV access
(3) O_2 @ 2 L/min, N_2O @ 2 L/min, and 1.25% isoflurane
(4) left uterine displacement

323. In a variable bypass vaporizer

(1) total gas flow passes through the vaporizing chamber
(2) the vaporizer output is less than the dial setting at very low flow rates
(3) the vaporizer output is independent of the composition of the carrier gas
(4) the vaporizer output is less than the dial setting at very high flow rates

324. Complications of a celiac plexus block include

(1) pneumothorax
(2) subarachnoid block
(3) postural hypotension
(4) retroperitoneal hematoma

325. Factors released in response to hypovolemia include

(1) aldosterone
(2) renin
(3) antidiuretic hormone
(4) atrial natriuretic peptide

326. Nerve injuries occurring during anesthesia are

(1) more frequent with long nerves
(2) always due to poor positioning
(3) more frequent with superficial nerves
(4) less common in the patient with ischemic disease

327. Airway resistance

(1) decreases with maximal inspiration
(2) increases with parasympathetic stim ulation
(3) increases with acetylcholine inhalation
(4) increases with inhalation of smoke

328. During surgery for correction of scoliosis, integrity of the spinal cord may be confirmed by

(1) performing an intraoperative wake up
(2) monitoring brainstem-evoked potentials
(3) monitoring somatosensory-evoked potentials
(4) monitoring the train-of-four on all four limbs

329. Agents in anesthetic practice that can cause bronchodilatation include

(1) ketamine
(2) atropine
(3) isoflurane
(4) thiopental

330. Therapy for neurogenic pulmonary edema includes

(1) reduction of intracranial hypertension
(2) α-adrenergic antagonists
(3) supportive respiratory care
(4) central nervous system depressants

331. True allergy is common to which of the following local anesthetics?

(1) lidocaine
(2) mepivacaine
(3) bupivacaine
(4) tetracaine

332. Mannitol reduces ICP because it decreases the volume of

(1) brain substance
(2) arterial blood
(3) brain intracerebral water
(4) venous blood

333. Postintubation laryngeal edema may require treatment with

(1) humidified oxygen
(2) intubation or tracheostomy
(3) nebulized racemic epinephrine 0.25 mL in 2 mL normal saline
(4) fluid restriction

334. Physiologic compensation for anemia includes

(1) increased plasma volume
(2) increased cardiac output
(3) increased levels of 2,3-diphosphoglycerate
(4) bradycardia

335. Herpetic whitlow

(1) is caused by the varicella zoster virus
(2) is usually acquired from a needle stick
(3) usually affects the CNS causing encephalitis
(4) may be effectively prevented by wearing gloves

336. Oxygen should be administered in the recovery room

(1) only to those who have had general anesthesia
(2) only to those with cyanosis
(3) never after laser surgery
(4) as determined by pulse oximetry

337. Which of the following agents will lead to a decrease in pulmonary vascular resistance in either acute or chronic pulmonary hypertension?

(1) inhaled nitric oxide
(2) epoprostenol
(3) sildenafil
(4) bosentan

338. Sickle cell formation is facilitated by

(1) hypoxemia
(2) fever
(3) acidosis
(4) halothane

339. Effects of fat embolism include

(1) hypoxia
(2) tachypnea
(3) interstitial pulmonary edema
(4) neurologic impairment

340. For patients with complex regional pain syndrome, the performance of a Bier block (IV regional block) with which of the following agents may be efficacious in relieving symptoms?

(1) bretylium
(2) hydralazine
(3) guanethidine
(4) methyldopa

341. Equalization of diastolic filling pressures (central venous, right ventricular diastolic, pulmonary artery diastolic, pulmonary capillary wedge) tends to occur in

(1) mitral stenosis
(2) pericardial constriction
(3) tricuspid regurgitation
(4) cardiac tamponade

342. Acute respiratory insufficiency will manifest itself by

(1) hypoxia
(2) dyspnea
(3) hypercapnia
(4) increased blood bicarbonate

343. A heated cascade humidifier

(1) can deliver 100% relative humidity at body temperature
(2) can overheat the airway resulting in a burn
(3) has an internal check valve
(4) should be filled with sterile water to minimize infection

344. Morbid obesity is associated with decreases in

(1) functional residual capacity
(2) residual volume
(3) total lung capacity
(4) closing capacity

345. Prior to initiating epidural blockade for pain relief during labor, which of the following is (are) desirable?

(1) prehydration with 500 to 1000 mL of balanced salt solution
(2) equipment available for resuscitation
(3) adequate IV access
(4) intramuscular ephedrine to prevent hypotension

346. The pulmonary capillary wedge pressure will be elevated with respect to the left ventricular end-diastolic pressure in the event of

(1) bradycardia
(2) high airway pressure
(3) tricuspid regurgitation
(4) mitral stenosis

347. The arteria radicularis magna or artery of Adamkiewicz

(1) is the largest of the radicular arteries supplying the spinal cord
(2) anastomoses with the anteroposterior spinal artery system
(3) is one of several radicular branches from the aorta
(4) always originates from the suprarenal aorta

348. Appropriate ways to assure proper placement of the endotracheal tube include

(1) chest radiograph
(2) bronchoscopic examination
(3) auscultation of the thorax
(4) end-tidal carbon dioxide capnography

349. The variable(s) that Ohm's law relate(s) is (are)

(1) resistance
(2) voltage
(3) current
(4) power

350. Transesophageal echocardiography provides information about

(1) intracardiac volumes
(2) intracardiac gas emboli
(3) regional myocardial contractility
(4) valvular function and regurgitation

Answers and Explanations

1. **(A)** Pacing will not improve this condition. A junctional rhythm can sometimes be corrected by means of overdrive pacing. *(1:862; 4:661; 5:1418; 6:72)*

2. **(C)** Pregnancy is associated with many laboratory deviations from "normal." Hgb and Hct decrease, platelets and most coagulation factors increase, blood urea nitrogen (BUN) and creatinine fall, and both serum HCO_3^- and PCO_2 fall resulting in little change in pH. *(1:1141–2, 4:1989–90, 5:2309–11)*

3. **(D)** Henry's law states that the amount of gas absorbed by a liquid is proportional to the pressure of the gas above the liquid, as long as there is no chemical reaction between the gas and the liquid. *(5:12)*

4. **(C)** The antithrombin III-heparin complex inactivates thrombin and the other proteases of the intrinsic clotting pathway. *(1:217; 3:1522–3; 4:1924)*

5. **(A)** At FRC, there is no pressure difference between the alveoli and atmosphere. Elastic forces of the lung and chest wall are balanced. The total pulmonary vascular resistance is at its lowest at FRC. Further ventilation is possible, since FRC is the volume that exists at the end of a normal tidal volume. The gradient between the alveoli and the intrapleural space will be nonzero, balancing the elastic tension of the lung. *(1:794–5; 4:1753; 5:681–3, 693)*

6. **(C)** The patient's caregivers have been noncompliant with asthma medications. He clearly is not at baseline nor optimized for an elective surgery. The patient should be referred to his pediatrician. He should be placed back on a maintenance dose of montelukast for a minimum of 2 weeks. Sedation and analgesia with local anesthesia is usually not a reasonable choice in this age group. *(2:49–50: 5:2382, 2853)*

7. **(D)** In the lithotomy position, the peroneal nerve is the most likely nerve to be injured, and the mechanism is usually pressure against the stirrups. The other nerves listed may be injured by compression due to flexion of the legs on the trunk. *(5:712)*

8. **(B)** Scleroderma is associated with many problems, but arterial dilatation is not one of them. The arterial problem is usually arterial constriction. Raynaud's phenomenon is frequently seen. *(1:510–1; 4:557, 563; 6:511–2)*

9. **(E)** Although PEEP may be necessary in the ventilation of the neurosurgical patient; its use should not be routine. The level of PEEP should be titrated to the need and effect. It is helpful to monitor ICP to ascertain possible deleterious effects. The head should be elevated. *(1:7163; 4:1617; 5:2141–2; 6:239, 241)*

10. **(D)** Halothane is soluble in rubber and, if one uses rubber delivery hoses, the agent can be absorbed into the rubber. Halothane can then have an effect on subsequent patients. The rubber does not change the composition of halothane. If high flow rates are used, the halothane will be washed out of the system faster. *(3:349–50; 4:1131; 5:142–3)*

11. **(C)** Advantages of continuous spinal anesthesia using microbore catheters include the ability to control the spread of the local anesthetic and a relatively low incidence of postdural puncture headache. While there were cases of catheters tearing during removal, they were probably related to inadequate flexion of the patient's spine and/or excessive force applied to the catheter during removal. These microbore catheters were withdrawn from the market due to a number of reported cases of cauda equina syndrome. *(1:695; 4:1380–1; 5:2320)*

12. **(A)** One gram of oxygen, molecular weight 32, is 0.03125 mole. Thus, by the ideal gas law, $V = nRT/P = (0.03125) \times (0.082) \times (273 + 20)/1$, i.e., $V = 0.751\ \text{L} = 751\ \text{mL}$. *(1:379; 4:1012–3)*

13. **(C)** Preeclampsia occurs after the 20th week of gestation and requires at least two of the following: systolic blood pressure greater than 140 or 30 mmHg above prepregnancy levels, diastolic blood pressure greater than 90 or 15 mmHg above prepregnancy levels, general edema, or proteinuria. It can involve all organ systems. Its etiology remains unknown. *(1:1152–4, 4:2011–2, 5:2329–33)*

14. **(D)** Pulse pressure widens in small distal vessels. *(1:673; 1:875; 4:811; 5:1274)*

15. **(C)** The afferent stimulus for the temperature-regulating center is an integrated input from multiple temperature-sensitive cells. Input signals from a variety of anatomic regions contribute including the skin surface, neuroaxis, and deep tissues. Afferent thermal sensing is neural, not hormonal. Muscle spindles do not provide thermal information. *(1:683–4, 1397; 4:2442–4; 5:1572–5)*

16. **(A)** The hardness of soda lime is due to the addition of silica. The hardness decreases the amount of dust that is present. The dust can cause channeling and increased resistance and can be carried throughout the system. *(1:582; 4:1046; 5:296)*

17. **(A)** Many drugs administered to the parturient cross the placenta and have neonatal effects ranging from fetal heart rate changes to neonatal depression. Opioids are the most commonly used agents during labor and may produce neonatal depression, which is related to the total dose and time interval from administration to delivery of the fetus. Inhalational agents rapidly cross but muscle relaxants have minimal transfer. *(1:1142–3, 4:1995–6)*

18. **(E)** No pharmacologic alternative meets both goals. *(1:915; 3:226; 4:191; 5:1991; 6:9)*

19. **(C)** Hurler's syndrome is associated with all of the options listed. Hypertelorism may be present, but it should not be a factor complicating the anesthetic. *(1:542–3; 4:224, 429; 6:466)*

20. **(C)** The centers located in the pons and medulla (brain stem) respond primarily to hydrogen ion concentration. Although this is related to the level of carbon dioxide, the primary stimulant is hydrogen ion. Chronic hypercapnia depresses these centers. *(1:796–7; 5:170–1, 1463)*

21. **(E)** Use of heparin in a patient with heparin-induced thrombocytopenia type II can lead to life-threatening thrombotic complications and is therefore not indicated. Low-molecular weight heparins can cross-react with heparin and should also be avoided. Warfarin should not be used until the thrombocytopenia has resolved because it can cause venous limb gangrene or multicentric skin necrosis. Tirofiban is an antiplatelet drug that acts by inhibition of the glycoprotein IIb/IIIa receptor. Lepirudin is a recombinant derivative of hirudin, which is administered intravenously. It is approved for the prevention of deep vein thrombosis in patients with heparin-induced thrombocytopenia. *(3:1525–6, 1536; 5:1116)*

22. **(A)** The hematocrit reaches the lowest value between 8 and 12 weeks in full-term infants. The MAC for inhalational agents is highest between 1 and 6 months of age. Blood volume is approximately 375 mL (70 to 80 mL/kg). Maintenance fluids are 20 mL/h. *(1:1198, 1201; 2:42–3, 134, 223; 5:2372–3, 2388–90)*

23. **(D)** The most likely cause of sudden apnea in this patient is the inadvertent subarachnoid

injection of local anesthetic. All of the other choices are potential side effects of a stellate ganglion block. *(1:862)*

24. **(C)** *(4:415; 5:2686)*

25. **(B)** The inferior wall of the left ventricle is supplied by the right coronary artery and is most apparent in leads II, III, and AVF. *(1:1507; 4:1672; 5:1390; 6:18)*

26. **(C)** A bacterial infection is likely to have responded to antibiotic therapy. A wheezy child may not be "asthmatic" but may have aspirated a foreign body. In a 2-year-old child this diagnosis should always be considered a possibility even if there is no clear history of aspiration. A child presenting with these symptoms and signs requires emergency therapy whatever the diagnosis. Agitation may be misinterpreted as emotional upset when it is due to serious underlying hypoxemia. If the child is stable, x-rays may be helpful in making the diagnosis and in identifying and localizing the foreign body; however, if the child is severely distressed, oxygen should be administered by facemask and immediate plans made for removal of the foreign body in the operating room. *(2:321–2; 4:464–5; 5:2853)*

27. **(A)** *(1:692; 4:1395)*

28. **(C)** FRC decreases approximately 33% after upper abdominal surgery recovering after 1 to 2 weeks. Peripheral surgery affects FRC little. *(1:1388–9; 4:2353; 5:2711–5)*

29. **(D)** Sevoflurane is not known to cause hepatotoxicity and it is not flammable. The agent produces good muscle relaxation, does not produce tachycardia, and may be a preferable agent in patients prone to myocardial ischemia. Sevoflurane is metabolized in the liver by the cytochrome P450 system. *(3:354–5; 4:1149–51)*

30. **(E)** At constant volume, the ratio of pressure to temperature (in K) is constant. Thus, at 155°F (68.3°C or 341.3 K), the pressure in the tank will be (1900 psi) × (341.3 K) / (293 K) = 2213 psi. *(1:379; 4:1013)*

31. **(D)** The difficult airway or failed intubation in obstetric anesthesia has potential for very serious consequences with two lives at risk. Everyone should have a protocol to follow, and many examples are available in the literature. The mother's life has priority, but every effort must be made to deliver a viable infant. Airway assessment should be reevaluated prior to anesthesia for cesarean section because the effects of labor can alter the airway. *(4:2006–8, 5:2325–6)*

32. **(B)** Hypotension rarely occurs with the administration of intrathecal morphine. *(3:1942)*

33. **(D)** At the level of the first rib, where local anesthetic is deposited for a supraclavicular block, the roots have fused to form the three trunks of the brachial plexus. *(1:726; 4:1415; 5:1738)*

34. **(D)** The A-gamma fiber is a medium-sized, myelinated fiber that serves as the efferent fiber to muscle spindles. *(1:415; 4:1297; 5:493)*

35. **(A)** α-Adrenergic agonists permit higher perfusion pressures for heart and brain during CPR. *(1:1494; 3:221; 4:662; 5:2942; 6:614)*

36. **(A)** Compliance refers to a change in volume per unit change in pressure. Compliance is the reciprocal of elastance. The reciprocal of conductance is resistance. No time units are involved since it is a static measurement. *(1:794–5; 4:1753; 5:1465–7)*

37. **(B)** The MAC for desflurane is about 5.5% and it is irritating to airways. It is insoluble in water and lipids and is metabolized to a lesser extent than other halogenated anesthetics. *(4:1124, 1148–9)*

38. **(B)** Double-lumen tubes have one lumen in the trachea and one lumen in a mainstem bronchus. The handedness of the tube reflects the bronchus intubated. This design principle applies to the Robertshaw and Carlens tubes, as well as the newer disposable PVC tubes. *(1:824–5; 4:1757; 5:1874–8)*

39. **(C)** Creatinine clearance may be estimated according to the following equation:

$$Cl_{Cr} \approx \frac{(140 - \text{age}) \times \text{weight}}{\text{Cr} \times 72}$$

where age is in years, weight is in kg, and Cr is in mg/dL. The values obtained at 20 and 85 years of age are approximately 117 and 53 mL/min, respectively. *(1:1015; 4:1168; 5:1499; 6:342–3)*

40. **(E)** One milliliter of desflurane liquid is 1.47 g or 0.00875 mol (1.47 g/168 g/mol). Thus, by the ideal gas law, $V = nRT/P = (0.00875) \times (0.082) \times (273 + 37) = 0.222$ L = 222 mL. *(1:379; 4:1021)*

41. **(E)** The head constitutes 10% of total body surface area in the adult, 13% in a 10- to 14-year-old child, 15% in a 5- to 9-year-old child, and about 19% in a 1- to 4-year-old child. *(2:523; 4:2167; 5:2874)*

42. **(B)** Thiopental is known to have an antianalgesic effect. If given in small doses for sedation to one who is experiencing severe pain, the pain may be aggravated. The drug is a poor choice for sedation in this type of patient. *(5:332)*

43. **(D)** With pressure control ventilation, the tidal volume is not set. Rather, the clinician sets an inspiratory pressure, respiratory rate, and inspiratory time. The tidal volume varies, depending on the patient's chest wall and lung compliance and airway resistance. *(1:1471; 5:2821–2)*

44. **(A)** Thiopental injection is followed by decreased blood pressure and increased heart rate. Coronary blood flow is increased. Stroke volume is decreased, as is cardiac output. *(3:322; 4:1218; 5:333–4)*

45. **(B)** Successful block of the celiac plexus will interrupt nociceptive stimuli from the pancreas as well as the stomach, liver, and other viscera. This block is highly effective if performed with neurolytic agents in those suffering from pancreatic carcinoma. It is less successful in the setting of nonmalignant chronic pancreatitis. *(3:1939)*

46. **(E)** Pulseless electrical activity (electromechanical dissociation) is organized electrocardiographic activity without evidence of myocardial contractions. *(1:1501; 4:661; 5:2944)*

47. **(B)** Ketamine administration is associated with changes in cardiovascular parameters similar to those seen with sympathetic stimulation. Therefore, the heart rate is increased, and there is an increase in cardiac output. *(3:346–7; 4:1218; 5:348)*

48. **(B)** The baroreceptor reflex (or carotid sinus reflex) originates in sensors located in the carotid bifurcation which reach the central processing center in the medulla via the glossopharyngeal nerve. The afferent stimulus is pulse pressure and slope of systolic pressure rise. Functionally, a decrease in blood pressure produces increases in cardiac contractility, heart rate, and peripheral vascular tone. Isoflurane depresses the reflex least. *(1:334; 5:738)*

49. **(C)** There are many types of arrhythmias seen with the second dose of succinylcholine, but the most frequent is nodal rhythm. The most threatening arrhythmia is sinus arrest, which, fortunately, is not common. *(4:781)*

50. **(B)** Electrical energy, in joules, is the product of the current in amperes and the time in seconds. *(1:143)*

51. **(B)** The hypotension that may be seen after atracurium administration is due to histamine release, which may be exaggerated in a patient with hypovolemia. The extent of histamine release is dependent on the dose and the speed of injection. Atracurium does not cause clinically significant ganglionic blockade. *(3:199–203; 5:511–3)*

52. **(E)** CDH occurs in 1 to 2 in 5000 live births. Approximately 90% of herniations occur through the foramen of Bochdalek, although herniations may occur through the substernal sinus (foramen of Morgani). Less than 1% of cases have bilateral herniations. Affected infants present with respiratory distress related to lung hypoplasia. Herniation of abdominal contents into the chest inhibits normal lung growth and leaves the abdomen scaphoid rather than distended. Bowel obstruction may occur and cause the presenting symptoms in 5% of cases. Cardiovascular abnormalities accompany CDH in about 23% of cases. Apart from congenital heart lesions, the increased intrathoracic components may also cause obstruction of the inferior vena cava and decreased preload, resulting in decreased cardiac output. *(2:304–5; 4:2066–7; 5:2355–6, 2396–7)*

53. **(B)** A supraclavicular block requires about 20 to 30 mL. Both the axillary and interscalene approaches require 40 mL or greater. A femoral-sciatic block requires about 20 mL for each of the two injections. Intercostal blocks require about 4 mL per injection; thus, to block three segments above and below the incision bilaterally requires about 48 mL. *(1:725–7, 731, 738–9; 4:1414, 1417, 1420, 1422, 1424, 1429; 5:1688, 1690, 1692, 1698, 1708, 1711)*

54. **(D)** This baby has a 25% chance of having sickle cell anemia. He is not anemic for his age and does not require a transfusion. Screening tests for HbS cannot differentiate between homozygous HbS patients and heterozygous HbS/HbA patients. Furthermore, even if this baby were homozygous for HbS, at 3 months of age he would have an amount of HbF (fetal hemoglobin) that would make a sickle crisis very unlikely. *(2:44–5; 4:466–7; 5:1112–3, 2868–9)*

55. **(C)** A laryngeal mask airway is usually inserted blindly. Its use requires an adequate depth of general anesthesia, so it cannot be used in the postoperative patient emerging from general anesthesia who has airway obstruction. It does not protect the airway against aspiration, and is a possible cause of, and not a treatment for, laryngospasm. Patients may be mechanically ventilated through a laryngeal mask airway. *(1:601–2; 4:1076; 5:1625–7)*

56. **(D)** Naloxone is considered to be a "pure" antagonist opioid. It acts as a competitive antagonist at mu, delta, and kappa opioid receptors. The other drugs listed have partial agonist actions at the mu receptor. *(3:599–604)*

57. **(C)** William T.G. Morton performed the first public demonstration of ether anesthesia at the Massachusetts General Hospital in 1846. The surgeon was Warren, the patient Abbott. Wells was a dentist who pioneered the use of nitrous oxide, and attempted a public demonstration of nitrous oxide in 1845, which was judged a failure. Long was one of the first to use ether, which he did in 1842. He unfortunately did not report his experiences until 1849. *(1:5; 5:14)*

58. **(B)** Blockade of sympathetic nerve fibers has been shown to be useful to relieve the pain in all the conditions listed, except for amputation. One would need sensory fiber blockade to provide anesthesia for amputation. *(4:1369; 5:1709, 2775)*

59. **(D)** An interscalene block should provide adequate anesthesia for surgery on the hand, arm, and shoulder. Blockade of the cervical plexus is common when larger volumes are injected, permitting surgery on the carotid artery. More commonly, however, when carotid endarterectomy is performed under regional anesthesia, a cervical plexus block is performed by making injections at the transverse processes of C2, C3, and C4. The axilla is innervated by the second and third thoracic nerves and is therefore located below the innervation of the brachial plexus. *(1:724–5; 4:1415; 5:1686–8)*

60. **(C)** Primidone is an anticonvulsant used to treat grand mal seizures. *(3:532–3)*

61. **(C)** Sodium etidronate is a phosphonate derivative used to treat Paget's disease of the bone and other disorders of bone mineralization. *(3:1734–5)*

62. **(A)** Gemfibrozil is used to lower serum cholesterol. *(3:993–4)*

63. **(C)** Zidovudine is an antiviral agent used to treat persons with AIDS. *(3:1353–6)*

64. **(B)** Fluoxetine is an antidepressant used to treat persons with manic depressive disorder. *(3:278, 451)*

65. **(D)** There is increased pulmonary blood flow in the other lesions. *(1:919; 2:359; 4:1714; 5:2009; 6:47)*

66. **(D)** Global CBF is 50 mL/100 g/min under normal physiologic conditions. Cortical areas rich in gray matter receive more flow, 75 to 80 mL/100 g/min, whereas subcortical, white matter rich areas receive 20 mL/100 g/min. *(1:746; 4:1608; 5:814; 6:238–9)*

67. **(A)** There appears to be no benefit to keeping the hemoglobin concentration in critically ill patients equal to or above 10 g/dL, unless they have active myocardial ischemia. A practice pattern where hemoglobin concentrations of 7 to 9 g/dL in patients without acute coronary syndromes appears to be safe and has the advantage of decreasing the number of transfusions administered to this patient population. All of the other choices are considered evidence-based strategies to decrease the length of mechanical ventilation and ICU stay, and decrease the morbidity and mortality related to nosocomial infections in the ICU. *(5:2791–4, 2797–9, 2803)*

68. **(C)** General anesthesia provides a rapid induction with less hypotension and less cardiovascular instability and a more secure airway once it is established. An increased risk of aspiration and the potential for a difficult intubation limit general anesthesia to situations in which regional anesthesia is not optimal, such as when there is a need for uterine relaxation. *(4:2006–7; 5:2325)*

69. **(B)** Closing volume measurement requires nitrogen washout or the use of a tracer gas. Its measurement is not possible by spirometry. *(1:804–6; 4:862; 5:695–7)*

70. **(C)** An incompetent tricuspid valve permits right ventricular pressure to be transmitted to the right atrium, causing increased amplitude of the **v** wave. Increased **a** waves occur in heart block and junctional rhythm, while a waves are absent in atrial fibrillation. *(1:675; 4:1673; 5:1299; 6:42)*

71. **(E)** Thiopental should not be a factor in the apnea after a time period needed to complete a thoracotomy. The muscle relaxants may be involved because of altered sensitivity; the patient may be exhibiting the Eaton-Lambert syndrome with muscle weakness due to the carcinoma. Hyperventilation may lead to apnea due to lack of carbon dioxide drive to ventilation. *(4:387; 5:1851; 6:527–8)*

72. **(C)** The principal extracellular anion is chloride. The most important extracellular cation is sodium. *(1:170; 4:946; 5:1764; 6:375)*

73. **(E)** Albumin is thought to have no risk of transmitting viral diseases such as hepatitis. There is a theoretical concern, not proven, that prion diseases (such as Creutzfeldt-Jakob disease) may be transmitted via transfusion of albumin. *(1:209–10; 5:1817–20, 1825)*

74. **(A)** The pin index safety system is one of the safety systems built into the compressed gas system to prevent delivery of the wrong gas. This system is effective at preventing the misconnection of compressed gas cylinders to the anesthesia machine. *(1:568; 4:1012; 5:276)*

75. **(E)** Propranolol is useful in the treatment of atrial flutter. Since it causes bradycardia, it is not useful in treatment of that disorder. It may aggravate bronchitis. It may mask the signs of hypoglycemia in diabetes millitus. *(1:314; 2:377; 3:948; 5:653; 6:82)*

76. **(B)** A double-lumen tube is used to isolate the two lungs. Of the listed diseases, only hemoptysis requires prevention of cross-contamination of the lungs. *(1:824; 4:1757; 5:1873)*

77. **(D)** Necrotizing enterocolitis is a disease, not an anomaly. It occurs predominantly in premature infants under 1500-g birth weight and the mortality is 10 to 30%. Associations include birth asphyxia, hypotension, systemic infections, early feeding, and umbilical vessel catheterization. Umbilical artery catheters are usually

replaced with peripheral arterial lines so that mesenteric blood flow is not compromised. Metabolic and hematologic abnormalities include hyperglycemia, thrombocytopenia, coagulopathy, and anemia. Infants usually lose vast quantities of fluid into the extracellular space leading to clinical shock and the need for cardiovascular support. *(1:1190, 1364; 2:308–9; 4:2068–9; 5:2864)*

78. **(C)** The frequency is in the 10^5 to 10^6-Hz range. *(1:157; 4:710; 5:3146)*

79. **(B)** The second stage of labor involves the distention of the vaginal vault and perineum. These impulses arise from the pudendal nerves, which are composed of lower sacral fibers. *(1:1146, 4:1999)*

80. **(B)** Procainamide is eliminated primarily by the kidneys. Some is metabolized in the liver, but the metabolite is excreted by the kidneys. It is useful orally, and decreases ventricular automaticity. It is resistant to hydrolysis by cholinesterase. *(3:963–4)*

81. **(B)** Ethacrynic acid has its effect on the ascending portion of the loop of Henle. The thiazides, e.g., hydrochlorothiazide, chlorthalidone, and benzthiazide, have their primary site of action in the distal convoluted tubule. Triamterene is a potassium-sparing diuretic that has its effect in the distal convoluted tubule. *(3:771–2, 775, 777–8)*

82. **(B)** The interaction of rocuronium and acetylcholine at the myoneural junction is one of competition for binding sites. As the relaxant occupies the receptors, acetylcholine cannot bind and have an effect. *(3:199)*

83. **(B)** Student's *t*-test allows one to compare the mean values of two populations. It allows the testing of hypotheses when the population variance is unknown. *(1:58; 5:888)*

84. **(D)** When meperidine, an acidic compound, and thiopental, a basic compound, are mixed in the same syringe or the same IV tubing, a precipitate is formed. This may completely plug an IV line. *(4:1211; 5:328)*

85. **(D)** Carbon dioxide, acidosis, diphosphoglycerate, and hyperthermia shift the oxyhemoglobin dissociation curve to the right. Nitrous oxide has no effect. *(1:202–3; 4:877; 5:1603, 1806; 6:472)*

86. **(C)** Rocuronium is metabolized and eliminated by the liver. It can be administered intramuscularly and by continuous infusion and does not cause significant histamine release or ganglionic blockade. *(1:430–1; 3:196, 198, 202–3; 5:511–2, 2378–9)*

87. **(D)** Catheter segments remaining in the epidural space are rarely of consequence. It is agreed that while one should inform the patient, no attempts should be made at retrieval. *(1:837)*

88. **(A)** Endogenous catecholamines are epinephrine, norepinephrine, and dopamine. Dobutamine is a synthetic drug. *(3:220–1, 225–7, 228–9)*

89. **(C)** Unlike halothane, isoflurane produces burst suppression of the EEG at 2 MAC. During carotid endarterectomy performed under isoflurane anesthesia, the cortical blood flow level at which EEG manifestations of ischemia are detected is lower than that of halothane. Neither agent abolishes autoregulation of CBF. Isoflurane, although producing less increase in CBF than halothane, is still a cerebral vasodilator. *(1:390, 941; 4:1624–7; 5:825–9)*

90. **(A)** Increased dead space may result from hypotension, short, rapid ventilations, and pulmonary embolus. Decreased cardiac output leads to increased dead space. *(1:802–3; 4:876; 5:697–8)*

91. **(E)** The parturient is at risk for aspiration because of decreased gastric emptying secondary to mechanical and hormonal factors. Hypotension and fetal distress occur secondary to aortocaval compression in the supine position. Airway management can be difficult due to rapid desaturation from a decreased FRC and capillary engorgement of the airway. *(1:1150; 4:2007)*

92. **(B)** The activated clotting time test is a measure of the activity of the intrinsic pathway. It is performed by mixing whole blood with diatomaceous earth and measuring the time to clot formation. A normal value is 90 to 120 seconds. The extrinsic pathway involves tissue thromboplastin (which is not added in the activated clotting time test) and factor VII (the factor most sensitive to warfarin). *(1:221; 3:1519; 4:931; 5:1339, 1973)*

93. **(A)** The flow rate through a tube of a smoothly flowing liquid is proportional to the pressure gradient and the fourth power of the radius of the tube. There is an inverse relationship to viscosity. Viscosity is a function of temperature; however, as temperature increases, viscosity of a liquid decreases, while viscosity of a gas increases. *(4:1034; 5:2842)*

94. **(B)** Milrinone is an inotrope that has little effect on heart rate. It causes some vasodilatation. Milrinone is a phosphodiesterase inhibitor, which causes an increase in cyclic AMP. *(3:927)*

95. **(E)** EA and TEF are often associated with other congenital abnormalities, in particular the VATER association (*v*ertebral abnormalities, imperforate *a*nus, TEF, *r*adial aplasia, and *r*enal abnormalities), cardiac defects, and duodenal atresia. Infants with TEF should be maintained prone or in the lateral position with a 30° head-up tilt to decrease the risk of pulmonary aspiration. *(1:1187–8; 2:302–4; 4:2069–71; 5:2355, 2396)*

96. **(D)** Lamotrigine is an anticonvulsant agent useful in the treatment of partial and secondarily generalized tonic-clonic seizures in adults, either as mono- or add-on therapy. It is also effective for the treatment of absence seizures. Its half-life is reduced to about 15 hours when pentobarbital, carbamazepine, or phenytoin is administered concomitantly. *(3:539–40)*

97. **(D)** The toxic CNS effects of local anesthetics are seen with lower doses and at lower blood levels than the cardiovascular effects. Some local anesthetics are much more toxic than others. For example, approximately one fifth as much bupivacaine is required to cause seizures as compared to procaine. High blood levels of local anesthetics increase conduction time resulting in prolongation of the PR and QRS intervals, and depression of the sinus node. *(4:1350–2; 5:592–4)*

98. **(D)** The glomerular capillaries originate from an afferent vessel. This is the mechanism that affects filtration. Bowman's capsule is the tubular part of the nephron in which the glomerulus lies. The capillaries are very thin to achieve easy filtration. *(1:1005; 5:777)*

99. **(E)** The usual maximum recommended doses of epinephrine given during halothane anesthesia are a total of 100 μg in 10 minutes or 300 μg in 1 hour. Increasing the concentration of epinephrine beyond 1:100,000 does not increase the vasoconstricting effect and increases the likelihood of toxicity due to excessive dose. *(4:1143, 1347)*

100. **(A)** Sensation to the sole of the fifth digit and its nail bed is supplied by the posterior tibial nerve, to the lateral aspect of the fifth toe by the sural nerve, and to the dorsal surface of the fifth toe by the superficial peroneal nerve. *(1:741; 5:1703–4)*

101. **(B)** Volatile anesthetics decrease hepatic blood flow—halothane to a greater degree than sevoflurane and isoflurane. There is little effect of etomidate or nitrous oxide on hepatic blood flow. *(4:1140, 1143, 1149, 1218; 5:214, 2209–12)*

102. **(B)** Because brain and blood are not compressible, as an intracranial tumor expands in size it displaces intracranial contents from the cranium. Initially, intracranial CSF is diminished in volume, then venous blood volume is diminished. Tumors increase brain water content by causing cerebral edema in the surrounding tissue. A decrease in arterial blood volume occurs when ICP exceeds arterial perfusion pressure. *(1:747–8; 4:1614–5; 5:2127–30; 6:235–6)*

103. **(D)** The Macintosh blade is manufactured in various sizes. It has a flat tip, a curved spatula, and a flange. *(1:607; 4:1081, 1083; 5:1631)*

104. **(E)** Halothane metabolism yields trifluoroacetic acid and chloride, fluoride, and bromide ions, as well as other organic compounds. *(3:349; 4:1144; 5:237–8)*

105. **(E)** Each of these components is that derived from 1 unit of fresh whole blood. *(1:231–2; 4:1491)*

106. **(E)** The patient who has been on chronic steroid therapy is at risk of experiencing acute adrenal insufficiency and should be given exogenous steroids in the perioperative period. Although the likelihood of such a stress reaction is small, the risk of administering supplemental steroids is very small and is routinely done in such patients. *(1:1129; 4:1588–91; 5:1039–40; 6:427)*

107. **(A)** The infant's glottis is at C2 and the minimum diameter is at the cricoid. In the adult, the glottis is at C4, and the tightest point is at the vocal cords. The infant's chest wall is incompletely ossified, and the airway caliber is smaller, thereby increasing resistance to airflow. *(1:596, 1095; 2:81–2; 4:440; 5:2384)*

108. **(A)** Vasopressin increases the permeability of the collecting ducts to water. Vasopressin is an antidiuretic hormone, secreted by the posterior pituitary. It acts primarily on the collecting tubules. *(1:1137; 5:793–4)*

109. **(D)** Meconium aspiration usually occurs in term babies and is rare in those weighing less than 2 kg at birth. Regular chest physical therapy and postural drainage are recommended to clear residual meconium from the lung. Long-term outcome is good, in terms of intellectual development and pulmonary function, unless asphyxia occurred in the perinatal period. When death occurs, it is often due to persistent fetal circulation. *(2:19; 4:2019, 2069, 5:2355)*

110. **(A)** Vital capacity includes tidal volume and inspiratory and expiratory reserve volumes but not all of FRC. FRC includes residual volume. *(1:804–5; 4:862; 5:693–4)*

111. **(A)** The reaction of carbon dioxide and soda lime yields calcium carbonate, sodium carbonate, and water. No hydrochloric acid is formed. *(1:582; 4:1046; 5:296)*

112. **(C)** Protamine is a positively charged protein isolated from salmon testes. It binds to negatively-charged heparin. If given in the absence of heparin or in doses exceeding those needed to neutralize circulating heparin, it has a mild anticoagulant effect of its own. *(1:229; 2:401; 3:1525; 4:932; 5:1340, 1982; 6:618)*

113. **(A)** Mannitol is an osmotic diuretic and is filtered at the glomerulus. It will acutely increase blood volume and therefore may be a hazard in congestive failure. *(3:767–8)*

114. **(D)** Sensation to the dorsal and palmar surfaces of the fifth finger is supplied by the ulnar nerve. *(1:729; 4:1420; 5:1693–4)*

115. **(A)** Atropine, lidocaine, and epinephrine can be given via an endotracheal tube if necessary. Calcium chloride is considered too irritating. *(1:1485; 2:265; 5:2944)*

116. **(B)** This patient is on the steep portion of the intracranial compliance curve. Since compensatory mechanisms have been exhausted, small increases in the volume of intracranial contents (such as an increase in blood volume due to coughing) will produce large increases in ICP. Reduction in brain volume by decreasing brain water content will move the patient toward position A. A Cushing's reflex in response to increased ICP is hypertension and bradycardia. *(1:747–8; 4:1614–5; 5:2127–30; 6:235–6)*

117. **(D)** Although this patient is approaching the steep portion of the intracranial compliance curve, he may not manifest signs of increased ICP. Unilateral mydriasis is often a sign of brain stem herniation or distortion from extremely high ICP. Hyperventilation, by producing cerebral arterial vasoconstriction, would reduce intracranial blood volume and pressure. Halothane, in contrast, increases blood flow by cerebral vasodilation and raises ICP. *(1:747–8; 4:1614–5; 5:2127–30; 6:235–6)*

118. **(D)** Clonidine is a centrally acting sympathetic inhibitor and may suppress some withdrawal symptoms. Chlorpromazine is ineffective for this indication. The other two options are both agonist-antagonist combinations and are likely to precipitate withdrawal. *(3:233–4, 400, 486, 601)*

119. **(E)** Initial resuscitative measures of the newborn include drying/warming the infant, suctioning of the mouth, and tactile stimulation. In the moderately depressed newborn (heart rate remains less than 100 bpm or persistent apnea) initial measures include the administration of 100% oxygen and bag mask ventilation with maintained pressures of 15 to 20 cm H_2O (higher initial pressures of 30 to 40 cm H_2O may be required to overcome high opening pressures). This is usually sufficient, however, if the heart rate remains below 100 bpm, further resuscitation may be required including chest compressions, fluids, and pharmacologic intervention. *(1:1162–3, 4:2017–9)*

120. **(B)** Coronary spasm in Prinzmetal's angina may respond to nitroglycerin. Dilation of venous capacitance vessels decreases preload. *(1:892; 3:852; 4:1553; 5:1947–51; 6:19)*

121. **(A)** Aerosol therapy is used to humidify inspired gases, deliver medications, and mobilize tracheobronchial secretions. While some drugs (e.g., *N*-acetylcysteine [Mucomyst]) may induce bronchospasm, this is not the intended effect. *(5:2816–7)*

122. **(D)** When halothane is used in sufficient doses to provide deliberate hypotension, myocardial blood flow decreases because of decreased oxygen demand due to the decrease in cardiac output. There is loss of autoregulation of CBF. Seizure activity is seen with enflurane, not halothane. *(3:349–50; 4:1142–3)*

123. **(B)** The serum sodium is decreased. There is an increase in circulating fluid, and the serum electrolytes are diluted. Hemolysis of red cells does not occur with 1.5% glycine because it is only slightly hyposmotic. *(1:1018–20; 4:1971–3; 5:2192–4)*

124. **(C)** In the patient in shock, there is peripheral vasoconstriction, and the amount of thiopental going to the brain is relatively greater. CBF is preferentially maintained. *(5:333)*

125. **(C)** The treatment of a postdural puncture headache includes reassurance and analgesics. In addition, the patient should have fluids and, if necessary, an epidural (not subarachnoid) blood patch. *(3:1394)*

126. **(A)** The patient with myxedema may have decreased myocardial contractility. Florid congestive heart failure generally occurs only in the presence of coexisting coronary artery disease. Myxedematous patients may also have psychosis, pericardial effusions, and urinary retention. Bradycardia is usually present. *(1:1122–3; 4:306–7; 5:1047; 6:418)*

127. **(E)** All of the factors cited may be involved in causing hypoxemia. Blood loss, cardiac output, and shunting are involved in oxygen transport; airway obstruction is involved in ventilation. *(1:800–4; 4:877; 5:707–14)*

128. **(A)** Ketamine anesthesia is associated with open eyes even at an adequate level of anesthesia, mydriasis, and nystagmus. Intraocular pressure is increased or unchanged. *(1:336–7; 3:346–7; 5:346–7)*

129. **(D)** The emergence reactions that are seen with ketamine include dreaming, hallucinations, and delirium. These effects are more common in women, and are less likely to be perceived as unpleasant by children. While ketamine produces amnesia for intraoperative events, the amnesia does not persist into the postoperative period, and many patients remember their dreams or hallucinations. The emergence reactions can be minimized by the concurrent administration of a benzodiazepine. *(1:336–7; 3:346–7; 5:346–7)*

130. **(E)** Diffusion hypoxia is due to an outpouring of both nitrous oxide, which displaces oxygen and leads to hypoxia, and carbon dioxide, which leads to hypoventilation. The latter effect may thus worsen the hypoxia in a patient breathing room air in the first minutes of emergence. *(1:387; 4:1135; 5:180)*

131. **(E)** Local anesthetics are affected by the pH of the surrounding tissue, which is lower in the presence of inflammation. The uncharged form of the drug penetrates the nerve membrane to block conduction from within the cell. Buffering the solution to an alkaline pH hastens the onset of action by favoring the uncharged form of the drug. *(3:368–72; 4:1306–10; 5:574–6)*

132. **(E)** Drug interaction may take place by all of the mechanisms noted. *(1:257–8; 3:54–6)*

133. **(E)** MAO inhibitors cause increased levels of catecholamines and serotonin in nerves and inhibit the metabolism of many exogenous compounds, such as ephedrine. Orthostatic hypotension is common as is a hypertensive crisis when tyramine-containing foods are eaten. Meperidine is metabolized via P450-catalyzed demethylation to normeperidine which is a CNS stimulant and which may cause hypertension and seizures. MAO catalyzes the metabolism of normeperidine. *(1:290–1; 4:263–4; 3:542; 5:2270)*

134. **(A)** Aortocaval compression decreases cardiac output and increases venous pooling, thus decreasing uterine blood flow which is proportional to perfusion pressure. Aortocaval compression can be prevented by placing the patient in the semilateral position with a wedge under the right hip. It becomes more common as the uterus becomes larger. *(1:1141, 4:1989)*

135. **(D)** Of the medications listed, only flumazenil is a benzodiazepine antagonist. *(3:412)*

136. **(A)** Verapamil causes negative inotropy, dromotropy, and chronotropy. Like all calcium entry blockers, it is a vasodilator. *(1:314–5; 3:857; 4:1561; 5:1122)*

137. **(E)** All of the options are true. Alveolar collapse must be prevented and treated when it occurs. Inhaled volume is more important than pressure. If maneuvers are to be effective, they must be used frequently and adequately. Sustained maximum inflation devices, such as incentive spirometers, are effective and may be placed at the bedside so that the patient can use them without the presence of a therapist. *(1:1387; 5:2818–9)*

138. **(E)** Pneumothorax is caused by a loss of the sealed, negative pressure in the pleural space. Gas tracking through fascial planes, a connection to the airways, or a connection through the chest wall can all cause a pneumothorax. Chest wall trauma can either produce a penetrating injury to the chest wall, or airway rupture, or a rib fracture, which in turn can puncture the lung. *(1:1270–1; 4:1782; 5:2479–80, 2713)*

139. **(D)** Atenolol is the only selective β_1-adrenergic antagonist listed. Nadolol and timolol are nonselective β-adrenergic antagonists. Labetalol has both nonselective β-adrenergic and selective α_1-adrenergic antagonistic actions. *(3:254–6)*

140. **(B)** A regulator is a device to allow the compressed gas in a cylinder to be delivered at a lower, constant, and useable pressure. It is not a device of measurement. *(1:568; 4:1014; 5:276)*

141. **(A)** Hypotension in these patients is treated with volume and vasoconstrictors (such as phenylephrine) lacking inotropic effects. *(1:896; 3:138; 5:1956; 6:120)*

142. **(C)** Methyldopa and enalapril are available in formulations which may be given intravenously. *(3:246, 823, 877–9, 892)*

143. **(A)** Newborn infants with significant congenital heart disease commonly present with either cyanosis or congestive cardiac failure. Cyanosis occurs when there is a significant degree of right-to-left shunting of blood. Ventricular septal defects usually result in a left-to-right shunt. The shunt may be reversed if pulmonary hypertension occurs, but this usually takes some years (Eisenmenger's syndrome). *(2:392–4; 4:1701, 1712–3; 5:2008–11)*

144. **(E)** All of the options are correct, although it is not known how significant the role of collateral ventilation is in humans. *(1:793–4; 5:692)*

145. **(B)** In a patient with congestive heart failure, the administration of an ACE inhibitor usually causes a decrease in preload and afterload and an increase in cardiac output and sodium excretion. *(1:317; 3:910; 5:659; 6:111)*

146. **(A)** Hypertension may be caused by many factors. In this case, the first three options are possible. Pheochromocytoma is possible if the tumor is part of the multiple endocrine neoplasia syndrome. Although hypocalcemia may occur during thyroidectomy if the parathyroid glands are also removed, hypotension will result. *(1:1130–1; 4:105; 5:1042; 6:430–1)*

147. **(B)** Omeprazole is an inhibitor of the H^+, K^+-ATPase responsible for the synthesis of hydrochloric acid in the stomach. It has no effect on gastric emptying or on the volume of gastric juice. It is metabolized by cytochrome P450 and inhibits the metabolism of some other medications by this enzyme. *(3:1007–9)*

148. **(A)** Heart sounds are diminished in intensity in tamponade. *(1:918; 5:1966; 6:136)*

149. **(A)** PEEP increases end-expiratory volume, so collapsed alveoli may be opened and airway closure prevented. PEEP also shifts extravascular water from the periphery to the lung hilum, thereby decreasing the alveolar diffusion distance. Decreased cardiac output is an effect of PEEP that decreases oxygen delivery. *(1:1468–9; 4:139; 5:2820)*

150. **(B)** Protocols for suctioning include preoxygenation and the use of a catheter about half the internal diameter of the tube. This allows air to pass down along the suction catheter and prevents the creation of a vacuum in the lungs when suction is applied. In addition, the suction should be applied for short periods (less than 10 seconds), and the patient should be manually ventilated with 100% oxygen at the end of the procedure. (*5:2819–20)*

151. **(A)** A halo device significantly limits neck motion necessary for direct laryngoscopy and necessitates alternative methods of airway management, including awake fiberoptic intubation. If the cervical spine is unstable, removal of the halo device is not recommended. The traumatized patient with a neck injury may have other associated injuries, including facial fractures and closed head injury. Anesthetizing the trachea is necessary before placement of the endotracheal tube to prevent coughing and straining. *(1:1109, 1258; 4:1646; 5:2152–3)*

152. **(D)** In patients with preexisting vitamin B_{12} deficiency even relatively short exposures to nitrous oxide can produce megaloblastic changes and a neuropathy. The neurologic findings characteristic of vitamin B_{12} deficiency are a bilateral peripheral neuropathy, which affects predominantly the lower extremities as well as an unsteady gait and diminished deep tendon reflexes. Nitrous oxide has been shown to irreversibly inactivate the enzyme methionine synthetase, which ultimately leads to impaired synthesis of phospholipids, myelin, and thymidine, which is an essential DNA base. Megaloblastic hematopoiesis and subacute combined degeneration of the spinal cord can ensue. *(3:1504–5; 5:257; 6:477–8)*

153. **(A)** Hypoventilation, low cardiac output, and V/Q mismatch are problems that are of most importance. Abnormal diffusion is rarely important in this situation, and is associated with lung diseases, such as sarcoidosis. *(1:800–4; 4:877; 5:707–14)*

154. **(B)** Hypoglycemia in the anesthetized patient may be difficult to detect. In the patient who is known to have diabetes, blood sugars should be determined. A glucose infusion will avoid hypoglycemia. Blood sugars can be determined even with an infusion running, but one should avoid drawing blood from the same extremity in which the infusion is running. *(4:1957–8; 5:911)*

155. **(E)** Because intracranial contents are not compressible, any increase in the mass of intracranial material will result in a rise in ICP unless the volume of another compartment decreases. Edema in tissue surrounding the tumor increases interstitial fluid volume and contributes to the rise in ICP. Cerebral blood volume may be either increased or decreased by tumor growth.

Hydrocephalus can result from tumor obstruction of CSF outflow from the brain. *(1:747–8; 4:1615; 5:2127–30; 6:235–8)*

156. (D) Stage 3 of general anesthesia is characterized by regular respiration, normal values for heart rate and blood pressure, absent lid reflex, and conjugate gaze. *(4:1136; 5:1228)*

157. (A) Oxygen toxicity occurs at high partial pressures of oxygen, which implies high FIO_2 at normal ambient pressures. If the FIO_2 is over 0.5, there is an increased risk of oxygen toxicity. Oxygen toxicity does occur more often in the patient who is critically ill. Damage is first manifest after several hours, and worsens with continued exposure. *(5:716–7)*

158. (B) Intraoperatively, the major causes of pacemaker failure include acute changes in potassium concentration, electromagnetic interference, and myocardial infarction. *(1:1514; 5:1425; 6:72)*

159. (D) In the presence of increased ICP, these patients may be vomiting and at risk for pulmonary aspiration. Therefore, preoperative sedation with opioid analgesics may be unsafe and a rapid sequence induction (despite the risks of further raising the ICP) may be the procedure of choice. Ketamine would be a poor choice of agent in this situation because it can lead to sudden massive intracranial hypertension. If IV access is a problem, inhalational induction with halothane and gentle cricoid pressure may be performed but is less desirable. Once induction is complete, the trachea intubated, and IV access secured, anesthesia may be maintained with IV agents or low concentrations of isoflurane. Of the commonly used potent volatile anesthetic agents, isoflurane appears to produce the smallest increase in CBF for the depth of anesthesia produced and is usually used freely in neurosurgical procedures up to a minimum alveolar concentration of about 1. *(2:497–500; 4:2105–6; 5:2469–73, 2856)*

160. (E) Pulmonary resistance is also increased by hyperinflation, sympathetic stimulation, and polycythemia. *(1:877–8, 4:1728; 5:683–9)*

161. (E) All of these pathways may cause shunting. Atelectasis can be treated by pulmonary toilet, CPAP or PEEP. A patent foramen ovale may respond to pharmacologic control. *(1:803–4; 4:877; 5:688–9)*

162. (B) Serum albumin and prothrombin time are true indicators of liver function because they measure the concentration or the effects of proteins synthesized by the liver. Determination of serum enzyme levels indicates the degree of inflammation or injury but does not correspond with functional deficit. *(1:1074–5; 4:340–2; 5:766; 6:305)*

163. (A) The precordial thump is not expected to cardiovert ventricular fibrillation, and it may precipitate ventricular fibrillation. The American Heart Association considers the maneuver to be an optional technique in a monitored arrest and a class IIb recommendation (acceptable and supported to be useful by fair to good evidence) when the patient is pulseless and a defibrillator is not immediately available. *(1:1485; 2:265; 4:648; 5:2937)*

164. (D) Total pulmonary compliance is volume divided by pressure. It involves the lung and the chest wall. There is hysteresis in the alveolar expansion, so hypoventilation and atelectasis will decrease compliance. Compliance is increased by surfactant. *(1:794–5; 4:1755; 5:1465–7)*

165. (E) All of these choices can cause hypercarbia. A pulmonary embolus or increased V_D/V_T will cause end-tidal CO_2 to fall while causing hypercarbia. *(1:800–4; 4:876; 5:178–9, 1439–44)*

166. (E) All of the above can be a consequence of anesthesia in the prone position. Care must be taken to protect the eyes. Cerebral perfusion can be impaired by excessive head rotation. The superficial location of the ulnar nerve makes it vulnerable to compression injury. Venous distention occurs as a consequence of abdominal compression. *(1:728)*

167. (E) All of these conditions or maneuvers will impede egress of venous blood from the head, either by occlusion of the jugular veins or by an increase in venous pressure. *(1:763; 4:1756–7; 5:2129–30, 2134–8; 6:239, 234)*

168. **(D)** Pressure-support ventilation is a pressure-limited, flow-cycled mode of ventilation. Thus, when the patient begins to inspire, the ventilator administers gas at a flow rate adequate to maintain a preset pressure. The tidal volume is thus dependent on the patient's pulmonary compliance. Inspiration ends when the flow of gas necessary to maintain this preset pressure declines to a preset value. This mode of ventilation is an improvement over synchronized intermittent mandatory ventilation (SIMV) as a means of partial ventilatory support, because the ventilator supports each inspiratory effort by the patient. *(1:1470; 5:2823–4)*

169. **(A)** Carbon dioxide is inversely proportional to ventilation, i.e., the higher the ventilation, the lower the CO_2. CO_2 will equilibrate with bicarbonate, acting as an acid. Ventilation will increase in response to endogenous or exogenous CO_2. *(1:168, 798–800; 4:875, 964; 5:178–9, 1600–1)*

170. **(A)** The net effect of dopamine is dose dependent. Dopaminergic receptors are activated by the lowest doses, while β-adrenergic, and then α-adrenergic, receptors are activated as the dose is increased. *(3:226–7)*

171. **(A)** Nonshivering thermogenesis is not an effective way to produce heat in adult humans. It does occur in neonates who do not shiver. Shivering increases oxygen consumption; therefore, it increases myocardial work. This can be harmful in elderly or otherwise compromised patients. Awakening can be delayed secondary to reduced perfusion and decreased drug biotransformation. *(1:1538)*

172. **(B)** In small doses, both methohexital and etomidate have been used to elicit seizures in patients with documented seizure disorders. Case reports have described propofol-induced seizures as a rare occurrence. Although during profound hypocapnia with high doses of enflurane human subjects exhibit seizure-like EEG activity, there is no documentation of elicitation of seizures. *(1:753; 4:1622–4; 5:352–3, 1095)*

173. **(E)** About two-thirds of filtered water is reabsorbed in the proximal tubule. Water reabsorption also occurs in the loop of Henle, the distal tubule, and the collecting ducts. *(1:1008; 5:781–3)*

174. **(E)** Increasing cardiac output causes the pulmonary pressure to increase, and the pulmonary vascular resistance to partially compensate by decreasing. Increasing pulmonary artery pressure decreases zone 1, where alveolar pressure is greater than arterial, and thus there are fewer unperfused alveoli. *(1:801–2; 4:879, 1752; 5:165–6)*

175. **(A)** Severe antepartum bleeding causes significant morbidity and mortality. Placental abnormalities such as placenta previa or abruptio placenta and, rarely, uterine rupture are the most common causes. If a vaginal examination is necessary, preparation for an emergency general anesthetic should be done, which includes two IV lines and blood products available in the operating room. If the patient is hypovolemic, thiopental will cause a greater decrease in blood pressure than will ketamine. *(1:1156; 4:2009)*

176. **(E)** Stridor occurring 14 hours after thyroidectomy and accompanied by signs of tetany is most likely due to hypocalcemia. This patient has an airway problem, and that should be given priority in management. The patient should be observed and intubated if necessary to protect the airway. *(1:1122; 4:1955; 5:2540; 6:415)*

177. **(E)** Etomidate is a substituted imidazole derivative. *(4:1212; 5:351)*

178. **(E)** Carboxyhemoglobin is involved in the transport of carbon monoxide. CO_2 exists as dissolved, bicarbonate and carbaminohemoglobin. *(1:1275; 4:963; 5:703)*

179. **(E)** Pierre Robin syndrome is characterized by mandibular hypoplasia and pseudomacroglossia (the tongue protrudes from the mouth but is not actually enlarged). It is associated with high arched and cleft palate but, unlike Treacher Collins syndrome, it is not associated with choanal atresia, vertebral abnormalities, or

cardiac defects. In the normal infant, the larynx is located higher in the neck compared with that of an older child or adult, resulting in acute angulation between the base of the tongue and the larynx. This anatomic relationship is further exaggerated in the Pierre Robin syndrome, thus making direct visualization of the larynx difficult and sometimes impossible. The midfacial hypoplasia of this syndrome results in the larynx being displaced posteriorly and the angulation between base of tongue and laryngeal inlet being even more acute. *(2:82, 119; 4:429, 460–1; 5:2386, 2853)*

180. **(C)** Nifedipine dilates blood vessels through its actions as a calcium channel blocker. *(1:315; 3:891–2; 4:1553; 5:1122)*

181. **(E)** Transdermal scopolamine is a muscarinic antagonist that can cross the blood-brain barrier. It is not recommended for the prophylaxis of emesis in children and can produce behavioral side effects. It has, however, been shown to be useful in adults if applied 12 hours before the operative procedure. *(3:171)*

182. **(D)** The patient with chronic renal failure who has missed a scheduled dialysis is expected to have uremia, hyperkalemia, hypervolemia, and metabolic acidosis. Platelet function is abnormal due to the uremia (which inhibits platelet aggregation) and a decrease in platelet factor III. Platelet count should not be abnormal unless there is another coexisting problem causing decreased platelet production or increased utilization. *(1:1010–1; 4:324–6; 5:2181–2; 6:346–8)*

183. **(C)** Intercostal blocks result in the highest plasma concentrations of local anesthetic, followed by caudal, epidural, and brachial plexus. *(4:1346; 5:591)*

184. **(B)** Once the skull is open and the dura mater incised, the brain is no longer confined within the cranium, and ICP is zero. *(1:747; 4:1615; 5:2127–30; 6:236)*

185. **(C)** Naloxone is a pure opioid antagonist, meaning that the drug has no opioid effect of its own. Butorphanol and levallorphan are both mixed opioid agonist-antagonists. Neostigmine and edrophonium are cholinesterase inhibitors. *(3:602–3)*

186. **(D)** The farad is the unit of capacitance. The ampere is the unit of current, the volt is the unit of voltage, the ohm is the unit of resistance, and the henry is the unit of inductance. *(4:702; 5:1210)*

187. **(C)** In the kidney position, the lower leg is flexed to give some stability to the position. The upper leg is kept straight. The arms are placed in a neutral position. The head is supported. If a kidney rest is used, it should be under the dependent iliac crest. *(1:609–10; 4:694–6)*

188. **(A)** All of the options are absolute contraindications except preexisting neurologic disease which may be a relative contraindication. Preexisting neurologic disease may be a contraindication, depending on the status and type of disease, in order to avoid confusion postoperatively between preexisting disease, obstetrical, and anesthetic complications. *(4:2000, 5:2320)*

189. **(A)** Umbilical vessel catheterization may be of great help in the resuscitation of the newborn infant. There are some complications with the technique, and one must weigh the benefits against the risks before undertaking the procedure. Bladder rupture is not a risk. AV fistulas have been reported, as have portal vein thrombosis, infection, hepatic necrosis, endocarditis, embolization of blood clot or air, and pulmonary infarction (resulting from a catheter introduced into the umbilical vein and migrating across the heart through the patent foramen ovale). *(2:747–51; 5:2356–7)*

190. **(C)** Milrinone is a vasodilating and inotropic agent thought to function through inhibiting phosphodiesterase and thereby elevating cyclic-AMP levels. It is not a catecholamine. *(3:927)*

191. **(C)** The toxicity associated with bupivacaine is due to blockade of sodium channels. The drug is an amide of long duration. It should not be used in concentrations above 0.5% for epidural anesthesia in pregnant patients. Toxicity may occur after injection other than intravenously. *(3:375; 5:595)*

192. **(B)** Blood should be warmed to 100°F (38°C). Hemolysis may occur if the blood is overheated. There is no need for bicarbonate administration since much of the acid load in a unit of stored blood is due to carbon dioxide. If the patient has a coexisting metabolic acidosis requiring therapy, it should be treated with bicarbonate, not lactate. There is approximately 18 mEq of acid per unit of blood. *(1:205–6, 229; 4:2147–8, 2421–2; 5:1814; 6:500–1)*

193. **(B)** In this scenario, the most likely cause is glycine toxicity. Glycine 1.5% is commonly used as an irrigating fluid. Toxicity is more likely to be observed if large amounts of irrigating fluid are used. Hyponatremia is also a complication associated with transurethral resection of the prostate. *(1:1019–20; 4:2055; 5:2192; 6:368–71)*

194. **(D)** Under normal conditions, intracranial arterial blood volume is about 7 to 8 mL and constitutes about 15% of the 50 mL of total intracranial blood volume. Conceptually, brain tissue can be divided into solid material (about 168 g or 12% of intracranial contents) and tissue water (1092 g or 78% of intracranial contents). Intracranial CSF volume is 75 mL. Although occupying the smallest volume within the cranium, cerebral blood volume is altered rapidly by physiologic and pharmacologic intervention. *(1:746, 762–3; 4:1615; 5:2129)*

195. **(A)** The correct ranking of local anesthetic toxicity, from most toxic to least toxic, is tetracaine > bupivacaine > lidocaine > chloroprocaine. *(1:458)*

196. **(C)** Boyle's law relates volume and pressure for isothermal processes. *(4:1013; 5:2667)*

197. **(D)** Cromolyn prevents release of histamine from mast cells. It has no direct effect on bronchioles, does not stimulate cyclic AMP, nor does it release catecholamines. *(3:742–3)*

198. **(C)** Preganglionic neurons are located in the thoracic and lumbar spinal cord (T1–L3) and synapse on postganglionic nerves either in paravertebral sympathetic ganglia or in plexi adjacent to organs of innervation. Except for sweat glands, which are cholinergic, target organ receptors are adrenergic. β-Adrenergic receptors mediate their effects through cyclic AMP; α-adrenergic receptors act via a more complex second messenger system involving G proteins. All preganglionic neurons are cholinergic. *(1:261–6; 4:723–6; 5:621–5)*

199. **(D)** Most dosage guidelines for pediatric patients have been derived from data extrapolated from adult studies. The dose of lidocaine is also contingent on the addition of vasoconstrictors; however, in a child the dose should not exceed a total dose of 7 mg/kg. This dose, as all doses, depends on the route and the speed of administration. If a dose of lidocaine is given intravenously, 7 mg/kg will be excessive. For infiltration of a laceration, the dose would be appropriate. *(2:639–40; 5:1728–9)*

200. **(D)** The preterm infant is particularly susceptible to maternally-administered drugs. They have immature enzyme systems, an incomplete bundle branch block, and decreased protein for binding. The preterm infant metabolizes both ester and amide local anesthetics and is actually less sensitive to the CNS toxicity of lidocaine than full-term infants. *(1:1199; 4:2017; 5:2394)*

201. **(C)** β-Adrenergic agonists increase AV conduction. *(1:271; 2:546; 3:228; 4:737; 5:1406–8; 6:77)*

202. **(A)** A bronchial blocker obstructs flow to one lung. It may have a small lumen suitable for suctioning (Univent tube) but still this is inadequate for ventilation or drainage of lavage fluid. *(1:824; 4:1757, 1778; 5:1883–7)*

203. **(A)** An increased FRC should improve oxygenation. FRC is usually decreased in the postsurgical period. The other options may all contribute to hypoxemia. *(1:1291; 4:2320; 5:2312)*

204. **(C)** Ventilation should be discontinued immediately following the discovery of an airway fire. The endotracheal tube should be clamped and disconnected from the circuit. Oxygen should be discontinued and the endotracheal tube removed. If any burning remnants remain, they should be extinguished. Following ventilation by mask, the patient should be reintubated and admitted to the ICU for observation. *(4:2264; 5:2580–1)*

205. **(C)** By the Henderson-Hasselbalch equation, $pH = pK + \log [HCO_3^-]/[H_2CO_3]$. $[H_2CO_3]$ may be replaced by $PaCO_2 \times 0.03$. With the values given, a bicarbonate concentration of 33.2 mEq/L is calculated. Therefore, it appears that this patient has, at baseline, a metabolic alkalosis as a compensatory mechanism for chronic respiratory acidosis. In the postoperative period, the respiratory acidosis has become worse, possibly as a result of medications which depress ventilation or the supplemental oxygen which has diminished this patient's hypoxic ventilatory drive. *(1:166–9; 5:1603, 2264; 6:391–2)*

206. **(E)** Renal failure has not been associated with amiodarone. All of the other options have been reported. *(1:907; 2:378; 3:956; 5:2221; 6:71)*

207. **(E)** The normal blood volume in an infant is 80 mL/kg. A premature infant has a higher volume per unit of weight. *(2:19–20, 236–7; 4:2059; 5:2390)*

208. **(D)** At 17°C, cerebral oxygen consumption is reduced to about 8% of the normothermic value and, while metabolic activity is decreased, cellular integrity is maintained. This accounts for the brain's tolerance to modest periods of cardiac arrest during hypothermia. Both CBF and electrical activity decrease during hypothermia. *(1:750; 2:504; 4:2370–4; 5:1977–9; 6:250)*

209. **(B)** Hetastarch is a synthetic colloid. The starch molecule is modified to resist metabolism by amylase to glucose monomers. The drug interferes with coagulation by diluting platelets and coagulation factors. The incidence of anaphylactoid reactions is about 1 in 1000. Hetastarch is eliminated primarily via the kidneys. *(4:990–1; 5:1787)*

210. **(D)** The peptide is released by the cardiac atria (and perhaps other organs) in response to intravascular volume expansion and tends to oppose the volume expansion. *(1:1088; 3:796; 5:737)*

211. **(C)** Metoclopramide increases gastric emptying but has no effect on the secretion of gastric acid. All of the other options have an effect on gastric acid secretion, some more efficiently than others. *(3:170, 1009–11, 1025–6)*

212. **(B)** Snow preferred chloroform to ether and administered chloroform to Queen Victoria during the birth of Prince Leopold. Suckling synthesized halothane. *(1:6–7; 5:44)*

213. **(C)** Ketamine causes increases in ICP, cerebral metabolism, and CBF. It is relatively contraindicated in patients with an intracranial mass or increased ICP. The other induction agents will decrease ICP. *(1:332–7; 4:1673; 5:820–5)*

214. **(D)** Arrhythmias and cardiovascular collapse are late signs of venous air embolism. The precordial Doppler in conjunction with a capnograph or pulmonary artery catheter will usually detect air before physiologic consequences occur. *(1:766–7; 4:1635–8; 5:2139; 6:243–4)*

215. **(E)** Bupivacaine is more cardiotoxic than other local anesthetics, but most arrests have been associated with the 0.75% solution, which is contraindicated in obstetrical anesthesia. Toxicity of local anesthetics includes cardiorespiratory depression and CNS depression followed by CNS excitability at higher toxic doses. Chloroprocaine is rapidly metabolized in maternal serum with a short half-life, and current preparations do not contain bisulfite, which has been implicated in neurotoxicity in the past. *(4:2002, 5:2322–3)*

216. **(A)** Etomidate has anticonvulsant activity, and it decreases intraocular pressure. There is no histamine release and it does not have any analgesic effects. Etomidate produces a higher incidence of postoperative nausea and vomiting than other IV induction agents. *(4:1216, 1219; 5:352–5)*

217. **(C)** $CMRO_2$ decreases with a fall in temperature. This decrease is quantitated as the Q_{10}, or changes in metabolic rate for a 10°C change in temperature. While the first Q_{10} is 2.2, the second Q_{10} is about 5; the difference between the first and second Q_{10} values is thought to be due to cessation of brain electrical activity. Temperature has no effect either on autoregulation or on oxygen extraction from blood. MAC decreases with a fall in temperature. *(1:750, 763; 4:259; 5:816–7)*

218. **(C)** The technique of saddle block spinal anesthesia may be used if outlet forceps, vacuum extraction, or extensive episiotomy repair are needed. Caudal anesthesia is rarely used because it is difficult to provide anesthesia higher than T10 and large volumes of local anesthetics may be required, increasing the risk of toxicity. Unintentional injection of the fetal head is a possible, though unusual, complication. Lumbar sympathetic block interrupts pain impulses from the uterus, cervix, and upper third of the vagina. *(1:1147, 4:2000–4)*

219. **(A)** The agents are listed in decreasing order of β-adrenergic to α-adrenergic activity. *(1:297; 4:662; 5:646–8)*

220. **(C)** Thiamine deficiency is common in malnourished persons, especially in those who rely on alcohol as a source of calories. The administration of glucose to such persons may cause symptoms of acute thiamine deficiency. *(6:641–2)*

221. **(A)** A successful axillary block will probably not provide adequate anesthesia for the upper arm. *(1:726–7; 4:1420; 5:1691)*

222. **(A)** The renal tubule has active transport processes for the tubular secretion of organic acids, such as glucuronides, and organic bases. Since the glucuronide derivative is likely to be much more polar than the parent compound, the rate of tubular reabsorption (which usually occurs by passive nonionic diffusion) will be less. *(3:11)*

223. **(C)** The V/Q ratio is higher at the apex (so PO_2 is higher). The blood flow and ventilation are greatest at the base. *(1:801–2; 4:1753; 5:679–83)*

224. **(D)** The correct ranking of the potencies of the volatile anesthetics, from most potent to least potent, is halothane > isoflurane > sevoflurane > desflurane. *(1:378; 4:1124)*

225. **(C)** Catecholamines are inactivated by MAO and COMT. Pseudocholinesterase inactivates succinylcholine and mivacurium, while acetylcholinesterase inactivates acetylcholine. *(1:271; 3:134–5; 4:748; 5:647)*

226. **(E)** Vincristine is commonly associated with the development of a peripheral neuropathy. *(5:1118; 6:588–9)*

227. **(E)** The child should have an induction that is most appropriate for his clinical condition. If there has been a large amount of blood lost, a rapid sequence induction with thiopental and succinylcholine may not be appropriate. Passing a nasogastric tube may not remove all of the stomach contents. The procedure should not be delayed to allow the stomach to empty. Once the injury has occurred, the stomach emptying probably stops, and the contents will still be there 6 hours later. The child should be allowed to awaken with the endotracheal tube in place and be extubated once protective reflexes are intact. *(2:188; 4:2107–8, 2150–4, 5:2384–5)*

228. **(A)** The interaction between meperidine and a MAO inhibitor may cause CNS stimulation manifested as delirium, hyperthermia, and convulsions. *(1:290–1; 4:263–4; 3:542; 5:2270)*

229. **(A)** Ranitidine is a histamine H_2 receptor antagonist and acts to increase the pH of gastric secretions. All of the other statements are true. *(4:2303)*

230. **(D)** Compliance becomes better as one moves down the lung. Dependent alveoli are smaller because of the weight of the tissue above, and therefore can expand more. The shape of the thorax and diaphragm enhances this effect *(1:794–5; 4:1755; 5:679–83)*

231. **(C)** Oxygen delivered at 2 L/min via a nasal cannula provides an FIO_2 of approximately

28% in a patient with a normal ventilatory pattern. *(5:2813)*

232. **(D)** At very high blood levels of ethanol, approximately 400 mg/dL in the nontolerant person, apnea is likely. The resulting hypoxemia may cause seizures and/or ventricular arrhythmias. Ethanol causes dose-dependent vasodilation; however, blood pressure is usually maintained (until hypoventilation leads to hypoxemia) because of reflexive increases in heart rate and cardiac output. *(1:1480; 6:641)*

233. **(C)** This patient would be classified as physical status III because she has a severe systemic illness with functional limitation. The proposed operation is not an emergency procedure, therefore "E" is not added to the physical status. *(1:474; 4:4; 5:2592)*

234. **(A)** Benztropine is a centrally acting anticholinergic agent used to treat Parkinson's disease and the extrapyramidal side effects of dopaminergic antagonists. All of the other drugs listed are used in various forms of epilepsy. *(3:522, 560)*

235. **(B)** A possible cause of hypotension is acute adrenal insufficiency in a patient with suppression of the pituitary-adrenal axis; however, more likely causes of hypotension during hip arthroplasty include hypovolemia due to blood loss, marrow or fat embolus, methyl methacrylate-induced vasodilation, excessive anesthetic depth, or an allergic reaction to an antibiotic or other medication. The appropriate therapy is dependent on the etiology. *(1:1129, 1380–3; 4:2129–32; 5:2415, 2716; 6:103, 426–7)*

236. **(E)** Of the anticholinergic agents, glycopyrrolate and scopolamine are better antisialagogues than atropine. The other medications listed are cholinesterase inhibitors that have increased salivation as a major side effect. *(1:560; 4:62)*

237. **(A)** The cricothyroid and transverse arytenoid muscles are the only muscles of the larynx that are not innervated by the recurrent laryngeal nerves. Their action is to tense and oppose the vocal cords. In practice, the degree of airway obstruction is variable, but voice and intubation can be impaired. The left recurrent laryngeal nerve has a longer course, and thus is more prone to injury. *(1:595; 5:2537–8)*

238. **(D)** Patients with mediastinal masses and superior vena cava obstruction can be difficult to manage. The trachea may be compressed and malacic. Airway patency may depend on position or intrinsic muscle tone. An appropriate anesthetic plan would preserve muscle tone and spontaneous ventilation until an airway past the area of obstruction is established. *(1:836; 4:1801; 5:1920–3)*

239. **(B)** The alveolar gas equation relates PAO_2 to FIO_2 as follows:

$$PAO_2 = FIO_2 \times (P_{atm} - P_{H_2O}) - \frac{PaCO_2}{RQ}$$

where P_{H_2O} is the value for saturated water vapor pressure at body temperature, and *RQ* is the respiratory quotient, which is normally approximately 0.8. Thus, in this patient, PAO_2 is calculated to be 307 mmHg. Since the normal alveolar-arterial oxygen difference is less than 10 mmHg, the PaO_2 should be approximately 300 mmHg. *(1:804; 5:1010; 1439)*

240. **(E)** The preeclamptic patient is at risk for all the complications noted, as well as pulmonary edema, congestive heart failure, thrombocytopenia, abnormal clotting studies, intravascular volume depletion, peripheral vasoconstriction, and extravascular sodium and water retention. Intracranial hemorrhage is the most common cause of death. *(1:1153, 4:2012)*

241. **(E)** Countershock will not improve AV conduction. Complete heart block may be treated by a cardiac pacemaker. *(1:1493; 2:265; 4:659; 5:2929–41)*

242. **(C)** Etomidate produces the highest incidence of postoperative vomiting after short procedures, propofol the lowest. *(1:336; 4:1219; 5:354)*

243. **(B)** The mode is the number that occurs most frequently in a sample. The median is the

number that divides a sample set into two equal parts. The mean is the sum of all the sample values divided by the number of values. *(1:56; 5:883)*

244. **(E)** Performance of a sciatic block does not rely on an arterial landmark. In performing a femoral nerve block, the femoral artery should be palpated; for an axillary block, the axillary artery; for a stellate ganglion block, the carotid artery; and for a supraclavicular block, the subclavian artery. *(1:726–7, 733–4, 738–9; 4:1415, 1418, 1422–3, 1464; 5:1690, 1692, 1698, 1708, 1754)*

245. **(C)** Postpartum hemorrhage, if secondary to retained placenta or uterine inversion, may require general anesthesia for uterine relaxation. Regional anesthesia may be beneficial for repair of lacerations. *(4:2010–1)*

246. **(A)** A has higher efficacy than B because the maximum response produced by A is greater than the maximum response produced by B. Because A and B are located at the same place along the x-axis, A and B are said to have the same potency. The dose at which the half-maximal response occurs, the ED_{50}, is the same for both A and B. It is likely that A and B act via the same receptor. *(3:39)*

247. **(B)** The effect of drug Y is to decrease the efficacy of X without shifting the dose-response relationship of X. This is the definition of a noncompetitive antagonist. It is likely that X and Y act via the same receptor. *(3:39–41)*

248. **(C)** The external noninvasive units are ventricular pacing devices. *(1:896; 4:817; 5:1084; 6:72)*

249. **(E)** This patient's parenteral nutrition solution is too concentrated to administer via a peripheral vein. The immediate concern is hypoglycemia resulting from the high circulating insulin level, which accompanies infusions of solutions high in glucose. The administration of a 10% glucose solution (the highest concentration recommended by peripheral vein) with concurrent regular monitoring of the blood glucose concentration would be the most appropriate temporary measure to take during the operation. *(1:1478; 5:1035; 6:453)*

250. **(B)** Pulmonary capillary wedge, left atrial, and aortic pressures will tend to decrease. *(1:1299; 4:820; 5:683; 5:686; 6:218)*

251. **(D)** Interleukins are proteins secreted by T cells which participate in the immune response. All of the other compounds are derivatives of the fatty acid, arachidonic acid. *(3:670–1, 688–9)*

252. **(B)** Hepatitis C, cytomegalovirus, and syphilis infections can be transmitted by blood transfusions, and infected patients should be considered able to transmit disease to health care workers who receive parenteral injuries with needle or surgical instruments contaminated with blood. Similarly, the virus of Creutzfeldt-Jakob disease is thought to be present in neural (and perhaps other) tissue in infected patients, and a parenteral injury with a contaminated instrument may be a method of transmission. The virus of hepatitis A is shed in feces, thus a possible mode of transmission to a health care worker is by the hand-to-mouth route after handling fecal-contaminated material. *(1:74–7, 210; 5:1819–20, 3157, 3160; 6:265, 300–1, 557, 565)*

253. **(B)** Administering succinylcholine to a patient with a massive burn may result in severe hyperkalemia; however, the pharmacokinetics of succinylcholine will not be altered (assuming the patient survives). Patients treated with inhibitors of cholinesterase (neostigmine, pyridostigmine, isoflurophate) or with atypical cholinesterase will experience a prolonged succinylcholine effect. *(1:421–4; 3:201–3; 4:767–70; 5:489–91)*

254. **(E)** The diaphragm has a central tendon and is innervated by the phrenic nerve, which arises from cervical nerves 3, 4, and 5, but principally the 4th. Slow-twitch fibers are endurance elements and do not fatigue with constant use. *(1:791)*

255. **(B)** Dextran is highly antigenic and many patients are sensitized even if they have not received dextran therapy in the past. In sensitized patients, dextran 1 binds to the mast cell receptors for dextran but is too small a molecule to bridge them; without bridging two (or more) receptors, the mast cell does not degranulate, and the receptors are blocked when the larger dextran 40 molecule is administered. *(5:1826)*

256. **(B)** There is no practical system for drug use accountability that can prevent the diversion of drugs by an anesthesia care provider. The other statements listed are correct. *(1:78–82; 4:2548; 5:3166)*

257. **(B)** Cromolyn sodium is used to prevent asthmatic attacks, but it is not a bronchodilator and is believed to prevent the immunologic causes of an asthmatic attack. All of the other agents are useful bronchodilators. *(3:742–3)*

258. **(C)** In flow through a tube, the Reynold's number depends on tube diameter but not length. A Reynold's number above 2000 predicts turbulent flow. *(5:1217)*

259. **(E)**. *(1:692; 4:1395)*

260. **(C)** Left-to-right shunts tend to increase mixed venous saturation. *(1:676; 4:350; 5:1331, 1453; 6:228)*

261. **(A)** The ICD is usually deactivated intraoperatively by means of a magnet. *(1:1515; 2:733; 5:1429)*

262. **(A)** In the Tec 6 vaporizer, desflurane is heated to 39°C, well above its boiling point of 23°C, yielding a pressure of desflurane vapor within the vaporizer of about 2 atm. The vaporizer may be filled while in use; the high pressure is transmitted to the bottle containing desflurane liquid forcing liquid out of the bottle and into the vaporizer. This vaporizer does not use the variable bypass principle because carrier gas never enters the vaporizing chamber but is mixed with desflurane vapor leaving the vaporizing chamber. *(1:577–8; 4:1030–3; 5:289–92)*

263. **(A)** Metaproterenol, isoetharine, and albuterol are selective β_2-adrenergic agonists. Metoprolol is a selective β_1-adrenergic antagonist. *(3:230, 255–6)*

264. **(C)** Ventilation under inhalation anesthesia is characterized by a decrease in tidal volume as the depth of anesthesia increases. Tachypnea partially compensates for the decreased tidal volume, but CO_2 levels do increase. Metabolic processes decrease, lowering O_2 consumption and CO_2 production. *(1:398–9; 4:1143; 5:706)*

265. **(B)** Droperidol is a potent antagonist at dopaminergic receptors and a weak antagonist at α-adrenergic receptors. *(3:490)*

266. **(A)** Acromegaly is due to the hypersecretion of growth hormone. Hypertrophy of skeletal and connective tissue, especially of the face and head, may make intubation difficult. Diabetes mellitus and hypertension are also common. The pituitary adenomas responsible for acromegaly usually do not secrete thyroid stimulating hormone (TSH) or cause hyperthyroidism. *(1:1137; 4:315; 5:1051–2; 6:436–7)*

267. **(E)** Effective postoperative analgesia can be obtained by the delivery of opioid drugs by oral, rectal, transdermal, sublingual injection, or infusion. In addition, opioids can be given in the epidural or intrathecal spaces. *(4:2137)*

268. **(B)** The irrigating fluids should be nonhemolytic and isosmolar. The composition should not be close to water, since water is hyposmotic. Electrolyte solutions should not be used because they conduct electricity and therefore interfere with the electrocautery. *(1:1018–9; 4:1972; 5:2192; 6:368)*

269. **(E)** Phenelzine, isocarboxazid, and tranylcypromine are the "classical" or nonselective MAO inhibitors and are used primarily in the treatment of depression. Selegiline is a selective MAO-B inhibitor used to treat Parkinson's disease. *(3:457, 556)*

270. **(E)** After subarachnoid hemorrhage ECG changes consistent with ischemia are common. A subpopulation of these patients do have elevated cardiac enzymes including CK, CK-MB, and troponin. Hydrocephalus can complicate the clinical picture and requires ventriculostomy for drainage or permanent shunting. Hypertension is often required to reverse neurologic deficits that develop as a result of vasospasm. *(1:769–72; 4:1640–6; 5:2147–9; 6:251–4)*

271. **(A)** Heparin inhibits the activated forms of factors IX, X, XI, XII, and thrombin. *(3:1522–3)*

272. **(A)** A PaO_2 of 30 mmHg corresponds to a saturation of approximately 60%. Dissolved oxygen is 0.003 mL O_2/100 mL blood per mmHg. The total assumes a hemoglobin-bound O_2 content of 20 mL O_2/100 mL blood when 100% saturation of 15 gm Hgb/100 mL blood. *(1:201; 5:699–700)*

273. **(A)** Patients should always be informed regarding the possible complications of muscle weakness, sphincter impairment, and sensory loss. Postneurolytic neuralgias can also occur. Seizures would be a rare and unexpected complication. *(3:1941)*

274. **(D)** Following a burn injury of this magnitude, vast amounts of fluid are lost from the circulation into the burned tissue, and thereafter are sequestered outside the circulation even in nonburned tissues. Albumin loss is usually at least twice the total body plasma content. Cardiac output is strikingly decreased immediately after injury because of the rapid reduction in circulating blood volume or the severe compressive effects of circumferential burns on the abdomen and chest, impairing venous return. This is despite a large increase in circulating catecholamines. Evaporative fluid losses are about 4 L/m^2/day. Pulse oximetry is not useful in monitoring oxygenation in carbon monoxide poisoning because carboxyhemoglobin produces an overestimation of oxygen saturation; the photodetector does not differentiate between oxyhemoglobin and carboxyhemoglobin. In contrast, transcutaneous oxygen analyzers are useful. *(2:522–6; 4:2165–78; 5:1793–4)*

275. **(C)** Scopolamine and diphenhydramine (an antihistamine) cross the blood-brain barrier and block central muscarinic receptors. The anticholinergic glycopyrrolate is a quaternary compound and does not cross the blood-brain barrier. Cimetidine is an antagonist at the histamine H_2 receptor. *(3:162, 169, 651, 1009)*

276. **(B)** Management of patients with cardiac disease should be done in association with a cardiologist and with an understanding of the pathophysiology of the specific lesion and drugs the patient may be taking. Hemodynamic stability often requiring invasive monitoring is sought. Central blood volume increases dramatically after delivery and heart rate and blood pressure may increase significantly with pain. Patients with cyanotic heart disease may not tolerate pregnancy well. *(1:1156–7, 4:2013–4, 6:664–8)*

277. **(B)** The vapor pressure is primarily a function of the temperature and increases with increased temperature. It is not affected by the atmospheric pressure above the liquid and is not correlated with the density of the liquid. *(1:379; 4:1019; 5:284–6)*

278. **(A)** Hypocarbia can inhibit hypoxic pulmonary vasoconstriction (HPV). High FIO_2 helps reduce shunt in low V/Q areas but along with dependent position can contribute to atelectasis. Higher airway pressures can counteract some atelectasis, but damage lung parenchyma and increase pulmonary vascular resistance. The fourth option is not true, since the high FIO_2 may cause vasodilation. *(1:831–3; 4:1767; 5:716–7, 1893–4)*

279. **(B)** All volatile anesthetics and depolarizing muscle relaxants (e.g., succinylcholine) are known to be triggering agents in patients susceptible to malignant hyperthermia. All other medications are considered safe. *(1:524; 4:2430; 5:1182)*

280. **(B)** Fibrillating tissue cannot be paced. *(1:1514; 4:817; 5:2932; 6:82)*

281. **(B)** The presence of nitrogen in alveolar gas aids in keeping the alveoli expanded. Absorption of oxygen from alveoli containing only oxygen may cause alveolar collapse. One hundred percent oxygen causes alveolar injury; initially there is inflammation and increased vascular permeability which progresses to pulmonary edema if 100% oxygen is continued. Blindness secondary to retrolental fibroplasia is common in newborns given 100% oxygen, but does not occur in adults. Cerebral edema may accompany oxygen therapy under hyperbaric conditions but does not occur at atmospheric pressure. *(5:716–7, 2676–8)*

282. (D) Currents less than 100 μA may cause ventricular fibrillation if applied directly to the heart. The line isolation monitor will sound an alarm, and the ground fault circuit interrupter will remove power from the circuit, only when the leakage current exceeds 5 mA. Connection of the equipment chassis to earth ground will divert the majority of the leakage current away from the patient. *(1:155–7; 4:708–9; 5:3142–4)*

283. (A) At 3 atm and an FIO_2 of 1.0, the amount of dissolved oxygen in plasma is adequate to supply tissues with oxygen. The half-life of carbon monoxide excretion is dependent on the PaO_2: at 1 atm and an FIO_2 of 0.21, it is approximately 214 minutes, and it decreases to approximately 43 minutes at an FIO_2 of 1.0. If the atmospheric pressure is increased to 2.5 atm, the half-life decreases further to approximately 19 minutes. The pulmonary toxicity of oxygen is a function of the PO_2 in the inspired gas: the toxicity of oxygen at an FIO_2 of 1.0 at 1 atm is the same as with an FIO_2 of 0.33 at 3 atm. Because the gas contained within pulmonary bullae is not in equilibrium with the external environment, during the decompression phase after hyperbaric therapy, the bullae may rupture. *(5:2671–2, 2676–80)*

284. (B) An Apgar score of 7 to 10 indicates a healthy infant, who cries after delivery, maintains tone and color, has a heart rate above 100 bpm, and requires only routine care. An infant with an Apgar of 4 to 6 is depressed, may not breathe immediately, and should be stimulated and have the airway cleared with a bulb and syringe. If the heart rate is less than 100 bpm the infant should be ventilated with a bag and mask at a rate of 60 bpm. Long inflations rather than short, fast ones are optimal. An infant with an Apgar of 0 to 3 is flaccid, apneic, pale, and unresponsive. The ones that do not respond to bag and mask ventilation should be intubated, and if the heart rate remains less than 100 bpm, chest massage is initiated at a rate of 120 bpm. *(1:1162; 2:2349, 2352–3; 4:1995)*

285. (A) Metolazone is a thiazide diuretic. All of the others are high-ceiling (or loop) diuretics. *(3:770, 774)*

286. (A) Neonates have 60 to 90% hemoglobin F, and it is not until they are about 6 months of age that the adult hemoglobin A:hemoglobin F ratio is achieved. Hemoglobin F has a high affinity for oxygen, and the oxygen dissociation curve is shifted to the left, resulting in decreased oxygen delivery to the tissues at a given oxygen tension. *(2:19–20; 4:432, 2146–8)*

287. (E) Despite increased blood flow accompanying the elevated metabolic rate during generalized seizures, cerebral acidosis develops. The mechanism underlying this phenomenon is unclear. Most general anesthetics suppress seizure activity, including barbiturates, benzodiazepines, and volatile anesthetics. Prolonged anesthesia with isoflurane has been used to treat refractory seizures. *(1:499–501; 5:596, 1094–5; 6:282–5)*

288. (D) A patient with mitral valve prolapse may be at increased risk of developing endocarditis during any procedure associated with bacteremia. Although perioperative arrhythmias are common, therapy should be directed at the specific arrhythmia once it occurs. While some clinicians favor long-term anticoagulation in such patients, the anticoagulation should be discontinued in the perioperative period to decrease the amount of surgical bleeding. A temporary transvenous pacemaker is not indicated because there is not an increased risk of intraoperative third-degree AV block. *(1:561; 2:421; 4:1595–6; 5:1080; 6:35)*

289. (E) Mucociliary flow is depressed by halothane, dry gases, cool temperatures, and hypoxia. *(1:400, 808–9; 5:163–5)*

290. **(E)** All of these drugs are mixed opioid agonist-antagonists. *(3:599–602; 4:1253–4)*

291. **(A)** There is often downward displacement of the J point during exercise. *(1:1511; 4:182; 5:1411)*

292. **(C)** The patient with anorexia nervosa has severe protein-calorie malnutrition, but the patient's dietary intake is usually well-balanced, although inadequate. Thus, serum albumin is usually normal and ascites nonexistent. Cardiomyopathy is common, as are hypokalemia, hypocalcemia, hypomagnesemia, and metabolic acidosis. *(5:1034; 6:451)*

293. **(B)** Anion gap may be estimated by subtracting the sum of the serum chloride and bicarbonate concentrations from the serum sodium concentration. A normal value is 10 to 12 mEq/L. The acidosis that occurs due to gastrointestinal loss of bicarbonate ion with diarrhea has a normal anion gap because it is accompanied by hyperchloremia. Diabetic ketoacidosis has a high anion gap because of the presence of increased levels of unmeasured anions such as acetoacetate and β-hydroxybutyrate. *(1:166–8; 4:965–6; 5:1607–8)*

294. **(C)** Risk factors for postoperative nausea and vomiting include female gender, young age, a history of motion sickness or postoperative nausea or vomiting, gastrointestinal pathology, and surgery on the eye, ear, or within the abdomen. All available antiemetics are competitive antagonists for neurotransmitters active at the chemoreceptor trigger zone in the medulla. The chemoreceptor trigger zone activates the emetic center in response to a variety of circulating compounds. The chemoreceptor trigger zone is on the peripheral side of the blood-brain barrier and thus is exposed to all compounds circulating in the blood. *(1:1395–6; 4:134–6, 2259–60; 5:2597–8)*

295. **(A)** PGE_1 is used to prevent closure of the PDA in cyanotic congenital heart disease and is a bronchodilator and vasodilator. *(1:774; 2:374; 3:679; 5:2838)*

296. **(E)** Horner's syndrome, anhydrosis, injection of the conjunctiva, nasal stuffiness, vasodilatation, a hoarse voice, and increased skin temperature are all signs of a successful stellate ganglion block. *(3:1431)*

297. **(B)** Hypoxemia and intracranial hypertension are common problems after closed head trauma. Securing the airway and mechanical ventilation may be required to maintain adequate oxygenation. Sedation, by decreasing $CMRO_2$, is useful to reduce ICP. Routine hyperventilation has been found to be detrimental to outcome and is no longer recommended. The recommended lower limit for coronary perfusion pressure (CPP) is 60 mmHg. *(1:777–80; 2:336, 568; 4:581–2; 5:2152–5; 6:254–6)*

298. **(A)** The pressure exerted by a vapor varies directly with temperature and inversely with volume. Agitation has no effect on pressure. *(1:379; 4:1012–3)*

299. **(B)** Contemporary flowmeters are variable orifice devices employing a float within a glass tube whose diameter increases from bottom to top. The pressure drop across the float is constant regardless of flow rate and depends only on the weight of the float. Calibration of a flowmeter depends on both the density and viscosity of the gas. *(1:570–2; 4:1015–7; 5:277–9)*

300. **(B)** Adenosine may normally regulate coronary blood flow. It may interrupt reentrant supraventricular tachycardia. It is not an adrenergic catechol. It is inactivated by adenosine deaminase. *(1:304; 2:281; 3:953; 4:1574; 5:2930–2)*

301. **(E)** Administration of humidified gases is of benefit to some patients. The risks are that the humidifier may become contaminated, that the body will have to expend heat to warm the humidified gases, that the patient will receive a water load, and that the airway resistance will increase. These are not insurmountable problems—the mist can be heated to avoid patient heat loss, the fluid load can be compensated for by decreasing other fluids, and contamination can be avoided with good technique. *(5:2816–7)*

302. **(A)** Catecholamine neurotransmitters are derived from phenylalanine via conversion to tyrosine. Once released into the synaptic cleft, the effect of norepinephrine is terminated by a variety of mechanisms, including reuptake into the presynaptic neuron either for enzymatic degradation by MAO or for recycling into synaptic vesicles. Norepinephrine can also diffuse into capillaries or be taken up by the effector cell. Nicotinic receptors are a subset of cholinergic receptors and do not bind catecholamines. *(1:268–9; 4:731; 5:630–1)*

303. (B) Losartan is a competitive antagonist of angiotensin II while enalapril is an antagonist of ACE. Prazosin is an α-adrenergic antagonist, while minoxidil is an arterial vasodilator. *(3:246, 822, 832–3, 887)*

304. (A) Closing volume is measured as the slow "phase IV" washout of nitrogen or tracer. Maximal efforts actually increase CV, presumably by dynamic compression of some bronchioles. CV increases with age. *(1:805, 1207; 4:510; 5:695–7)*

305. (A) Fluphenazine, chlorpromazine, and haloperidol are used to treat schizophrenia. Doxepin is an antidepressant. *(3:487–9)*

306. (B) Sickling in such patients is increased by fever and hypoxia. Tissue hypoxia may be increased if regional blood flow is reduced due to hypovolemia or vasoconstriction (which may accompany hypothermia). The administration of supplemental oxygen to raise PaO_2 decreases the likelihood of sickling. *(1:506; 4:297, 466–7; 5:1113, 2423; 6:478–80)*

307. (B) Metoclopramide has antagonistic effects on dopamine (D_2) as well as 5-HT_3 receptors. Its main pharmacologic action is 5-HT_4 receptor activation. *(3:1023, 1025–6; 4:2326)*

308. (B) The pressure gradient requested is the transthoracic pressure gradient. Transpulmonary pressure is the gradient from alveoli to pleura and transmural is the gradient from pleura to ambient. Esophageal balloon pressure is a measure of pleural pressure. Pulmonary artery occlusion pressure (or wedge pressure) is a measure of left atrial pressure. *(1:794, 801; 5:689)*

309. (A) The Venturi mask uses oxygen flow to entrain air and regulate the oxygen concentration. The FIO_2 is independent of the patient's minute ventilation. *(5:2813)*

310. (D) Captopril inhibits conversion of angiotensin I to angiotensin II and thus decreases arteriolar and venous tone. *(1:317; 3:821–3; 4:1552; 5:793)*

311. (A) It is a common misconception that anesthetics abolish the coupling between blood flow and metabolic requirements. The mechanism underlying this relationship remains unclear. Although volatile anesthetics increase flow, a constant ratio of flow to metabolism is maintained. *(1:745–6; 4:1621; 5:814–6)*

312. (E) The optimal PEEP level depends on the degree of hypoxemia. One need not go to high levels of PEEP if the degree of hypoxemia is minimal. The type of lung disease also will affect the amount of PEEP used. The patient with emphysema and blebs in the lung will not benefit from PEEP and is more prone to barotrauma. Hydration and cardiac status are also important, since the patient with hypovolemia and poor contractility may not be able to withstand any increase in intrathoracic pressure. On the other hand, these patients may be helped by an increase in oxygenation. Careful titration is required. *(1:1469; 5:2820–1)*

313. (E) Cardiac β-adrenergic receptors are mainly of the β_1-adrenergic class. Bronchi and peripheral vascular smooth muscle bear mainly β_2-adrenergic receptors. *(1:275–6; 4:731; 5:636)*

314. (B) Exposure of the probe to extraneous light and the presence of red-absorbing dyes in the circulation can give erroneous pulse oximetry values. Tachycardia will not alter the accuracy of the saturation value as long as the oximeter is still able to record the proper pulse. Helium has no effect on pulse oximetry. *(1:670–1; 4:92; 5:1450–2)*

315. (C) Fentanyl is more lipophilic than morphine and is less likely to cause delayed onset of respiratory depression. It has a more rapid onset than morphine and is enhanced by the addition of local anesthetics. *(5:1432)*

316. **(B)** MAC-BAR (blockade of autonomic response) is defined as the minimum alveolar concentration necessary to prevent a rise in blood catecholamine levels after skin incision. This value therefore cannot be determined by observation but requires the assay of blood for catecholamine levels. It is about 50% higher than MAC and is much higher than MAC-awake for all anesthetics. *(1:389; 4:1138; 5:1239, 1242)*

317. **(C)** Patients with Down's syndrome and rheumatoid arthritis are at high risk for atlantoaxial instability. Patients with ankylosing spondylitis may have their cervical vertebrae fused in flexion. Such patients may have unstable cervical spines after attempts to extend the neck during intubation. *(1:597; 4:222, 428; 5:2409–11, 1099)*

318. **(C)** Infants of up to 50 weeks postconceptual age are at risk of postoperative apnea, even if they have no history of previous apneic episodes. Therefore, they are usually kept in the hospital for the first postoperative night for close apnea monitoring and are not operated on as ambulatory surgery patients. Halothane has been shown to decrease chemoreceptor sensitivity and is therefore implicated in postoperative apnea, but with appropriate monitoring its use is not contraindicated. Spinal anesthesia can be used as the sole technique for this procedure, but should not be combined with a sedative agent because of the risk of postoperative apnea. Infants have a relatively rapid uptake of volatile agents because of decreased blood-gas partition coefficients, decreased MAC, increased cardiac output per kilogram body mass, and relatively high blood flow to the brain. *(1:1190–1; 2:45–8, 644–7; 4:431, 2238–9; 5:2383, 2398)*

319. **(A)** Pulmonary embolus increases dead space. Changes in cardiac output will change oxygen delivery, altering mixed venous oxygen content, and thus shunt. High FIO_2 reduces increases in PaO_2 from low V/Q regions. An atelectatic region has perfusion with no ventilation, contributing to shunt. *(1:803–4; 5:683–9)*

320. **(B)** The rostral portion of the brain receives its blood supply via the internal carotid arteries; the posterior circulation is supplied by the vertebral arteries which form the basilar system. Vertebral arteries arise from the subclavian arteries. Although anastomoses can exist between branches of the external and internal carotid arteries, these usually do not contribute a major supply of blood. The occipital artery is entirely extracranial. *(1:856; 6:246–7)*

321. **(B)** Labetalol blocks α-adrenergic and β-adrenergic receptors and tends to lower blood pressure without compensatory tachycardia. *(1:312; 4:746; 5:657)*

322. **(C)** Adequate IV access, administration of a nonparticulate antacid, and left uterine displacement are mandatory. After denitrogenation, a rapid sequence induction is performed with cricoid pressure, followed by intubation and maintenance with 50% N_2O and 0.5 MAC isoflurane. After delivery, anesthesia is deepened with 70% nitrous and opioids and benzodiazepines. Isoflurane may be decreased or discontinued if uterine atony is present. Small doses of neuromuscular blocking agents may be used as needed. *(1:1150; 4:2007)*

323. **(C)** In a variable bypass vaporizer, the stream of carrier gas is split and only a portion is saturated with the vapor of the volatile agent. Adjusting the dial on the vaporizer changes the ratio of carrier gas entering the vaporizing chamber in relation to total gas flow. Vaporizer output is nearly constant over a wide range of flow rates, but is less than the dial setting at both very low and high flow rates. Alteration of the composition of the carrier gas also changes the vaporizer output. *(1:574–7; 4:1024–5; 5:285–9)*

324. **(E)** All of the choices listed are possible complications of a celiac plexus block. Hypotension can occur when a patient assumes an upright position secondary to the blockade of visceral vasoconstrictor fibers. *(4:1433)*

325. **(A)** Renin is released by the kidney in response to sympathetic stimulation or hypotension. Renin generates angiotensin I, which is converted to angiotensin II, a stimulator of aldosterone release. Hypovolemia leads to ADH release while hypervolemia releases atrial natriuretic peptide. *(1:282–3, 876–7; 4:1593; 5:795–6, 1487, 1765; 6:373, 425)*

326. **(B)** The nerves that are most subject to injury are those that are superficial and run a long course. The ulnar nerve is an example. The patient with ischemic disease, e.g., arteriosclerosis or diabetes mellitus, is more subject to nerve injury. One must also remember that not all postoperative nerve injuries are due to positioning. Some may be due to injections or direct damage to the nerve. *(1:716–8; 5:706–7)*

327. **(E)** Resistance decreases with inspiration because the airways dilate. The other factors constrict airways. *(1:795–6; 4:238; 5:690–1)*

328. **(B)** Both an intraoperative wake up, where the patient is asked to move each limb, and the monitoring of somatosensory-evoked potentials, test the integrity of the spinal cord. Brainstem-evoked potentials may be used during certain intracranial procedures, especially those in the posterior fossa. The train-of-four monitor only evaluates the function of a peripheral nerve and the neuromuscular junction. *(1:1112–3; 4:2121; 5:2418–20)*

329. **(A)** Ketamine has sympathomimetic activity that can reverse bronchospasm. Atropine causes bronchodilatation by its muscarinic antagonism, blocking the action of the vagus on bronchial smooth muscle. All the potent inhalation agents block the bronchial response to mediators of bronchoconstriction, although in spontaneously ventilating patients, the respiratory depression will cause lower lung volumes. Therefore ventilation will occur at a less compliant region of the pressure-volume curve. Thiopental has no significant activity on bronchial tone. *(1:287–8, 337, 400; 4:238, 1145, 1217; 5:347–8, 659–60, 1124)*

330. **(E)** Neurogenic pulmonary edema is thought to involve massive sympathetic discharge from injured brain in response to intracranial hypertension. *(1:780; 4:256; 6:208)*

331. **(D)** Allergy is common with local anesthetics in the ester class, such as tetracaine, and extremely rare with local anesthetics in the amide class. *(4:1353; 5:596–7)*

332. **(B)** Because mannitol is an osmotic diuretic that does not cross the blood-brain barrier, water is drawn out of the brain intracellular compartment. This ultimately reduces the volume of brain substance. *(1:762–3; 4:247–8, 1632; 5:2134; 6:237–8)*

333. **(A)** Postintubation laryngeal edema is more common in children than adults because of their relatively small airway size and usually manifests as a brassy cough and inspiratory stridor. Humidified oxygen and racemic epinephrine are frequently the only measures required, but if obstruction is severe and persistent, intubation or tracheostomy must be considered. Fluid restriction is not appropriate. *(1:1386, 4:2319; 5:1650)*

334. **(A)** Myocardial depressant effects of anesthetics may be exaggerated in patients with increased cardiac output at rest as compensation for anemia. *(1:504; 4:35; 5:701, 1112; 6:471–2)*

335. **(D)** Herpetic whitlow is a painful infection of a finger caused by the herpes simplex virus. It is most commonly acquired from contact with herpetic lesions or respiratory secretions containing virus. The wearing of gloves is highly protective. *(1:71–2; 4:422; 5:3161; 6:564)*

336. **(D)** Most patients recovering from an anesthetic are in need of supplemental oxygen. Cyanosis is a late finding (and may not be present in some states of hypoxia). Laser ignition only is a risk at the time of the actual application. *(1:1390; 4:2320; 5:2712)*

337. **(E)** All of the agents mentioned will lead to a decrease in pulmonary vascular resistance, each through a different mechanism. Inhaled nitric oxide is used for the treatment of acute pulmonary hypertension and acts through stimulation of soluble guanylate cyclase with resultant vascular smooth muscle relaxation.

IV prostacyclin (epoprostenol, PGI_2) induces relaxation of smooth muscle by stimulating the production of cyclic AMP and can also be administered via inhalation. Other drugs in this category are inhaled iloprost and subcutaneous treprostinil. Sildenafil is a phosphodiesterase type 5 inhibitor. These drugs have an acute pulmonary vasodilator effect, which is due to enhancement of nitric oxide-mediated pulmonary vasodilation. Bosentan is an orally active endothelin antagonist used for the treatment of primary pulmonary hypertension. *(3:680; 5:685–6; 6:130–2)*

338. **(A)** Hyperthermia increases the rate of sickling, but hypothermia is also undesirable because of peripheral vasoconstriction and stasis. *(1:506; 4:365; 5:1113, 2536; 6:478–9)*

339. **(E)** Fat emboli cause a mechanical blockage, and then endothelial damage from free fatty acid breakdown products. The pulmonary and cerebral vascular beds can be involved. *(1:1114–5; 4:295; 5:2425; 6:175–6)*

340. **(B)** Bretylium and guanethidine may have efficacy because they deplete the sympathetic nerve terminal of norepinephrine. Hydralazine is a vasodilator and has no specific effects on the sympathetic nervous system. Methyldopa acts in the central nervous system to decrease sympathetic tone. *(1:1592; 5:1385–7)*

341. **(C)** The filling pressures are elevated and tend to equalize in pericardial constriction and pericardial tamponade. *(1:918; 5:1316; 5:1966)*

342. **(A)** Respiratory insufficiency will affect oxygenation and CO_2 elimination. Dyspnea will occur if the patient is conscious and the respiratory drive is active. Bicarbonate retention is a slower response to elevated CO_2 by the kidney. *(1:167, 1385; 5:1603)*

343. **(E)** A heated cascade humidifier can deliver 100% relative humidity at body temperature or at an elevated temperature, causing a burn. There is an internal check valve which prevents humidified gas from entering the anesthesia machine. This valve requires that the humidifier be properly connected or else obstruction to flow will exist. Contamination of the water in the humidifier can cause a respiratory tract infection. *(4:446, 2101:5:1585)*

344. **(B)** The morbidly obese have lower values for FRC and TLC. Since RV and CC remain unchanged, normal tidal ventilation may occur at lung volumes below CC. *(1:1035; 4:509; 5:1027–33; 6:446–7)*

345. **(A)** Prehydration with 500 to 1000 mL of salt solution decreases the incidence of hypotension, but prophylactic ephedrine is not recommended. Adequate equipment and supplies for resuscitation and IV access are mandatory. *(4:2000; 5:2326)*

346. **(C)** Airway pressure elevates the wedge pressure with respect to the left atrial pressure. *(1:898; 4:814; 5:1314)*

347. **(A)** There are several radicular arteries that branch from intercostal and lumbar arteries to anastomose with the anteroposterior spinal artery system to supply the spinal cord with blood. The artery of Adamkiewicz is the largest of these 4 to 10 radicular branches. Its origin may be either supra- or infrarenal. *(1:691; 4:1868; 5:2088)*

348. **(E)** CO_2 can demonstrate respiratory gas exchange, but the correct position within the tracheobronchial tree can only be determined by bronchoscopy, x-ray, or (less reliably) auscultation. *(1:609; 4:1094)*

349. **(A)** Ohm's law states that the current in a circuit is equal to the voltage divided by the resistance. *(1:143; 4:701; 5:1209)*

350. **(E)** Two-dimensional slices of the heart are imaged in real time. The transesophageal probe provides image acquisition without interference with chest surgery. Some cardiac structures, such as the sometimes thrombus-laden left atrial appendage, are better seen from the esophagus than from chest surface positions.

Cardiac chamber sizes and motion are readily appreciated, and air is very echogenic. Much information about heart valves is apparent from their echo images, but further information is provided by means of the Doppler effect, used to provide "color-flow" images of jets of blood passing through stenotic or leaking valves or through septal defects. *(1:886; 4:829; 5:1363–85; 6:28)*

References

1. Barash PG, Cullen BF, Stoelting RK, eds. *Clinical Anesthesia,* 4th ed. Philadelphia, PA: Lippincott Williams & Wilkins; 2001.
2. Coté CJ, Ryan JF, Todres ID, Goudsouzian NG, eds. *A Practice of Anesthesia for Infants and Children,* 3rd ed. Philadelphia, PA: W.B. Saunders; 2001.
3. Hardman JG, Limbird LE, eds. *Goodman and Gilman's The Pharmacological Basis of Therapeutics,* 10th ed. New York: McGraw Hill; 2001.
4. Longnecker DE, Tinker JH, Morgan GE, eds. *Principles and Practice of Anesthesiology,* 2nd ed. St. Louis, MO: Mosby Year Book; 1997.
5. Miller RD, ed. *Miller's Anesthesia,* 6th ed. Philadelphia, PA: Churchill Livingstone; 2005.
6. Stoelting RK, Dierdorf SF. *Anesthesia and Co-Existing Disease,* 4th ed. Philadelphia, PA: Churchill Livingstone; 2002.

Index

The numbers following the topic refer to the question numbers. Numbers preceded by "t" refer to questions in Chapter 17, the practice test.

B

Q

R